PROGRESS IN CLINICAL AND BIOLOGICAL RESEARCH

Series Editors

Nathan Back	Vincent P. Eijsvoogel	Kurt Hirschhorn	Sidney Udenfriend
George J. Brewer	Robert Grover	Seymour S. Kety	Jonathan W. Uhr

RECENT TITLES

Vol 227: **Advances in Chronobiology**, John E. Pauly, Lawrence E. Scheving, *Editors.* Published in two volumes.

Vol 228: **Environmental Toxicity and the Aging Processes,** Scott R. Baker, Marvin Rogul, *Editors*

Vol 229: **Animal Models: Assessing the Scope of Their Use in Biomedical Research,** Junichi Kawamata, Edward C. Melby, Jr., *Editors*

Vol 230: **Cardiac Electrophysiology and Pharmacology of Adenosine and ATP: Basic and Clinical Aspects,** Amir Pelleg, Eric L. Michelson, Leonard S. Dreifus, *Editors*

Vol 231: **Detection of Bacterial Endotoxins With the Limulus Amebocyte Lysate Test,** Stanley W. Watson, Jack Levin, Thomas J. Novitsky, *Editors*

Vol 232: **Enzymology and Molecular Biology of Carbonyl Metabolism: Aldehyde Dehydrogenase, Aldo-Keto Reductase, and Alcohol Dehydrogenase,** Henry Weiner, T. Geoffrey Flynn, *Editors*

Vol 233: **Developmental and Comparative Immunology,** Edwin L. Cooper, Claude Langlet, Jacques Bierne, *Editors*

Vol 234: **The Hepatitis Delta Virus and Its Infection**, Mario Rizzetto, John L. Gerin, Robert H. Purcell, *Editors*

Vol 235: **Preclinical Safety of Biotechnology Products Intended for Human Use,** Charles E. Graham, *Editor*

Vol 236: **First Vienna Shock Forum,** Günther Schlag, Heinz Redl, *Editors.* Published in two volumes: Part A: *Pathophysiological Role of Mediators and Mediator Inhibitors in Shock.* Part B: *Monitoring and Treatment of Shock.*

Vol 237: **The Use of Transrectal Ultrasound in the Diagnosis and Management of Prostate Cancer,** Fred Lee, Richard McLeary, *Editors*

Vol 238: **Avian Immunology,** W.T. Weber, D.L. Ewert, *Editors*

Vol 239: **Current Concepts and Approaches to the Study of Prostate Cancer,** Donald S. Coffey, Nicholas Bruchovsky, William A. Gardner, Jr., Martin I. Resnick, James P. Karr, *Editors*

Vol 240: **Pathophysiological Aspects of Sickle Cell Vaso-Occlusion,** Ronald L. Nagel, *Editor*

Vol 241: **Genetics and Alcoholism,** H. Werner Goedde, Dharam P. Agarwal, *Editors*

Vol 242: **Prostaglandins in Clinical Research,** Helmut Sinzinger, Karsten Schrör, *Editors*

Vol 243: **Prostate Cancer,** Gerald P. Murphy, Saad Khoury, Réne Küss, Christian Chatelain, Louis Denis, *Editors.* Published in two volumes: Part A: *Research, Endocrine Treatment, and Histopathology.* Part B: *Imaging Techniques, Radiotherapy, Chemotherapy, and Management Issues*

Vol 244: **Cellular Immunotherapy of Cancer,** Robert L. Truitt, Robert P. Gale, Mortimer M. Bortin, *Editors*

Vol 245: **Regulation and Contraction of Smooth Muscle,** Marion J. Siegman, Andrew P. Somlyo, Newman L. Stephens, *Editors*

Vol 246: **Oncology and Immunology of Down Syndrome,** Ernest E. McCoy, Charles J. Epstein, *Editors*

Vol 247: **Degenerative Retinal Disorders: Clinical and Laboratory Investigations,** Joe G. Hollyfield, Robert E. Anderson, Matthew M.LaVail, *Editors*

Vol 248: **Advances in Cancer Control: The War on Cancer—15 Years of Progress,** Paul F. Engstrom, Lee E. Mortenson, Paul N. Anderson, *Editors*

Vol 249: **Mechanisms of Signal Transduction by Hormones and Growth Factors,** Myles C. Cabot, Wallace L. McKeehan, *Editors*

Vol 250: **Kawasaki Disease,** Stanford T. Shulman, *Editor*

Vol 251: **Developmental Control of Globin Gene Expression,** George Stamatoyannopoulos, Arthur W. Nienhuis, *Editors*

Vol 252: **Cellular Calcium and Phosphate Transport in Health and Disease,** Felix Bronner, Meinrad Peterlik, *Editors*

Vol 253: **Model Systems in Neurotoxicology: Alternative Approaches to Animal Testing,** Abraham Shahar, Alan M. Goldberg, *Editors*

Vol 254: **Genetics and Epithelial Cell Dysfunction in Cystic Fibrosis,** John R. Riordan, Manuel Buchwald, *Editors*

Vol 255: **Recent Aspects of Diagnosis and Treatment of Lipoprotein Disorders: Impact on Prevention of Atherosclerotic Diseases,** Kurt Widhalm, Herbert K. Naito, *Editors*

Vol 256: **Advances in Pigment Cell Research,** Joseph T. Bagnara, *Editor*

Vol 257: **Electromagnetic Fields and Neurobehavioral Function,** Mary Ellen O'Connor, Richard H. Lovely, *Editors*

Vol 258: **Membrane Biophysics III: Biological Transport,** Mumtaz A. Dinno, William McD. Armstrong, *Editors*

Vol 259: **Nutrition, Growth, and Cancer,** George P. Tryfiates, Kedar N. Prasad, *Editors*

Vol 260: **EORTC Genitourinary Group Monograph 4: Management of Advanced Cancer of Prostate and Bladder,** Philip H. Smith, Michele Pavone-Macaluso, *Editors*

Vol 261: **Nicotine Replacement: A Critical Evaluation,** Ovide F. Pomerleau, Cynthia S. Pomerleau, *Editors*

Vol 262: **Hormones, Cell Biology, and Cancer: Perspectives and Potentials,** W. David Hankins, David Puett, *Editors*

Vol 263: **Mechanisms in Asthma: Pharmacology, Physiology, and Management,** Carol L. Armour, Judith L. Black, *Editors*

Vol 264: **Perspectives in Shock Research,** Robert F. Bond, *Editor*

Vol 265: **Pathogenesis and New Approaches to the Study of Noninsulin-Dependent Diabetes Mellitus,** Albert Y. Chang, Arthur R. Diani, *Editors*

Vol 266: **Growth Factors and Other Aspects of Wound Healing: Biological and Clinical Implications,** Adrian Barbul, Eli Pines, Michael Caldwell, Thomas K. Hunt, *Editors*

Vol 267: **Meiotic Inhibition: Molecular Control of Meiosis,** Florence P. Haseltine, *Editor*

Vol 268: **The Na^+,K^+-Pump,** Jens C. Skou, Jens G. Nørby, Arvid B. Maunsbach, Mikael Esmann, *Editors*. Published in two volumes: Part A: *Molecular Aspects.* Part B: *Cellular Aspects.*

Vol 269: **EORTC Genitourinary Group Monograph 5: Progress and Controversies in Oncological Urology II,** Fritz H. Schröder, Jan G.M. Klijn, Karl H. Kurth, Herbert M. Pinedo, Ted A.W. Splinter, Herman J. de Voogt, *Editors*

Vol 270: **Cell-Free Analysis of Membrane Traffic,** D. James Morré, Kathryn E. Howell, Geoffrey M.W. Cook, W. Howard Evans, *Editors*

Vol 271: **Advances in Neuroblastoma Research 2,** Audrey E. Evans, Giulio J. D'Angio, Alfred G. Knudson, Robert C. Seeger, *Editors*

Vol 272: **Bacterial Endotoxins: Pathophysiological Effects, Clinical Significance, and Pharmacological Control,** Jack Levin, Harry R. Büller, Jan W. ten Cate, Sander J.H. van Deventer, Augueste Sturk, *Editors*

Vol 273: **The Ion Pumps: Structure, Function, and Regulation,** Wilfred D. Stein, *Editor*

Vol 274: **Oxidases and Related Redox Systems,** Tsoo E. King, Howard S. Mason, Martin Morrison, *Editors*

Vol 275: **Electrophysiology of the Sinoatrial and Atrioventricular Nodes,** Todor N. Mazgalev, Leonard S. Dreifus, Eric L. Michelson, *Editors*

Please contact the publisher for information about previous titles in this series.

The Na^+,K^+-Pump
Part B: Cellular Aspects

THE Na^+,K^+-PUMP
Part B: Cellular Aspects

Proceedings of the 5th International Conference on Na^+,K^+-ATPase, Held at Fuglsø Conference Center, Denmark, June 14–19, 1987

Editors

Jens C. Skou
Jens G. Nørby
Arvid B. Maunsbach
Mikael Esmann
University of Aarhus
Aarhus, Denmark

ALAN R. LISS, INC. • NEW YORK

Address all Inquiries to the Publisher
Alan R. Liss, Inc., 41 East 11th Street, New York, NY 10003

Copyright © 1988 Alan R. Liss, Inc.

Printed in the United States of America

Under the conditions stated below the owner of copyright for this book hereby grants permission to users to make photocopy reproductions of any part or all of its contents for personal or internal organizational use, or for personal or internal use of specific clients. This consent is given on the condition that the copier pay the stated per-copy fee through the Copyright Clearance Center, Incorporated, 27 Congress Street, Salem, MA 01970, as listed in the most current issue of "Permissions to Photocopy" (Publisher's Fee List, distributed by CCC, Inc.), for copying beyond that permitted by sections 107 or 108 of the US Copyright Law. This consent does not extend to other kinds of copying, such as copying for general distribution, for advertising or promotional purposes, for creating new collective works, or for resale.

Library of Congress Cataloging-in-Publication Data

International Conference on Na^+,K^+-ATPase (5th : 1987 : Denmark)
    The Na^+,K^+-pump : proceedings of the 5th International Conference on Na^+,K^+-ATPase held at Fuglsø Conference Center, Denmark, June 14–19, 1987 / editors, J.C. Skou . . . [et al.].
    p. cm. — (Progress in clinical and biological research ; v. 268A-B)
    Includes bibliographies and indexes.
    Contents: pt. A. Molecular aspects — pt. B. Cellular aspects.
ISBN 0-8451-5118-5 (set). ISBN 0-8451-5400-1 (pt. A).
ISBN 0-8451-5401-X (pt. B)
    1. Biological transport, Active—Congresses. 2. Sodium in the body—Congresses. 3. Potassium in the body—Congresses.
4. Adenosine triphosphatase—Congresses. 5. Cell membranes — Congresses. I. Skou, J.C. II. Title. III. Series.
QH509.I575 1987   88-585
574.87′5—dc19   CIP

Contents

Contributors and Participants xiii
Contents of Part A xxv
Preface
Jens C. Skou, Jens G. Nørby, Arvid B. Maunsbach, and Mikael Esmann xxix
Acknowledgments xxxi

THE CONCEPT OF ACTIVE TRANSPORT

Overview: The Development of the Concept of Active Transport
Hans H. Ussing 3

BIOSYNTHESIS AND EXPRESSION

Overview: Biosynthesis, Membrane Insertion and Maturation of Na^+,K^+-ATPase
Kaethi Geering 19

In Vitro Biosynthesis of the β-Subunit of the Na^+, K^+-ATPase in the Brine Shrimp
Lee Ann Baxter-Lowe, James M. Yohanan, and Lowell E. Hokin 35

Comparative Properties of Alpha and Alpha Plus Subunits in Adult Rat Brain and Heart
Isabelle Berrebi-Bertrand, Jean-Michel Maixent, and Lionel G. Lelievre 43

Sorting of Newly Synthesized Na^+, K^+-ATPase in Polarized Epithelial Cells
Michael J. Caplan, H. Clark Anderson, George E. Palade, and James D. Jamieson 51

Quantitative Analysis of Sodium Pump-Specific mRNA From Human Endothelial (HeLa) and Canine Kidney (MDCK) Cell Cultures
Christopher P. Cutler, Gordon Cramb, and Joseph F. Lamb 59

Tissue-Specific Differences in the Expression of the Genes Encoding the Alpha and Beta Subunits of Rat Na^+, K^+-ATPase
Gregory G. Gick, Faramarz Ismail-Beigi, and Isidore S. Edelman 65

In Vitro Expression of the Alpha and Beta Subunits of the Na^+, K^+-ATPase
M. Gilmore-Hebert, R.W. Mercer, J.W. Schneider, and E.J. Benz 71

Membrane Insertion of α and β Subunits of Human Na^+, K^+-ATPase
Haruo Homareda, Kiyoshi Kawakami, Kei Nagano, and Hideo Matsui 77

Development of a Heterologous Gene Expression System for the Na^+,K^+-ATPase Subunits in the Yeast *Saccharomyces cerevisiae*
Burton Horowitz and Robert A. Farley 85

The Appearance of Na^+, K^+-ATPase Isozymes in Rat Neuronal Cell Cultures
Nobuo Inoue, Hideo Matsui, Hiroko Tsukui, and Hiroshi Hatanaka 91

Application of Gene Transfer and Gene Mapping Techniques to Functional Analysis of the Na^+, K^+-ATPase
Rachel B. Kent, Dorothy A. Fallows, Janet Rettig Emanuel, Peter A. Lalley, Robert Levenson, and David E. Housman 97

Multiple Forms of the Na^+, K^+-ATPase: Their Genes and Tissue Specific Expression
Jerry B. Lingrel, Regina M. Young, and Marcia M. Shull 105

Identification of a Third Form of Na^+, K^+-ATPase Catalytic Subunit in Rat Brain by Photoaffinity Labeling
Joseph M. Lowndes, Arnold E. Ruoho, and Mabel Hokin-Neaverson 113

Molecular Cloning and Characterization of α-Subunit Isoforms of the Na^+,K^+-ATPase
Robert W. Mercer, Jay W. Schneider, and Edward J. Benz, Jr. 119

Characterization and Expression of a cDNA Clone for the Beta Subunit of Rat Brain Na^+, K^+-ATPase
Koichiro Omori, Kyoko Omori, Mary Flanagan, Vandana A. Desai, Nuran Nabi, Yuji Sugita, Dipak Haldar, Jane M. Sherman, David D. Sabatini, and Takashi Morimoto 127

Intraindividual Tissue-Specific Variations in Organization of Genes Coding for the Na^+, K^+-ATPase Subunits
E.D. Sverdlov, N.E. Broude, G.S. Monastyrskaya, V.E. Sverdlov, A.V. Grishin, N.S. Akopyantz, K.E. Petrukhin, and N.N. Modyanov 135

Increased *In Vivo* Synthesis of Retinal Na^+ Pump Isoforms After Pre-Treatment With Citrate Buffer
Susan C. Specht, Enid Del Valle, and Sonia Castro Hernández 143

Two Forms of the Na^+, K^+-ATPase Catalytic Subunit Detected by Their Immunological and Electrophoretic Differences. Alpha and Alpha+ in Mammalian Heart
Kathleen J. Sweadner and Sima K. Farshi 149

The Effect of Growth in Monensin or Low Potassium on Internal Sodium, Alpha Sub-Unit mRNA and Sodium Pump Density in Human Cultured Cells
Shiela Unkles, Trudi McDevitt, Carol Voy, James R. Kinghorn, Gordon Cramb, and Joseph F. Lamb 157

Effects of Tunicamycin on the Cellular Expression and Structural Arrangement of Na^+,K^+-ATPase
D. Zamofing, B.C. Rossier, and K. Geering 163

SECONDARY ACTIVE TRANSPORT, Na^+,K^+-HOMEOSTASIS, AND KIDNEY FUNCTION

Overview: Physiological Role of the Na^+-K^+ Pump
Claude Lechene 171

Overview: Maintenance of Na^+, K^+-Homeostasis by Na^+, K^+-Pumps in Striated Muscle
Ole M. Sejersted 195

Overview: Role of Na^+-K^+-ATPase in Kidney Function
Adrian I. Katz 207

Comparison of the Maximum Capacity for Active Sodium-Potassium Transport in the Left and Right Ventricle of Mammalian Heart
Tamás Bányász, Tibor Kovács, and János Somogyi 233

Is the Na^+,K^+-Pump Capacity in Skeletal Muscle Inadequate During Sustained Work?
Torben Clausen and Maria Elisabeth Everts 239

Na^+,K^+-ATPase Co-Distributes With Ankyrin and Spectrin in Renal Tubular Epithelial Cells
Michael Kashgarian, Jon S. Morrow, Harald G. Foellmer, Andrea S. Mann, Carol Cianci, and Thomas Ardito 245

Na^+,K^+-ATPase Concentration in Skeletal Muscle: Quantification, Regulation, and Significance
Keld Kjeldsen, Maria E. Everts, and Aage Nørgaard 251

Quantification of the Na^+,K^+-Pump in Heart Muscle by Measurement of 3-O-Methylfluorescein Phosphatase Activity
Jim S. Larsen, Keld Kjeldsen, and Aage Nørgaard 257

Alpha Subunits (α and α^+) Isoforms of the Na^+,K^+-ATPase in Dog Heart. Alteration of α^+ in Ischemia
Jean-Michel Maixent, Pierre Birkui, Simone Fenard, and Lionel G. Lelievre . . 263

The Na^+,K^+-Pump in Human Myocardium: Quantification in Normal Subjects and Patients With Suspected Cardiomyopathy
Aage Nørgaard and Keld Kjeldsen 269

REGULATION OF Na^+,K^+-ATPase: HORMONES, INHIBITORS, AND INOTROPIC ACTION

Overview: Hormonal Regulation of Na^+,K^+-ATPase
G.G. Gick, F. Ismail-Beigi, and I.S. Edelman 277

Overview: Physiological Inhibitors of Na^+,K^+-ATPase: Concept and Status
Garner T. Haupert, Jr. 297

Overview: Role of the Na^+, K^+-ATPase in the Cardiotonic Action of Cardiac Glycosides
Arnold Schwartz, Gunter Grupp, Earl Wallick, I.L. Grupp, and W.J. Ball, Jr. . . 321

Aldosterone and Sodium Induce Kidney Na^+,K^+-ATPase *In Vitro* by Two Different Mechanisms
Catherine Barlet-Bas and Alain Doucet 339

Regulation of Cardiac Glycoside Receptors With Different Affinities for Cardiac Glycosides in Cultured Rat Heart Cells
H.J. Berger, K. Werdan, and E. Erdmann 345

Pertussis Toxin Modulates Dopamine Inhibition of Na^+,K^+-ATPase Activity in Rat Proximal Convoluted Tubule Segments
Alejandro Bertorello and Anita Aperia 353

Loss of Na^+,K^+-ATPase During Sheep Reticulocyte Maturation
Rhoda Blostein and Eva Grafova 357

The Effect of External Na^+ and Ca^{2+} on the Rate of Ouabain Binding to Resealed Human Red Blood Cell Ghosts
H. Harm Bodemann and Joseph F. Hoffman 365

Time Course of Changes in Na^+,K^+-Pump Concentration and Passive Na^+,K^+-Fluxes in Skeletal Muscle After Administration of Thyroid Hormone
Maria E. Everts, Torben Clausen, and Keld Kjeldsen 371

Na^+,K^+-ATPase Isozymes in Rat Tissues: Differential Sensitivities to Sodium, Vanadate, and Dihydroouabain
Gudrun Feige, Thomas Leutert, and Alain De Pover 377

Effects of Gossypol on the Activity of Rabbit Kidney Na^+,K^+-ATPase and the Functions of Human Erythrocyte Membrane
Fu Yun-Feng, Zhang Shi-Lian, Lu Zhen-Min, and Wang Wei 385

Interaction Between Palytoxin and Purified Na^+,K^+-ATPase
E. Grell, E. Lewitzki, and D. Uemura 393

Control of the Sodium Pump by Liponucleotides and Unsaturated Fatty Acids: Side-Dependent Effects in Red Cells
W.-H. Huang, Z. Xie, S.S. Kakar, and A. Askari 401

Binding of Ouabain to the Na^+-K^+-ATPase in Oocytes of *Xenopus laevis* is Voltage-Independent
Andreas V. Lafaire, Bernd Schweigert, and Wolfgang Schwarz 409

An Endogenous Inhibitor of Na^+,K^+-ATPase Isolated From Human Plasma Inhibits the Acid Pump of the Stomach
Sven Mårdh 417

Two Molecular Forms of Na^+, K^+-ATPase in the Ferret Heart and Developmental Changes in Digitalis Sensitivity
Yuk-Chow Ng and Tai Akera 423

Two-Sided Functional Na^+, K^+-ATPase-Liposomes for Characterizing the Permeability and Side of Action of Pump Inhibitors
H.G. Rey, P. Meda, and B.M. Anner 429

Effect of Primaquine on the Topology of Na^+, K^+-ATPase and the Receptor for Asialoglycoproteins
G.J. Strous, P. van Kerkhof, A. van Bokhoven, A.L. Schwartz, and
J.J.H.H.M. de Pont 437

Non-Esterified Fatty Acids and the Circulating Inhibitor of Na^+, K^+-ATPase
H.G.P. Swarts, J.A.H. Timmermans, F.M.A.H. Schuurmans Stekhoven,
J.J.H.H.M. de Pont, S.J. Graafsma, and T.A. Thien 443

The Decisive Role of Na^+, K^+-ATPase in Stimulation Process of T-Lymphocytes
Márta Szamel, Mária Kövecses, P. Csermely, L. Szollár, Margarete Goppelt,
K. Resch, and J. Somogyi 449

Membrane Phosphorylation is Involved in the Inhibitory Effect of Micromolar Ca^{2+} on Rat Myometrial Na^+ K^+-ATPase
A. Turi and J. Somogyi 457

Effects of Aldosterone on Na^+, K^+-ATPase Transcription, mRNAs, and Protein Synthesis, and on Transepithelial Na^+ Transport in A6 Cells
F. Verrey, E. Schaerer, P. Fuentes, J.P. Kraehenbuhl, and B.C. Rossier 463

Inotropic and Toxic Actions of Several Cardiac Steroids in Sheep Cardiac Tissues
J. Andrew Wasserstrom and David E. Farkas 469

Effect of Apamin on the Na^+, K^+ Pump and Na^+, K^+-ATPase
Hana Zemková, Jan Teisinger, and František Vyskočil 477

Index 485

Contributors and Participants

Norma C. Adragna-Lauf*, Department of Physiology and Biophysics, Wright State University School of Medicine, Dayton, OH 45401-0927

Tai Akera, Department of Pharmacology and Toxicology, Michigan State University, East Lansing, MI 48824 **[423]**

N.S. Akopyantz, M.M. Shemyakin Institute of Bioorganic Chemistry, USSR Academy of Sciences, Moscow, USSR **[135]**

Jens P. Andersen*, Institute of Physiology, University of Aarhus, 8000 Aarhus C, Denmark

Olaf Sparre Andersen*, Department of Physiology and Biophysics, Cornell University Medical College, New York, NY 10021

H. Clark Anderson, Department of Cell Biology, Yale University School of Medicine, New Haven, CT 06510 **[51]**

Beatrice M. Anner*, Department of Pharmacology, Geneva University Medical Center, CH-1211 Geneva 4, Switzerland **[429]**

Hans-Jürgen Apell*, Department of Biology, University of Konstanz, D-7750 Konstanz, Federal Republic of Germany

Anita Aperia*, Department of Pediatrics, St. Göran's Children's Hospital, 112 81 Stockholm, Sweden **[353]**

Thomas Ardito, Departments of Pathology, Physiology, and Medicine, Yale University, New Haven, CT 06510 **[245]**

Amir Askari*, Department of Pharmacology, Medical College of Ohio, Toledo, OH 43699 **[401]**

W.J. Ball, Jr., Departments of Pharmacology and Cell Biophysics, University of Cincinnati College of Medicine, Cincinnati, OH 45267-0575 **[321]**

Ernst Bamberg*, Max-Planck-Institut für Biophysik, 6000 Frankfurt am Main, Federal Republic of Germany

Tamás Bányász, Department of Physiology, Medical University of Debrecen, H-4012 Debrecen, Hungary **[233]**

Catherine Barlet-Bas, Laboratoire de Physiologie Cellulare, College de France, 75231 Paris, France **[339]**

The numbers in brackets are the opening page numbers of the contributors' articles.
*Participants are indicated by an asterisk.

xiv / Contributors and Participants

Lee Ann Baxter-Lowe*, Department of Pharmacology, University of Wisconsin Medical School, Madison, WI 53706 **[35]**

Luis Beaugé*, División de Biofísica, Instituto de Investigación Médica Mercedes y Martín Ferreyra, 5000 Córdoba, Argentina

Edward J. Benz, Jr., Departments of Internal Medicine and Human Genetics, Yale University School of Medicine, New Haven, CT 06510 **[71,119]**

H.J. Berger*, Medizinische Klinik I, Klinikum Grosshadern, University of Munich, D-8000 München 70, West Germany **[345]**

Isabelle Berrebi-Bertrand*, INSERM: U127, Hôpital Lariboisière, Université Paris, 75010 Paris 7, France **[43]**

Alejandro Bertorello*, Department of Pediatrics, St. Göran's Children's Hospital, 112 81 Stockholm, Sweden **[353]**

Pierre Birkui, INSERM U 141, Hôpital Lariboisière, Université Paris, 75010 Paris 7, France **[263]**

Rhoda Blostein*, Montreal General Hospital Research Institute, Montreal, Quebec H3G 1A4, Canada **[357]**

H. Harm Bodemann*, Städtisches Krankenhaus, Sindelfingen, Federal Republic of Germany **[365]**

Alexander A. Boldyrev*, Department of Biochemistry, Moscow State University, 119899 Moscow, USSR

Rolf T. Borlinghaus*, Department of Biology, University of Konstanz, D-7750 Konstanz, Federal Republic of Germany

C.-H. Brogren*, Department of Animal Physiology and Biochemistry, Royal Veterinary and Agricultural University, DK-1871 Copenhagen, Denmark

N.E. Broude, M.M. Shemyakin Institute of Bioorganic Chemistry, USSR Academy of Sciences, Moscow, USSR **[135]**

Michael J. Caplan*, Department of Cell Biology, Yale University School of Medicine, New Haven, CT 06510 **[51]**

J.D. Cavieres*, Department of Physiology, Leicester University, Leicester LE1 7RH, England

Carol Cianci, Departments of Pathology, Physiology, and Medicine, Yale University, New Haven, CT 06510 **[245]**

Torben Clausen*, Institute of Physiology, University of Aarhus, 8000 Aarhus C, Denmark **[239,371]**

Flemming Cornelius*, Institute of Biophysics, Aarhus University, DK-8000 Aarhus C, Denmark

Gordon Cramb*, Department of Physiology and Pharmacology, Molecular Genetics Unit, University of St. Andrews, Fife, England KY16 9TS **[59,157]**

P. Csermely, Institute of Biochemistry I, Semmelweis University Medical School, Budapest 8, Hungary **[449]**

Christopher P. Cutler*, Department of Physiology and Pharmacology, University of St. Andrews, Fife, England KY16 9TS **[59]**

Enid Del Valle, Department of Pharmacology and Laboratory of Neurobiology, University of Puerto Rico School of Medicine, San Juan, Puerto Rico 00936 **[143]**

Leopoldo de Meis*, Instituto de Ciências Biomédicas, Departamento de Bioquímica, Universidade Federal do Rio de Janeiro, Ilha do Fundão, Rio de Janeiro 21910, Brasil

Contributors and Participants / xv

Victor V. Demin*, Shemyakin Institute of Bioorganic Chemistry, USSR Academy of Sciences, 117871 Moscow, USSR

J.J.H.H.M. de Pont*, Department of Biochemistry, University of Nijmegen, Nijmegen 6525 EZ, The Netherlands [437,443]

Alain De Pover*, Research Department, Pharmaceutical Division, Ciba-Geigy Ltd., CH-4002 Basel, Switzerland [377]

Vandana A. Desai, Department of Cell Biology, New York University School of Medicine, New York, NY 10016 [127]

Paul De Weer*, Department of Cell Biology and Physiology, Washington University School of Medicine, St. Louis, MO 63110

Alain Doucet*, Laboratoire de Physiologie Cellulare, College de France, 75231 Paris, France [339]

Philip B. Dunham*, Department of Biology, Syracuse University, Syracuse, NY 13244

Isidore S. Edelman*, Department of Biochemistry and Molecular Biophysics, Columbia University, New York, NY 10032 [65,277]

E. Erdmann*, Medizinische Klinik I, Klinikum Grosshadern, University of Munich, D-8000 München 70, West Germany [345]

Mikael Esmann*, Institute of Biophysics, University of Aarhus, DK-8000 Aarhus, Denmark [xxix]

Maria Elisabeth Everts*, Institute of Physiology, University of Aarhus, 8000 Aarhus C, Denmark [239,251,371]

Dorothy A. Fallows, Center for Cancer Research, Massachusetts Institute of Technology, Cambridge, MA 02139 [97]

David E. Farkas, University of Illinois Medical School, Chicago, IL 60612 [469]

Robert A. Farley*, Department of Physiology and Biophysics, University of Southern California School of Medicine, Los Angeles, CA 90033 [85]

Sima K. Farshi, Department of Physiology, Harvard Medical School, Boston, MA 02114 [149]

Gudrun Feige, Research Department, Pharmaceutical Division, Ciba-Geigy Ltd., CH-4002 Basel, Switzerland [377]

Simone Fenard, Laboratoire Nativelle, 91160 Longjumeau, France [263]

Klaus Fendler*, Max-Planck-Institut für Biophysik, 6000 Frankfurt am Main, Federal Republic of Germany

Mary Flanagan, Department of Cell Biology, New York University School of Medicine, New York, NY 10016 [127]

Harald G. Foellmer, Departments of Pathology, Physiology, and Medicine, Yale University, New Haven, CT 06510 [245]

Bliss Forbush III*, Department of Cellular and Molecular Physiology, Yale University School of Medicine, New Haven, CT 06510

P.A. George Fortes*, Department of Biology, University of California, San Diego, La Jolla, CA 92093

Harry Fozzard*, University of Chicago Hospital, Chicago, IL 60637

Jeffrey P. Froehlich*, National Institute on Aging, National Institutes of Health, Francis Scott Key Medical Center, Baltimore, MD 21224

Yun-Feng Fu*, Department of Biochemistry, Institute of Experimental Medicine, Hebei Academy of Medical Sciences, Shijiazhuang, Hebei Province, China [385]

P. Fuentes, Institute of Pharmacology and Biochemistry, University of Lausanne, CH-1005 Lausanne, Switzerland **[463]**

Yutaka Fukuda*, c/o Dr. Anita Aperia, St. Görans Children's Hospital, S-11281 Stockholm, Sweden

Yoshihiro Fukushima*, Laboratory of Active Transport, National Institute for Physiological Sciences, Okazaki 444, Japan

Patricio J. Garrahan*, Instituto de Química y Fisicoquímica Biológicas, Facultad de Farmacia y Bioquímica, Junin 956, 1113 Buenos Aires, Argentina

Kaethi Geering*, Institut de Pharmacologie, Université de Lausanne, CH-1005 Lausanne, Switzerland **[19,163]**

Gregory G. Gick*, Departments of Biochemistry and Molecular Biophysics, Columbia University, New York, NY 10032 **[65,277]**

Maureen Gilmore-Hebert*, Department of Internal Medicine, Yale University School of Medicine, New Haven, CT 06510 **[71]**

I.M. Glynn*, Physiological Laboratory, University of Cambridge, Cambridge CB2 3EG, England

Margarete Goppelt, Division of Molecular Pharmacology, Department of Pharmacology and Toxicology, Medical School of Hannover, 3000 Hannover, Federal Republic of Germany **[449]**

S.J. Graafsma, Department of Biochemistry, University of Nijmegen, 6500 Nijmegen, The Netherlands **[443]**

Eva Grafova, Montreal General Hospital Research Institute, Montreal, Quebec H3G 1A4, Canada **[357]**

E. Grell*, Max-Planck-Institute for Biophysics, Frankfurt, Federal Republic of Germany **[393]**

Charles M. Grisham*, Department of Chemistry, University of Virginia, Charlottesville, VA 22901

A.V. Grishin, M.M. Shemyakin Institute of Bioorganic Chemistry, USSR Academy of Sciences, Moscow, USSR **[135]**

Gunter Grupp, Departments of Physiology and Medicine, University of Cincinnati College of Medicine, Cincinnati, OH 45267-0575 **[321]**

I.L. Grupp, Departments of Pharmacology and Cell Biophysics, University of Cincinnati College of Medicine, Cincinnati, OH 45267-0575 **[321]**

Robert B. Gunn*, Department of Physiology, Emory University School of Medicine, Atlanta, GA 30322

Dipak Haldar, Department of Biological Science, St. John's University, Flushing, NY 11439 **[127]**

Otto Hansen*, Institute of Physiology, University of Aarhus, DK-8000 Aarhus C, Denmark

Yukichi Hara*, Department of Biochemistry, Tokyo Medical and Dental University, Tokyo 113, Japan

Hiroshi Hatanaka, Department of Neuroscience, Mitsubishi-Kasei Institute of Life Sciences, Machida, Tokyo 194, Japan **[91]**

Garner T. Haupert, Jr.*, Renal Unit, Massachusetts General Hospital and Harvard Medical School, Boston, MA 02114 **[297]**

Contributors and Participants / xvii

Yutaro Hayashi*, Department of Biochemistry, Kyorin University School of Medicine, Mitaka, Tokyo 181, Japan

Hans Hebert*, Department of Medical Biophysics, Karolinska Institutet, S-10401 Stockholm, Sweden

Sonia Castro Hernandez, Department of Pharmacology and Laboratory of Neurobiology, University of Puerto Rico School of Medicine, San Juan, Puerto Rico 00936 **[143]**

Gudrun Hippler-Feldtmann*, Physiologisch-Chemisches Institut, Abteilung Molekularbiologie, Universität Hamburg, D-2000 Hamburg 13, Federal Republic of Germany

Ann S. Hobbs*, National Institute of Neurological and Communicative Disorders and Stroke, National Institutes of Health, Bethesda, MD 20892

Joseph F. Hoffman*, Department of Cellular and Molecular Physiology, Yale University School of Medicine, New Haven, CT 06510 **[365]**

Lowell E. Hokin, Department of Pharmacology, University of Wisconsin Medical School, Madison, WI 53706 **[35]**

Mabel Hokin-Neaverson, Department of Psychiatry, University of Wisconsin, Madison, WI 53706 **[113]**

Haruo Homareda*, Department of Biochemistry, Kyorin University School of Medicine, Mitaka, Tokyo 181, Japan **[77]**

Burton Horowitz, University of Southern California School of Medicine, Los Angeles, CA 90033 **[85]**

David E. Housman, Center for Cancer Research, Massachusetts Institute of Technology, Cambridge, MA 02139 **[97]**

W.-H. Huang*, Department of Pharmacology, Medical College of Ohio, Toledo, OH 43699 **[401]**

Nobuo Inoue*, Department of Biochemistry, Kyorin University School of Medicine, Mitaka, Tokyo 181, Japan **[91]**

Faramarz Ismail-Beigi, Department of Medicine, Columbia University, New York, NY 10032 **[65, 277]**

James D. Jamieson, Department of Cell Biology, Yale University School of Medicine, New Haven, CT 06510 **[51]**

Einar Jebens*, Arbeidsfysiologisk Institut, Box 8159, Oslo 1, Norway

Peter Leth Jørgensen*, Institute of Physiology, Aarhus University, DK-8000 Aarhus C, Denmark

S.S. Kakar*, Department of Pharmacology, Medical College of Ohio, Toledo, OH 43699 **[401]**

Jack H. Kaplan*, Department of Physiology, University of Pennsylvania School of Medicine, Philadelphia, PA 19104-6085

Steven J.D. Karlish*, Biochemistry Department, Weizmann Institute of Science, Rehovot 76100, Israel

Michael Kashgarian*, Department of Pathology, Yale University, New Haven, CT 06510 **[245]**

Adrian I. Katz*, Department of Medicine, The University of Chicago, Pritzker School of Medicine, Chicago, IL 60637 **[207]**

Kiyoshi Kawakami, Department of Biology, Jichi Medical School, Minamikawachi, Tochigi 329-04, Japan **[77]**

Masaru Kawamura*, Department of Biology, University of Occupational and Environmental Health, Kitakyusyu 807, Japan xviii / Contributors and Participants

Linda J. Kenney*, Department of Physiology, University of Pennsylvania School of Medicine, Philadelphia, PA 19104; present address: Departments of Cellular and Molecular Physiology and Internal Medicine, Division of Hematology, Yale University School of Medicine, New Haven, CT 06510

Rachel B. Kent*, Center for Cancer Research, Massachusetts Institute of Technology, Cambridge, MA 02139 [97]

James R. Kinghorn, Department of Physiology and Pharmacology, Molecular Genetics Unit, University of St. Andrews, Fife, England KY16 9TS [157]

Keld Kjeldsen*, Medical Department, State University Hospital, DK-2100 Copenhagen, Denmark [251, 257, 269, 371]

Irena Klodos*, Institute of Biophysics, University of Aarhus, DK-8000 Aarhus C, Denmark

Hermann Koepsell*, Max-Planck-Institut für Biophysik, D-6000 Frankfurt 70, Federal Republic of Germany

Jens Kort*, Department of Molecular Biology, University of Hamburg, D-2000 Hamburg 13, Federal Republic of Germany

Tibor Kovacs*, Department of Physiology, Medical University of Debrecen, 4012 Debrecen, Hungary [233]

Maria Kovecses, Institute of Biochemistry I, Semmelweis University Medical School, H-1444 Budapest 8, Hungary [449]

J.P. Kraehenbuhl, Institute of Pharmacology and Biochemistry, University of Lausanne, CH-1005 Lausanne, Switzerland [463]

Andreas V. Lafaire, Max-Planck-Institut für Biophysik, D-6000 Frankfurt, Federal Republic of Germany [409]

Peter A. Lalley, Institute for Medical Research, Bennington, VT 05201 [97]

Joseph F. Lamb*, Department of Physiology and Pharmacology, Molecular Genetics Unit, University of St. Andrews, Fife, England KY16 9TS [59,157]

Jim S. Larsen*, Institute of Physiology, University of Aarhus, DK-8000 Aarhus C, Denmark [257]

Lars Larsson*, Department of Pediatrics, St. Görans Childrens' Hospital, S-11281 Stockholm, Sweden

Stefan C.H. Larsson*, Department of Developmental Physiology, St. Görans Barnkliniker, S-11281 Stockholm, Sweden

Søren Lassen*, Department of Cell Biology, Institute of Anatomy, University of Aarhus, DK-8000 Aarhus C, Denmark

Peter K. Lauf*, Department of Physiology and Biophysics, Wright State University School of Medicine, Dayton, OH 45401-0927

Peter Läuger*, Department of Biology, University of Konstanz, D-7750 Konstanz, Federal Republic of Germany

Claude Lechene*, Laboratory of Cellular Physiology and National Electron Probe Resource for Analysis of Cells, Harvard Medical School and Brigham and Women's Hospital, Boston, MA 02115 [171]

Lionel G. Lelievre*, INSERM U 127, Hôpital Lariboisière, Université Paris, 75010 Paris 7, France [43, 263]

Contributors and Participants / xix

Thomas Leutert, Research Department, Pharmaceutical Division, Ciba-Geigy Ltd., CH-4002 Basel, Switzerland [377]

Robert Levenson, Department of Cell Biology, Yale University School of Medicine, New Haven, CT 06510 [97]

E. Lewitzki, Max-Planck-Institute for Biophysics, Frankfurt, Federal Republic of Germany [393]

Jerry B. Lingrel*, Department of Microbiology and Molecular Genetics, University of Cincinnati College of Medicine, Cincinnati, OH 45267-0524 [105]

A.G. Lowe*, Department of Biochemistry, University of Manchester, Manchester M13 9PL, England

Joseph M. Lowndes*, Departments of Physiological Chemistry and Psychiatry, University of Wisconsin, Madison, WI 53706 [113]

Zhen-Min Lu, Department of Biochemistry, Institute of Experimental Medicine, Hebei Academy of Medical Sciences, Shijiazhuang, Hebei Proivince, China [385]

Jean-Michel Maixent*, Laboratoire Nativelle, 91160 Longjumeau, France [43, 263]

Andrea S. Mann, Departments of Pathology, Physiology, and Medicine, Yale University, New Haven, CT 06510 [245]

Sven Mårdh*, Department of Medical and Physiological Chemistry, Biomedical Centre, Uppsala University, S-751 23 Uppsala, Sweden [417]

Reinaldo Marin*, Department of Cellular and Molecular Physiology, Yale University School of Medicine, New Haven, CT 06510; present address: Instituto Venezolano de Investigaciones Cientificas, Caracas 1020A, Venezuela

Anthony Martonosi*, Department of Biochemistry, SUNY Health Sciences Center, State University of New York at Syracuse, Syracuse, NY 13210

Flemming Østerby Mathiasen*, Department of Cell Biology, Institute of Anatomy, University of Aarhus, DK-8000 Aarhus C, Denmark

Hideo Matsui*, Department of Biochemistry, Kyorin University School of Medicine, Mitaka, Tokyo 181, Japan [77, 91]

Arvid B. Maunsbach*, Department of Cell Biology, Institute of Anatomy, University of Aarhus, DK-8000 Aarhus C, Denmark [xxix]

Trudi McDevitt*, Department of Physiology and Pharmacology, Molecular Genetics Unit, University of St. Andrews, Fife, England KY16 9TS [157]

P. Meda, Department of Histology, Geneva University Medical Center, CH-1211 Geneva 4, Switzerland [429]

Maria Luisa Melzi*, Department of Developmental Physiology, St. Görans Hospital, S-112 81 Stockholm, Sweden

Robert W. Mercer*, Department of Cell Biology and Physiology, Washington University School of Medicine, St. Louis, MO 63110 [71, 119]

N.N. Modyanov*, M.M. Shemyakin Institute of Bioorganic Chemistry, USSR Academy of Sciences, Moscow, USSR [135]

Manijeh Mohraz*, Department of Cell Biology, New York University School of Medicine, New York, NY 10016

Jesper Vuust Møller*, Institut for Medicinsk Biokemi, University of Aarhus, DK-8000 Aarhus C, Denmark

G.S. Monastyrskaya, M.M. Shemyakin Institute of Bioorganic Chemistry, USSR Academy of Sciences, Moscow, USSR [135]

Takashi Morimoto*, Department of Cell Biology, New York University School of Medicine, New York, NY 10016 [127]

Jon S. Morrow, Department of Pathology, Yale University, New Haven, CT 06510 [245]

Nuran Nabi, Department of Cell Biology, New York University School of Medicine, New York, NY 10016; present address: Colgate Palmolive Company, Edison, NJ 08854 [127]

Kei Nagano, Department of Biology, Jichi Medical School, Minamikawachi-machi, Tochigi 329-04, Japan [77]

Makoto Nakao*, Department of Biochemistry, Tokyo Medical and Dental University, Tokyo 113, Japan

Toshiko Nakao*, Department of Biochemistry, Tokyo Medical and Dental University School of Medicine, Tokyo 113, Japan

Yuk-Chow Ng*, Department of Pharmacology and Toxicology, Michigan State University, East Lansing, MI 48824 [423]

Jens G. Nørby*, Institute of Biophysics, University of Aarhus, DK-8000 Aarhus C, Denmark [xxix]

Aage Nørgaard*, Institute of Physiology, University of Aarhus, DK-8000 Aarhus C, Denmark [251, 257, 269]

Toshiko Ohta*, Department of Biology, Jichi Medical School, Minamikawachi Tochigi, 329-04 Japan

Koichiro Omori, Department of Cell Biology, New York University School of Medicine, New York, NY 10016; present address: Department of Physiology, Kansai Medical University, Osaka, Japan [127]

Kyoko Omori, Department of Cell Biology, New York University School of Medicine, New York, NY 10016; present address: Department of Physiology, Kansai Medical University, Osaka, Japan [127]

George E. Palade, Department of Cell Biology, Yale University School of Medicine, New Haven, CT 06510 [51]

Carlos H. Pedemonte*, Department of Physiology, University of Pennsylvania School of Medicine, Philadelphia, PA 19104-6085

K.E. Petrukhin*, M.M. Shemyakin Institute of Bioorganic Chemistry, USSR Academy of Sciences, 117871 Moscow, USSR [135]

Igor W. Plesner*, Institute of Chemistry, University of Aarhus, DK-8000 Aarhus C, Denmark

Liselotte Plesner*, Institute of Biophysics, University of Aarhus, DK-8000 Aarhus C, Denmark

Carlomaria Polvani*, Montreal General Hospital Research Institute, Montreal, Quebec H3G 1A4, Canada

Thomas Pressley*, Department of Biochemistry, Columbia University, New York, NY 10032

Fulgencio Proverbio*, Department of Cellular and Molecular Physiology, Yale University School of Medicine, New Haven, CT 06510; present address: Instituto Venezolano de Investigaciones Cientificas, Caracas 1020A, Venezuela

Contributors and Participants / xxi

Ruth Rasch*, Department of Cell Biology, Institute of Anatomy, University of Aarhus, DK-8000 Aarhus C, Denmark

Kurt R.H. Repke*, Energy Conversion Unit, Central Institute of Molecular Biology, Academy of Sciences of GDR, 115 Berlin-Buch, German Democratic Republic

K. Resch, Division of Molecular Pharmacology, Department of Pharmacology and Toxicology, Medical School of Hannover, 3000 Hannover, Federal Republic of Germany **[449]**

Janet Rettig Emanuel, Department of Cell Biology, Yale University School of Medicine, New Haven, CT 06510 **[97]**

Helene G. Rey*, Department of Pharmacology, Geneva University Medical Center, CH-1211 Geneva 4, Switzerland **[429]**

Jacqueline A. Reynolds*, Department of Physiology, Duke University Medical Center, Durham, NC 27710

Donald E. Richards*, Physiological Laboratory, University of Cambridge, Cambridge CB2 3EG, England

Joseph D. Robinson*, Department of Pharmacology, SUNY Health Sciences Center, Syracuse, NY 13210

B.C. Rossier, Institute of Pharmacology and Biochemistry, University of Lausanne, CH-1005 Lausanne, Switzerland **[163]**

Paul M. Rowe*, Laboratory of Neurochemistry, National Institute of Neurological and Communicative Disorders and Stroke, National Institutes of Health, Bethesda, MD 20892

Arnold E. Ruoho, Department of Pharmacology, University of Wisconsin, Madison, WI 53706 **[113]**

David D. Sabatini, Department of Cell Biology, New York University School of Medicine, New York, NY 10016 **[127]**

John R. Sachs*, Department of Medicine, State University of New York at Stony Brook, Stony Brook, NY 11794

E. Schaerer, Institute of Pharmacology and Biochemistry, University of Lausanne, CH-1005 Lausanne, Switzerland **[463]**

Georgios Scheiner-Bobis*, Institut für Biochemie und Endokrinologie, Justus-Liebig-Universität Giessen, D-6300 Giessen, Federal Republic of Germany

Jay W. Schneider*, Department of Human Genetics, Yale University School of Medicine, New Haven, CT 06310 **[71,119]**

Wilhelm Schoner*, Institut für Biochemie und Endokrinologie, Justus-Liebig-Universität Giessen, D-6300 Giessen, Federal Republic of Germany

F.M.A.H. Schuurmans Stekhoven*, Department of Biochemistry, University of Nijmegen, Nijmegen, 6525 EZ, The Netherlands **[443]**

A.L. Schwartz, Edward Mallinckrodt Departments of Pediatrics and Pharmacology, Washington University School of Medicine, Division of Pediatric Hematology-Oncology, Children's Hospital, St. Louis, MO 63178 **[437]**

Arnold Schwartz*, Departments of Pharmacology and Cell Biophysics, University of Cincinnati College of Medicine, Cincinnati, OH 45267-0575 **[321]**

Wolfgang Schwarz*, Max-Planck-Institut für Biophysik, D-6000 Frankfurt, Federal Republic of Germany **[409]**

Bernd Schweigert, Max-Planck-Institut für Biophysik, D-6000 Frankfurt, Federal Republic of Germany **[409]**

Ole M. Sejersted*, Department of Physiology, National Institute of Occupational Health, 0033 Oslo 1, Norway **[195]**

Amar K. Sen*, Department of Pharmacology, Faculty of Medicine, University of Toronto, Toronto, Ontario M5S 1A8, Canada

Mohammed Iqbal Sheikh*, Institute of Medical Biochemistry, University of Aarhus, DK-8000 Aarhus C, Denmark

Jane M. Sherman, Department of Cell Biology, New York University School of Medicine, New York, NY 10016 **[127]**

Gary E. Shull*, Department of Microbiology and Molecular Genetics, University of Cincinnati Medical School, Cincinnati, OH 45267-0524

Marcia M. Shull*, Department of Microbiology and Molecular Genetics, University of Cincinnati College of Medicine, Cincinnati, OH 45267-0524 **[105]**

Jens Christian Skou*, Institute of Biophysics, Aarhus University, DK-8000 Aarhus C, Denmark **[xxix]**

Elisabeth Skriver*, Department of Cell Biology, Institute of Anatomy, University of Aarhus, DK-8000 Aarhus C, Denmark

Margareta Söderholm*, Department of Medical Physics, Karolinska Institutet, S-10401 Stockholm, Sweden

Janos Somogyi*, Department of Chemistry and Biochemistry I, Semmelweis University Medical School, H-1444 Budapest 8, Hungary **[233,449,457]**

Susan C. Specht*, Department of Pharmacology and Laboratory of Neurobiology, University of Puerto Rico School of Medicine, San Juan, Puerto Rico, 00936 **[143]**

Wilfred D. Stein*, Department of Biochemistry, Hebrew University, Jerusalem 91904, Israel

Marcia Steinberg*, Department of Pharmacology, State University of New York Health Science Center at Syracuse, Syracuse, NY 13210

G.J. Strous, Laboratory for Cell Biology, University of Utrecht, 3511 Utrecht, The Netherlands **[437]**

Yuji Sugita, Department of Cell Biology, New York University School of Medicine, New York, NY 10016; present address: Department of Pharmaceutical Sciences, Showa University, Tokyo, Japan **[127]**

E.D. Sverdlov, M.M. Shemyakin Institute of Bioorganic Chemistry, USSR Academy of Sciences, Moscow, USSR **[135]**

V.E. Sverdlov, M.M. Shemyakin Institute of Bioorganic Chemistry, USSR Academy of Sciences, Moscow, USSR **[135]**

H.G.P. Swarts*, Department of Biochemistry, University of Nijmegen, 6500 Nijmegen, The Netherlands **[443]**

Kathleen J. Sweadner*, Pappas Laboratories, Neurosurgical Service, Massachusetts General Hospital, Boston, MA 02114 **[149]**

Marta Szamel, Institute of Biochemistry I, Semmelweis University Medical School, H-1444 Budapest 8, Hungary **[449]**

Contributors and Participants / xxiii

L. Szollár, Institute of Pathophysiology, Semmelweis University Medical School, H-1444 Budapest 8, Hungary [449]

Charles Tanford*, Department of Physiology, Duke University Medical Center, Durham, NC 27710

Kazuya Taniguchi*, Department of Pharmacology, School of Dentistry, Hokkaido University, Sapporo 060, Japan

Jan Teisinger, Institute of Hygiene and Epidemiology, 10000 Prague 10, Czechoslovakia [477]

T.A. Thien, Departments of Biochemistry and Internal Medicine, University of Nijmegen, 6500 Nijmegen, The Netherlands [443]

J.A.H. Timmermans, Department of Biochemistry, University of Nijmegen, 6500 Nijmegen, The Netherlands [443]

Daniel C. Tosteson*, Department of Physiology, Harvard Medical School, Boston, MA 02115

Hiroko Tsukui, Department of Neuroscience, Mitsubishi-Kasei Institute of Life Sciences, Machida, Tokyo 194, Japan [91]

Agnes Turi*, Department of Chemistry and Biochemistry I, Semmelweis University Medical School, H-1444 Budapest 8, Hungary [457]

D. Uemura, Faculty of Liberal Arts, Shizuoka University, Ohya, Shizuoka, Japan [393]

Shiela Unkles, Department of Physiology and Pharmacology, Molecular Genetics Unit, University of St. Andrew's, Fife, England KY16 9TS [157]

Hans H. Ussing*, Institute of Biological Chemistry A,, University of Copenhagen, 2100 Copenhagen, Denmark [3]

A. van Bokhoven, Department of Biochemistry, University of Nijmegen, 6500 Nijmegen, The Netherlands [437]

Harry van der Hijden*, Department of Biochemistry, University of Nijmegen, 6500 HB Nijmegen, The Netherlands

P. van Kerkhof, Laboratory for Cell Biology, University of Utrecht, 3511 Utrecht, The Netherlands [437]

F. Verrey*, Institute of Pharmacology and Biochemistry, University of Lausanne, CH-1005 Lausanne, Switzerland [463]

Bente Vilsen*, Institute of Physiology, University of Aarhus, DK-8000 Aarhus C, Denmark

Carol Voy, Department of Physiology and Pharmacology, Molecular Genetics Unit, University of St. Andrews, Fife, England KY16 9TS [157]

František Vyskočil, Institute of Physiology, Czechoslovak Academy of Sciences, 14220 Prague 4, Czechoslovakia [477]

Bonnie A. Wallace*, Department of Chemistry and Center for Biophysics, Rensselaer Polytechnic Institute, Troy, NY 12180

Earl Wallick, Departments of Pharmacology and Cell Biophysics, University of Cincinnati College of Medicine, Cincinnati, OH 45267-0575 [321]

Wei Wang, Department of Biochemistry, Institute of Experimental Medicine, Hebei Academy of Medical Sciences, Shijiazhuang, Hebei Province, China [385]

J. Andrew Wasserstrom*, Department of Medicine, Northwestern University Medical School, Chicago, IL 60611 [469]

K. Werdan, Medizinische Klinik I, Klinikum Grosshadern, University of Munich, D-8000 München 70, West Germany [345]

Z. Xie, Department of Pharmacology, Medical College of Ohio, Toledo, OH 43699 [401]

Kai-yuan Xu*, Department of Chemistry, University of California, San Diego, La Jolla, CA 92093

Atsunobu Yoda*, Department of Pharmacology, University of Wisconsin Medical School, Madison, WI 53706

James M. Yohanan, Department of Pharmacology, University of Wisconsin Medical School, Madison, WI 53706 [35]

Regina M. Young, Department of Microbiology and Molecular Genetics, University of Cincinnati Colege of Medicine, Cincinnati, OH 45267-0524 [105]

Daniele Zamofing*, Institut de Pharmacologie, Université de Lausanne, CH-1005 Lausanne, Switzerland [163]

Hana Zemková*, Institute of Physiology, Czechoslovak Academy of Sciences, 14220 Prague 4, Czechoslovakia [477]

Shi-Lian Zhang, Department of Biochemistry, Institute of Experimental Medicine, Hebei Academy of Medical Sciences, Shijiazhuang, Hebei Province, China [385]

Reinhard Zibirre*, Physiologisch-Chemisches Institut, Abteilung Molekularbiologie, Universität Hamburg, D-2000 Hamburg, Federal Republic of Germany

Y.S. Zou*, Department of Biochemistry, University of Nijmegen, Nijmegen 6525 EZ, The Netherlands

Contents of Part A

AMINO ACID SEQUENCE AND STRUCTURE

Overview: Amino Acid Sequences of the α and β Subunits of the Na^+,K^+-ATPase / *Gary E. Shull, Regina M. Young, Jeannette Greeb, and Jerry B. Lingrel*

Overview: Structural Basis for Coupling of E_1-E_2 Transitions in α/β-Units of Renal Na^+,K^+-ATPase to Na^+,K^+-Translocation / *Peter Leth Jørgensen*

Overview: Three-Dimensional Structure and Topography of Membrane Bound Na^+,K^+-ATPase / *Arvid B. Maunsbach, Elisabeth Skriver, Margareta Söderholm, and Hans Hebert*

Sequence Analysis of Exposed Domains of Membrane-Bound Na^+, K^+-ATPase. Model of Transmembrane Arrangement / *N.M. Arzamazova, E.A. Arystarkhova, N.M. Gevondyan, N.M. Luneva, R.G. Efremov, N.A. Aldanova, V.A. Nesmeyanov, and N.N. Modyanov*

Spatial Structure and Membrane Organization of Pig Kidney Na^+,K^+-ATPase / *Victor V. Demin, Alexander A. Barnakov, and Alexander P. Kuzin*

The Third Type of Alpha-Subunit of Na^+,K^+-ATPase / *Yukichi Hara, Osamu Urayama, Kiyoshi Kawakami, Hiroshi Nojima, Hideaki Nagamune, Toshiyuki Kojima, Toshiko Ohta, Kei Nagano, and Makoto Nakao*

Three-Dimensional Structure of Protomeric Renal Na^+,K^+-ATPase / *H. Hebert, E. Skriver, M. Söderholm, and A.B. Maunsbach*

Localization of Tryptic and Chymotryptic Cleavage Sites in α-Subunit of Na^+,K^+-ATPase / *Peter L. Jørgensen and John H. Collins*

Structure and Function of the β Subunit of (Na^+,K^+) ATPase / *Masaru Kawamura and Shunsuke Noguchi*

Three-Dimensional Structure of Na^+,K^+-ATPase and a Model for the Oligomeric Form and the Mechanism of the Na^+,K^+ Pump / *Manijeh Mohraz and P.R. Smith*

Structure of Na^+,K^+-ATPase: The Locations of ATP Binding Regions, Trypsin Accessible Regions and Membrane-Embedded Regions / *Toshiko Ohta, Masasuke Yoshida, Masaru Kawamura, and Kei Nagano*

Antibodies to Synthetic Peptides as Probes of the Structure of the α-Subunit of the Na^+,K^+-ATPase / *Paul M. Rowe, W. Timothy Link, Anup K. Hazra, Philip G. Pearson, and R. Wayne Albers*

Structural Studies of Na^+,K^+-ATPase Subunits / *B.A. Wallace, D.E. Elstein, J. Salon, and T. DiNolfo*

An Inquiry Into the Binding of Cr-ATPa by $(Na^+ + K^+)$-ATPase / *Kai-yuan Xu*

OLIGOMERIC STRUCTURE: PHYSICOCHEMICAL AND KINETIC EVIDENCE

Overview: The Oligomeric Structure of the Na^+,K^+ Pump Protein / *Jacqueline A. Reynolds*

Overview: Ligand Binding Sites of $(Na^+ + K^+)$-ATPase: Nucleotides and Cations / Amir Askari

Characterization of Pig Kidney Na^+,K^+-ATPase by Antibodies / C.-H. Brogren, O. Hansen, and J. Jensen

Association of Biochemical Functions With Specific Subunit Arrangements in Purified Na^+,K^+-ATPase / J.D. Cavieres

Target-Residues of the Active Site Affinity Modification are Different in E_1 and E_2 Forms / K.N. Dzhandzhugazyan, S.V. Lutsenko, and N.N. Modyanov

A Revised Boundary Lipid Count for Na^+,K^+-ATPase From *Squalus acanthias* / Mikael Esmann, Kalman Hideg, and Derek Marsh

Distances Between 5-Iodoacetamidofluorescein and the ATP and Ouabain Sites of (Na^+,K^+)-ATPase Determined by Fluorescence Energy Transfer / P.A. George Fortes and Richard Aguilar

High-Performance Gel Chromatography of Active Solubilized Na^+,K^+-ATPase Maintained by Exogenous Phosphatidylserine / Yutaro Hayashi, Kunihiro Mimura, Hideo Matsui, and Toshio Takagi

Properties of the Na^+,K^+, and ATP Binding Sites of the Ouabain-Complexed $(Na^+ + K^+)$-ATPase: Implications for the Mechanism of Ouabain Action / S.S. Kakar, W.-H. Huang, and A. Askari

The Catalytic Subunits Cooperate in $(Na^+ + K^+)$-ATPase: Demonstration With $MgPO_4$ and MgATP Complex Analogues / Georgios Scheiner-Bobis, Engelbert Buxbaum, and Wilhelm Schoner

CONFORMATIONAL TRANSITION, ION OCCLUSION AND PHOSPHOINTERMEDIATES

Overview: Occluded Ions and Na^+,K^+-ATPase / Bliss Forbush III

Overview: The Phosphointermediates of Na^+,K^+-ATPase / Jens G. Nørby and Irena Klodos

Pools of (Na^+,K^+)-ATPase Isoforms in Unborn and Adult Rats Can Be Detected by Their Sensitivity to Strophanthidin, but Have Similar Reactivity to Na^+,K^+ and ATP / Gustavo Blanco and Luis Beaugé

Mechanism of Substrate Specificity of the Na^+,K^+-Pump / Alexander A. Boldyrev, Olga D. Lopina, and Irina A. Svinukhova

Isolation of Two Isoforms of Na^+,K^+-ATPase From *Artemia salina*: Kinetic Characterization / N. Cortas and I.S. Edelman

Determination of Ox Brain Na^+,K^+-ATPase Isoforms and Their Phosphorylated Intermediates by Analysis of Anthroylouabain Binding and Its Inhibition by Oligomycin / P.A.G. Fortes, Liselotte Plesner, and Igor W. Plesner

Amine Base-Dependent Inactivation of Na^+,K^+-ATPase: Two Different Aspects / Yoshihiro Fukushima

Complex Time Dependence of Phosphoenzyme Formation and Decomposition in Electroplax Na^+,K^+-ATPase / Ann S. Hobbs, R. Wayne Albers, and Jeffrey P. Froehlich

ATP Synthesis by Na^+,K^+-ATPase. Effect of K^+ on the ADP-Dependent Dephosphorylation of Phosphoenzymes / Irena Klodos

Does ATP Affect the Interconversion and the Dephosphorylation of the Phosphoenzymes of the Na^+,K^+-Pump? / Irena Klodos and Jens G. Nørby

Interaction of the Na^+,K^+-ATPase with H_2DIDS / Carlos H. Pedemonte and Jack H. Kaplan

Exchange and Hydrolysis Kinetics of Na^+,K^+-ATPase / *Liselotte Plesner and Igor W. Plesner*

Dephosphorylation Schemes for the Na^+,K^+-ATPase Reconsidered in Light of Allosteric Sites for Na^+ / *Joseph D. Robinson*

A Comparison Between the Empirical Behaviour of the Na^+,K^+-ATPase and the Predictions of the Albers-Post Model / *Roland C. Rossi and Patricio J. Garrahan*

Phosphorylation of $(Na^+ + K^+)$-ATPase. Stimulation and Inhibition by Substituted and Unsubstituted Amines / *F.M.A.H. Schuurmans Stekhoven, H.G.P. Swarts, G.K. 't Lam, Y.S. Zou, and J.J.H.H.M. de Pont*

Pb^{2+} Activated ATP-Phosphorylation as a Tool to Determine the E1/E2 Ratio of Na^+,K^+-ATPase / *H.G.P. Swarts, H.A.H.T. Zwartjes, F.M.A.H. Schuurmans Stekhoven, and J.J.H.H.M. de Pont*

Structural Changes in Na^+,K^+-ATPase Estimated by Intrinsic and Extrinsic Fluorescence Probes / *Kazuya Taniguchi, Kuniaki Suzuki, Toshio Sasaki, Hiroyuki Tosa, and Eiji Shinoguchi*

Quenching of the Tryptophan Fluorescence of Na^+,K^+-ATPase With Acrylamide / *Patricia A. Tyson and Marcia Steinberg*

Occlusion of $^{22}Na^+$ and $^{86}Rb^+$ in Membrane-Bound and Soluble Protomeric α/β-Units of Na^+,K^+-ATPase / *Bente Vilsen, Jens P. Andersen, Janne Petersen, and Peter L. Jørgensen*

Effect of Cytoplasmic K^+ on ADP- and K^+-Sensitive EP of Reconstituted Na^+,K^+-ATPase Proteoliposomes / *Atsunobu Yoda and Shizuko Yoda*

TRANSPORT MECHANISM:MOLECULAR ASPECTS

Overview: Mechanism of Free Energy Coupling Between ATP Hydrolysis and Ion Transport / *Charles Tanford*

Overview: Stoichiometry and Voltage Dependence of the Na^+/K^+ Pump / *Paul De Weer, David C. Gadsby, and R.F. Rakowski*

Overview: The Coupling of Enzymatic Steps to the Translocation of Sodium and Potassium / *I.M. Glynn*

Symmetric Active Transport in Cholate-Dialysed Liposomes Containing Randomly Oriented Sodium Pumps / *B.M. Anner, M. Moosmayer, and H.G. Rey*

Na^+,K^+-ATPase in Artificial Lipid Vesicles: Potential Dependent Transport Rates Investigated by a Fluorescence Method / *Hans-Jürgen Apell and Beate Bersch*

Fast Charge Translocations Associated With Partial Reactions of Na^+,K^+-ATPase Induced by ATP Concentration-Jump / *Rolf T. Borlinghaus, Hans-Jürgen Apell, and Peter Läuger*

Non-Equivalent Cytoplasmic Na^+ Sites and Their Susceptibility to Transmembrane Interaction From Extracellular Na^+ / *Flemming Cornelius and Jens Christian Skou*

The Na^+/K^+ Pump of LK Sheep Erythrocytes: Modulation by a Membrane Antigen / *Philip B. Dunham*

Pump Currents Generated by Renal $Na^+ K^+$-ATPase and Gastric $H^+ K^+$-ATPase on Black Lipid Membranes / *Klaus Fendler, Harry van der Hijden, Georg Nagel, J.J.H.H.M. de Pont, and Ernst Bamberg*

Effects of Ca^{2+} in the Presence and Absence of Mg^{2+} on the Cytoplasmic Aspect of the Na^+/K^+ Pump in High K^+ Sheep Red Cells / *Hiroshi Fujise and Peter K. Lauf*

Charge Transfer by the Na^+/K^+ Pump / *Steven J.D. Karlish, Rivka Goldschleger, Yosepha Shahak, and Ada Rephaeli*

The Vectorial Effect of Ligands on the Occluded Intermediate in Red Cell Sodium Pump Transport / *Linda J. Kenney and Jack H. Kaplan*

Properties of Two Types of HeLa Low K^+ (LK) Mutants With Respect to K^+ Transport / *Jens Kort and Gebhard Koch*

Two Different Types of ATP-Dependent Anion Coupled Na^+ Transport are Mediated by the Human Red Blood Cell and Na^+/K^+ Pump / *Reinaldo Marin and Joseph F. Hoffman*

Interaction of External Na^+ With the Na^+-ATPase. A Byproduct Inhibition? / *Carlos Humberto Pedemonte*

Proton Effects on the Sodium Pump / *Carlomaria Polvani and Rhoda Blostein*

Functional Consequences of the Membrane Pool of ATP Associated With the Human Red Blood Cell Na^+/K^+ Pump / *Fulgencio Proverbio, David G. Shoemaker, and Joseph F. Hoffman*

The Na^+ Pump Reaction Mechanism is Ping-Pong With Respect to Na^+ and K^+ / *John R. Sachs*

How the Energy From ATP Hydrolysis is Used to Render Ion Pumping Effective / *Wilfred D. Stein*

COMPARISON OF Na^+,K^+-ATPase, H^+,K^+-ATPase, and Ca^{2+}-ATPase

Overview: H^+,K^+-ATPase: Na^+,K^+-ATPase's Stepsister / *J.J.H.H.M. de Pont, M.L. Helmich-de Jong, A.T.P. Skrabanja, and H.T.W.M. van der Hijden*

Overview: Subunit Interaction and Conformational States. Ca^{2+}-ATPase and Na^+,K^+-ATPase Compared / *Jens P. Andersen and Bente Vilsen*

Overview: Energy Transduction in Biological Membranes. Role of Solvation Energy / *Leopoldo de Meis*

NMR Studies of Complexes of Gd^{3+} With Nucleotides and Substrate Analogues on Sarcoplasmic Reticulum Ca^{2+}-ATPase / *Mary R. Klemens and Charles M. Grisham*

Preface

The idea of a sodium pump in the cell membrane was introduced by R.B. Dean in 1941 in a paper titled, "Theories of electrolyte equilibrium in muscle" (Biol. Symp. 3:331–348). Referring to experiments by L.A. Heppel (1938), by C.L.A. Schmidt (1939), and by H.B. Steinbach (1940) Dean concluded "that the muscle can actively move potassium and sodium against concentration gradients . . . [but] this requires work. Therefore there must be some sort of a pump possibly located in the fiber membrane, which can pump out the sodium or, which is equivalent, pump in the potassium."

In the nearly half century that has passed since Dean's proposal, extensive work has been done by scientists all over the world to understand how a biological system can convert chemical energy from the hydrolysis of adenosine triphosphate (ATP) into a movement of cations against electrochemical gradients. For thirty years it has been known that the Na^+,K^+-pump is identical with membrane-bound Na^+, K^+-ATPase, and a large amount of information has been collected about the system. The increase in knowledge has revealed the complexity of the system, and thus far it has not been possible to determine the nature of molecular events that lead to coupling between the chemical reaction and the translocation reaction.

To facilitate the exchange of information, five international conferences have been held on the subject, the first in 1973 in New York, sponsored by the New York Academy of Sciences; the second in 1978 in connection with Aarhus University, Denmark. Since then it has been a tradition to hold a conference every third year: in 1981 at Yale University, New Haven, CT; in 1984 in Cambridge, England; and in 1987 again in connection with Aarhus University, Denmark.

The 1987 meeting was held June 14–19 at Fuglsø Conference Center in Denmark. The conference was organized by F. Cornelius, M. Esmann, I. Klodos, J.G. Nørby, J.C. Skou (Institute of Biophysics), and A.B. Maunsbach (Department of Cell Biology, Institute of Anatomy), University of Aarhus, Denmark. On behalf of the organizers we thank all who took care of the practical arrangements, and those who helped to run the conference. We wish to express our special thanks to Edith Moldt and Jytte Kragelund for their excellent secretarial assistance.

This volume, Part B, concerns the cellular and physiological aspects of the Na^+, K^+-pump presented at the conference. It contains the papers dealing with biosynthesis and molecular biology; the physiological regulation of Na^+, K^+-transport; Na^+, K^+-homeostasis; the role of Na^+, K^+-transport in secondary active transport in specialized tissues like the kidney, skeletal muscle, and heart muscle; and the Na^+, K^+-pump as receptor for cardiac glycosides. The volume begins with a historical review of the concept of active transport.

The accompanying volume, Part A, presents the papers on the molecular aspects of the Na^+, K^+-pump: Na^+, K^+-amino acid sequence, structure and the enzymatic and transport mechanism of Na^+, K^+-ATPase. Part A also contains a section where this system is compared with other transport ATPases.

The two volumes give an overview of the present status in the field and also of the problems to be solved before it is possible to understand the molecular events underlying the coupling between the chemical reaction and transport.

Jens C. Skou
Jens G. Nørby
Arvid B. Maunsbach
Mikael Esmann

Acknowledgments

On behalf of all the participants, the organizing committee of the Fifth International Conference on Na^+,K^+-ATPase, held at Fuglsø June 14–19, 1987, acknowledge the generous financial support from: The Danish Ministry of Education, University of Aarhus Research Foundation, The Danish Medical Research Council, The Danish Science Research Council, The Novo Foundation, The Leo Foundation, The M. Møller Foundation, The M.C. Holst Foundation, and the following companies: Provinsbanken A/S, Handelsbanken A/S, Sparekassen SDS, Nordisk Gentofte A/S, Upjohn, Janssenpharma A/S, Packard Instruments, and Ceres Breweries A/S.

The Concept of Active Transport

OVERVIEW:
THE DEVELOPMENT OF THE CONCEPT OF ACTIVE TRANSPORT

Hans H. Ussing

Institute of Biological Chemistry A,
University of Copenhagen, 2100 Copenhagen Ø,
Denmark

When I was asked to lecture on the development of the concept of active transport I soon realized that a proper coverage of the subject could not be given in 45 minutes. Therefore, being both a participant in and an observer of that development I decided to tell the story as it looked from my own vantage point. A more extensive treatment of the subject can be found in two earlier reviews (Ussing, 1949b, Ussing et al., 1960).

As long as physiology has existed, physiologists have been aware of the fact that living organisms can secrete and absorb substances in a way which could not be expected in a "dead" system. For more than half a century such processes have been designated as active transfer or active transport and the vast majority of scientists implicitly assumed that, somehow, they were driven by metabolic energy. However, when it came to deciding which processes should be rightly considered as active transport, the opinions have differed and scientific battles have been fought for over a century.

Active Transport Explained Away

When, in 1935, I started as research assistant under professor August Krogh, it was toward the end of a long period of attempts to explain away apparent cases of active transport. Some will remember from the history of respiratory physiology the controversy between two outstanding Danish physiologists, Kristian Bohr and his pupil August Krogh, concerning the exchange of respiratory gasses in the lung. Bohr had come to the conclusion that the lung must transfer oxygen from the air to the blood by a secretory process, an opinion which he shared with many physiologists, for instance J.B. Halldane. Krogh, however, was able to demonstrate that the transport of respiratory gasses could always be explained as a result of simple diffusion. With respect to the distribution of electrolytes between cells and their surroundings, a large minority believed it to be passive, but that the colloids of the cells had a high affinity for potassium and a low affinity for sodium and chloride ions. The majority assumed the cells to be tight to most electrolytes, with the exception that excitable cells were permeable to potassium ions.

The Isotope Age

There were conflicting observations which did not fit these two simple cell models and more were to come. Toward the end of the thirties isotopic tracers for sodium, potassium, and chloride became available, and it became clear that the supposedly tight cells exchanged their contents of the monovalent ions mentioned more or less rapidly with their surroundings. A last ingenious attempt to explain the electrolyte distribution in striated muscle without resorting to active transport was made by Boyle and Conway (1941). They made the assumption that the fibre membrane was tight to phosphate esters, protein anions, and sodium ions, but permeable to chloride and potassium ions. If, initially, the fibres had a low content of sodium and a high content of phosphates and proteins, then a double Donnan equilibrium would be established, so that the nondiffusible cellular anions would be matched by the diffusible potassium ions, whereas the ensuing negative cellular potential would keep chloride low in the cytoplasm. This model beautifully predicted cell volume, membrane potential, and electrolyte composition when the electrolyte composition of the medium was altered. However, already at the time the Boyle - Conway theory was advanced, there was ample evidence that muscle fibres can lose and regain sodium ions (see f.ex. Fenn (1938) and that the fibre sodium can exchange with radioactive sodium added to the perfusion medium (Heppel, 1940). For further references, see Ussing (1949b, Using et al., 1960). In order to explain the apparently conflicting observations, Dean (1941) proposed that the fibre membrane was provided with a "sodium pump" which expelled sodium as fast as it entered.

The existence of active sodium transport already had been demonstrated beyond any reasonable doubt in connection with the osmoregulation of freshwater animals. Thus Krogh had found salt-depleted frogs to perform a net uptake of sodium as well as chloride from extremely dilute solutions (Krogh, 1937, 1938) and he had also shown that sodium uptake could take place in the absence of any net uptake of chloride and vice versa. For further references to the early literature on active transport by aquatic animals, see Krogh (1939). He also had initiated a study by aid of isotopes of active ion uptake by giant plant cells (see Holm-Jensen et al., 1944). At the time of his retirement 1945 he asked me to head a small team with the object of exploiting the possibilities for the use of isotopes, produced by The Niels Bohr Institute, for studies of active transport. At the time my main interest was to study protein synthesis with the aid of deuterium-labelled amino acids, but I found it imperative to give the electrolyte studies a good start.

In his Croonian lecture in 1946 Krogh had pointed out that in many cases the large differences between the concentration of substances in cells and their surroundings must be upheld by a simultaneous active transport in one direction and passive diffusion in the other. In such cases the measurement of the rate of exchange of the substance across the cell membranes, as measured with a proper isotope under steady state conditions, would simultaneously give the rate of active transport and the rate of passive leak. Quite naturally we decided to study the sodium transport in frog sartorius muscle which was the object used by Conway and his group. Right from the outset it was clear that there was no a priori reason to consider the distribution of potassium and chloride ions between fibre and medium to be due to active transport of these ions. Their distribution was amply explained by Donnan distribution as it was proposed by Boyle and Conway.

"Exchange Diffusion" May Mimic Active Transport

By loading frog sartorius muscles with ^{24}Na in ^{24}Na Ringer and following the rate of washout we were able to separate the rate of washout of the interspaces from the rate of exchange across the fibre membranes, and thus we could calculate the sodium flux through the membrane (which ought to be equal to the pumping rate of the sodium pump). At the same time, since we knew the sodium concentration in the fibres as well as the membrane potential, we could calculate the minimum work necessary to perform the active transport. It turned out that more than the total free energy made available by metabolism was required in order to pump sodium at the rate found. It then occurred to me (Ussing, 1947) that thermodynamics does not prohibit a strict one-to-one exchange of an ion, even if its electrochemical potential is different on the two sides of the membrane. Since I had no other choice, I proposed as an explanation for our findings that, besides active transport of sodium, the membrane must also possess a mechanism for what was named "exchange diffusion".

The Isolated Frog Skin Performs Active Na$^+$ Transport

In principle the same argument might be correct for all cases where isotopic tracers were used for measuring steady state exchanges. Studies of ion exchange between cells and their surroundings now were faced by a double set of uncertainties. In the first place, as already mentioned, several powerful schools did not believe in the cell membrane as a diffusion barrier but preferred to think of the skewed distribution of certain ion species to be due to specific adsorption by the colloids of the cytoplasm. In any case the determination of activity coefficients of ions in the cytoplasm presented unsolved problems. And now the possible interference of exchange diffusion made it even more difficult to estimate the true rate of active transports. Under these circumstances I decided to direct our efforts toward systems where a manifest net transport of ions was taking place. We therefore chose the frog skin as experimental object. Already in 1935 Huf had demonstrated net inward transport of chloride through isolated frog skin when it was bathed with Ringer solution on both sides; and, as mentioned above, Krogh had demonstrated that salt-depleted frogs will take up sodium and chloride from very dilute solutions. The frog skin is known to maintain an electric potential difference between the inside and outside solutions of up to 170 mV (inside plus), as long as the outside bathing solution contains sodium ions. Thus the inward transport of sodium ions has to be active. The chloride transport, on the other hand, might at least partly be passive, driven by the potential difference.

The Flux Ratio as a Criterion for Active Transport

Only a transport against the combined effects of potential and concentration differences should be considered active. This was the definition implicit in our work with sartorius muscle, and this definition was explicitly expressed by Rosenberg (1948) who was working on ionic exchanges in red cells.

But now our problem was: what part, if any, of a measured isotope flux is due to active transport, what part to exchange diffusion, and what part to passive electrodiffusion? It was obviously necessary to find a criterion for simple passive diffusion of an electrolyte. In the case of a thin homogeneous membrane the problem had been solved by Goldman (1944) who used the constant field assumption to integrate the flux equation. But we studied multilayered epithelia where it would be senseless to assume a constant electric field. Even today, with the help of computers, the problem is insolvable, because conditions have to be known along the whole transport path. However, after some attempts, I realized that all the unknowns vanish if one calculated the flux ratio instead of the individual fluxes. In any system with only electrodiffusion, the ratio between forward and backward fluxes are determined solely by the difference in electrochemical potential between the outside and inside bathing solutions (Ussing, 1949a). At the same time the difference between the forward and backward flux is a measure of the net transport. Thus, whereas the flux measured in only one direction contained virtually no information concerning net transport and driving force, the measurement of flux-pairs determined both these parameters. At that time, before the development of commercial flame spectrometers, a precise determination of net transports was just as difficult as estimation of driving forces.

Based on the flux theories outlined above we were now able to show that the net transport of sodium ions through the frog skin was almost 100 per cent active, whereas chloride showed ideal passive behaviour.

Single-file Diffusion May Mimic Active Transport

It looked as if it would be possible to characterize a transport as being active solely on the basis of the flux ratio: if the forward flux of a substance divided by the backward flux was larger than that calculated from its electrochemical potentials on the two sides of the "membrane", the active transport must be responsible, wholly or in part. As long as we had to consider only electrodiffusion, exchange difusion, and active transport, the method would have been applicable, even for a net transport going from a higher to a lower electrochemical potential of the ion. However, in 1955 Hodgkin and Keynes discovered the "single file" phenomenon for potassium ions passing the membrane of a giant squid nerve fibre: the net flux of K^+ was always in the direction of the electrochemical potential gradient, but the flux ratio was much larger than expected. Indeed one had to raise the calculated value to the third power to obtain the flux ratios measured. According to the theory developed by the two authors, the potassium pathway has three sites in single file, so that all of them have to be occupied in succession by any potassium ion crossing the membrane. The relationship mentioned means that for potassium, at least, an abnormally high flux ratio during "downhill" transport does not necessarily indicate active transport. So far, to my knowledge, single file behaviour has not been found for other "biological ions", but the mere fact that the phenomenon exists, calls for caution in the evaluation of downhill transports.

The Short-circuit Criterion for Active Transport

The finding that sodium transport was active, whereas net chloride transport was driven by the transepithelial potential difference, made it tempting to propose that the potential difference across the skin was solely due to the active inward transport of the positive charges of sodium.

In order to prove that, we developed the short-circuit technique (Ussing & Zerahn, 1951). At the time, it was already known that, if the skin was bathed with Ringer on both sides, it was possible to draw current when the two bathing solutions were connected via reversible electrodes. The opinions differed with respect to the origin of the potential. One hypothesis was that respiratory CO_2 was split in hydrogen and bicarbonate ions which left the skin in opposite directions. Another hypothesis was that it was an oxydation reduction potential. A rough calculation showed me that, if the potential could be totally short-circuited, the current would be of the same order of magnitude as the net sodium flux. Even the best reversible electrodes had too high resistance for total short-circuit, and to that came the resistance of the bathing solutions. However, by putting a battery and a variable resistance in series with a skin separating identical Ringer solutions, a complete short-circuit was achieved. It then turned out that the short-circuit current was exactly equal to the net sodium transport. Thus the latter must be the sole reason for electric asymmetry of the skin. Control experiments showed that, during short-circuit, the net chloride transport was nil. The existence of active sodium transport had been demonstrated beyond any reasonable doubt.

Without many noticing it, the acceptance of the short-circuiting technique as a means of demonstrating active transport had subtly changed the definition of this phenomenon. During short-circuiting there is not a transport from a lower to a higher electrochemical potential, but the system is considered as an electric element with an electromotive force and an internal resistance. Under the assumption that we were dealing with a specific and reversible chemical reaction, and that both forward and backward reaction used the same chemical pathway, the electromotive force could be calculated as $(RT/F)\ln(J_{in}/J_{out})$ where J_{in} and J_{out} are the inward and outward sodium flux, respectively. The assumption that we were dealing with a specific and reversible reaction may seem quite natural today (see Glynn and Lew, 1970) but it was not so at the time. When I presented our first isotope experiments at a meeting at Cold Spring Harbor (Ussing, 1948) and gave the arguments for a chemical membrane process which could distinguish sodium from potassium, the grand old man of physical chemistry, L. Michaëlis stood up and said that it was impossible. Sodium and potassium were so similar chemically that my working hypothesis must be wrong. Two days later, however, he called and asked me to pay him a visit at his laboratory in New York. He wanted to show me something. It turned out that he had just received a reprint on the cation binding of EDTA (which had recently been synthesized), and from the paper it was evident that Na was bound more strongly than was potassium. "So you may have something", he said. Three years later I applied for a grant from the Carlsberg Foundation for the support of further studies of active sodium transport. The head of the Foundation at that time was the well known physico-chemist Niels Bjerrum. He did not believe that our working hypothesis could be correct (again due to the similarity between sodium and potassium), but he gave us a yearly grant of 10.000 Danish crowns so that we could find out where we made the mistake. As it turned out there was no mistake, but I may mention in passing that our early values for the electromotive force of the sodium transport were too low due to a small paracellular shunt path for sodium and other small ions. With a new isotope technique (Sten-Knudsen & Ussing, 1981), which permits simultaneous determination of the fluxes through the cellular and the paracellular pathway for sodium, we have found the electromotive force of the "pump" to be about 200 mV in the agreement with the assumption that the process is reversible (see Ussing et al., 1981; Eskesen & Ussing, 1985).

Stoichiometric Relation to Metabolism Criterion for Active Transport

The assumption of active transport being due to specific chemical reactions in the cell membrane suggests that a stoichiometric relationship should exist between active transport and metabolism. Indeed, such a relationship may be proposed as a criterion of active transport (compare Kedem, 1961). In 1954, Lundegårdh reported that a one-to-one relationship existed between monovalent anions taken up by plant roots and the number of electrons passing through their respiratory chain. Conway (1954) proposed a similar relationship between sodium ions and respiratory electrons in connection with active sodium transport in animal cells and tissues.

Clearly, the identity between net active sodium

transport and short-circuit current made the frog skin an ideal object for studying the relationship between respiration and active transport, and two of my former associates, K. Zerahn and Alex Leaf, independently studied the increase in oxygen consumption seen during short-circuiting. The result was in both studies (Zerahn, 1956) and (Leaf & Renshaw, 1957) that 18 sodium ions were transported for each molecule of oxygen consumed in excess of the "resting" consumption. This finding, which was later confirmed for sodium transport in many other objects, was very important. In the first place it showed that there must be a stoichiometric relationship between transport and respiration, but moreover, it disproved a hypothesis, mentioned above, that active sodium transport meant a transport of one sodium ion for each electron passing through the respiratory chain. The correct number was 4.5 sodium ions for one electron, or, in other words, three sodium ions per ATP. As an alternative to a direct coupling between electron flow and sodium transport, we favoured a role of ATP, because, in early experiments, we had found dinitrophenol to abolish the active Na^+ transport. We also knew that Ouabain inhibited the short-circuit current of the frog skin (Koefoed-Johnsen, 1957), but I dare say that none of us had the audacity to propose that the ATPase was the sodium pump, so Skou's paper (1957) on the sodium potassium ATPase came as a godsend. We were immediately convinced that the enzyme from crab nerve and the sodium pump of frog skin were the same thing.

Until now I have been talking about the sodium pump, but by 1957 we also were convinced that the sodium pump of frog skin was a sodium/potassium pump. By studying the ion selectivity of the apical and basolateral membranes of the frog skin we had realized (Ussing & Koefoed-Johnsen, 1956; Koefoed-Johnsen & Ussing, 1958) that the apical membrane had a sodium-selective conductance, whereas the basolateral membrane was the seat of the sodium pump and a potassium conductance. Thus, if we placed the sodium/potassium exchange pump, then known from red cells, in the basolateral membrane of the frog skin, potassium would recycle and the net result of the operation of the whole assembly would be the same as that produced by a "pure" sodium pump. For a period of several years many epitheliologists preferred the "pure" sodium pump concept, because the sodium/potassium pump hypothesis demanded a stoichiometric relationship between sodium transport and potassium recycling. This discussion has now subsided; especially convincing evidence in favour of the sodium/potassium pump was provided by Robert Nielsen (1979) from our institute, who demonstrated the coupling between sodium and potassium pumping in frog skin beyond any reasonable doubt, and found the coupling ratio of 3 Na^+ for two K^+ as one would expect for the Na^+/K^+ ATPase.

Solvent Drag May Mimic Active Transport

In the foregoing we have assumed that ions passing through an epithelial cell are only influenced by electrochemical potential gradients and active transport. It is a well known fact, however, that many epithelia perform net water transport. In our effort to circumscribe the concept of active transport, we also had to take into account the possible effect of water flow on the movement of ions. One still had to consider seriously the fluid-circuit hypothesis advocated by Ingraham et al., 1938. The idea was that fluid from the intestinal contents was somehow pumped through the cells, and pure water pumped back again. With such a mechanism there was no reason to find a strict relationship between net transfer of water and of solutes. We now wanted to find out to what extent such a solvent drag influences the sodium transport through the frog skin. Just as it was the case for simple electrodiffusion, the problem is simplified if we consider the effect of solvent drag on the flux ratio (Koefoed-Johnsen and Ussing, 1953). To make a long story short, we found no evidence for fluid circulation in amphibian skin. With identical solutions on the two sides of the skin, small test molecules (smaller or the same size as sodium) only exhibited net transport if there was a net transport of water (Ussing and Andersen, 1955). Only if water was forced inward by an osmotic gradient, a significant solvent drag effect was seen. In "tight epithelia" like frog skin epithelium, solvent drag in general and fluid circuit in particular can normally be disregarded. However, if transport through leaky epithelia is studied, one should always keep in mind that paracellular solute flow may accomplish solvent drag on the solutes.

"Secondary Active Transport"

Socalled "leaky epithelia" like those of small intestine and proximal kidney tubules, perform net transport not only of sodium, but also of many organic and inorganic substances. I cannot help mentioning that in 1958-59 we had as a visitor dr. T.Z. Csaky who studied intestinal uptake of the non-metabolizable methyl glucose, using the isolated small intestine of the toad as test object. He found that the apparent active transport of this substance was strictly dependent on the presence of sodium ions in the intestinal fluid, whereas the nature of the anion was immaterial. The transport did not depend on any net transport of water (Csaky and Thale, 1960). I wondered if it still could be a case of solvent drag. On that occasion we missed the train. A few years later Crane (1962) solved the riddle: certain sugars undergo cotransport with sodium ions in a strictly coupled process. Since then numerous cases of cotransport (or "symport") have been described, most of them with sodium ions as an obligatory participant. Other types of strictly coupled transport depend on two (or more) reactants sharing an exchange diffusion (also called "antiport") system. It is now common usage to designate cases where the primary transport of sodium ions is coupled with transport of another substance via symport or antiport as secondary active transport (Kedem, 1961). The term should be reserved for cases where a strict stoichiometric relationship has been established, whereas transport due to solvent drag and an electric potential difference still is considered passive. As an example of secondary active transport I may mention the very recent finding (Ussing, 1988) that the active transport of lithium through the frog skin (Zerahn, 1955) is absolutely dependent on the presence of sodium ions in the basolateral bath, and seems to be brought about by a one-to-one exchange of Li^+ and Na^+ via a basolateral Na^+/H^+ exchanger which accepts Li^+ instead of H^+. Na^+ is thus recycled and an equivalent amount of Li^+ is undergoing net transport. In this case both of the primarily transported ions, Na^+ and K^+, are recycled and do not show any net transport, whereas the secondarily transported ion Li^+ undergoes bona fide active transport. The example shows that it is much easier to define active transport theoretically than to demonstrate it in practice.

REFERENCES

Boyle PJ, Conway EJ (1941). Potassium accumulation in muscle and associated changes. J Physiol 100: 1-63.

Conway EJ (1954). Some aspects of ion transport through membranes. In Brown R, Danielli JF (eds): "Active transport and secretion", Cambridge: Cambridge University Press, pp 297-324.

Csáky TZ, Thale M (1960). Effect of ionic environment on intestinal sugar transport. J Physiol 151: 59-65.

Crane RK (1962). Hypothesis for mechanism of intestinal active transport of sugars. Fed Proc 21: 891-895.

Dean RB (1941). Theories of electroyte equilibrium in muscle. Biol Symp 3: 331-348.

Eskesen K, Ussing HH (1985). Determination of the electromotive force of active sodium transport in frog skin epithelium (Rana temporaria) from presteady-state flux ratio experiments. J Membrane Biol 86: 105-111.

Fenn WO (1938). Factors affecting the loss of potassium from stimulated muscles. Am J Physiol 124: 213-229

Glynn IM, Lew VL (1970). Synthesis of adenosine triphosphate at the expense of downhill cation movements in intact human red cells. J Physiol (London) 207: 393-402.

Goldman DE (1944). Potential, impedance and rectification in membranes. J Gen Physiol 27: 37-60.

Heppel LA (1940). The diffusion of radioactive sodium into the muscles of potassium-depleted rats. Am J Physiol 128: 449-454.

Hodgkin AL, Keynes RD (1955). The potassium permeability of a giant nerve fibre. J Physiol (London) 128: 61-88.

Holm-Jensen J, Krogh A, Wartiovaara V (1944). Some experiments on the exchange of potassium and sodium between single cells of Characeae and the bathing fluid. Acta Bot Fenn 36: 1-22.

Huf E (1935). Versuche über den Zusammenhang zwischen Stoffwechsel, Potentialbildung und Funktion der Froschhaut. Arch Ges Physiol 235: 655-673.

Ingraham RC, Peters HC, Visscher M (1938). On the movement of materials across living membranes against concentration gradients. J Phys Chem 42: 141-150.

Kedem O (1961). Criteria of active transport. In Kleinzeller A, Kotyk A (eds): "Membrane transport and metabolism", New York: Academic Press.

Koefoed-Johnsen V (1957). The effect of g-strophanthin (ouabain) on the active transport of sodium through the isolated frog skin. Acta Physiol Scand 42, suppl 145: 87-88.

Koefoed-Johnsen V, Ussing HH (1953). The contributions of diffusion and flow to the passage of D_2O through living membranes. Acta Physiol Scand 28: 60-76.

Koefoed-Johnsen V, Ussing HH (1958). The nature of the frog skin potential. Acta Physiol Scand 42: 298-308.

Krogh A (1937). Osmotic regulation in the frog (R. esculenta) by active absorption of chloride ions. Scand Arch Physiol 76: 60-74.

Krogh A (1938). The active absorption of ions in some fresh water animals. Z vergleich Physiol 25: 335-350.

Krogh A (1939). Osmotic regulation in aquatic animals. Cambridge: The University Press.

Krogh A (1946). The active and passive exchange of inorganic ions through the surface of living cells and through living membranes generally. Proc Roy Soc B: 131-200.

Leaf A, Renshaw A (1957). Ion transport and respiration of isolated frog skin. Biochem J 65: 82-90.

Levi H, Ussing HH (1948). The exchange of sodium and chloride across the fibre membrane of the isolated frog sartorius. Acta Physiol Scand 16:232-249.

Lundegårdh, H (1954). Anion respiration symposium. Soc exp Biol 8: 262-296.

Nielsen R (1979). A 3 to 2 coupling of the Na-K pump in frog skin disclosed by the effect of Ba. Acta Physiol Scand 107: 189-191.

Rosenberg T (1948). On accumulation and active transport in biological systems. Acta Chem Scand 2: 14-33.

Skou JC (1957). The influence of some cations on adenosine-triphosphatase from periferal nerves. Biochim Biophys Acta 23: 394-401.

Sten-Knudsen O, Ussing HH (1981). The flux ratio equation under non-stationary conditions. J Membrane Biol 63: 233-242.

Ussing HH (1947). Interpretation of the exchange of radiosodium in isolated muscle. Nature 160: 262.

Ussing HH (1948). The use of tracers in the study of active ion transport across animal membranes. Cold Spring Harbor Symposium. Quant Biol 13: 193-200.

Ussing HH (1949a). The distinction by means of tracers between active transport and diffusion. Acta Physiol Scand 19: 43-56.

Ussing HH (1949b). Transport of ions across cellular membranes. Physiol Rev 29: 127-155.

Ussing HH (1988). Odd behavior of Li in frog skin ion transport. J Comp Physiol Biochem (in press).

Ussing HH, Andersen B (1955). The relation between solvent drag and active transport of ions. Proc 3rd Int Congress Biochem Biophys, Brussels. New York: Academic Press.

Ussing HH, Eskesen K, Lim J (1981). The flux ratio transient as a tool for separating pathways in epithelia. In MacKnight ADC, Leader JB: "Epithelial ion and water transport", New York: Raven Press, pp 257-264.

Ussing HH, Koefoed-Johnsen V (1956). Nature of the frog skin potential. Abstracts Com 20th Int Physiol Congress, Brussels: Vol. 2, p 511.

Ussing HH, Kruhøffer P, Hess Thaysen J, Thorn NA (1960). The alkali metal ions in biology . Handbuch der experimentellen Pharmakologie 13, Berlin: Springer-Verlag.

Ussing HH, Zerahn K (1951). Active transport of sodium as the source of electric current in the short-circuited isolated frog skin. Acta Physiol Scand 23: 110-127.

Zerahn K (1955). Studies on the active transport of lithium in the isolated frog skin. Acta Physiol Scand 33: 347-358.

Zerahn K (1956). Oxygen consumption and active sodium transport in the isolated and short-circuited frog skin. Acta Physiol Scand 36: 300-318.

Biosynthesis and Expression

OVERVIEW:

BIOSYNTHESIS, MEMBRANE INSERTION AND MATURATION OF Na,K-ATPase

K. Geering

Institut de Pharmacologie,
Rue du Bugnon 27,
CH-1005 Lausanne, Switzerland

INTRODUCTION

Several characteristics of the Na,K-ATPase render the study of its biosynthesis and cellular expression particularly interesting . We are dealing with an enzyme which is composed of two heterologous subunits, a catalytic alpha-subunit and a glycosylated beta-subunit (for review, see Jørgensen 1986). Proteolytic cleavage studies on the purified enzyme (Jørgensen et al., 1982), as well as sequence data (Kawakami et al., 1985; Noguchi et al., 1986; Ovchinnikov et al., 1986; Shull et al., 1985, 1986) indeed suggest that each subunit has a particular membrane topology. The beta-subunit spans the membrane once and the alpha-subunit six to eight times and both subunits expose their N-terminal to the cytoplasmic side. Most of the mass of the alpha-subunit is at the cytoplasmic surface of the membrane while most of the mass of the beta-subunit is at the extracellular surface containing several glycosylation sites. In the last few years, particular interest has been given to the questions where the two enzyme subunits are synthesized and how they are inserted into the membrane in their correct orientation.

The two subunits are encoded for by two independent mRNA species (Geering et al., 1985). Mechanisms must therefore exist which synchronize the synthesis of the two subunits either at the transcriptional or post- transcriptional level to permit the expression of a minimal functional enzyme unit consisting of an alpha-beta protomer (Brotherus et al., 1983).

In view of the fact that all major catalytic functions, namely ATP hydrolysis and conformational transitions upon cation binding, or else interaction with cardiac glycosides, can be assigned to the alpha-subunit (for review, see Jørgensen, 1986) and that no functional role for the beta-subunit could yet be established, the question arises whether the assembly of the two subunits has any consequences for the functional integrity of the enzyme. An interesting hypothesis would be that the beta-subunit, instead of assuming proper catalytic functions, rather plays a role in the biosynthesis process of the enzyme, e.g. by assisting the alpha-subunit to become correctly and stably organized in the membrane. In the same line of thoughts, it might be asked whether the fact that the beta-subunit is a glycoprotein and thus subjected to co- and post-translational modifications is important for the enzyme to acquire its functional properties or else to be targeted to the plasma membrane. In this review, I would like to summarize the present state of knowledge on some of these questions and to reveal the unresolved or contradictory problems which should now be addressed and hopefully soon be elucidated with the help of recently developed molecular technologies.

BIOSYNTHESIS, MEMBRANE INSERTION AND ASSEMBLY OF THE alpha-AND beta-SUBUNITS OF Na, K-ATPase

The site of synthesis and the mode of membrane insertion of the two subunits of Na,K-ATPase have been matter of debate in the last few years. Certain reports indeed favour the hypothesis that either the alpha-subunit (Hiatt et al., 1984; Sabatini et al., 1981) or else the beta-subunit (Fisher et al., 1984) are made in their entirety on free polysomes and that the resulting soluble precursors would ultimately be posttranslationally inserted into the ER membrane. Although many membrane proteins destined for mitochondria, chloroplasts, peroxisomes or prokaryotic membranes follow this synthesis pathway (for review, see Wickner and Lodish, 1985; Rapoport, 1986) no eukaryotic protein ultimately located in the plasma membrane has yet been found to behave in this fashion. It is generally agreed that these latter proteins are integrated into the ER membrane at least in part in the course of their synthesis on ribosomes bound to the ER membrane (for recent review, see Rapoport, 1986).

The hypothesis of posttranslational membrane insertion of the alpha-subunit was based on the observation that free polysomes produced more alpha-subunits in a reticulocyte lysate than bound polysomes (Hiatt et al., 1984, Sabatini et al., 1981). In addition, in one report (Hiatt et al., 1984) the 96 KD alpha-polypeptide was recovered in a soluble form even when rough microsomes were present during translation. The membrane associated form immunoprecipitated with an anti alpha-serum from a microsomal pellet exhibited instead a molecular mass of 135 KD. These results were interpreted in the sense that the integration of the alpha-subunit in vitro is dependent on the presence in the membrane of a smaller peptide presumably the beta-subunit, the association of which would give rise to the higher molecular mass of the immunoprecipitated membrane-associated product. In addition, in the same publication the authors presented evidence that a soluble precursor existed in pulse labelled intact cells, the amount of which decreased during increasing chase time, in parallel with an increase in membrane associated forms.

The hypothesis that the beta-subunit might function as a receptor for the newly synthesized soluble alpha-subunit is appealing but would presuppose that an important pool of unassembled alpha- and beta- subunits exists in the intact cell. This seems however not to be the case since alpha- and beta-subunits could be coprecipitated by a monoclonal antibody against the beta-subunit, as stoichiometric complexes, rapidly after synthesis and incorporation of labelled precursor into the two subunits was linear with time of pulse (Fambrough, 1983). These data are incompatible with a delayed assembly of the two subunits. In addition, recent experiments by Caplan et al. (1986) suggest that the soluble precursor of alpha-subunit observed in the intact cell is actually an experimental artefact and represents a component of small non-pelletable membrane vesicles which can be generated by harsh cell disruption procedures.

Finally, the in vitro data obtained with bound and free polysomes are in contradiction to our own observations which indicate that both alpha- and beta-subunits translated from mRNA, enriched on sucrose gradients, are cotranslationally inserted into ER microsomes in a reticulocyte lysate (Geering et al., 1985). Soluble alpha-subunit recovered in a 100000 xg supernatant was indeed only observed when rough microsomes were absent during translation. On the other hand, more than

95% of the immunoprecipitable material was associated with the microsomal pellet when ER membranes are present during translation. Integration into membranes was documented by alkali resistance of the alpha-subunit as well as by the production of a membrane protected and thus trypsin resistant fragment. Finally, the alpha-subunit was no longer able to integrate into membranes when synthesis was completed.

In our hands, the beta-subunit behaved in a similar fashion in the reticulocyte lysate and cotranslational membrane insertion was documented by the concomitant coreglycosylation of the polypeptide (Geering et al., 1985). These data are not in agreement with the observation made by Fisher et al. (1984) that mRNA isolated from free polysomes of brine shrimp nauplii produced more beta-subunits in a reticulocyte lysate than mRNA from bound polysomes. In our opinion, it is conceivable that these results were artifactually produced due to instability of the mRNA. Indeed, increasing amounts of beta-subunits were synthesized from bound polysomes when increasing concentrations of ribonuclease inhibitors were used during polysome preparation.

A classic insertion pathway of the beta-subunit is further supported by the fact that it depends strongly on the interaction with certain cytosolic and membrane components which have been found to be implicated in the synthesis and membrane insertion of all membrane proteins originating in the ER or else of secretory proteins (for review, see Hortsch and Meyer, 1986). The general model predicts that the signal recognition particle (SRP) binds to a signal sequence on the nascent polypeptide made on free ribosomes. The complex of ribosome, mRNA, nascent chain and SRP then attaches to the ER membrane via the SRP receptor or docking protein and the translocation process is initiated. In some instances, the interaction of SRP to the signal sequence leads to a translation arrest of polypeptide synthesis which is only released after binding of the complex to the ER membrane (for review, see Hortsch and Meyer, 1984). This translation arrest is thought to strictly couple membrane translocation to nascent chain elongation since the polypeptide cannot be completed in the cytoplasm. N-terminal signal sequences are generally cleaved when the protein emerges at the luminal side of the ER while internal signal sequences might become part of a membrane-spanning domain anchoring the polypeptide in the membrane. Finally, as soon as an acceptor asparagine residue reaches the ER lumen, carbohydrates are added to future N-linked glycoproteins (for review, see Kornfeld and Kornfeld, 1985).

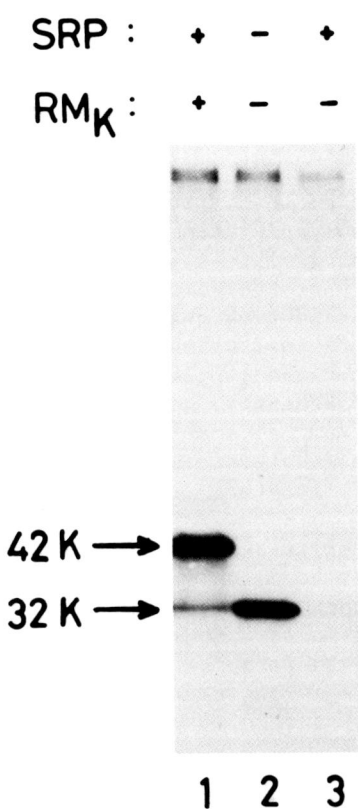

Figure 1. SRP-dependent membrane insertion of the β-subunit of Na,K-ATPase. mRNA fractions coding for the β-subunit were translated in the wheat germ system in the absence or presence of signal recognition particle (SRP) and rough microsomes (RM$_K$) followed by immunoprecipitation of the synthesis products with anti β-subunit serum. For further explanation, see text.

As shown in Fig 1, the beta-subunit of Na,K-ATPase followed this synthesis pathway when mRNA, isolated from cultured amphibian kidney cells (A6 cells), was translated in the wheat germ system.

In the absence of SRP and rough microsomes, wheat germ ribosomes produced soluble unglycosylated beta-subunits (Fig.1, lane 2). In the presence of SRP, elongation was arrested and thus no completed beta-subunit was produced (Fig. 1, lane 3). Finally, in the presence of SRP and microsomes, synthesis was restored through the interaction of the complex with the docking protein on the microsomal membranes and the beta-subunit became coreglycosylated due to its insertion across the membrane (Fig. 1, lane 1). Trypsin treatment of the microsomes released a proteolytic fragment of about 2 KD (Geering et al., 1985) suggesting that the completed beta-subunit has only a short cytoplasmic extension while the bulk of the protein is at the extracytoplasmic side, e.g. in the membrane vesicle lumen. Recently, sequence data confirmed the existence of a short N-terminal domain on the beta-subunit exposed to the cytoplasm (Shull et al., 1986; Otha et al., 1986).

Unfortunately, a similar SRP-dependent membrane integration could not yet be demonstrated for the alpha-subunit mainly because this large subunit is very inefficiently translated in the wheat germ system in which such experiments must be performed. The use of recombinant DNA technology to produce large quantities of pure mRNA will hopefully aid in the near future to realize such experiments.

To terminate the controversy on the insertion mechanism of the two subunits of Na,K-ATPase, I would like to stress the fact that in general the question of co-or posttranslational membrane insertion becomes more and more irrelevant. New data from in vitro studies indeed indicate that protein translocation is not obligatory coupled to protein synthesis (for review, see Zimmermann and Meyer, 1986). In fact, small proteins such as prepromelittin can cross the membrane posttranslationally in the absence of any targeting element (Müller and Zimmermann, 1987). In addition, an increasing number of larger SRP and docking protein dependent proteins, e.g. IgG light chain, prolactin precursor (Ainger and Meyer, 1986) and glucose transporter (Mueckler and Lodish, 1986) are described which can be translocated either late in their synthesis or in the completed state. Nobody claims that these data indeed reflect the in vivo situation but since in all these latter examples efficient posttranslational membrane insertion in vitro can only occur when the polypeptide is still associated to the ribosome, the data can be interpreted in the sense that in vivo, the cellular components implicated in protein translocation, e.g. SRP, docking protein and ribosomes all serve a main function, namely the prevention of a folding of a big nascent polypeptide which would impede the contact of the protein with the cellular translocation machinery. The question of co- or posttranslational membrane insertion of a particular protein will thus probably be reduced in the future to a largely kinetic argument, e.g. how long can a nascent polypeptide chain maintain an unfolded state.

On the basis of these results, we might presume that the beta-subunit with 1 transmembrane segment, the N-terminal in the cytoplasm and its large glycosylated extracytoplasmic domain (Ohta et al., 1986) might have rather stringent requirements for membrane translocation. It could well follow the insertion model lately proposed for the asialoglycoprotein receptor, another such type II transmembrane protein. Holland and Drickamer (1986) propose that in these proteins, an internal signal sequence which eventually becomes the transmembrane domain is recognized by SRP when it emerges from the ribosomes. Translation is temporarily inhibited until the complex is brought to the ER membrane where SRP binds to its receptor. Synthesis is restored and the C-terminal is cotranslationally translocated and finally released through the membrane while the N-terminal domain remains in the cytoplasm. During this step, the beta-subunit of Na,K-ATPase would in addition acquire its coresugars.

The alpha-subunit of Na,K-ATPase might follow a different insertion pathway which has recently been proposed for the glucose transporter, another multispanning membrane protein (Mueckler and Lodish, 1986). As stated by the authors, this model could apply to all proteins which as the alpha-subunit, have no cleavable signal sequence or large extracytoplasmic domains and whose C- and N-termini are positioned in the cytoplasm. A key aspect of this model is that ribosomes implicated in the synthesis of such proteins never interact directly with the ER membrane. SRP-docking protein interaction would indeed be necessary for initiating membrane integration of a few internal signal sequences during chain elongation but hydrophobic loop structures nearby the signal sequences might fold into the membrane posttranslationally.

Although this model could well explain the fact that in vitro the alpha-subunit can insert into membranes later than the beta-subunit (Geering et al., 1985), the possibility exists that the alpha-subunit follows still another insertion pathway in vivo. In fact, the question remains open whether the beta-subunit influences not only the membrane topology of the finished alpha-protein as might be expected but also its membrane insertion. This argument brings us to the question of where and how the two enzyme subunits assemble. Fambrough (1983) has studied this question in pulse labelled myogenic cell cultures by using a monoclonal antibody against the beta-subunit which coprecipitated associated alpha-subunit in non-denatured enzyme preparations; he produced evidence that the assembly process occurs very rapidly either during or else soon after the synthesis of the two subunits.

Such data indeed suggest that the synthesis of the two subunits must be strictly coordinated and the question then arises whether actually the membrane integration or even the synthesis of the two subunits depend on each other. At first sight, this appears not to be the case. In vitro, synthesis and membrane insertion of the alpha-subunit are indeed possible in the absence of concomitant synthesis of the beta-subunit, and vice versa (Geering et al., 1985). On the other hand, however, we cannot estimate the efficiency of these processes in vitro compared to the in vivo situation and the question thus arises whether the in vitro data indeed reflect what actually happens in the intact cell. To answer this question, it is important to compare the structural features of the newly synthesized subunits in the two experimental systems and to study eventual maturation processes of the enzyme occuring after the translational step.

MATURATION OF Na,K-ATPase

The structural processing of the beta-subunit during its intracellular transport was studied in chick sensory neurons (Tamkun and Fambrough, 1986) and in cultured amphibian cells (Geering et al., 1985). In both cell types, the use of tunicamycin to inhibit cotranslational coreglycosylation revealed a non-glycosylated membrane-bound peptide of 32 KD which was identical to the synthesis product obtained in the reticulocyte lysate in the absence of rough microsomes (Geering et al., 1985). These data agree with the notion that the beta-subunit has no cleavable signal sequence as predicted by sequence data (Shull et al., 1986). In the absence of tunicamycin, the beta-subunit is processed as other N-linked glycoproteins (for review, see Kornfeld and Kornfeld, 1985). It acquires coresugars of the high mannose type during synthesis in vivo (Geering et al., 1985; Tamkun and Fambrough, 1986) as well as in vitro in the presence of rough microsomes (Geering et al., 1985) which are susceptible to endoglycosidase H (endo H) treatment. With the commonly used microsomes, no further processing occurs in vitro, while in the intact cell, mannoses are trimmed from the coresugars of the beta-subunit during its transport from the ER to the Golgi and complex type sugars are added when the beta-subunit reaches a trans Golgi compartment. The time required for the beta-subunit to adopt this endo H resistant mature form and thus to be transported from the site of synthesis to the trans Golgi is between 45 and 60 min (Geering et al., in press, Tamkun and Fambrough, 1986).

In different species, the beta-subunit expresses different glycosylation patterns (Geering et al., 1985); this fact might either be due to a different number of aspargine residues available for coreglycosylation or else to different complex type glycosylation. In addition, the beta-subunit can be different in different tissues of the same species (Jørgensen, 1982, Sweadner and Gilkeson, 1985). As shown in Fig. 2 in Xenopus laevis, the mature beta-subunits of the brain or of oocytes, migrated differently on SDS PAGE than the ones of the kidney or the heart. Characterization of cDNAs from brain and kidney indicate that the different forms of the beta-subunit in the same species are not due to the existence of different proteins but rather to a different glycosylation of the same protein in different cell types (Young et al., 1987).

The structural and functional maturation of the alpha-subunit during its intracellular transport was assessed by using a pulse chase protocol combined with controlled trypsinolysis (Geering et al., in press). Since alpha- and beta-subunits rapidly assemble after their synthesis (Tamkun and Fambrough, 1986) it can be assumed that during a pulse chase experiment alpha- and beta-subunits are in the same cellular compartment.

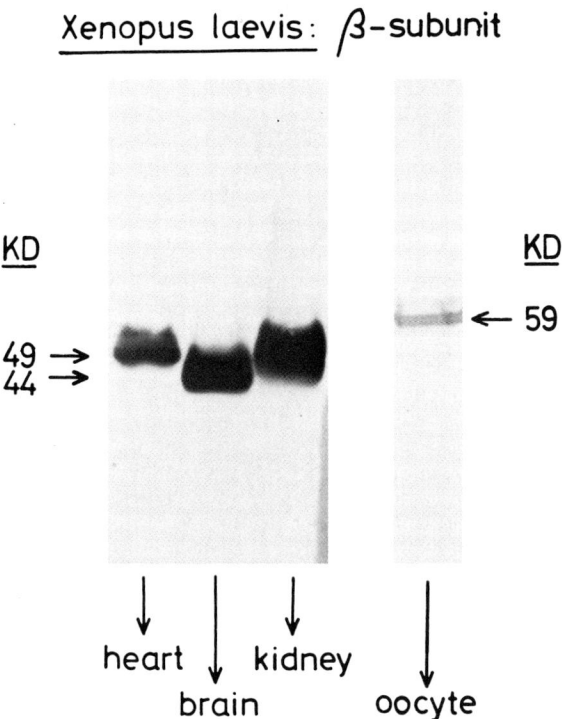

Figure 2. Differently glycosylated β-subunits in different tissues of Xenopus laevis. Shown are immunoblots of cell homogenates revealed with anti β-subunit serum.

On the basis of beta-subunit processing, it was therefore possible to define an early transport compartment namely the ER, an intermediate compartment comprising trans ER and cis Golgi and a late transport compartment beyond the trans Golgi. The newly synthesized alpha-subunit in an early transport compartment showed to be highly trypsin-sensitive and the proteolytic product produced was identical to the one obtained from the alpha-subunit synthesized in vitro in the presence of microsomes (Geering et al., 1985). On the other hand, alpha-subunit assessed in intermediate or late transport compartments was much more trypsin resistant and had acquired one of its functional properties namely to respond to Na$^+$ and K$^+$ by a conformational change (Geering et al., in press). As defined by Jørgensen (1975) on the purified enzyme, this property was reflected in a different tryptic pattern of the alpha-subunit in the presence of Na$^+$ and K$^+$. Thus these data indicate that the alpha-subunit is subjected to a structural reorganization early during intracellular transport which might be a prerequisite for the functional maturation of the enzyme. Evidence that alpha-subunit is transformed from an inactive to an active form was also provided by the observation that only about 10 minutes after synthesis, the alpha-subunit becomes competent to bind ouabain in an ATP dependent fashion (Caplan et al., 1985).

The molecular mechanisms which underly this maturation process are not yet known. An appealing hypothesis would be that posttranslational membrane integration of certain domains of the alpha-subunit and/or rearrangement of the large cytoplasmic domain might be responsible for the increase in trypsin resistance and the acquisition of functional competence. Since the alpha-subunit made in vitro in the absence of concomitant synthesis of the beta-subunit resembles closely to the alpha-subunit in an early transport compartment in the intact cell (Geering et al., 1985) it is tempting to speculate that the inactive alpha-subunit is not yet associated to the beta-subunit and that perhaps the assembly process of the two subunits marks the functional maturation of the enzyme.

Experimental proof for this hypothesis is difficult to obtain in the intact cell. Indeed, in all in vivo systems so far tested, alpha-and beta-subunits are expressed in parallel and it has not yet been possible to selectively suppress the synthesis of one or the other subunit and thereby to reveal its inherent functions. In this context, the recent data obtained by Takeyasu et al. (in press) are revealing. They introcuded cDNA coding for chick beta-subunit into a mouse cell and found that the avian beta-subunit became expressed in the plasma membrane of the mouse cell and was complexed to the alpha-subunit of murine Na,K-ATPase. Despite the high number of hybrid molecules expressed, the mouse cell did not gain ouabain sensitivity characteristic of avian cells. It was concluded that the ouabain binding site located on the alpha-subunit is not greatly affected by the interaction of beta- to alpha-subunit. At present, we indeed know very little on the functional role of the beta-subunit or its glycomoiety in the catalytic turnover of the enzyme. Nevertheless, on the basis of the discussed results and some recent data published in this volume (Zamofing et al.), I would like to present in conclusion a working model which proposes some specific roles for the beta-subunit at several steps of the intracellular processing of the Na,K-ATPase.

CONCLUSIONS

As discussed in detail in this review, it is fairly sure that the alpha- and the beta-subunits become inserted into the ER membrane during their synthesis and that the beta-subunit acquires its coresugars during this process. Interestingly, when coreglycosylation of the beta-subunit was inhibited by tunicamycin, the newly synthesized pools of both beta- and alpha-subunits were specifically decreased by 70 %, compared to controls (Zamofing et al., in this volume). Since the alpha-subunit is not a glycoprotein and other nonglycosylated membrane proteins were not affected by tunicamycin treatment, this result suggests that the synthesis of the two subunits is strictly coordinated and that the linking factor in the synthesis might be the beta-subunit itself or else another glycoprotein. It is conceivable that the synthesis synchronization of the two polypeptides might favour the documented rapid association after synthesis of the two subunits (Fambrough, 1983). The assembly in turn could trigger the structural and functional maturation of the alpha-subunit which occurs also soon after synthesis (Caplan et al., 1983; Geering et al., in press). In addition, the acquisition of coresugars of the beta-subunit seems to play a certain role in the structural maturation of the enzyme. Zamofing et al. (in this volume) could indeed show that inhibition of coreglycosylation of the beta-subunit led to a destabilization e.g. a higher trypsin sensitivity not only of the newly synthesized beta- but also of the alpha-subunit. On the other hand, however, inhibition of coreglycosylation prevented neither assembly of the two subunits nor their transfer to the plasma membrane (Tamkun and Fambrough, 1986). Finally, although we do not yet know whether the non-glycosylated enzyme becomes functionally active, recent data suggest that acquisition of coresugars is sufficient for the enzyme to perform cation dependent conformational changes (Geering et al., in press) or ouabain inhibitable ATP hydrolysis (Zamofing et al., in preparation).

Many questions concerning the biosynthesis, the assembly and the functional role of the beta-subunit in the maturation process of the enzyme remain yet open. Hopefully, with the help of recombinant DNA technology which has recently become available, we should soon be able to elucidate some of these problems.

REFERENCES

Ainger KJ, Meyer DI (1986). Translocation of nascent secretory proteins across membranes can occur late in translation. EMBO J 5:951-955.

Brotherus JR, Jacobsen L, Jørgensen PL (1983). Soluble and enzymatically stable (Na^++K^+)-ATPase from mammalian kidney consisting predominantly of protomer alpha-beta-units: Preparation, assay and reconstitution of active Na^+, K^+ transport. Biochim Biophys Acta 731:290-303.

Caplan MJ, Palade GE, Jamieson JD (1985). Cell surface expression and activation of newly synthesized Na,K-ATPase in MDCK cells. The Sodium Pump: Glynn I, Ellory C, ed.:147-151.

Caplan MJ, Palade GE, Jamieson JD (1986). Newly synthesized Na,K-ATPase alpha-subunit has no cytosolic intermediate in MDCK cells. J Biol Chem 25:2860-2865.

Fambrough DM (1983). Studies on the Na^+-K^+ ATPase of skeletal muscle and nerve. Cold Spring Harbor Symp Quant Biol 48:297-304.

Fisher JA, Baxter-Lowe LA, Hokin LE (1984). Site of synthesis of the alpha and beta subunits of the Na,K-ATPase in brine shrimp Nauplii. J Biol Chem 259:14217-14221.

Geering K, Meyer DI, Paccolat MP, Kraehenbühl JP, Rossier BC (1985). Membrane insertion of alpha-and beta-subunits of Na^+,K^+-ATPase. J Biol Chem 260:5154-5160.

Geering K, Kraehenbühl JP, Rossier BC. Maturation of the catalytic alpha subunit of Na,K-ATPase during intra-cellular transport. J Cell Biol, in press.

Hiatt A, McDonough AA, Edelman IS (1984). Assembly of the (Na^+ + K^+)-adenosine triphosphatase: Post-translational

membrane integration of the alpha subunit. J Biol Chem 259:2629-2635.
Holland EC, Drickamer K (1986). Signal recognition particle mediates the insertion of a transmembrane protein which has a cytoplasmic NH_2 terminus. J Biol Chem 261:1286-1292.
Hortsch M, Meyer DI (1984). Pushing the signal hypothesis: What are the limits? Biol Cell 52:1-8
Hortsch M, Meyer DI (1986). Transfer of secretory proteins through the membrane of the endoplasmic reticulum. Int Rev Cytol 102:215-242.
Jørgensen PL (1975). Purification and characterization of (Na^+, K^+)-ATPase. V. Conformational changes in the enzyme. Transitions between the Na-form and the K-form studied with tryptic digestion as a tool. Biochim Biophys Acta 401:399-415.
Jørgensen PL (1982). Mechanism of the Na^+, K^+ pump: Protein structure and conformations of the pure (Na^++K^+)-ATPase. Biochim Biophys Acta 694:27-68.
Jørgensen PL, Karlish SJD, Gitler C (1982). Evidence for the organization of the transmembrane segments of (Na,K)-ATPase based on labeling lipid-embedded and surface domains of the alpha-subunit. J Biol Chem 257:7435-7442.
Jørgensen PL (1986). Structure, function and regulation of Na,K-ATPase in the kidney. Kidney Int 29:10-20.
Kawakami K, Noguchi S, Noda M, Takahashi H, Ohta T, Kawamura M, Nojima H, Nagano K, Hirose T, Inayama S, Hayashida H, Miyata T, Numa S (1985). Primary structure of the alpha-subunit of *Torpedo californica* (Na^++K^+)ATPase deduced from cDNA sequence. Nature 316:733-736.
Kornfeld R, Kornfeld S (1985). Assembly of asparagine-linked oligosaccharides. Ann Rev Biochem 54:631-664.
Mueckler M, Lodish HF (1986). The human glucose transporter can insert posttranslationally into microsomes. Cell 44:629-637.
Müller G, Zimmermann R (1987). Import of honeybee pre-promelittin into the endoplasmic reticulum: structural basis for independence of SRP and docking protein. EMBO J 6:2099-2107.
Noguchi S, Noda M, Takahashi H, Kawakami K, Ohta T, Nagano K, Hirose T, Inayama S, Kawamura M, Numa S (1986). Primary structure of the beta-subunit of *Torpedo californica* (Na^++K^+)-ATPase deduced from the cDNA sequence. FEBS Lett 196:315-320.
Ohta T, Yoshida M, Nagano K, Hirano H, Kawamura M (1986). Structure of the extra-membranous domain of the beta-subunit

of (Na,K)-ATPase revealed by the sequences of its tryptic peptides. FEBS Lett 204:297-301.
Ovchinnikov Yu.A, Modyanov NN, Broude NE, Petrukhin KE, Grishin AV, Arzamazova NM, Aldanova NA, Monastyrskaya GS, Sverdlov ED (1986). Pig kidney Na^+, K^+-ATPase: Primary structure and spatial organization. FEBS Lett 201:237-245
Rapoport TA (1986). Protein translocation across and integration into membranes. CRC Crit Rev Biochem 20:73-137.
Sabatini D, Colman D, Sabban E, Sherman J, Morimoto T, Kreibich G, Adesnik M (1981). Mechanisms for the incorporation of proteins into the plasma membrane. Cold Spring Harbor Symp Quant Biol 46:807-818.
Shull GE, Schwartz A, Lingrel JB (1985). Amino-acid sequence of the catalytic subunit of the $(Na^+ + K^+)$ATPase deduced from a complementary DNA. Nature 316:691-695
Shull GE, Lane LK, Lingrel JB (1986). Amino-acid sequence of the beta-subunit of the $(Na^+ + K^+)$ ATPase deduced from a cDNA. Nature 321:429-431.
Sweadner KJ, Gilkeson RC (1985). Two isozymes of the Na,K-ATPase have distinct antigenic determinants. J Biol Chem 260:9016-9022.
Takeyasu K, Tamkun MM, Siegel NR, Fambrough DM. Expression of hybrid $(Na^+ + K^+)$-ATPase molecules after transfection of mouse Ltk- cells with DNA encoding the beta-subunit of an avian brain sodium pump. J Biol Chem, in press.
Tamkun MM, Fambrough DM (1986). The $(Na^+ + K^+)$-ATPase of chick sensory neurons: Studies on biosynthesis and intracellular transport. J Biol Chem 261:1009-1019.
Wickner WT, Lodish HF (1985). Multiple mechanisms of protein insertion into and across membranes. Science 230:400-407.
Young RM, Shull GE, Lingrel JB (1987). Multiple mRNAs from rat kidney and brain encode a single Na^+, K^+-ATPase beta subunit protein. J Biol Chem 262:4905-4910.
Zamofing D, Rossier BC, Geering K (1988). Effects of tunicamycin on the cellular expression and structural arrangement of Na,K-ATPase. This volume.
Zimmermann R, Meyer DI (1986). 1986: A year of new insights into how proteins cross membranes. TIBS 11:512-515.

IN VITRO BIOSYNTHESIS OF THE ß-SUBUNIT OF THE Na,K-ATPase IN THE BRINE SHRIMP

Lee Ann Baxter-Lowe, James M. Yohanan, and Lowell E. Hokin

Department of Pharmacology, University of Wisconsin Medical School, Madison, Wisconsin 53706

INTRODUCTION

The Na,K-ATPase plays a major role in cellular function; thus, it is vital to understand the molecular mechanisms of its biosynthesis and regulation. One model system which has proved to be useful for studying these processes is the brine shrimp, Artemia, which demonstrates a marked increase in Na,K-ATPase activity and biosynthetic rate during the first 24 hr of development of rehydrated cysts (Conte et al., 1977; Peterson et al., 1982).

In order to understand the mechanism of biosynthesis of the Na,K-ATPase in the brine shrimp, the sites of synthesis of the α- and ß-subunits were previously determined by measuring the distribution of $mRNA_\alpha$ and $mRNA_\beta$ in preparations of free and membrane-bound polysomes (Fisher et al., 1984). Isolation of intact polysomes required that samples contain a ribonuclease inhibitor, vanadyl ribonuclease complex (VRC). However, it was observed that the distribution of $mRNA_\beta$ between the free and membrane-bound polysome fractions varied as a function of VRC concentration. It was concluded that the ß-subunit was synthesized on free polysomes, because the VRC concentration curve suggested that in the absence of VRC, $mRNA_\beta$ would be found in the free polysome fraction (Fisher et al., 1984). In contrast, studies conducted in other species suggested synthesis of the ß-subunit on membrane-bound polysomes (Sabatini et al.; 1981, Hiatt et al., 1984; Geering et al., 1985).

The difficulties related to high ribonuclease levels in the brine shrimp have been circumvented in this study by examining the synthesis, glycosylation, and membrane insertion of the ß-subunit in a cell-free translation system. The development of such in vitro systems will set the stage for examining many important questions concerning the biosynthesis of the Na,K-ATPase, including the mechanisms of insertion of the subunits into the membrane, assembly of the subunits into holoenzyme, and post-translational modifications of the subunits.

RESULTS AND DISCUSSION

Cell-Free Translation Products

The biosynthesis and membrane insertion of the ß-subunit of the Na,K-ATPase of brine shrimp was studied in a cell-free translation system supplemented with microsomal membranes. Total polysomal RNA was translated in a reticulocyte lysate system, and the products were immunoprecipitated with antibodies raised against highly purified mature ß-subunit. The immunoprecipitated products were separated by electrophoresis on SDS-polyacrylamide gels and detected by fluorography. If translation was carried out in the absence of microsomal membranes, immunoprecipitates contained a protein which migrated as a single, broad band on 12% acrylamide gels (Fig. 1). The apparent molecular weight of this protein was $33,000 \pm 1,000$ (n=7). This broad band could be resolved into two bands by carrying out electrophoresis on gels containing a higher concentration of bisacrylamide (data not shown).

The above results are reminiscent of the observations of Geering et al. (1985), who also observed two immunoprecipitable forms of the ß-subunit synthesized in translation mixtures supplied with mRNA isolated from A6 cells. Possible explanations for the two forms of primary translation products include partial degradation of the primary translation product, premature termination of translation, synthesis of isoforms of the subunits, or impurities in immunoprecipitates.

When microsomal membranes were included in the cell-free translation mixture, a band of an average apparent molecular weight of $37,000 \pm 1,000$ (n=7) was also present in the immunoprecipitates (Fig. 1).

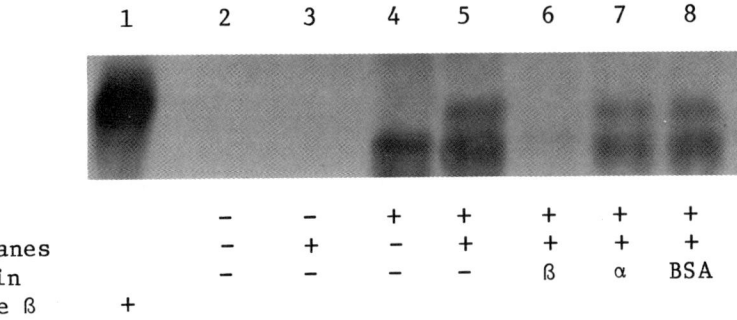

Figure 1. Cell-free translation and immunocompetition. Cell-free translation products were immunoprecipitated as described previously (Fisher et al., 1984), separated by electrophoresis through a 12% polyacrylamide gel (Hokin et al., 1973), and detected by fluorography. Mature ß-subunit (Peterson and Hokin, 1980), which was externally labelled with ^{14}C, is shown in lane 1. Immunoprecipitable cell-free translation products are shown in lanes 2-8, as indicated in the figure. Immunocompetition was performed by removing four aliquots from a pool of four translation reactions and adding the following nonradioactive proteins prior to addition of antibody: lane 5, no addition; lane 6, 10 μg ß-subunit; lane 7, 30 μg α-subunit; and lane 8, 100 μg bovine serum albumin.

Immunocompetition was utilized to investigate the identity of all bands. Addition of highly purified ß-subunit to samples prior to the addition of antibodies resulted in displacement of all radioactive bands (Fig. 1, lane 6). In contrast, addition of much larger quantities of purified α-subunit of BSA had no effect (Fig. 1, lanes 7 and 8). The selective displacement with purified ß-subunit indicates that all of the immunoprecipitated cell-free translation products are forms of the ß-subunit.

Glycosylation

Since the mature form of the ß-subunit is glycosylated (Peterson and Hokin, 1980), it was reasonable to expect that the larger molecular weight cell-free translation product might be a glycosylated primary translation product. This possibility was investigated by incubation of the cell-free translation products with endoglycosidase H (Fig. 2), which catalyzes the hydrolysis of the chitobiose core of high mannose oligosaccharides. After incubation with endoglycosidase H, the protein with an apparent molecular weight of 37,000 was reduced in size and its migration was indistinguishable from that of the primary cell-free translation product (Fig. 2). This indicates that the larger cell-free translation product is a ß-subunit which has been glycosylated.

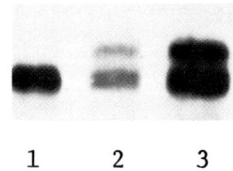

1   2   3

Figure 2. Endoglycosidase digestion of cell-free translation products. Cell-free translations were carried out in the presence of microsomal membranes, and the mixture was divided into three aliquots of 54 µl. Each aliquot was adjusted to a final volume of 160 µl containing 2% SDS (w/v), 0.78 mg/ml Aprotinin, 2 µM PMSF, 0.28 M Tris (pH 6.0). Samples were incubated with endoglycosidase H (lane 1), incubated without endoglycosidase H (lane 2), or stored at -20°C (lane 3). After these treatments were completed, ß-subunits were immunoprecipitated, separated by electrophoresis on 12% acrylamide gels, and detected by fluorography, as described in Fig. 1.

N-linked glycosylation of the primary cell-free translation product was achieved in the *in vitro* system, and the difference between the apparent molecular weights of the primary and glycosylated forms of the ß-subunit would indicate the addition of about two oligosaccharides. Analysis of the amino acid sequences of ß-subunits from several other species has revealed the presence of three to four potential N-glycosylation sites (Kawakami et al., 1986; Mercer et al., 1986; Noguchi et al., 1986; Ovchinnikov et al., 1986; Shull et al., 1986; Brown et al., 1987). It is possible that one or two of these sites have not been conserved or utilized in the brine shrimp. This would be consistent with the observation that the ß-subunit of the brine shrimp contains less carbohydrate than that detected in other species (Peterson and Hokin, 1980). Sequencing of a cDNA encoding the ß-subunit of the brine shrimp is currently in progress, and the amino acid sequence deduced from the nucleotide sequence may help to resolve this question.

The glycosylation of the primary translation product was not quantitative. Attempts to increase the glycosylation of the primary translation product by changing the amount of microsomal membranes in the cell-free translation mixtures were unsuccessful (data not shown).

Membrane Insertion

Since glycosylation occurs after the protein has crossed the microsomal membrane, the presence of oligosaccharides indicates that the ß-subunit is inserted into the membrane. This was confirmed by experiments in which all of the glycosylated form of the ß-subunit remained associated with microsomal membranes after alkaline extraction, which results in exclusive retention of integral membrane proteins (data not shown).

Evidence for co-translational insertion into the membrane was provided by examining the biosynthesis of the ß-subunit in a synchronized translation system. The synchronized cell-free translation system was achieved by carrying out translation in a single, large-scale reaction for 2 min, followed by addition of an inhibitor of initiation of translation, 7-methyl-guanosine 5-phosphate. Aliquots were removed from the large-scale reaction after various intervals between 5 and 120 min. Examination of the immunoprecipitable ß-subunit in these samples indicated that 12 to 15 min were required for synthesis of full-length ß-subunit (data not shown). This corresponds to a rate of synthesis of about 20 to 25 amino acids incorporated into ß-subunit/min incubation.

In order to examine whether the completed ß-subunit is inserted into the membrane, microsomal membranes were added after 30 min of translation (data not shown). No glycosylated subunit could be detected, suggesting that the newly synthesized and full-length ß-subunit was not inserted into the membrane.

The temporal requirement for the presence of membranes was investigated by adding microsomal membranes to aliquots removed from a synchronized cell-free translation system after 5 to 30 min of translation. After addition of membranes, samples were incubated for a total of 120 min, immunoprecipitated, and applied to acrylamide gels (data not shown). Glycosylation of the ß-subunit was observed in all samples which contained incomplete ß-subunits (up to 8.5 min of translation). However, addition of membranes after 12 min, a time point at which most of the ß-subunit is full-length, resulted in very little or no glycosylated product.

Maximal levels of glycosylation were observed if membranes were added to translation mixtures during the first 8.5 min of translation. If the rate of synthesis of the ß-subunit is linear, about 70% of the peptide is completed at this time. Therefore, it can be estimated that glycosylation can be achieved when at least 70% of the ß-subunit is synthesized prior to association with membranes.

The temporal relationship between biosynthesis of the peptide chain and its association with the membranes has also been studied in another model system, amphibian bladder (Geering et al., 1985). In contrast to the present study, that study showed that insertion of the ß-subunit into the membrane could only be achieved if microsomal membranes were present at very early stages of the synthesis of the ß-subunit. We have no explanation for the discrepancy at this time, partly because too little is known about the molecular mechanisms involved in the insertion of proteins into membranes.

This study has shown that the ß-subunit of brine shrimp is co-translationally inserted into microsomal membranes, where it is glycosylated. The data indicate that an earlier study which concluded that the ß-subunit of the Na,K-ATPase of the brine shrimp was synthesized on free polysomes (Fisher et al., 1984) was probably incorrect. The early study was complicated because VRC altered the distribution of mRNA$_ß$ between polysome fractions. Since most of the ß-subunit was observed in the free polysome fraction at the lowest concentrations of VRC, it was suggested that in the absence of VRC, mRNA$_ß$ would be associated with the free polysome fraction, indicating synthesis of the ß-subunit on free polysomes.

The in vitro studies reported here clearly indicate that the ß-subunit can be synthesized on membrane-bound polysomes. Since the in vitro studies also show that the ß-subunit can still be inserted into membranes and glycosylated when up to 70% of the peptide chain is completed, it is also possible that the appearance of mRNA$_\beta$ in the free polysome fraction was due to synthesis of a large portion of the ß-subunit prior to association with the membranes.

This study also demonstrates the potential usefulness of cell-free translation systems in the investigation of the molecular mechanisms of biosynthesis of the Na,K-ATPase, particularly in situations where results generated by in vivo systems may be difficult to interpret.

ACKNOWLEDGEMENTS

This work was supported by grants from the National Institutes of Health (GM 33850 and AM 07389) and the National Science Foundation (ACM-8120635). We wish to thank Karen Wipperfurth for her dedication and skill in the preparation of this manuscript.

REFERENCES

Brown TA, Horowitz B, Miller RP, McDonough AA, Farley RA (1987). Molecular cloning and sequence analysis of the $(Na^+ + K^+)$-ATPase ß-subunit from dog kidney. Biochim Biophys Acta 912:244-253.

Conte FP, Droukas PC, Ewing RD (1977). Development of sodium regulation and de novo synthesis of Na+K-activated ATPase in larval brine shrimp, Artemia salina. J Exp Zool 202:339-362.

Fisher JA, Baxter-Lowe LA, Hokin LE (1984). Site of synthesis of the α- and ß-subunits of Na,K-ATPase in brine shrimp nauplii. J Biol Chem 259:14217-14221.

Geering K, Meyer DI, Paccolat M-P, Kraehenbuhl J-P, Rossier BC (1985). Membrane insertion of α- and ß-subunits of Na^+,K^+-ATPase. J Biol Chem 260:5154-5160.

Hiatt A, McDonough A, Edelman IS (1984). Assembly of the $(Na^+ + K^+)$-adenosine triphosphatase. Post-translational membrane integration of the α-subunit. J Biol Chem 259:2629-2635.

Hokin LE, Dahl JL, Deupree JD, Dixon JF, Hackney JF, Perdue JF (1973). Studies on the characterization of the sodium-potassium transport adenosine triphosphatase. X. Purification of the enzyme from the rectal gland of Squalus acanthias. J Biol Chem 248:2593-2605.

Kawakami K, Nojima H, Ohta T, Nagano K (1986). Molecular cloning and sequence analysis of human Na,K-ATPase ß-subunit. Nucl Acids Res 14:2833-2844.

Mercer RW, Schneider JW, Savitz A, Emanual J, Benz EJ Jr, Levinson R (1986). Rat-brain Na,K-ATPase ß-chain gene: primary structure, tissue-specific expression, and amplification in ouabain-resistant HeLa C$^+$ cells. Mol Cell Biol 6:3884-3890.

Noguchi S, Noda M, Takahashi H, Kawakami K, Ohta T, Negano K, Hirose T, Inayama S, Kawamura M, Numa S (1986). Primary structure of the ß-subunit of Torpedo californica (Na$^+$ + K$^+$)-ATPase deduced from the cDNA sequence. FEBS Lett 196:315-320.

Ovchinnikov YA, Modyanov NN, Broude NE, Petrukhin KE, Grishin AV, Arzamazova NM, Aldanova NA, Monastyrskaya GS, Sverdlov ED (1986). Pig kidney Na$^+$,K$^+$-ATPase. Primary structure and spatial organization. FEBS Lett 201:237-245.

Peterson GL, Churchill L, Fisher JA, Hokin LE (1982). Structural and biosynthetic studies on the two molecular forms of the (Na$^+$ + K$^+$)-activated adenosine triphosphatase large subunit in Artemia salina nauplii. J Exp Zool 221:295-308.

Peterson GL, Hokin LE (1980). Improved purification of brine-shrimp (Artemia salina) (Na$^+$ + K$^+$)-activated adenosine triphosphatase and amino-acid and carbohydrate analyses of the isolated subunits. Biochem J 192:107-118.

Sabatini D, Coleman D, Sabban E, Sherman J, Morimoto T, Kreibich G, Adesnik M (1981). Mechanisms for the incorporation of proteins into the plasma membrane. Cold Spring Harbor Symp Quant Biol 46:807-818.

Shull GE, Lane, LK, Lingrel JB (1986). Amino acid sequence of the ß-subunit of the (Na$^+$ + K$^+$)-ATPase deduced from a cDNA. Nature 321, 429-431.

Walter P, Blobel G (1983). Preparation of microsomal membranes for cotranslational protein translocation. Methods Enzymol 96:84-93.

COMPARATIVE PROPERTIES OF ALPHA AND ALPHA PLUS SUBUNITS IN ADULT RAT BRAIN AND HEART.

Isabelle BERREBI-BERTRAND, Jean-Michel MAIXENT, Lionel G. LELIEVRE.

Laboratoire Nativelle, (J.M.M.) 1 chemin Saulxier, 91160 Longjumeau, France. INSERM : U 127, (I.B.B., L.G.L.) Hopital Lariboisière, Université Paris 7, 41 Bld de la Chapelle, 75010 Paris, France

INTRODUCTION

The heterogeneity of the alpha subunit of the Na,K-ATPase is now well established and it has been demonstrated in many tissues : heart, brain, adipocytes, kidney, and species : rat, mouse, dog and ferret (Lingrel and Young, 1987 ; Sweadner, 1979, 1987 ; Charlemagne et al., 1986 ; Shull et al., 1987 ; Lytton et al., 1985 ; Siegel et al., 1986 ; Maixent et al., 1987 ; Ng et al., 1987).

The main two discriminatory criteria for such an heterogeneity were the differencies in gel mobility and in sensitivity to digitalis (Sweadner KJ (1979)). It is generally accepted that the alpha chain has an apparent molecular weight of 95 Kda and exhibits a low sensitivity to digitalis (K_D values in the range of 10^{-5} - 10^{-4} M). The alpha plus subunit (apparent molecular weight : 98 Kda) exhibits a relatively high sensitivity to ouabain, 10^{-8} - 10^{-7} M.

However, there is no experimental evidence that the so-called alpha plus or alpha forms were identical in different tissues from the same species.

Our objective was to compare the respective properties of alpha and alpha plus subunits present in membranes from adult rat brain and heart.

Both the respective sensitivities of Na,K ATPase activities to ouabain and the respective rates of ouabain release from the brain and cardiac enzyme forms were used to determine a putative tissue specificity of these subunits.

Our results show that active alpha plus subunit in brain was different from that found in heart. Conversely, both rat brain and heart expressed the same alpha isoform.

MATERIAL AND METHODS

1) Isolation of Membranes From Adult Rat Brain.

Membrane fractions enriched in Na,K-ATPase activity were purified from adult (6 weeks) Wistar rat brains according to the method of Sweadner KJ (1979). The only modification of the above protocol was to buffer all the solutions by addition of 30 mM imidazole-HCl pH 7,2. The final pellet enriched in Na,K-ATPase activity was resuspended in the same buffer and kept frozen at -70°C. The protein content was determined by the method of Lowry OH et al. (1951).

2) Isolation of Membranes From Adult Rat Heart

Sarcolemmal vesicles were isolated according to the published procedure of (Mansier P and Lelievre LG (1982) ; Mansier P et al. (1983). The sarcolemma - enriched fraction was kept frozen in samples (about 100 ul) at -70°C.

3) Sensitivity of Na,K-ATPase To Ouabain

In order to reveal full enzyme activity in brain and cardiac preparations, the membrane vesicles were submitted to a treatment with sodium dodecyl-sulfate (SDS) (0,2 mg/mg of protein, 30 min at 20°C (Mansier P et al. (1983)) to make them permeable to substrates and ligands before Na,K ATPase assays.

The Na,K-ATPase activity was determined using a coupled assay method as previously described (Thomas R et al (1979), Lelièvre LG et al (1986)) in the absence or presence of various concentrations of ouabain (1 nM to 2 mM). The activity was measured by continuously recording NADH oxydation in an ATP regenerating assay medium. Each cell contained (final volume 0,6 ml) : 100 mM NaCl, 5,86 mM KCl, 4 mM $MgCl_2$, 4 mM ATP, 40 mM Imidazole/HCl (pH 7.4), 0.4 mM $NADH_2$, 2 mM phosphoenolpyruvic acid, 3.5 units pyruvate kinase and 5 units lactate dehydrogenase.

The enzymatic reaction was initiated by addition of 0,1 to 2 ug microsomal proteins. When linearity was reached (5 minutes) various amounts of ouabain were added. Inhibition percentage was calculated by comparing the activities in the presence or absence of ouabain after correcting for the ouabain insensitive ATPase activity measured in the presence of 2 mM ouabain.

4) Release of Ouabain From The Enzyme Isoforms

Dissociation of ouabain from the enzyme forms was started by a 100-fold dilution of the incubation medium in the same buffer but without ouabain. Thirty microgramms of membrane vesicles were incubated for 30 minutes at 37°C in 30 ul of the assay medium containing either 10 nM or 100 uM ouabain. A period of time of 30 minutes was chosen to ensure the equilibrium ATPase-ouabain has been reached. Ouabain bound to the enzyme dissociated via a first-order process when chased by a 100 - fold dilution.

The interactions between ouabain and the specific high-affinity sites are characterized by fast association sites, the association is slow, whereas the dissociation process is fast.

The dissociation processes could be accurately determined for both types of sites in each type of preparation. Indeed, the time course of relief from inhibition paralleled the release of ouabain from the enzyme. Since the dissociation is a first-order reaction, the relief from inhibition may be evaluated by the t1/2 value, which represents the time required to recover 50% of the Na,K ATPase activity.

RESULTS AND DISCUSSION

I - Sensitivities to Digitalis

As shown in Table 1, in both tissues, the alpha plus isoforms exhibited the same __apparent__ high affinities in a range from 30 to 60 nM. Regarding the alpha chain (of low affinity) in rat brain membranes, the dose-response curves to ouabain exhibited a complex pattern that revealed a large heterogeneity in the K_D values. Indeed, the computed values of the apparent affinity varied from 0.86 to 100 uM (Table 1). This might be due to the presence of a third alpha isoform (of intermediate affinity) as demonstrated by Schull et al. (1986). The complex pattern of the active low affinity forms was not observed in rat cardiac sarcolemma. Thus, the sensitivities of the alpha form to ouabain did not seem to be the same in the two organs. All these observations were also true when digitoxigenin instead of ouabain was used.

Table I. Computed apparent affinities of high (alpha plus) and low (alpha) sensitivity enzyme forms for ouabain.

	brain	heart
Alpha plus	60 $\pm$ 20 nM	29 $\pm$ 11 nM
Alpha	0.86 $\pm$ 100 uM	18 $\pm$ 16 uM

The values are mean $\pm$ S.E. of duplicate experiments performed with 12 membrane preparations from either 6 rat brains or 6 rat hearts.

Specific activities of the Na,K-ATPase used here were 135 $\pm$ 15 and 105 $\pm$ 16 umol inorganic phosphate liberated per hour per mg of protein in rat brain and heart, respectively, when the enzyme preparations were treated by SDS.

II - Dissociation of Ouabain From High and Low Sensitivity Na,K ATPase Forms.

Brain and cardiac alpha plus isoforms largely differed in their respective rate of ouabain release (Table 2). The half time for dissociation was 6.07 $\pm$ 0.7 minutes in brain preparations whereas it was two fold slower in rat heart 11.6 $\pm$ 1.2 minutes (significantly different $p < 0.01$). Thus, the difference in the dissociation rate constants between adult brain and heart, as reflected by the t1/2 values, supposes that the K_D observed in the two tissues (Table 1) were not similar. The factor of two existing between the dissociation rate constants was also found in the K_D values (60 $\pm$ 20 and 30 $\pm$ 11 for brain and heart, respectively). Note that the enzyme form with the slowest drug release had the lowest K_D values, confirming the assumption of Wellsmith and Lindemayer (1980).

As previously shown with Na,K ATPases in rat cardiac hypertrophy (Charlemagne et al (1986), the measurement of the dissociation rate constant is a better criterion than K_D values to discriminate enzyme isoforms. This is confirmed here.

Under the experimental conditions used here to characterize the low affinity forms, i.e. a shift in ouabain concentration from 100 to 1 uM, the t1/2 values calculated for the alpha chains were very similar in both brain and cardiac preparations (Table 2). This high homology of the alpha chains have been already suggested by Sweadner (1979) on the basis of the similarity of the K_D values.

Table 2. Dissociation rate constants expressed as half-time for ouabain dissociation (t1/2 in minutes).

isoforms	brain	heart
alpha plus	6.7 ± 0.7	11.6[o] ± 1.2
alpha	1 ± 0.1	1.05 ± 0.07

[o] significantly different $p < 0.01$

As a conclusion, it is clear that the heterogeneity of the alpha subunits of the Na,K-ATPase reflects different physiological roles, a common one for the alpha subunit in both brain and heart and a tissue specific action for alpha plus. Indeed, it has been shown that in dog and rat hearts (Maixent et al., (1987) ; Lelièvre et al., (1986)) the alpha plus chain was the inotropic site. However, the tissue specific role of the brain alpha plus remains unknown.

REFERENCES

Charlemagne D, Maixent JM, Preteseille M, Lelievre LG (1986). Ouabain binding sites and (Na$^+$,K$^+$)-ATPase activity in rat cardiac hypertrophy. J Biol Chem, 261:185.

Lelievre LG, Maixent JM, Lorente P, Mouas C, Charlemagne D, Swynghedauw B (1986). Prolonged responsiveness to ouabain in hypertrophied rat heart : physiological and biochemical evidence. Am J Physiol 251:H923.

Lingrel JB, Young RM (1987). Tissue distribution of mRNAs encoding the α isoforms and β subunit of rat Na$^+$,K$^+$-ATPase. This issue.

Lowry OH, Rosenbrough NJ, Farr A, Randall RJ (1951). Protein measurement with the folin phenol reagent. J Biol Chem 193:265.

Lytton J, Lin JC, Guidotti G (1985). Identification of two molecular forms of (Na$^+$,K$^+$)-ATPase in rat adipocytes. J Biol Chem 260:10075.

Maixent J-M, Birkui P, Fenard S, Lelievre LG (1987). Alpha subunits (α and α$^+$) isoforms of the Na$^+$,K$^+$-ATPase in dog heart. alterations in ischemia. This issue.

Mansier P, Lelievre LG (1982). Ca^{2+}-free perfusion of rat heart reveals a (Na$^+$ + K$^+$)ATPase form highly sensitive to ouabain. Nature 300:535.

Mansier P, Charlemagne D, Rossi B, Preteseille M, Swynghedauw B, Lelievre LG (1983). Isolation of impermeable inside-out vesicles from an enriched sarcolemma fraction of rat heart. J Biol Chem 258:6628.

Ng YC, Akera T (1987). Two molecular forms of Na$^+$,K$^+$-ATPase in the ferret heart and developmental changes in digitalis sensitivity. This issue.

Shull GE, Greeb J, Lingrel JB (1986). Molecular cloning of three distinct forms of the Na$^+$,K$^+$ATPase subunit from rat brain. Biochemistry 25:8125.

Shull GE, Greeb J, Lingrel JB (1987). Amino acid sequence of the α and β subunits of the Na$^+$,K$^+$ATPase. This issue.

Siegel GJ, Desmond T, Ernst SA (1986). Immunoreactivity of ouabain - dependent phosphorylation of (Na^+,K^+) - Adenosetriphosphate catalytic subunit doublets. J Biol Chem 261:13766.

Sweadner KJ (1979). Two molecular forms of (Na^+,K^+)-stimulated ATPase in brain. J Biol Chem 254:6060.

Sweadner KJ (1987) Multiple forms of the Na,K-ATPase catalytic subunit detected by their immunological and electrophoretic differences. Alpha and alpha (+) in heart. This issue.

Thomas R, Allen J, Pitts BJR, Schwartz A (1979). Cardenolic analogs. An explanation for the unusual properties of AX 22241. Eur J Pharmacol 53:227.

Wellsmith NV, Lindenmayer GE (1980). Two receptor forms for ouabain in sarcolemma enriched preparations from canine ventricle. Circ Res 47:710.

SORTING OF NEWLY SYNTHESIZED Na,K-ATPase IN POLARIZED EPITHELIAL CELLS

Michael J. Caplan, H. Clark Anderson, George E. Palade and James D. Jamieson

Department of Cell Biology, Yale University School of Medicine, New Haven, Ct. 06510

INTRODUCTION

The plasma membranes of transporting epithelial cells are specially adapted to the task of vectorial solute transport. The surface membranes of such cells are divided into two distinct domains, which are composed of markedly different lipid and protein components, face different environments and are separated from one another by intercellular tight junctions (Simons and Fuller, 1985). In most of these systems, the apical membrane is equiped for passive transport while the basolateral domain possesses the cell's entire complement of Na,K-ATPase (Ernst and Mills, 1980). This configuration permits the Na,K-ATPase to generate transepithelial sodium gradients which can then be exploited to drive the unidirectional absorption or secretion of solutes. Little is currently known about the cellular machinery which produces this architectural anisotropy.

Polarized epithelial cells must be equiped with mechanisms which target newly synthesized plasma membrane proteins such as the Na,K-ATPase to one or the other cell surface domain and retain them there following delivery. It can be further concluded that the newly synthesized proteins must themselves possess information which can be interpreted by the cellular sorting and transport apparatus as a signal which dictates these proteins' apical or basolateral delivery. The manner in which this sorting signal is woven into a protein's structure has yet to be elucidated. The nature and subcellular location of the device which reads this signal and mediates a protein's segregation into the appropriate delivery pathway is also currently unclear.

Following their synthesis, plasma membrane proteins are carried from their site of co-translational membrane insertion at the rough endoplasmic reticulum to the Golgi complex and subsequently to the cell surface (Lodish et al., 1981). Throughout their transit, the proteins are subjected to a characteristic program of post-translational modifications (Palade, 1975). Clearly, targeting of the proteins of polarized epithelial cells to one or the other plasmalemmal domain must occur somewhere in the course of this maturation pathway.

For the purposes of experimental investigation, it is useful to divide the various models of membrane protein sorting into three distinct classes (Simons and Fuller, 1985). Models of the first class predict that sorting of newly synthesized polypeptides is completed prior to their arrival at the cell surface. Since the membrane proteins would be moving unidirectionally from an intracellular sorting site to a single cell surface domain, this scheme can be said to represent "vectorial" sorting. The second type of model requires that newly synthesized plasmalemmal proteins be inserted randomly into the two cell membrane domains. Following this "shotgun" delivery, an endocytic process would mediate the removal of apical proteins from the basolateral domain and vice versa, presumably carrying the errant proteins to their appropriate destinations. The final type of model also invokes a sorting event which follows cell surface delivery. In this formulation, however, all apical and basolateral membrane proteins would be carried to a single plasmalemmal domain. Those polypeptides which found themselves in the appropriate cell surface membrane would be retained, while proteins belonging to the other surface domain would be transported to it via a transcytotic process. This model differs from the previous one in that only a single class (rather than all) of the newly synthesized plasma membrane proteins are initially mis-sorted. It should also be noted that the first model is distinct from the second two in its assertion that apical proteins should at no time be incorporated into the basolateral membrane and vice versa.

Previous experiments aimed at distinguishing among these three models have made use of the spike glycoproteins of enveloped viruses. Certain enveloped viruses bud from infected epithelial cells with polarity (Rodriguez-Boulan and Sabatini, 1978). This polarity of budding reflects the preferential pre-budding distribution of viral spike glycoproteins. By following the post-synthetic processing of viral membrane proteins, several investigators have demonstrated that these polypeptides are subject to the first--that is, the vectorial, model of sorting (Matlin and Simons, 1984; Misek et al., 1984; Pfeiffer et al., 1985; Rindler et al., 1985). It must be noted, however, that these experiments followed the sorting of proteins which are not native to epithelial cells and furthermore employed cells infected with cytopathic viruses. Thus, the applicability of their findings to the pathways pursued by endogenous membrane proteins in unmodified epithelia needed to be verified.

We have, therefore, undertaken experiments designed to examine the sorting of newly synthesized Na,K-ATPase. Our protocols make use of the N-azidobenzoyl (NAB) derivative of ouabain (Forbush et al., 1978) and anti-ouabain antibodies to monitor the appearance of newly synthesized sodium pumps at the two surface domains of cultured polarized renal epithelial cells (MDCK cells). We find that the Na,K-ATPase is delivered directly to the basolateral membrane without appearing, even briefly, at the apical surface. These results strongly suggest that Na,K-ATPase is sorted vectorially to the basolateral membrane and that this sorting must occur in an intracellular organelle (Caplan et al., 1986).

EXPERIMENTAL DESIGN

The Madin Darby canine kidney (MDCK) cell line is derived from distal nephric epithelium and retains many of the differentiated characteristics of its in situ ancestors. The cells grow as a polarized monolayer and are interconnected by junctional complexes. The MDCK cell line is a rich source of Na,K-ATPase, with each cell's 1,000,000 sodium pumps restricted to its basolateral cell surface (Louvard, 1980). The cells can be grown on permeable filter supports, thus permitting simultaneous and independent access to the apical and basolateral plasmalemmal domains (Richardson and Simmons, 1979).

We have shown that MDCK cells grown on Nuclepore polycarbonate filters mounted in two-compartment chambers form confluent monolayers which exhibit the characteristic morphology (Caplan et al., 1986). We have further demonstrated that the sodium pump is localized exclusively to the basolateral membrane of filter-grown cells. The intercellular tight junctions of filter-grown monolayers are sufficiently impermeable to substantially retard the transepithelial passage of small molecules. When the medium bathing the apical surface is supplemented with [^{3}H]-ouabain or [^{14}C]-inulin, less than 3% of the added radioactivity can be detected in the basolateral medium after a 3 h incubation.

The n-azidobenzoyl (NAB) derivative of ouabain binds to and inhibits the Na,K-ATPase with the same affinity as its parent compound (Forbush et al., 1978). When exposed to UV irradiation, bound NAB-ouabain is covalently incorporated with high efficiency into the α-subunit. We have shown that the 100 kDa α-subunit is the sole labeled species when MDCK cells are incubated with [^{3}H]-NAB-ouabain (kind gift of B. Forbush III), photolyzed and analyzed by SDS-PAGE (Caplan et al., 1986). This ouabain-labeled protein can be used as a substrate for immunoprecipitation with anti-ouabain antibodies (kindly provided by D. Louvard and B. Rossi). Both the labeling and the immunoprecipitation could be completely inhibited through the addition of non-radioactive ouabain.

On the basis of these findings, filter-grown MDCK cells, nonradioactive NAB-ouabain and anti-ouabain antibodies were employed in a protocol designed to determine whether newly synthesized Na,K-ATPase is sorted vectorially to the basolateral membrane or whether it appears transiently at the apical surface. MDCK monolayers on filters were pulse labeled for 20 min with [^{35}S] methionine. During the ensuing 90 min chase period NAB-ouabain was present in the medium bathing either the apical or the basolateral surface. The cells were then exposed to UV light, harvested from the filters by scraping, and used to prepare a crude membrane fraction which was solubilized and subjected to immunoprecipitation with the anti-ouabain antibody. Immunoprecipitates were analyzed by SDS-PAGE followed by fluorography. It is important to note that only Na,K-ATPase molecules synthesized during the pulse period will be radiolabeled and hence detectable in the fluorograms. Furthermore, only those sodium pumps which have been exposed at the cell surface to a compartment containing NAB-ouabain will be available for recognition by the anti-ouabain antibody. We have previously shown that 90 min is sufficient for the majority of Na,K-ATPase synthesized during a 20 min period to reach the cell surface and that these sodium pumps are competent to bind ouabain before they arrive at the cell surface (Caplan et al., 1985). Thus, this protocol should allow us to monitor the delivery of the Na,K-ATPase to both MDCK plasmalemmal domains throughout the course of its post-translational processing.

RESULTS AND DISCUSSION

When pulse labeled MDCK cell monolayers were exposed to NAB-ouabain at their basolateral surfaces, the 100 kDa α-subunit band could be readily detected in fluorograms of anti-ouabain immunoprecipitates. In contrast, little or no α-subunit was immunoprecipitable from cells which were presented solely with apical NAB-ouabain. Treatment of monolayers with 2mM EDTA results in the dissolution of intercellular tight junctions. Under these conditions, NAB-ouabain present in the apical medium was capable of interacting with newly synthesized Na,K-ATPase, since radio-labeled α-subunit was demonstrable in anti-ouabain immunoprecipitates.

These results are consistent with the idea that newly synthesized Na,K-ATPase is sorted intracellularly and is delivered vectorially to the basolateral membrane. Consequently, it never appears at the apical surface and is never available to NAB-ouabain present in the apical medium. Dissolution of the intercellular junctions permits apical NAB-ouabain to gain access to the cache of basolateral Na,K-ATPase. Before this conclusion can be accepted, however, the limit of sensitivity of these methods must be established.

In order to determine the lower limit of detectability for our NAB-ouabain labeling and antibody techniques, membranes from pulse labeled monolayers exposed to basolateral NAB-ouabain were serially diluted prior to anti-ouabain immunoprecipitation. Densitometric analysis of the resulting fluorograms revealed that less than 7% of the pre-dilution signal could be reliably detected. Next, it was necessary to show that the time resolution of our protocol is sufficiently fine to detect a cohort of newly synthesized sodium pumps which might have only an extremely brief residence in the apical membrane. We therefore measured the rate and stability of ouabain binding to MDCK cell Na,K-ATPase.

Under the conditions employed in the NAB-ouabain experiments, ouabain binding was fully saturated in less than one minute. During the course of a subsequent 90 min incubation, less than 30% of the bound ouabain dissociated from the Na,K-ATPase. Furthermore, this dissociation rate was not affected by adjusting the pH of the incubation medium to 5.0. Thus, if newly synthesized Na,K-ATPase were to appear on the apical surface it should bind NAB-ouabain very rapidly when this compound is present in the apical medium. Furthermore, NAB-ouabain bound during exposure on the apical surface should remain associated with the Na,K-ATPase throughout the remainder of the 90 min chase period, after which photolysis would render the bond irreversible. Even if the transit from the apical to the basolateral membrane were to occur via an acidic transcytotytic vesicle, any NAB-ouabain-sodium pump complexes formed at the apical surface should survive and thus be detectable in the immunoprecipitates. In light of these findings, it would appear that less than 1/10 of the cell's newly synthesized Na,K-ATPase arrives (and escapes detection) at the apical plasma membrane during its passage to the basolateral cell surface.

If newly synthesized Na,K-ATPase is delivered vectorially to the basolateral membrane, without appearing even briefly at the apical surface, then this protein must be sorted and targeted during its passage through an intracellular organelle. Experiments performed on the viral spike glycoproteins indicate that newly synthesized membrane proteins destined for insertion in opposite plasmalemmal domains can occupy the same transmost cisterna of the Golgi apparatus (Fuller et al., 1985). Sorting of apical from basolateral proteins must, therefore, occur at or after this late Golgi compartment and prior to arrival at the cell surface. Recent immunocytochemical evidence suggests that that the transmost cisterna is in fact a network (the trans Golgi network) which may be specialized to accomodate sorting functions (Griffith and Simons, 1986).

The mechanisms which mediate this sorting process are as yet unknown. It is interesting to note that the interactions which are required to direct newly synthesized Na,K-ATPase to the basolateral surface of MDCK cells must differ, at least in part, from those which function to target secretory proteins for basolateral release. Sorting of basolateral secretory proteins appears to require the participation of an intracellular acidic compartment, since weak bases (which raise the pH of such compartments) induce the random secretion of these polypeptides (Caplan et al., 1987). In contrast, sodium pump sorting is unaffected by the presence of weak bases (Caplan et al., 1986). The lack of an obligate involvement of intracellular acidic compartments in sodium pump sorting suggests that this process is not encompassed by models which invoke the participation of a pH-dependent sorting receptor. A fascinating proposal which may account for some aspects of Na,K-ATPase sorting is suggested by recent evidence that the Na,K-ATPase can interact directly with ankyrin, an element of the cytoskeleton (Nelson and Veshnock, 1987; Kashgarian et al., this volume). It has been shown that a web of ankyrin and fodrin lies immediately adjacent to the cytoplasmic surface of the basolateral membrane of MDCK cells (Nelson and Veshnock, 1986; Kashgarian et al., this volume). Thus, the formation of a complex with ankyrin may aid in targeting the newly synthesized Na,K-ATPase to this plasmalemmal domain. Future research on the molecular requirements for sodium pump sorting should be greatly facilitated by the availabilty of genetic probes for the α-subunit as well as for the β-subunit, whose synthesis, processing and targeting were not directly investigated in our system.

References

Caplan M, Palade G, Jamieson J (1985). Cell surface expression and acivation of newly synthesized Na,K-ATPase in MDCK cells. in Glynn I, Ellory C (eds): "The Sodium Pump," Cambridge: The Company of Biologists, Ltd. pp147-151.

Caplan M, Anderson H, Palade G, Jamieson J (1986). Intracellular sorting and polarized cell surface delivery of Na,K-ATPase, an endogenous component of MDCK cell basolateral plasma membranes. Cell 46:623-631.

Caplan M, Stow J, Newman A, Madri J, Anderson H, Farquhar M, Palade G, Jamieson J (1987). Polarized sorting of secreted proteins is pH-dependent. Nature in press.

Ernst S, Mills J (1980). Autoradiographic localization of tritiated ouabain-sensitive sodium pump sites in ion transporting epithelia. J Histochem Cytochem 28:72-77.

Forbush B, Kaplan J, Hoffman J (1978). Characterization of a new photoaffinity derivative of ouabain: labeling of the large polypeptides and of a proteolipid component of the Na,K-ATPase. Biochemistry 17:3667-3675.

Fuller S, Bravo R, Simons K (1985). An enzymatic assay reveals that proteins destined for the apical and basolateral domains of an epithelial cell line share the same late Golgi compartments. EMBO J 4:297-307.
Griffith G, Simons K (1986). The trans Golgi network: sorting at the exit site of the Golgi complex. Science 234:438-443.
Lodish H, Braell W, Schwartz A, Strous G, Zilberstein A (1981). Synthesis and assembly of membrane and organelle proteins. Int Rev Cytol Suppl 12:247-309.
Louvard D (1980). Apical membrane aminopeptidase appears at sites of cell-cell contact in cultured epithelial cells. Proc Nat Acad Sci 777:4132-4136.
Matlin K, Simons K (1984). Sorting of a plasmalemma glycoprotein occurs before it reaches the cell surface in cultured epithelial cells. J Cell Biol 99:2131-2139.
Misek D, Bard E, Rodriguez-Boulan E (1984). Biogenesis of epithelial cell polarity: intracellular sorting and vectorial exocytosis of an apical plasma membrane glycoprotein. Cell 39:537-546.
Nelson W, Veshnock P (1986). Dynamics of membrane skeleton (fodrin) organization during development of polarity in Madin-Darby canine kidney epithelial cells. J Cell Biol 103:1751-1765.
Nelson W, Veshnock P (1987). Ankyrin binding to Na,K-ATPase and implications for the organization of membrane domains in polarized cells. Nature 328:533-536.
Palade G (1975). Intracellular aspects of the process of protein synthesis. Science 189:347-358.
Pfeiffer S, Fuller S, Simons K (1985). Intracellular sorting and basolateral appearance of the G protein of vesicular stomatitis virus in MDCK cells. J Cell Biol 101:470-476.
Richardson J, Simmons N (1979) Demonstration of protein asymmetries in the plasma membrane of cultured renal (MDCK) epithelial cells by lactoperoxidase-mediated iodination. FEBS Lett 105:201-204.
Rindler M, Ivanov I, Plesken H, Sabatini D (1985). Polarized delivery of viral glycoproteins to the apical and basolateral plasma membranes of MDCK cells infected with temperature sensitive viruses. J Cell Biol 100:136-151.
Rodriguez-Boulan E, Sabatini D (1978). Asymmetric budding of viruses in epithelial monolayers: a model system for study of epithelial polarity. Proc Nat Acad Sci 75:5071-5075.
Simons K, Fuller S (1985). Cell surface polarity in epithelia. Ann Rev Cell Biol 1:295-340.

QUANTITATIVE ANALYSIS OF SODIUM PUMP-SPECIFIC mRNA FROM
HUMAN ENDOTHELIAL (HeLa) AND CANINE KIDNEY (MDCK) CELL
CULTURES.

Christopher P. Cutler, Gordon Cramb and Joseph F. Lamb.

Department of Physiology and Pharmacology,
Bute Medical Buildings, University of
St. Andrews, Fife, U.K., KY16 9TS.

INTRODUCTION

Several methods are available for the isolation of intact messenger RNA (mRNA) from tissue or cultured cell extracts. Extraction procedures employ the use of strong denaturants such as guanidinium salts to denature all cell protein including active RNAses, followed by separation of RNA from other cell constituents and purification of mRNA by affinity chromatography. During these extraction procedures a certain percentage of RNA is inevitably lost as a result of hydrolysis by intact RNAses or by non-specific absorption to glass or denatured proteins. Therefore accurate quantification of mRNA from tissue or cell extracts still remains a problem and, as a result, estimates of specific mRNA concentrations are usually reported qualitatively by comparison to a control sample or standard. We have attempted to develope a method for the quantitative isolation of sodium pump mRNA from cultured cells which would enable us to directly compare results between as well as within different experiments. The method is adapted from the total RNA extraction described by Cathala, et al. (1983).

METHODS

RNA Isolation

HeLa or MDCK cell cultures (40-60 x 10^6 cells) were extracted with a lysis solution comprising 4.5M guanidine thiocyanate, 1.43M ß-mercaptoethanol, 45mM tris pH 7.5, 9mM EDTA and 5000-10000 cpm [^{3}H]-rabbit globin mRNA as an internal standard to estimate recovery of total or poly A$^+$ (adenylated) mRNA. Total RNA was then selectively precipitated by the addition of 6.25 vol. 4M LiCl overnight at 4°C and isolated by centrifugation. The pellet was then washed with

3M LiCl, and finally resuspended in 10mM tris, 1mM EDTA, and 1% SDS pH 7.5. The extract was incubated with Proteinase K (200ug/ml), extracted twice in phenol-chloroform (1:1 v/v) and finally precipitated by addition of 0.1 vol. of 3M sodium acetate and 2.5 vol. of 95% ethanol.

Radiolabelling of Globin mRNA.

Rabbit globin mRNA was purchased from Bethesda Research Laboratories (BRL). The labelling reaction was carried out using the enzyme Poly A Polymerase (BRL). This almost exclusively adds ATP nucleotides to the 3' end of RNA molecules. The reaction was carried out at 37°C in the presence of ^{3}H-ATP (Amersham International) as described by Sippel (1973), with the exception that 5uM cordycephin 5'-triphosphate was added to prevent over extension of the poly A tails. The reaction was terminated by precipitation of the RNA after 3 hours with 2.5 vol. ethanol and 0.1 vol. 3M sodium acetate.

Radiolabelled poly A$^+$ globin mRNA was recovered by centrifugation and repurified by separation on a oligo (dT) cellulose column (Maniatis et al., 1982), (fig 1). The ethanol precipitate from the radiolabelling mixture was resuspended in

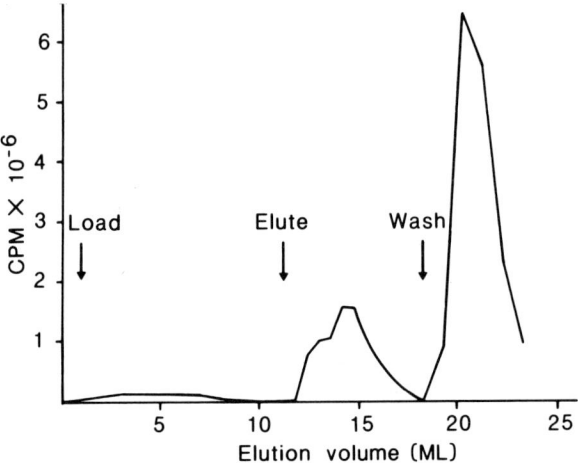

Figure 1. Purification of ^{3}H-labelled rabbit globin mRNA by adsorption and specific elution from oligo (dT) cellulose.

0.25ml of loading buffer (0.5M NaCl, 1mM EDTA, 0.1% SDS and 20mM Tris pH 7.5) and applied to a 1ml column of oligo (dT) cellulose. The column was washed with loading buffer and ^{3}H-labelled mRNA was eluted by reducing the [NaCl] to zero - elution buffer. The column was regenerated with 0.1M NaOH and 5mM EDTA to remove non-specifically adsorbed radiolabel.

Detection of Sodium Pump alpha-Subunit mRNA.

Total RNA was separated by formaldehyde gel electrophoresis, blotted onto nitrocellulose and hybridised with a ^{32}P-labelled cDNA probe for the alpha subunit of the sodium pump. Autoradiographic analysis indicated the presence of only one band of around 27S (fig. 2).

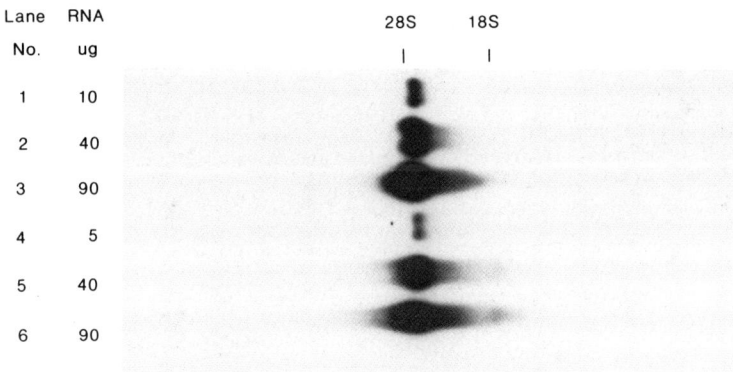

Figure 2. Northern Blot of total RNA probed with ^{32}P-labelled sodium pump alpha subunit cDNA. RNA (5-90ug) was extracted from HeLa or MDCK cells and electrophoresed in formaldehyde/ agarose gels and blotted onto nitrocellulose. Filters were probed with ^{32}P-labelled cDNA for the alpha subunit of the pump.

Total RNA and Poly A$^+$ mRNA (purified by oligo (dT) cellulose affinity chromatography) was serially diluted and filtered through nitrocellulose using a Dot Blot Manifold. Filters were baked, prehybridised, and then hybridised (Hames and Higgins, 1985) with a ^{32}P-labelled Na,K-ATPase cDNA probe (Donated by Prof. R. Levenson; Emanuel et al., 1986) of known specific activity. Blots were washed at 55°C with buffers of decreasing ionic strength. The filter was exposed to preflashed X-ray film at -70°C with intensifying screens. Blots were scanned by densitometry and peak areas integrated by computer. A standard series of radioactive probe dilutions of known cpm are filtered to relate silver grain density to count rate.

RESULTS AND CONCLUSIONS

To determine the recovery of total or messenger RNA from cell extracts 10,000 dpm of [^{3}H]-RNA (total HeLa cell RNA obtained from BRL) or [^{3}H]-rabbit globin mRNA was added to the guanidine thiocyanate/ß-mercaptoethanol extraction solution and the radioactivity appearing in the extracts was monitored throughout the different stages of the purification. The extent of recovery of radiolabelled total RNA or rabbit globin mRNA applied in tracer amounts to the original extraction of cell suspensions are shown in Tables no. 1 and 2.

Table 1. Losses of ^{3}H-RNA during extraction. Values represent the percentage loss of radiolabelled RNA (total HeLa cell RNA and globin mRNA) at each stage and the final recovery of RNA.

Purification stage	HeLa cell ^{3}H-RNA (n=2)	Globin ^{3}H-mRNA (n=18)
4M LiCl precipitation	15-20	15-25
3M LiCl precipitation	1-2	0.2-2.5
phenol extractions	0.5-2.5	0.6-3.0
ethanol precipitation	<0.1	<0.1
TOTAL RNA RECOVERY	25-50	60-80

Table 2. Losses of globin ^{3}H-mRNA during oligo (dT) column purification (see text for details). Values represent the percentage loss or total recovery of poly A$^+$ mRNA from the column.

Purification stage	Percentage radioactivity added (n=3)
Eluate during sample application (poly A$^-$ RNA)	1.0-2.5
Eluate from NaOH column regeneration	2.5-8.0
ethanol precipitation	<0.5
TOTAL POLY A$^+$ mRNA RECOVERED	77-93

Since only one mRNA species hybridised to the labelled probe all quantitative analysis of mRNA were carried out using a dot blot manifold to prevent possible extra losses of mRNA during Northern blotting. Duplicate samples of total RNA or isolated mRNA were double diluted with 20 X SSC (3M Na Cl and 0.3M Na citrate) and vacuum filtered through pre-soaked nitrocellulose. The stringency of hybridisation, pre-hybridisation and washing conditions were investigated (fig.3) with similar conditions to that of Northern Blot being adopted.

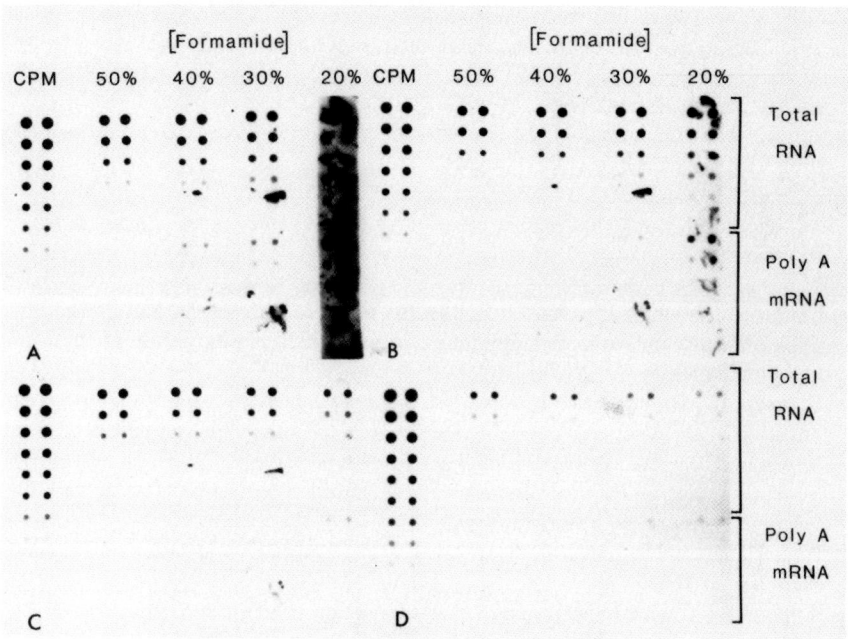

Figure 3. Effect of stringency of hybridisation on RNA dot blots. ^{32}P-labelled cDNA probe for the alpha subunit of the pump was hybridised with increasing stringency to total and mRNA dot blots. The stringency was altered by varying the formamide concentration (20-50%) of the incubations and by altering the ionic strength of the wash buffers. All washes were carried out for 30 min. at 55°C in A). 2 x SSC, B). as A plus 1 x SSC, C). as B plus 0.5 x SSC, or D). as C plus 0.1 x SSC. Radioactive standards (cpm) represent a duplicate series of filtered radioactive probe samples ranging from 250 to 5 cpm.

By relating the intensity of RNA dots on a autoradiograph to the standard dots, the cpm (counts per minute) per dot can be determined. The specific activity (cpm per ug) of the ^{32}P-labelled Na,K-ATPase DNA probe can be found by Cerenkov counting. The number of DNA molecules per ug can be calculated. Therefore, the number of probe molecules bound during hybridisation per dot can be calculated. Assuming a hybridisation ratio of 1:1 (DNA: Na,K-ATPase mRNA), the number of mRNA molecules bound per dot is determined. Adjustments can then be made for losses during extraction by accounting for the recovery of ^{3}H-RNA added and figures expressed in terms of mRNA molecules per cell. The efficiency of hybridisation can be checked by hybridising with a ^{32}P-labelled DNA probe for globin and counting the ^{3}H-radioactivity blotted on the filter. Using this method sodium pump mRNA species can be quantified from cell cultures grown under various conditions which are known to alter the density of pump sites on the membrane.

ACKNOWLEDGEMENTS

The authors would like to thank the British Heart Foundation for financial support. C.P.C. is in receipt of an MRC studentship.

REFERENCES

Cathala G, Savouret JF, Mendez B, Best BL, Karin M, Martial JA and Baxter JD (1983). A method for isolation of intact, translationally active ribonucleic acid. DNA 2: 329-335.

Emanuel JR, Garetz S, Schneider J, Ash J, Benz EJ Jnr, Levenson R (1986). Amplification of DNA sequences coding for the Na,K-ATPase alpha subunit in ouabain-resistant C$^+$ cells. Mol Cell Biol 6: 2476-2481.

Maniatis T, Fritsch EF and Sambrook J (1982). (eds.) Molecular Cloning - a Laboratory Manual, Cold Spring Harbor.

Hames BD and Higgins SJ (1985). (eds.) Nucleic Acid Hybridisation - a Practical Approach, IRL Press.

Sippel AE (1973). Purification and characterization of adenosine triphosphate: Ribonucleic acid adenyltransferase from Escherichia coli. Eur J Biochem 37: 31-40.

TISSUE-SPECIFIC DIFFERENCES IN THE EXPRESSION OF THE GENES ENCODING THE ALPHA AND BETA SUBUNITS OF RAT NA,K-ATPASE

Gregory G. Gick, Faramarz Ismail-Beigi and Isidore S. Edelman

Departments of Biochemistry and Molecular Biophysics (G.G.G., I.S.E.) and of Medicine (F.I.B.), Columbia University, New York, New York 10032, U.S.A.

INTRODUCTION

The abundance and specific activity of Na,K-ATPase varies significantly in a tissue between species and with respect to tissues within a single species (Bonting et al., 1962). In a normal rat, for example, the activity of Na,K-ATPase (V_{max}) in the kidney is 5 to 7-fold greater than that of liver (expressed per unit DNA) (Ismail-Beigi and Edelman, 1971; Gick et al., unpublished observations). The cloning of cDNAs encoding the alpha and beta subunits of the Na,K-pump has provided the tools to investigate the mechanisms contributing to tissue-specific differences in the expression of this enzyme. We present here an analysis of rat kidney and liver Na,K-ATPase gene expression at the level of mRNA abundance and corresponding rates of alpha and beta gene transcription.

METHODS

Na,K-ATPase mRNA alpha and beta content.

Replicate samples of rat liver (37 ug) and kidney (6 ug) total RNA were fractionated by electrophoresis through a 1% agarose gel containing 6% formamide, and transferred to nitrocellulose paper and prehybridized and hybridized as described previously (Chaudhury et al., 1987). Equal amounts of rat brain Na,K-ATPase alpha-specific (Schneider et al., 1985) and beta-specific (Mercer et al., 1986) cDNA probes labeled by nick-translation to approximately the same specific activity were used in Northern blot analysis. After hybridization, the blots were washed four times for 15 min each at 50°C in 250 ml of 0.1X SSC containing 0.1% SDS (1X SSC contains 0.15 M NaCl, 0.015 M sodium citrate, pH 7.0). Autoradiography was performed at -70 C for 18 and 72 h durations for alpha and beta blots, respectively. The relative intensities of Na,K-ATPase mRNA alpha and beta were measured by densitometry, and corrected for differences in the time of exposure. A RNA to DNA ratio of 1.0 and 3.2 in rat kidney and liver, respectively, was used to calculate the relative abundance of Na,K-ATPase mRNA alpha and beta per unit DNA (Loeb and Yeung, 1975; Gick et al., unpublished observations).

Na,K-ATPase alpha and beta gene transcription.

Nuclei were purified from rat liver and kidney as described (Schibler et al., 1983). Elongation reactions were initiated with an equivalent number of nuclei (20 ug DNA) and RNA isolated utilizing minor modifications of previous methods (McKnight and Palmiter, 1979). In this study, pBR322 and cDNA plasmids encoding full-length sheep mRNA alpha and beta (Shull et al., 1985; Shull et al., 1986) were immobilized on nitrocellulose paper by a modification of the method described by Kafatos et al. (1979). Immobilized plasmid DNA was prehybridized for 18 hr at 45 C and nuclear RNA hybridized for 4 days with agitation. Filters were washed and incubated with RNase for 30 min. Hybridized, labeled RNA was released from the filters and radioactivity associated with pBR322 filters was subtracted from alpha and beta cDNA filters to yield specific in vitro transcription.

RESULTS AND DISCUSSION

Northern blots of total RNA from rat liver and kidney were hyridized with Na,K-ATPase alpha and beta-specific cDNA probes (Fig. 1). Autoradiography of RNA blots hybridized to cDNA alpha revealed a single band characteristic of mRNA alpha migrating at 26-27 S in both kidney and liver RNA (Schneider et al., 1985; Chaudhury et al., 1987). Hybridization of a replicate Northern blot of kidney and liver RNA with cDNA beta revealed four bands with migration rates corresponding to 22, 20, 18 and 17 S. The predominant expression of the 22 S beta mRNA species in kidney RNA, and the approximate equal abundance of 22 and 20 S mRNA beta forms in liver, is consistent with previous descriptions of the tissue-specific expression of Na,K-ATPase beta mRNAs (Mercer et al., 1986; Young et al., 1987; Young and Lingrel, 1987).

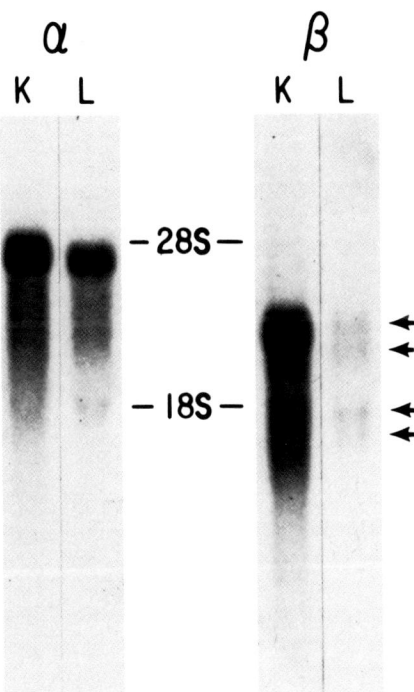

Figure 1. Northern blots of kidney (K) and liver (L) total RNA probed with Na,K-ATPase cDNA alpha and beta.

Densitometric analysis of the autoradiographs indicated that the relative abundance of kidney mRNA alpha and beta was 5 and 36-fold, respectively, greater than those of the corresponding liver Na,K-pump mRNAs (expressed per unit DNA) (Table 1).

Table 1. Na,K-ATPase Activity, mRNA Abundance and Transcription Rate in Kidney and Liver.

Tissue	Na,K-ATPase Activity umol Pi/h/ ug DNA	mRNA Abundance (per unit DNA) alpha	beta	Transcription Rate (cpm/min/ug DNA) alpha	beta
Kidney	0.330*	40**	36	0.6±0.1***	0.4±0.1
Liver	0.076	9	1	0.6±0.1	0.4±0.1

* Calculated from Ismail-Beigi and Edelman (1971) and kidney and liver mg protein/ug DNA ratios of 51 and 80, respectively (Loeb and Yeung, 1975; Gick et al., unpublished observations).

** Abundance of Na,K-ATPase mRNA's are normalized to liver mRNA beta.

*** n=3, mean ± S.E.M.

The greater abundance of kidney mRNA alpha strongly correlates with the higher kidney Na,K-ATPase activity relative to liver (Ismail-Beigi and Edelman, 1971). The difference in mRNA beta abundance, however, is significantly greater than the difference in enyzme activity, suggesting the potential involvement in the liver of translational and/or post-translational mechanisms in the determination of the abundance of functional Na,K-pumps.

An estimate of the relative abundance of Na,K-ATPase mRNAs within a tissue was also made, as the cDNA probes were approximately the same size and specific activity (Table 1). The abundances of kidney mRNA alpha and beta were equivalent. In contrast, the abundance of liver mRNA alpha was 9-fold greater than that of mRNA beta. The striking discrepancy in the abundance of the alpha and beta messages in liver suggests additional levels of complexity in the biogenesis of hepatic Na,K-ATPase. The low abundance of liver mRNA beta may reflect either the absence of this subunit in the liver enzyme or the presence of variant beta polypeptides encoded by mRNAs with sufficient

nucleotide sequence divergence to preclude hybridization to the rat brain cDNA beta (Hubert et al., 1986).
Alternatively, if we are detecting mRNA beta forms in liver which give rise to polypeptides associated with Na,K-ATPase alpha subunits with the expected 1:1 stoichiometry, then the discrepancy in subunit mRNA abundances may reflect the involvement of translational and/or post-translational mechanisms in the accumulation of liver Na,K-ATPases.

In vitro run-on transcription assays were employed to investigate the molecular mechanisms of tissue-specific differences in the abundance of Na,K-ATPase mRNAs (Table 1). Transcription analysis of isolated kidney and liver nuclei revealed similar rates of alpha and beta gene transcription. In the kidney, the near equivalence of alpha and beta gene transcription correlates with similar abundances in alpha and beta messages. In contrast, in liver, gene transcription is greater than predicted from the abundance of liver mRNA alpha and beta relative to kidney. Thus, kidney Na,K-ATPase mRNAs may be more stable or are processed and/or transported at a higher rate than liver mRNA alpha and beta. Furthermore, in the liver, mRNA alpha content is significantly greater than mRNA beta while alpha and beta gene transcription rates are comparable, thus suggesting regulation of liver Na,K-ATPase mRNAs at a post-transcriptional level.

REFERENCES

Bonting SL, Caravaggio LL, Hawkins NM (1962). Studies on sodium-potassium-activated adenosine triphosphatase. Arch Biochim Biophys 98:413-419.

Chaudhury S, Ismail-Beigi F, Gick GG, Levenson R, Edelman IS (1987). Effect of thyroid hormone on the abundance of Na,K-adenosine triphosphatase alpha-subunit messenger ribonucleic acid. Mol Endocrinology 1:83-89.

Hubert JJ, Schenk DB, Skelly H, Leffert HL (1986). Rat hepatic (Na^+,K^+)-ATPase: alpha-subunit isolation by immunoaffinity chromatography and structural analysis by peptide mapping. Biochemistry 25:4156-4163.

Ismail-Beigi F, Edelman IS (1971). The mechanism of the calorigenic action of thyroid hormone. J Gen Physiol 57:710-722.

Kafatos FC, Jones CW, Estratiadis A (1979). Determination of nucleic acid sequence homologies and relative concentrations by a dot hybridization procedure. Nucl Acids Res

7:1541-1552.

Loeb JN, Yeung LL (1975). Synthesis and degradation of ribosomal RNA in regenerating liver. J Exp Med 142:575-587.

McKnight GS, Palmiter RD (1979). Transcriptional regulation of the ovalbumin and conalbumin genes by steroid hormones in chick oviduct. J Biol Chem 254:9050-9058.

Mercer RW, Schneider JW, Savitz A, Emanuel J, Benz Jr EJ, Levenson R (1986). Rat-brain Na,K-ATPase beta-chain gene: primary structure, tissue-specific expression, and amplification in ouabain-resistant HeLa C^+ cells. Mol Cell Biol 6:3883-3890.

Schibler U, Hagenbuchle O, Wellauer PK, Pittet AC (1983). Two promoters of different strengths control the transcription of the mouse alpha-amylase gene amy-1^a in the parotid gland and the liver. Cell 33:501-508.

Schneider JW, Mercer RW, Caplan M, Emanuel R, Sweadner KJ, Benz Jr EJ, Levenson R (1985). Molecular cloning of rat brain Na,K-ATPase alpha-subunit cDNA. Proc Natl Acad Sci USA 82:6357-6361.

Shull GE, Schwartz A, Lingrel JB (1985). Amino-acid sequence of the catalytic subunit of the (Na^+-K^+) ATPase deduced from a complementary DNA. Nature 316:691-695.

Shull GE, Lane LK, Lingrel JB (1986). Amino-acid sequence of the beta-subunit of the (Na^++K^+) ATPase deduced from a cDNA. Nature 321:429-431.

Young RM, Shull GE, Lingrel JB (1987). Multiple mRNAs from rat kidney and brain encode a single Na^+,K^+-ATPase beta subunit protein. J Biol Chem 262:4905-4910.

Young RM, Lingrel JB (1987). Tissue distribution of mRNAs encoding the alpha isoforms and beta subunit of rat Na^+,K^+-ATPase. Biochem Biophys Res Comm 145:52-58.

IN VITRO EXPRESSION OF THE ALPHA AND BETA SUBUNITS OF THE Na,K-ATPase

M. Gilmore-Hebert[1], R. W. Mercer[3*], J. W. Schneider[2], and E.J. Benz[1,2]. Departments of Internal Medicine[1], Human Genetics[2] and Cellular and Molecular Physiology[3], Yale University School of Medicine, New Haven CT USA. *Present address: Department of Cell Biology and Physiology, Washington University School of Medicine, St. Louis, MO USA.

INTRODUCTION

The Na^+,K^+-ATPase generates the flux of Na^+ and K^+ ions across the plasma membranes of animal cells by the hydrolysis of ATP. The enzyme is a heterodimeric protein consisting of alpha (97-100 KDa) and beta (55KDa) subunits. It is responsible for the regulation of cell volume, the generation of Na^+ gradients which drive the transport of glucose, amino acids and Ca^{++} ions, and the electrochemical gradients in nerve and muscle tissue. Na^+,K^+ ATPase is widely distributed and is known to display structural and ouabain-binding heterogeneity in several tissues, including brain and skeletal muscle (Lytton 1985).

Two isoforms of the alpha subunit of the Na^+,K^+ ATPase have been previously characterized by biochemical and immunological techniques. They are designated alpha and alpha+. The alpha+ isoform can be distinguished from alpha by its slower mobility on $NaDod\ SO_4$ - PAGE (Sweadner 1979), its higher affinity for ouabain and Na^+ ions (Sweadner 1985), greater sensitivity to reducing agents (Sweadner 1979), inactivation by pyrithiamin (Akera et al., 1986) and higher resistance to trypsin digestion. These distinct biochemical properties suggest that each isoform may fulfill a specialized cellular requirement for ion transport.

A third alpha subunit isoform has recently been identified using c-DNA cloning by our group and Schull et al., (1986) designated alpha 3. Northern and in situ hybridization data (Schneider et al., 1987) strongly suggest that alpha 3 is the major isoform in rat brain and fetal heart yet this novel isoform has not been previously detected in either tissue by analysis of the membrane proteins.

In order to characterize and compare the biochemical and immunological properties of this novel isoform to alpha and alpha +, we used an in vitro transcription-translation system to synthesize alpha 3 protein. The alpha 3 translation product associated with microsomes, and shared antigenic determinants with alpha and alpha + proteins.

METHODS

Cell-Free Protein Synthesis. Templates for in vitro translation were generated by transcription of vectors which contained the entire coding regions of alpha 1, alpha +, alpha 3 and beta c-DNAs. The transcription reactions were performed with reagents provided in the Riboprobe Gemini II System (Promega Biotec). For translations three ul of a 20 ul transcription reaction (150 ng of RNA) were directly added to a 30 ul rabbit reticulocyte lysate translation reaction (Promega-Biotec) supplemented with ^{35}S-methionine (Amersham). For membrane association, translation reactions were supplemented with dog pancreatic microsomes (Amersham). Immuno-precipitations were performed using Kl" antisera (generous gift of K. Sweadner) (Sweadner and Gilkeson, 1985) as previously described (Caplan et al., 1986). The proteins were analyzed by NaDod SO_4-PAGE (5 or 7.5% polyacrylamide) followed by enhanced flourography of the dried gel.

RESULTS

Figure 1 shows the results of an experiment in which synthetic alpha 1, alpha + and alpha 3 isoform m-RNAs were separately added to a rabbit reticulocyte lysate in the presence of dog pancreatic microsomes. After 60 minutes of incubation, the reaction mixture was centrifuged and the microsomal pellets were collected. The supernatant and microsomal pellet fractions were either analyzed directly on 5% NaDod SO_4-PAGE or the microsomal associated proteins were immunoprecipitated with an antisera raised against purified rat alpha subunit and then subsequently analyzed on 5% NaDod SO_4 PAGE.

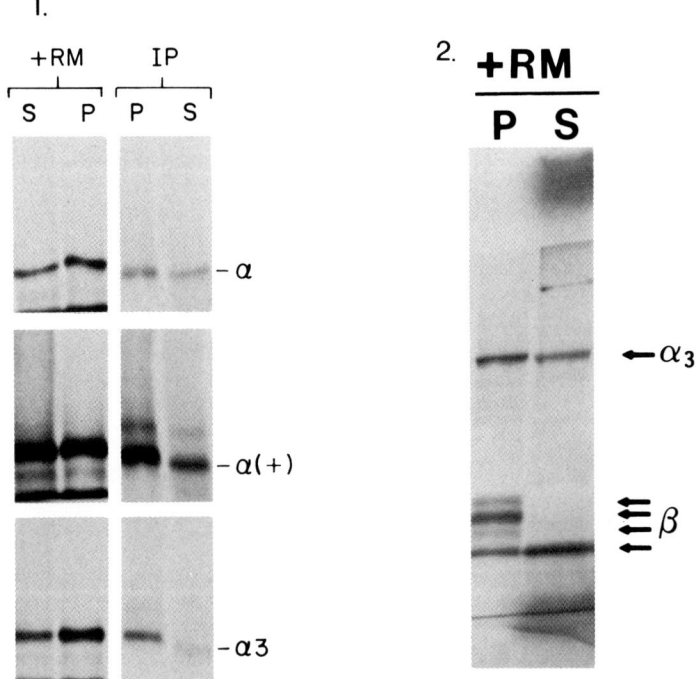

Figure 1. Cell Free Synthesis, Membrane Association and Immunoprecipitation of Na^+,K^+-ATPase Proteins.
[RM] ^{35}S-labeled Na^+,K^+ ATPase alpha polypeptides were generated by translation of synthetic mRNAs in rabbit reticulocyte lysate supplement with rough microsomes. Approximately 50% of the protein associated with the membrane pellet (P) while 50% remained in the membrane-free supernatant (S). [IP] Synthetic $Na+,K^+$-ATPase alpha isoform proteins were individually immunoprecipitated with an anti rat alpha antibody. Greater than 50% of the isoform protein was found in the pellet (P) versus that which remained in the supernatant (S).

Figure 2. Cell-Free Synthesis and Membrane Association of alpha 3 and beta.
^{35}S-labeled alpha 3 and beta polypeptides were co-translated in a reticulocyte lysate in the presence of microsomes. Approximately 50% of the alpha 3 and beta protein associates with the membranes (P) while 50% remains in the supernatant. The beta subunit is glycosylated as indicated by its shift to a higher molecular weight (55kDa).

Radiolabeled alpha 1, alpha + and alpha 3 proteins each associates with the microsomal pellet which suggest integration into the membrane. All three isoform proteins were immunoprecipitated with a Na^+,K^+-ATPase antibody suggesting that they share common idiotypes.

Figure 2 shows a similar experiment in which alpha 3 m-RNA was co-translated with beta m-RNA in the presence of pancreatic microsomes. The microsomal pellet and supernatant were analyzed on 7.5% NaDod SO_4-PAGE. Both alpha 3 and beta proteins are associated with the membrane pellet. The glycosylated 55KDa form at beta is only seen in the microsomal pellet fraction indicating the addition of carbohydrate to the protein core. Addition of the microsomal membranes had no apparent effect on the molecular weight of the alpha 3 protein indicating that like alpha and alpha + proteins it is not glycosylated. The amount of alpha and beta protein associated with the microsomes is directly proportional to the amount of microsomes added to the translation reaction suggesting an SRP dependent association.

Attempts to demonstrate enzymatic activity in the microsomal membranes containing co-translated alpha isoforms and beta have been unsuccessful. The failure may be due to the lack of proper association between the alpha and beta subunits within the membranes or the lack of other organelle components required to process the subunits into an active conformation (Caplan et al., 1986, Geering et al., 1985).

Figure 3 shows in vitro synthesized immunoprecipitated alpha 1, alpha + and alpha 3 proteins analyzed on 7.5% NaDod SO_4-PAGE gel. The alpha 3 protein migrates like alpha +, slower than alpha 1 in this gel system. The alpha 3 protein was also precipitated with an alpha + specific antibody "A2" (generous gift of K. Sweadner). Thus alpha 3 has features similar to both alpha 1 and alpha +.

SUMMARY

Our analysis of the cloned alpha 3 protein strongly suggests that this c-DNA represents a bona fide Na^+,K^+ ATPase isoform. Its similarity to alpha + may have made its detection in tissues by gel migration or immunoreactivity difficult.

Expression of an enzymatically active alpha 3 beta Na^+K^+ATPase either in a completely in vitro system or in a heterologous tissue culture system will clearly establish the biochemical properties of this isoform. Development of alpha 3 specific immunochemical probes will allow a proper assessment of its in vivo expression.

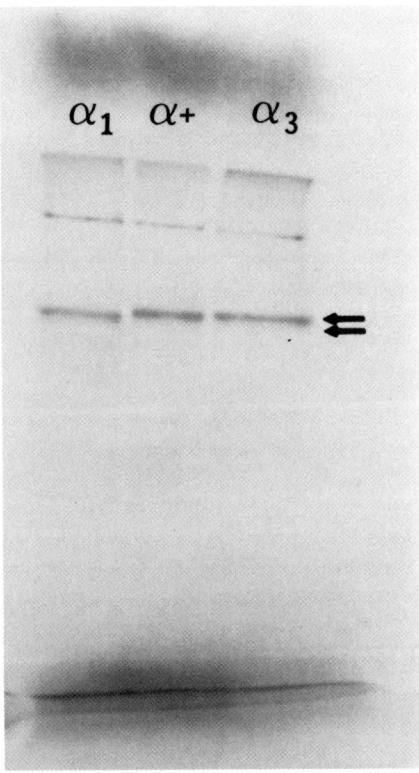

Figure 3. Higher Resolution of Isoform Peptides.
^{35}S-label and Na$^+$K$^+$-ATPase alpha polypeptides were generated in vitro. Protein associated with the microsomal membranes was subsequently immunoprecipitated and analyzed on a 7.5% NaDod SO$_4$-PAGE.

ACKNOWLEDGMENTS

This work was supported by NIH grants #1P01-DK 28376 and #T32 HL-07262.

REFERENCES

Akera T, Ng Y, Hadley R, Katano Y, Brody TM (1986). High affinity and low affinity ouabain binding sites in rat heart. Eur J Pharm 132:137-146.

Caplan MJ, Anderson HC, Palade GE, Jamieson JD (1986). Intracellular sorting and polarized cell surface delivery of Na,K-ATPase, an endogenous component of MDCK cell basolateral plasma membranes. Cell 46:623-631.

Geering K, Meyers PI, Paccolat MP, Krachenbuhl JP, Rossier BC (1985). Membrane insertion of α and β - subunits of the Na^+,K^+-ATPase. J Biol Chem 260:5154-5160.

Lytton J (1985). Insulin affects the sodium affinity of the rat adipocyte (Na^+,K^+)ATPase. J Biol Chem 260:10075-10080.

Schneider JW, Mercer R, Benz EJ Jr. (1987). Identification by molecular cloning of isoforms of the α-subunit of the Na,K-ATPase expressed in brain and heart. Clin Res 35:5905A.

Schull GE, Greeb J, Lingrel JB (1986). Molecular cloning of three distinct isoforms of the Na^+,K^+-ATPase α-subunit from rat brain. Biochemistry 25:8125-8132.

Sweadner KJ (1979). Two molecular forms of (Na^++K^+)-stimulated ATPase in brain. J Biol Chem 254:6060-6067.

Sweadner KJ, Gilkeson RC (1985). Two isozymes of Na,K-ATPase have distinct antigenic determinants. J Biol Chem 260:9016-9022.

Sweadner KJ (1985). Enzymatic properties of separated isozymes of the Na,K-ATPase. J Biol Chem 260:11508-11513.

MEMBRANE INSERTION OF α AND β SUBUNITS OF HUMAN Na^+,K^+-ATPase

Haruo Homareda, Kiyoshi Kawakami, Kei Nagano and Hideo Matsui

Department of Biochemistry, Kyorin University School of Medicine, Mitaka, Tokyo 181 (H.H., H.M), and Department of Biology, Jichi Medical School, Minamikawachi, Tochigi 329-04 (K.K., K.N), Japan

INTRODUCTION

Na^+,K^+-ATPase is an intrinsic membrane protein composed of α and β subunits(Jørgensen, 1982). Elucidation of the membrane insertion of these subunits is essential for understanding the molecular assembly of Na^+,K^+-ATPase. We have already deduced the amino acid sequences of the α and β subunits from the cloned cDNAs of HeLa cells (Kawakami et al., 1986 a,b). From the hydropathy profiles of the amino acid sequences, it has been assumed that the α subunit has at least six transmembrane segments and the β subunit has only one. Since these subunits do not have a cleavable signal sequence, it has been predicted that the signals for membrane insertion of these subunits are included in the transmembrane segments. To determine if this is true, we have constructed various deletion mutants from cloned cDNA encoding the α and β subunits of human Na^+,K^+-ATPase and examined the insertion of their translation products into microsomal membranes. The present data show that the first to fourth transmembrane segments from the N-terminal in the six transmembrane segments (TM1 to TM6) of the α subunit include the membrane insertion signals and at least a 16 amino acid stretch in the transmembrane segment of the β subunit is essential for membrane insertion of β subunits.

EXPERIMENTAL PROCEDURES

<u>Construction of deletion plasmids.</u> The cloning and sequencing of the cDNAs of HeLa Na^+,K^+-ATPase α and β

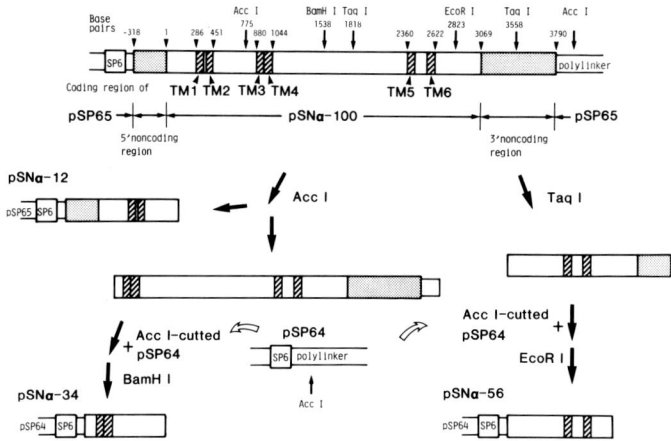

Fig. 1. Construction of deletion plasmids of the α subunit. pSP64 and pSP65 are plasmid vectors. SP6 designates the promoter site of SP6 RNA polymerase, and TM1 to TM6 designate the first to sixth transmembrane segments of the α subunit from the N-terminal.

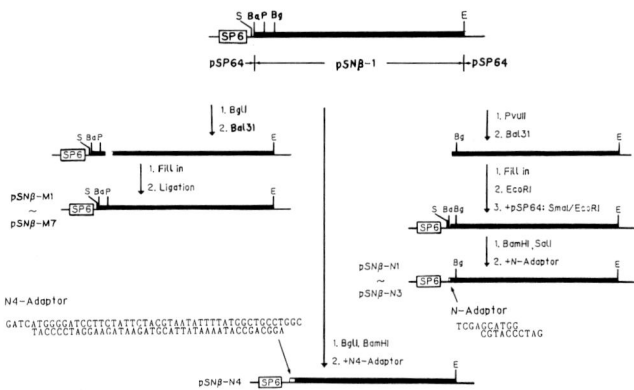

Fig. 2. Construction of deletion plasmids of the β subunit. Ba, Bg, E, H, P and S designate BamHI, BglI, EcoRI, HhaI, PvuII and SalI sites, respectively.

subunits were described elsewhere (Kawakami et al., 1986 a, b). Outlines of the methods for constructing the deletion plasmids of these subunits are given in Figs. 1 and 2.

In vitro transcription. Three μg of linearized plasmid DNA was incubated with 40 mM Tris-HCl(pH7.5), 6mM $MgCl_2$, 2mM spermidine, 0.5mM each of ATP, CTP, GTP and UTP, 1mM m^7GpppG; 10mM dithiothreitol, 1 unit/μl RNasin, and 7.5 units of SP6 RNA polymerase, in a total volume of 25 μl, for 80 min at 37°C. Nucleic acid was precipitated with ethanol and redissolved in 20 μl H_2O.

In vitro translation and membrane insertion of translation products. Outline of the method is summarized in Fig. 3.

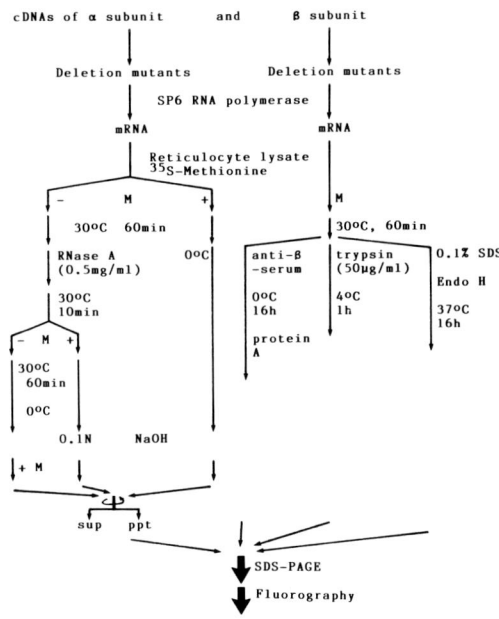

Fig. 3. Outline of the methods for in vitro translation and membrane insertion of translation products. M and Endo H designate canine pancreatic microsomal membranes and Endo-β-N-acetylglucosaminidase H, respectively.

RESULTS AND DISCUSSION

Biosyntheses and membrane insertions of the α subunit and its deletion mutants

pSNα-100, which is the cDNA encoding the entire region of human Na$^+$,K$^+$-ATPase α subunit, was transcribed by SP6 RNA polymerase and the resultant mRNA was translated in reticulocyte lysate with [^{35}S]methionine. Fluorographic patterns after SDS-polyacrylamide gel electrophoresis (SDS-PAGE) of the translation products are shown in Fig. 4. The translation product with molecular weight (MW) of 104 kDa was observed in the presence of dog pancreatic microsomal membranes (M) (lane 1). This MW was consistent with that of the purified Na$^+$,K$^+$-ATPase α subunit (Jørgensen, 1982). To determine whether this product was inserted into membranes,

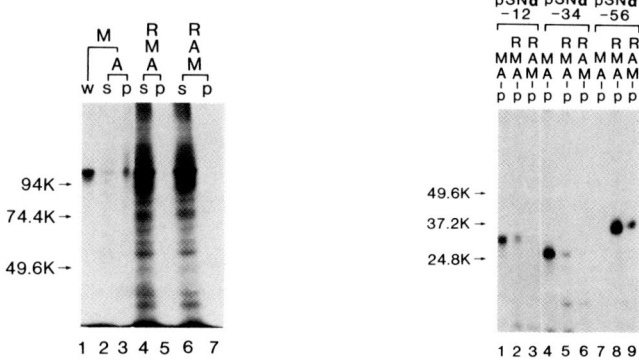

Fig. 4. (left) Biosynthesis and membrane insertion of the α subunit. pSNα-100 was transcribed as described in the "Experimental Procedure", and the synthesized mRNA was translated under the conditions indicated in the figure. M, R, A, w, s and p designate pancreatic microsomal membranes, RNase A, alkali treatment, untreated translation mixture, supernatant and pellet of translation mixture, respectively.

Fig. 5. (right) Biosyntheses and membrane insertions of polypeptides with two transmembrane segments in six ones of the α subunit. pSNα-12, pSNα-34 and pSNα-56 were transcribed and the synthesized mRNAs were translated under the conditions indicated in the figure.

the translation mixture was treated with 0.1N NaOH prior to centrifugation for separation of M from the mixture (lanes 2-7), because alkali treatment is known to release the polypeptides peripherally associated to the surface of the membranes or those secreted from the membrane vesicles (Mostov et al., 1981). When M was added at the start of the translation reaction (lanes 2 and 3), a dense band was observed in the precipitated M fraction (lane 3). When M was added after the translation reaction followed by RNase A treatment for inhibiting further synthesis of the translation products (lanes 4 and 5) or added after the alkali treatment for examining nonspecific precipitation of the translation products (lanes 6 and 7), the band was observed in only the supernatants (lanes 4 and 6). These results suggest that the α subunit is inserted into membranes cotranslationally.

pSNα-12, pSNα-34 and pSNα-56, which are the cDNAs encoding the regions including TM1 and TM2, TM3 and TM4, and TM5 and TM6, respectively, were expressed in the same transcription-translation system as that used for pSNα-100 (Fig. 5). When pSNα-12 was used as a template, a translation product with MW of 31 kDa was precipitated with M which was added to the translation mixture at the start of the translation reaction (lane 1). The observed MW was the same as the value calculated from the amino acids sequence which was deduced from the sequence of pSNα-12. When M was added to the translation mixture after the RNase A treatment, a faint band was observed in the precipitated M fraction (lane 2). However, when M was added to the mixture after the alkali treatment, no band was observed (lane 3). When pSNα-34 was used as the template, a result similar to that with pSNα-12 was observed (lanes, 4-6). The MW of the translation product, i.e., 26 kDa, was well consistent with the calculated value from pSNα-34, i.e, 30 kDa. Therefore, the membrane insertion signals are included in the polypeptide with TM1 and TM2 and that with TM3 and TM4, which can be inserted into membranes posttranslationally as well as cotranslationally.

When pSNα-56 was used as a template, the translation product was not precipitated with M (lane 7), while it was precipitated when M was added after RNase A treatment (lane 8) or added after the alkali treatment (lane 9). Although the MW of this translation product was 36 kDa which was also consistent with the calculated value from pSNα-56, i.e., 39 kDa, appearance of the polypeptide in lanes 8 and 9 showed the possibility that it was not precipitated by insertion into M but was nonspecifically precipitated by the addition of RNase A. This possibility was supported by the experimental result that the polypeptide was precipitated by addition of RNase A to the translation mixture without M (data not shown). These results show that the membrane insertion signals are not included in TM5 and TM6.

Biosyntheses and membrane insertions of the β subunit and its deletion mutants

pSNβ-1, which is the cDNA encoding the entire region of the human Na^+,K^+-ATPase β subunit, was expressed in the same transcription-translation system as that used for the α subunit. The main translation product with a MW of 37 kDa was observed in the absence of M (Fig. 6A, lane 1; Fig. 6B, lane 4) and when M was added after RNase A treatment (Fig. 6B, lane 3). The addition of M at the start of the translation reaction produced a translation product with a 52 kDa MW (Fig. 6A, lane 3; Fig. 6B, lane 2). This

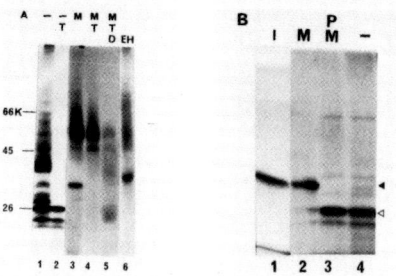

Fig. 6. Biosynthesis and membrane insertion of the β subunit. pSNβ-1 was transcribed, and the synthesized mRNA was translated under the conditions indicated in the figure. T, M, D, EH and I designate trypsin, pancreatic microsomal membranes, detergent (NP-40), endo-β-N-acetylglucosaminidase H and anti-β subunit serum, respectively. PM designates the addition of M to the translation mixture after incubation at 30°C for 30min. ◄ and ◁ indicate the glycosylated and non glycosylated translation products, respectively.

increase in MW was due to the glycosylation in the presence of M, since digestion of the product with endo-β-N-acetylglucosaminidase H decreased the MW of the product from 52 to 34 kDa (Fig. 6A, lane 6). The translation product with the MW of 52 kDa was insensitive to trypsin (Fig. 6A, lane 4), contrary to the 34 kDa one in the absence of M and the one with a 52 kDa MW in the presence of detergent (Fig. 6A, lanes 2 and 5). The 52 kDa product was immunoprecipitated with anti-Na^+,K^+-ATPase β subunit serum (Fig.6B, lane 1). From these results, it is clear that the 52 kDa product was a complete β subunit, which was inserted into the membranes cotranslationally.

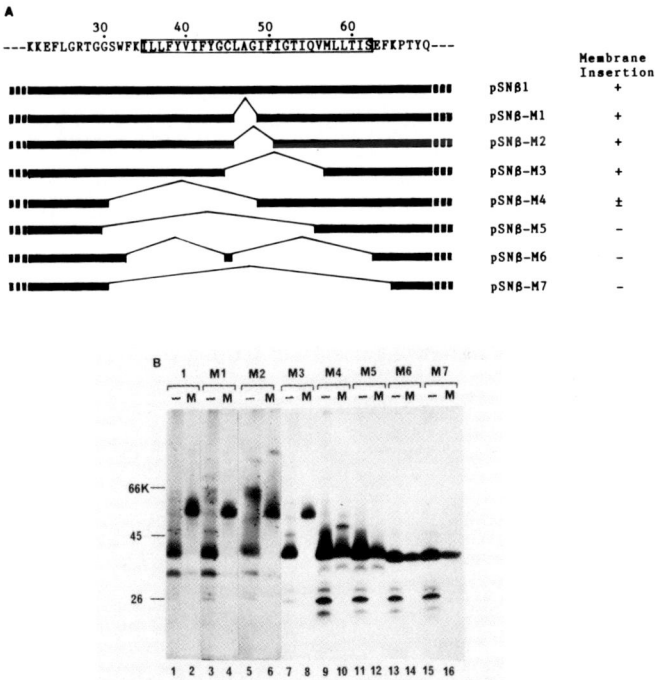

Fig. 7. Membrane insertion of the β subunit deleted its transmembrane segment to various extent. pSNβ-M1 to pSNβ-M7 were transcribed and the synthesized mRNAs were translated in the absence or presence of M. A: summary of result shown in B. A group of amino acids in box is the transmembrane segment. B: fluorographic pattern after SDS-PAGE of the translation products.

pSNβ-M1 to pSNβ-M7, which are cDNAs in which the sequences encoding the transmembrane region of the β subunit have been deleted to various extents, were expressed in the presence or absence of M. The membrane insertion of the translation products were judged from the increase in MW to about 50 kDa due to glycosylation in the presence of M. The results are shown in Fig. 7. The translation products of pSNβ-M1 to pSNβ-M4 lacking the coding regions for up to 12 amino acids in the transmembrane segment, which was assumed to be composed of 28 amino acids (Kawakami et al., 1986 b), could be inserted into the membranes. Therefore, the remaining 16 hydrophobic amino acids, in which a specific sequence may not be included, is suggested to be the minimal requirement for the membrane insertion process of the β subunit.

The N-terminal domain preceding the transmembrane segment of the β subunit has several positively charged amino acids that might have some role in the membrane insertion process. Therefore, pSNβ-N1 to pSNβ-N4, which are cDNAs in which the sequences coding for the region corresponding to the N-terminal domain have been deleted to various extents, were expressed in the presence or absence of M. All of the translation products showed increases in MW to about 50 kDa in the presence of M (data not shown). This result indicates that the N-terminal domain preceding the transmembrane segment is not required for the membrane insertion process of the β subunit.

REFERENCES

Jørgensen P L (1982). Mechanism of the Na^+,K^+ pump : protein structure and conformations of the pure (Na^++K^+)-ATPase. Biochim Biophys Acta 694 : 27.
Kawakami K, Ohta T, Nojima H, Nagano K (1986). Primary structure of the α-subunit of human Na,K-ATPase deduced from cDNA sequence. J Biochem 100 : 389.
Kawakami K, Nojima H, Ohta T, Nagano K (1986). Molecular cloning and sequence analysis of human Na^+,K^+-ATPase β-subunit. Nucl Acids Res 14 : 2833.
Mostov K E, DeFoor P, Fleisher S, Blobel G (1981). Co-translational membrane integration of calcium pump protein without signal sequence cleavage. Nature 292 : 87.

DEVELOPMENT OF A HETEROLOGOUS GENE EXPRESSION SYSTEM FOR THE Na,K-ATPase SUBUNITS IN THE YEAST SACCHAROMYCES CEREVISIAE

Burton Horowitz and Robert A. Farley
University of Southern California School of Medicine
Los Angeles, CA 90033 (USA)

ABSTRACT

cDNA fragments coding for the α and β subunits of the Na,K-ATPase were separately ligated into the yeast expression vector YEp1PT in both the sense (YEpNKA(+)) and anti-sense (YEpNKA(-)) orientations with respect to the promoter. The recombinant plasmids were intoduced into Saccharomyces cerevisiae strain UT4 by transformation. Total RNA from the transformed strains was isolated and analyzed by Northern hybridization. The resulting autoradiogram revealed strong signals indicative of a high level of transcriptional expression of both subunits in both orientations of the cDNA. 35S-Methionine labeled extracts were immunoprecipitated with antibodies specific for the β subunit. A β subunit translation product was produced from YEpβNKA(+) but not from YEpβNKA(-). Experiments to detect an α specific translation product are in progress.

INTRODUCTION

The Na,K ATPase is composed of two noncovalently linked polypeptide subunits, a catalytically active α subunit (M_r=112000), and a glycosylated β subunit (M_r=35000). The structure of the α subunit has been extensively studied and the role of the α subunit in Na,K ATPase catalytic activity is well documented. The function of the β subunit polypeptide however is unknown. There is currently a controversy concerning the biosynthesis and mechanism of assembly of the Na,K-ATPase in the plasma membrane. Hiatt et al. (1984) suggested that the insertion of the α subunit into the plasma membrane requires the presence of the β subunit already incorporated into the membrane. On the other hand, results obtained by Geering et al. (1985) indicate that the α

subunit is inserted into the plasma membrane late during its translation and independent of the β subunit. One potential problem that may compromise the interpretation of both of these studies is that they were carried out in cell free systems utilizing microsomal membranes. Microsomes contain endogenous Na,K-ATPase α and β subunits (Mircheff, 1983), which make evaluation of the roles of the individual subunits difficult. The fact that the data conflict also adds to the ambiguity.

A more direct approach to determine the role of the β subunit in Na,K ATPase assembly, conformation, or function is to express the genes encoding the protein subunits individually in an in vivo system. One major obstacle to this, however, is the fact that all animal cells contain endogenous Na,K ATPase. Endogenous Na,K ATPase would make detection and analysis of a heterologous polypeptide ambiguous. The yeast Saccharomyces cerevisiae is an attractive alternative to animal cell expression systems for the heterologous expression and analysis of the genes encoding the α and β subunits of the Na,K ATPase. Heterologous gene expression in yeast is well established and is becoming a popular method for expressing eukaryotic genes (see Hitzeman et al., 1985 for review). Large scale membrane purification procedures are relatively simple and perhaps most important, Na,K ATPase activity is not detectable in yeast, thereby eliminating the problem of detecting heterologous Na,K ATPase expression with a background of endogenous activity. In addition to elucidating the role of the β subunit in Na,K ATPase assembly and activity, the development of an expression system for the subunits of Na,K ATPase in yeast will provide the means to further analyze the structure and mechanism of the sodium pump through manipulations such as site directed mutagenesis.

RESULTS

The cDNA for sheep kidney α subunit of the Na-K, ATPase was obtained from Dr. Jerry Lingrel, U. Cinncinati (Shull et al.,1985). Bal 31 exonuclease was employed to remove 151bp from the 5' untranslated region of the cDNA. The extent of exonuclease digestion was monitored by nucleotide sequencing. The yeast expression vector, YEp1PT was obtained from Dr. Ronald Hitzeman at Genentech Inc. This plasmid has a unique Eco RI restriction site placed between the 3-phosophoglycerate kinase (PGK) promoter and the yeast termination signal sequence. The trimmed cDNA fragment was ligated to Eco RI linkers and inserted at the YEp1PT unique Eco RI site. YEp1PT has a pBR322 based origin of replication; therefore routine manipulations can be accomplished in E. coli. Transformants were screened for a plasmid containing the cDNA inserted into YEp1PT in the sense orientation with respect to the PGK promoter (YEpαNKA(+)) and a plasmid containing the cDNA inserted in the anti-sense orientation (YEpαNKA(-)). The recombinant plasmids were introduced into Saccharomyces cerevisiae strain UT4 (Etcheverry, 1985) by transformation (Hinnen et al., 1978).

cDNA encoding the β subunit of the Na,K-ATPase from dog kidney was isolated from a λgt11 cDNA library constructed in this laboratory (Brown et al., 1987). The two fragments comprising the full length cDNA were ligated together and the fidelity of that construction was confirmed by restriction enzyme analysis. The full length cDNA was then deleted of all but 20bp of 5' untranslated DNA by a combination of Exonuclease III and Mung Bean nuclease digestion. The cDNA fragment was ligated to YEp1PT in the sense (YEpβNKA(+)) and anti-sense (YEpβNKA(-)) orientations and these plasmids were introduced into Saccharomyces cerevisae strain UT4 by transformation.

Total RNA from the transformed strains was isolated and analyzed by Northern Hybridization (Thomas, 1980). The resulting autoradiogram (Fig 1) revealed a strong signal at a molecular weight of approximately 3.8kb in lanes containing RNA from both YEpαNKA(+) and YEpαNKA(-). This RNA contains approximately 500bp of 3' mRNA added to the heterologous transcript by the expression plasmid. The strain containing YEpβNKA(+) yielded mRNA which hybridized to the β probe at a molecular weight of 1.6kb. RNA from YEpNKA(-) also hybridized to the β probe at a molecular weight of 1.6kb, however, the majority of β specific RNA hybridized at a lower molecular weight suggesting the existence of a transcriptional stop signal in the anti-sense orientation.

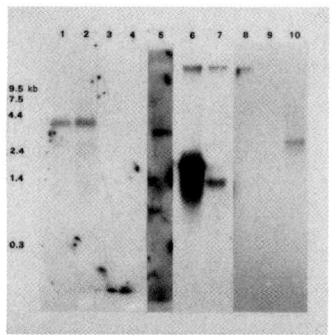

Figure 1. Northern hybridization analysis of total RNA prepared from yeast strains containing recombinant plasmids.

Lane 1: RNA from YEpαNKA(+)-transformed UT4 cells. Lane 2: RNA from YEpαNKA(-)-transformed UT4. Lane 3: RNA from YEp1PT (no insert)-transformed UT4. Lane 4: RNA from yeast strain UT4 without plasmid. Lane 5: Poly A mRNA from dog kidney α subunit. Lanes 1-5 were hybridized to a 32P-labeled 1.4kb dog kidney α subunit probe representing the carboxy terminal end of the polypeptide. Lane 6: RNA from YEpβNKA(+)-transformed UT4 cells. Lane 7: RNA from YEPβNKA(-)-transformed UT4 cells. Lane 8: RNA from YEP1PT-transformed UT4 cells. Lane 9: RNA from yeast strain UT4 without plasmid. Lane 10: Poly A mRNA from dog kidney. Lanes 6-10 were hybidized to a 32P-labeled 0.8kb dog kidney β subunit probe representing the carboxy terminal end of polypeptide and 225bp of 3'-nontranslated sequence.

35S-Methionine labeled extracts were immunoprecipitated with antibodies (JK594 obtained from Jack Kyte, U. of California at San Diego) specific for the α and β subunits. The yeast strain YEpβNKA(+) yielded β specific translation product which has a molecular weight of approximately 40kd (Fig. 2). This corresponds to the molecular weight of core glycosylated β subunit derived from MDCK cells. Strain YEpβNKA(-) does not express a β specific translation product. Experiments to detect an α specific translation product are in progress.

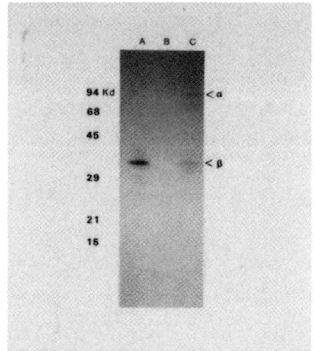

Figure 2. Immunoprecipitation of 35S-methionine labeled lysates from yeast cells transformed with recombinant plasmids.

Lane A: Extract prepared from YEpβNKA(+)-transformed UT4 cells. Lane B: Extract prepared from 35S-methionine labeled YEpβNKA(-)-transformed UT4 cells. Lane C: Extract prepared from 35S-methionine labeled MDCK cells. All immunoprecipitations were performed after preadsorption of the extracts with non-immune rabbit serum. The precipitating antibody used was form AHE594, an ammonium-sulfate precipitate of antisera to Na,K-ATPase holoenzyme, and obtained from J. Kyte (UC San Diego).

ACKNOWLEDGEMENT

This work was supported by NIH Grant GM28673, NSF Grant DMB-8613999, American Heart Association, Greater Los Angeles Affiliate Award 809 IG3, and a USC Faculty Research and Innovation Fund Award. This work was done during the tenure of an Established Investigatorship of the American Heart Association and with funds contributed in part by the Greater Los Angeles Affiliate. The expression plasmid YEp1PT was generously provided by Dr. R. Hitzman (Genentech), and the cDNA for sheep kidney α subunit was from Dr. J. Lingrel (U. Cincinnati).

REFERENCES

Brown TA, Horowitz B, Miller RP, McDonough AA, Farley RA (1987). Molecular cloning and sequence analysis of the (Na+K)-ATPase β subunit from dog kidney. Biochim Biophys Acta 912:244-253.

Etcheverry MT (1984). Amplification of the copper-chelatin gene in yeast. In Simon M, Herskowitz I (eds): "Genome Rearrangement. Proceedings of the UCLA Symposium" New York: Alan R. Liss, pp221-232.

Geering K, Meyer DI, Paccolat MP, Kraehenbuhl JP, Rossier B (1985). Membrane insertion of α- and β-subunits of Na,K-ATPase. J Biol Chem 260:5154-5160.

Hiatt A, McDonough AA, Edelman IS (1984). Assembly of the (Na+K)ATPase. J Biol Chem 259:2629-2635.

Hinnen A, Hicks JB, Fink GR (1978). Transformation of yeast. Proc Natl Acad Sci 75:1929-1933.

Hitzeman RA, Cristina CY, Hagie FE, Lugovoy JM, Singh A (1985). Yeast: An alternative organism for foreign protein production. In Bolon A (ed): "Recombinant DNA products: Insulin, Interferon, and Growth Hormone," New York, Cold Spring Harbor Laboratory, pp 47-65.

Mircheff AK (1983). Empirical strategy for analytical fractionation of epithelial cells. Am J Physiol 244 (Gastrointest Liver Physiol 7) 347-G356.

Shull GE, Schwartz A, Lingrel JB (1985). Amino-acid sequence of the catalytic subunit of the (Na+K)ATPase deduced from a complementary DNA. Nature 316:691-695.

Thomas PS (1980). Hybridization of denatured RNA and small DNA fragments transferred to nitrocellulose. Proc Natl Acad Sci USA 77:5201-5205.

THE APPEARANCE OF Na^+,K^+-ATPase ISOZYMES IN RAT NEURONAL CELL CULTURES.

Nobuo Inoue, Hideo Matsui, Hiroko Tsukui, and Hiroshi Hatanaka

Dept. of Biochemistry, Kyorin Univ. Sch. of Med., Mitaka, Tokyo 181, (N.I., H.M.), Dept. of Neuroscience, Mitsubishi-Kasei Inst. of Life Sciences, Machida, Tokyo 194, (H.T., H.H.), Japan.

INTRODUCTION

Na^+,K^+-ATPase plays an important role in neurons to generate and maintain the resting membrane potential. Two molecular forms, $\alpha = \alpha I$ and $\alpha(+) = \alpha II$, of the enzyme are known to be present in brain (Sweadner, 1979). During the course of development, the total enzyme activity increases and the ratio of the αII form of the enzyme to the αI form increases (Specht, 1984; Schmitt and McDonough, 1986). These results suggest the presence of the isozymes in neurons and that there is a change of the form of the enzyme during neuronal differentiation.

In the present study, we cultured two types of nerve cells: primary cultured rat cerebral neurons and clonal PC12h cells derived from rat pheochromocytoma. Thus, we have been able to investigate changes in the activity and in the forms of the enzyme in the cells, both during the in vitro maturation of the cerebral neurons in culture and during the differentiation of PC12h cells to neuron-like cells induced by nerve growth factor (NGF). The activities of Na^+,K^+-ATPase and the uptake of K^+ (or Rb^+) in the cells were measured, and the molecular form was classified either as a digitalis-resistant (presumed αI) form or a digitalis-sensitive (presumed αII) form, according to the differences in the affinities for digitalis inhibition (Sweadner, 1979; Inoue et al., in press).

MATERIALS AND METHODS

The dissociated cerebral cells were prepared from 17 day old rat fetuses (Arimatsu and Hatanaka, 1985; Hatanaka and Tsukui, 1986). The cells were cultured in a medium which contained 5% heat-inactivated horse serum, 5% pre-colostrum newborn calf serum and a 90% 1 : 1 mixture of Dulbecco's modified Eagle's and Ham's F12 media. The cells were treated by cytosine arabinoside at a concentration of 1 µM, and almost all of the surviving cells were neurons which were positive against anti-neurofilament monoclonal antibody. The clonal PC12h cells, derived from a pheochromocytoma of rat adrenal medulla (Greene and Tischler, 1976; Hatanaka, 1981), were cultured in the same medium that was used for cerebral neurons, in the absence or presence of 50 ng/ml NGF (Inoue and Hatanaka, 1982). The crude particulate fractions obtained from the cultured cells and fetal cerebra were assayed for Na^+,K^+-ATPase activity. The standard assay mixture contained 50 mM imidazole-HCl, pH 7.5, 100 mM NaCl, 10 mM KCl, 4 mM $MgCl_2$, 1 mM EDTA, 2 mM ATP and samples in the presence and absence of strophanthidin. The reaction was allowed to proceed at 37°C for 10 min after preincubation for 1 min. Na^+,K^+-ATPase activity was determined to be the difference between the activities in the absence and presence of 10 mM strophanthidin. Potassium (or rubidium) uptake into cells was measured in a solution containing 130 mM NaCl, 0.5 mM KCl (or RbCl), 1.8 mM $CaCl_2$, 5 mM glucose, 5 mM Hepes-imidazole, pH 7.3, and a trace amount of ^{42}KCl (or $^{86}RbCl$) in the presence and absence of ouabain. The reaction was allowed to proceed at 25°C for 11 min after preincubation for 30 min. Seven mM ouabain was used to inhibit K^+ (or Rb^+) uptake by Na^+,K^+-ATPase. ^{42}K was milked from a ^{42}Ar-^{42}K generator, developed by Dr. H. Morinaga of Technischen Universität München.

RESULTS AND DISCUSSION

The Na^+,K^+-ATPase activity of the crude particulate fractions of rat 17 embryonic day cerebra was 0.12 ± 0.01 (n=8; µmol Pi/min/mg protein). The pattern of strophanthidin inhibition of enzyme activity showed that almost all of the enzyme was a digitalis-resistant form (Fig. 1), which was presumed to be the αI form (Sweadner, 1979; Inoue et al., in press). This indicates that almost all the Na^+,K^+-ATPase of immature cerebral neurons of age E17 was the αI form. However, detection of the minute presence of the αII form was of considerable apparent difficulty.

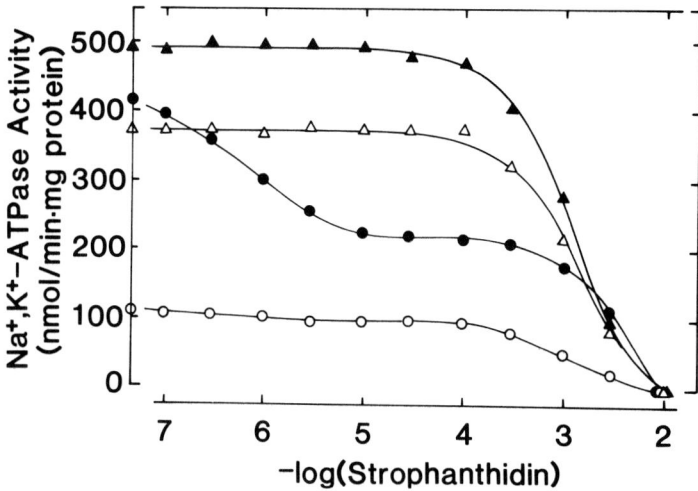

Figure 1. Strophanthidin inhibition of rat Na^+,K^+-ATPase activities. (O) Cerebra from 17 day-old fetuses; (●) cerebral neurons cultured for 13 days; PC12h cells cultured without NGF (△) and with 50 ng/ml NGF for 10 days (▲).

During the course of the culture day, an increase was observed of the total Na^+,K^+-ATPase activities of the cultured neurons, and as well, a digitalis-sensitive form, which was presumed to be the αII form (Sweadner, 1979; Inoue et al., in press) of the enzyme appeared and increased (manuscript in preparation). By day 13 in culture, the total enzyme activity of the neurons rose to 0.38 ± 0.02 (n=11), and the strophanthidin and ouabain inhibition patterns of the enzyme activity and K^+ uptake, respectively, clearly showed the existence of two components corresponding to the αI and αII forms of the enzymes (Fig. 1 and 2). The activity of the αII form overtook that of the αI form by day 13 in culture (Fig. 1). These results indicate that immature cerebral neurons contain only very minute amounts indeed of the αII form, but do possess readily-detectable amounts of the αI form of the enzyme, that the neurons contain gradually increasing levels of the αII form of the enzyme during the course of in vitro maturation, and that the mature neurons contain both αI and αII forms of the enzymes. The presence of αI and αII forms of the enzymes in mature rat cerebral neurons deduced from this study, is

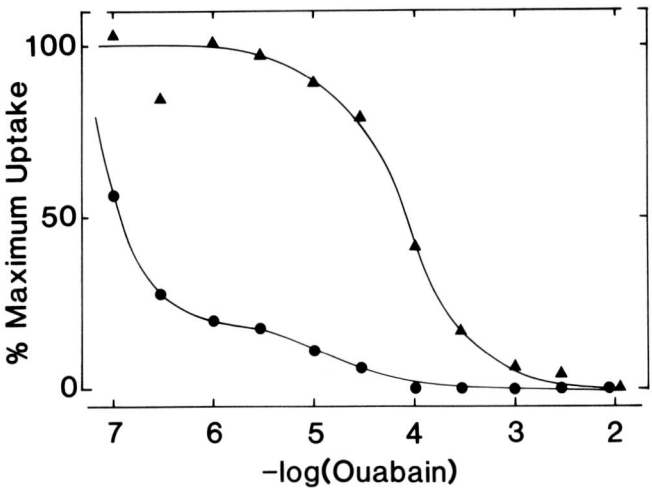

Figure 2. Ouabain inhibition of K^+ uptake into rat cerebral neurons and Rb^+ uptake into PC12h cells. The maximal values for uptake are 10.0 (nmol K^+/min/mg protein) for neurons cultured for 13 days (●), and 15.6 (nmol Rb^+/min/mg protein) for PC12h cells cultured with 50 ng/ml NGF for 8 days (▲).

consistent with the reports that αI and αII forms are present in rat retinal neurons (Specht and Sweadner, 1984; McGrail and Sweadner, 1986) and that ouabain-sensitive and resistant types of the enzyme activities are present in cultured and separated rat cerebellar neurons (Atterwill et al, 1984).

In PC12h cells, which have been known to undergo differentiation to neuron-like cells in culture in the presence of NGF (Greene and Tischler, 1976; Hatanaka, 1981; Inoue and Hatanaka, 1982), Na^+,K^+-ATPase activity of the crude particulate fractions increased from 0.37 ± 0.02 (n=19) to 0.55 ± 0.02 (n=20; μmol Pi/min/mg protein), when cultured with NGF for 5 to 11 days. The increase of the enzyme activity by NGF was prevented by the addition of specific anti-NGF antibodies (Inoue et al, in press). The molecular form of the enzyme, however, was only digitalis-resistant and was presumed to be αI, both in the NGF-untreated and in the NGF-treated cells (Fig. 1 and 2).

These results indicate that the αI form of the enzyme is induced, but the αII form enzyme is not induced in PC12h cells, during the differentiation to neuron-like cells by NGF. This result differs from that shown above in the case of cultured cerebral neurons. Although the reason for this discrepancy is unknown, one possible explanation is that PC12h cells were differentiated by NGF to cells which are similar to sympathetic neurons (Green and Tischler, 1976) which are thought not to contain the αII form of the enzyme (Sweadner, 1979; Inoue et al., in press).

Recently, a third type of α subunit, αIII, has been found in rat brain by cDNA analysis (Shull et al., 1986; Hara et al., 1987). However, the enzyme comprising the αIII subunit, the αIII form of the enzyme, has not yet been demonstrated as an enzyme protein nor has its enzyme activity yet been demonstrated. Because the affinity of the αIII form of the enzyme for digitalis inhibition is unknown, this form enzyme was not considered in the present study. However, if it is in fact digitalis-resistant and is present in cultured cells, the αIII form of the enzyme most likely was regarded as being the αI form of the enzyme in this study. In contrast, if it is digitalis-sensitive and is present in cultured cells, the αIII form of the enzyme most likely was regarded as the αII form of the enzyme. Interestingly, RNAs encoding all three isoforms of the α subunit have been shown to be present in cultured neurons, according to our recent analysis through the use of cDNA probes for the three different types of α subunit (manuscript in preparation).

ACKNOWLEDGMENT

We thank Dr. H. Morinaga and The Japan Radioisotope Association for their generous supply of the ^{42}Ar-^{42}K generator. This work was supported in part by Grants-in-Aid for Scientific Research from the Ministry of Education, Science and Culture of Japan.

REFERENCES

Arimatsu Y, Hatanaka H (1985). Estrogen treatment enhances survival of cultured fetal rat amygdala neurons in a defined medium. Develop. Brain Res. 26: 151-159.

Atterwill CK, Cunningham VJ, Balázs R (1984). Characterization of Na^+,K^+-ATPase in cultured and separated neuronal and glial cells from rat cerebellum. J. Neurochem. 43: 8-18.
Greene LA, Tischler AS (1976). Establishment of a noradrenergic clonal line of rat adrenal pheochromocytoma cells which respond to nerve growth factor. Proc. Natl. Acad. Sci. USA. 73: 2424-2428.
Hara Y, Urayama O, Kawakami K, Nojima H, Nagamune H, Kojima T, Ohta T, Nagano K, Nakao M (1987). Primary structures of two types of alpha-subunit of rat brain Na^+,K^+-ATPase deduced from cDNA sequences. J. Biochem. 102: 43-58.
Hatanaka H (1981). Nerve growth factor-mediated stimulation of tyrosine hydroxylase activity in a clonal rat pheochromocytoma cell line. Brain Res. 222: 225-233.
Hatanaka H, Tsukui H (1986). Differential effects of nerve growth factor and glioma-conditioned medium on neurons cultured from various regions of fetal rat central nervous system. Develop. Brain Res. 30: 47-56.
Inoue N, Hatanaka H (1982). Nerve growth factor induces specific enkephalin binding sites in a nerve cell line. J. Biol. Chem. 257: 9238-9241.
Inoue N, Matsui H, Hatanaka H. Nerve growth factor induces Na^+,K^+-ATPase in a nerve cell line. J. Neurochem. (in press).
McGrail KM, Sweadner KJ (1986). Immunofluorescent localization of two different Na,K-ATPase in the rat retina and in identified dissociated retinal cells. J. Neurosci. 6: 1272-1283.
Schmitt CA, McDonough AA (1986). Developmental and thyroid hormone regulation of two molecular forms of Na^+-K^+-ATPase in brain. J. Biol. Chem. 261: 10439-10444.
Shull GE, Greeb J, Lingrel JB (1986). Molecular cloning of three distinct form of the Na^+,K^+-ATPase α-subunit from rat brain. Biochem. 25: 8125-8132.
Specht SC (1984). Development and regional distribution of two molecular forms of the catalytic subunit of the Na,K-ATPase in rat brain. Biochem. Biophys. Res. Commun. 121: 208-212.
Specht SC, Sweadner KJ (1984). Two different Na,K-ATPase in the optic nerve: Cells of origin and axonal transport. Proc. Natl. Acad. Sci. USA 81: 1234-1238.
Sweadner KJ (1979). Two molecular form of (Na^++K^+)-stimulated ATPase in brain. Separation, and difference in affinity for strophanthidin. J. Biol. Chem. 254: 6060-6067.

APPLICATION OF GENE TRANSFER AND GENE MAPPING TECHNIQUES TO FUNCTIONAL ANALYSIS OF THE NA,K-ATPASE

Rachel B. Kent, Dorothy A. Fallows, Janet Rettig Emanuel, Peter A. Lalley, Robert Levenson, David E. Housman

Center for Cancer Research, Massachusetts Institute of Technology, Cambridge, MA 02139 (R.B.K., D.A.F., D.E.H.); Department of Cell Biology, Yale University School of Medicine, New Haven, CT 06510 (J.R.E., R.L.); Institute for Medical Research, Bennington, VT 05201 (P.A.L.)

INTRODUCTION

The techniques of molecular biology provide a powerful set of approaches to understanding the structure and function of the Na,K-ATPase. We have focused on gene transfer and gene mapping to provide new insight into the functional relationship among the three distinct isoforms of the rodent Na,K-ATPase α subunit. Structural differences among the multiple α subunit isoforms have been suggested to be responsible for the observed tissue differences in Na,K-ATPase sensitivity to ouabain in the rat (Sweadner, 1979). Species-specific differences in ouabain sensitivity between rodent and primate cells are presumably due to analogous differences in α subunit structure. The isolation of cDNA clones for three isoforms of the rat Na,K ATPase α subunit (Shull et al., 1986; Herrera et al., 1987) served as a point of departure for the studies described here. Nucleotide sequence comparison indicates that while the three α subunit isoforms share highly conserved regions of structure (greater than 85% overall amino acid sequence homology), significant differences in codon usage occur indicating that the multiple α subunit isoforms are likely to be encoded by different genes. To investigate the relationship among these three genes, we performed the gene transfer and gene mapping experiments described here.

CHROMOSOME-MEDIATED TRANSFER OF OUABAIN RESISTANCE

To identify murine genomic DNA sequences encoding gene products responsible for the relative ouabain resistance of rodent cells, we performed chromosome-mediated gene transfer experiments (Fallows et al., 1987). Metaphase chromsomes were prepared from mouse cells and transferred to African green monkey kidney CV-1 cells. These cells were then selected for survival in 10^{-6} M ouabain, a dose normally lethal to monkey cells. Colonies of cells resistant to this dose of ouabain were not observed in control dishes. However, monkey cells which were the recipients of mouse metaphase chromosomes did give rise to ouabain resistant colonies. To determine which mouse genes had been transferred to the monkey cells, we hybridized genomic DNA isolated from these ouabain resistant colonies to cDNA probes specific for each rat α subunit isoform, and to a cDNA for the rat β subunit. Results are shown in Table 1. Clones arising in the same transfection plate were given the same letter designation. A difference in letter designation between two clones therefore denotes their independent derivation. The transfer of the mouse α_1 subunit gene is observed in all 16 transferents. This result is consistent with the view that the mouse α_1 subunit participates in the formation of a Na,K ATPase enzyme molecule which is much more resistant to ouabain than the primate enzyme. The mouse genes encoding the α_2, α_3 and β subunits were not detected in any of the transferents. It therefore would appear that these genes are either incapable of conferring ouabain resistance to primate cells in this protocol or do so at a frequency substantially lower than the α_1 gene.

In order to assess the extent of ouabain resistance in transferents containing murine Na,K-ATPase α_1 subunit DNA sequences, we measured the survival of four independent transferents (A2, C1, D1, G2) at 10^{-6}, 10^{-5}, 10^{-4}, and 10^{-3} M ouabain (Fallows et al., 1987). The survival of each of the four transferent clones at each ouabain concentration tested was the same as the original mouse NIH 3T3 donor line, indicating that these clones are true transferents for murine ouabain resistance.

MOUSE NA,K-ATPASE α_1 SUBUNIT cDNA CONFERS OUABAIN RESISTANCE

In the chromosome transferents a substantial amount of rodent DNA was transferred to the recipient CV-1 cells in

TABLE 1. Chromosome Transfer of the a_1 Isoform of the Mouse Na,K-ATPase a Subunit into CV-1 Cells Correlates with Acquisition of Ouabain Resistance.

Cell type	Hybridization to mouse genes			
	Atpa-1 (a_1)	Atpa-2 (a_2)	Atpa-3 (a_3)	Atpb (β)
NIH 3T3	+	+	+	+
CV-1	−	−	−	−
Transferent				
A1, A2				
B1				
C1, C2, C3				
D1, D2	+	−	−	−
G1, G2, G3				
H1				
I'1, I2				
K1, K'1				

addition to the Na,K-ATPase a_1 subunit gene. This consideration makes it impossible to definitively conclude that transfer of the rodent a_1 subunit gene is responsible for species-specific differences in ouabain sensitivity. To resolve this issue, we tested the capacity of a cDNA encoding the mouse a_1 subunit to directly confer ouabain resistance to CV-1 cells (Kent et al., 1987a).

A 3.6 kilobase (kb) cDNA encoding the full-length murine a_1 subunit was isolated from a mouse pre-B cell cDNA library and inserted into the eukaryotic expression vector pSV2 (Figure 1). Insertion of the 3.6 kb cDNA in the sense orientation generated plasmid pSV2a3.6 which was introduced into ouabain-sensitive CV-1 cells as described in Table 2. Ouabain-resistant cells obtained from transfection of plasmid pSV2a3.6 into CV-1 cells grew rapidly in the presence of ouabain at concentrations as high as 10^{-4} M. No ouabain-resistant clones were observed in control CV-1 cells transfected with pSV2 DNA or in mock transfected cells. No ouabain-resistant colonies were produced by transfection of plasmid pSV2a3.0 which had a deletion of approximately 250 bp encoding the COOH-terminus of the a_1 subunit and 350 bp of the 3' untranslated region. This result indicates that a

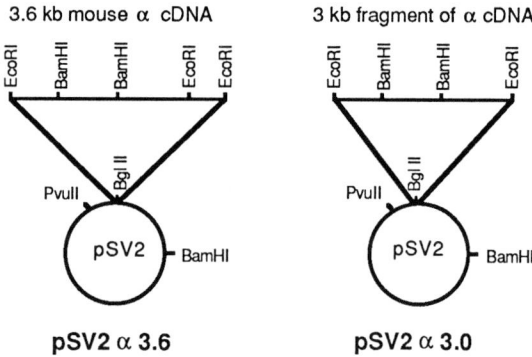

Figure 1. Schematic representation of mouse Na,K-ATPase α subunit cDNA clone and insertion into the vector pSV2. The cDNA contains two internal BamHI sites and one internal EcoRI site. BclI linkers were added to the 3.6 kb partial EcoRI digestion fragment or to the 3.0 kb EcoRI fragment for ligation to pSV2 DNA. The pSV2 vector was constructed from pSV2DHFR by excising the DHFR insert and adding BglII linkers. In the diagram of pSV2 the area between PvuII and BglII contains the SV40 origin and promoter and the area between BglII and BamHI includes the SV40 small T antigen splice site and polyA addition site.

Table 2. Efficiency of Transfection of Mouse Na,K ATPase α_1 Subunit cDNA Clones into Ouabain-sensitive Monkey Cells.

DNA Transfected	Colonies/μg DNA/10^6 Cells		
	Expt. 1	Expt. 2	Expt. 3
pSV2α3.6	200	337	650
pSV2α3.0	ND	ND	0
pSV2	0	0	ND
None	0	0	ND

CV-1 cells were exposed to a calcium phosphate precipitate of plasmid pSV2α3.6, pSV2α3.0, or pSV2, or to calcium phosphate without DNA. Cells were subcultured 25 h later and, after 4 h, ouabain was added to a final concentration of 10^{-6} M. Cells were maintained under selection for two weeks after transfection, then colonies were fixed and stained. ND, not determined.

cDNA encoding a complete polypeptide chain is required for biological activity in this assay.

To confirm that the ouabain-resistant phenotype resulted from the introduction of the murine α_1 subunit cDNA construct, genomic DNA from ouabain-resistant transfectants was analyzed (Figure 2). The complete correlation between the presence of α_1 subunit cDNA in all transfectants and the high level of ouabain resistance of the transfected cells strongly supports the view that the α_1 cDNA indeed confers ouabain resistance to recipient CV-1 cells. Equal intensity of hybridization to an 11 kb BamH1 fragment, corresponding to the endogenous monkey α_1 subunit gene, was observed in all cell lines tested including untransfected ouabain sensitive CV-1 cells. Amplification of the endogenous monkey α_1 subunit gene therefore does not contribute to the ouabain resistance phenotype of the transfectants. In addition no significant differences in total α_1 subunit mRNA expression levels were detected between the ouabain resistant transfectants and ouabain sensitive controls, indicating that the ouabain resistance of the transfectants was not due to a major increase in expression level of the α_1 subunit gene (Kent et al., 1987a).

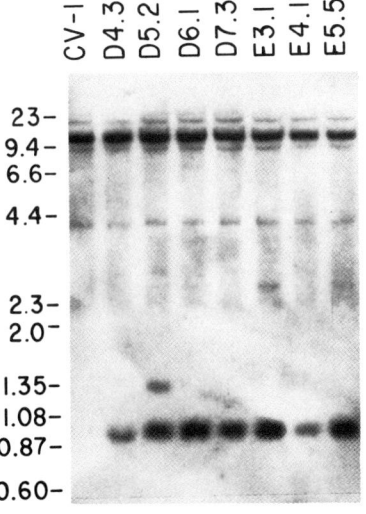

Figure 2. Southern analysis of genomic DNA from cells transfected with pSV2α3.6. Genomic DNA (10 μg) from clones selected in 10^{-6} M ouabain and from CV-1 cells was digested with BamHI. Digestion products were electrophoresced in a 1% agarose gel containing 0.04 M Tris acetate (pH 8), 0.002 M EDTA, and transferred to a hybridization membrane (Zetabind, AMF Cuno). The blot was hybridized to 10^7 c.p.m. ^{32}P-labelled 0.95 kb internal BamHI fragment DNA from the mouse α cDNA clone. Size markers, HindIII-digestion products of bacteriophage lambda DNA and HaeIII-digestion products of ϕX174 replicative form DNA.

The transfer of ouabain resistance by murine Na,K-ATPase α_1 subunit cDNA demonstrates that differential ouabain sensitivity between primate and rodent cells can be directly ascribed to primary sequence differences between the primate and rodent α_1 subunit genes. These results also suggest that the murine α_1 subunit is capable of association with the monkey β subunit to form a functional Na,K-ATPase.

ASSIGNMENT OF NA,K-ATPASE SUBUNIT GENES TO MOUSE CHROMOSOMES

To further investigate the molecular basis for Na,K-ATPase isoform diversity, we determined the chromosomal positions of the mouse genes encoding α and β subunits of the enzyme (Kent et al., 1987b). Rat cDNA probes for three distinct isoforms of the α subunit and for the β subunit were used in two types of genomic DNA analyses. The segregation of genomic DNA sequences in mouse/hamster somatic cell hybrids was used to identify a chromosomal assignment for each gene. The segregation of restriction fragment length polymorphisms (RFLPs) in N_1 backcross progeny of a Mus musculus-Mus spretus mating was then used to confirm chromosome assignment. Using this approach, we established linkage between each ATPase subunit gene and another marker previously assigned to the same chromosome. The α_1 subunit (Atpa-1) gene expressed in all tissues examined to date is located on chromosome 3. The α_2 subunit gene (Atpa-2), designated αIII by Shull et al. (1986) and expressed predominantly in brain and fetal heart, is on chromosome 7. The α_3 subunit gene (Atpa-3), designated α^+ by Shull et al. (1986), and the β subunit gene (Atpb) both exhibit complex tissue-specific patterns of expression and are located on chromosome 1. These two genes are a considerable distance from each other on the chromosome however. The genomic distribution of α subunit genes indicates that they form a multigene family. The chromosomal mapping of Na,K-ATPase subunit genes reveals their dispersion in the murine genome, therefore the existence of a common cis-acting mechanism regulating expression of the subunit genes examined in this study is ruled out.

The difference in ouabain sensitivity between rodent and primate cells had been previously assigned to mouse chromosome 3 in somatic cell genetic experiments (Kozak et al., 1979). Thus the functional studies suggesting a primary role of the α_1 subunit in differential ouabain sensitivity presented earlier in this report are consistent with and lend further support to the gene mapping studies.

DISCUSSION

The molecular genetic approaches presented here provide new strategies for addressing questions of fundamental interest in the relationship between structure and function of the Na,K ATPase. For example, an extension of the gene transfer approach in which chimeric cDNAs between ouabain resistant and ouabain sensitive α subunit genes are tested for ouabain resistance should lead to a definitive identification of the domain in the α subunit directly determining ouabain resistance. Site specific mutagenesis should then provide definitive resolution of this issue. Extension of these approaches to other aspects of enzymatic function should be of equally great interest.

In a broader context, understanding the entire repertoire of responses of the Na,K ATPase to intracellular signals is more likely given the availability of molecular genetics. At present, we have little insight into the requirements for three alternative isoforms of the α subunit. Differences in intracellular distribution of the isoforms hint at underlying functional differences. The gene mapping studies presented here confirm at least one functional difference among the isoforms. Many others are likely to exist. The ability to reintroduce the genes for each isoform into appropriate target cells suggests a valuable line of approach to this set of issues.

The extension of the approaches presented here should also permit insight into interaction among the polypeptides which form the Na,K-ATPase. Experiments to determine levels of α and β subunit mRNA expression have shown that coordinate expression of the subunit genes does not occur in some mammalian tissues (Young et al., 1987; Emanuel et al., 1987). Although no direct expression studies of the β subunit are presented here, it is clear that these approaches can be extended to the interaction between the subunits of the enzyme. Cellular assembly processes are likely to be necessary for the introduction of α and β subunits into the membrane environment. The extension of the genetic approach described here will also be useful in analyzing the cellular processes which lead to insertion of the enzyme into the correct membrane context.

The identification of genetic abnormalities in Na,K-ATPase function in experimental organisms or man could lead to a better understanding of the normal functions of this enzyme. Gene mapping studies of the type presented here provide the opportunity to rapidly determine whether a genetic abnormality is likely to be due to a defect in function of one of the genes encoding a component of the Na,K-ATPase.

REFERENCES

Emanuel JR, Garetz S, Stone L, Levenson R (1987). Differential expression of Na,K-ATPase α and β subunit genes in rat tissues and cell lines. Proc Natl Acad Sci USA, in press.

Fallows D, Kent RB, Nelson DL, Emanuel JR, Levenson R, Housman DE (1987). Chromosome-mediated transfer of the murine Na,K-ATPase alpha subunit confers ouabain resistance. Mol Cell Biol 7:2985-2987.

Herrera VL, Emanuel JR, Ruiz-Opazo N, Levenson R, Nadal-Genard B (1987). Three differentially expressed Na,K-ATPase α subunit isoforms: structural and functional implications. J Cell Biol, in press.

Kent RB, Emanuel JR, Ben Neriah Y, Levenson R, Housman DE (1987a). Ouabain resistance conferred by expression of the cDNA for a murine Na^+,K^+-ATPase α subunit. Science 237:901-903.

Kent RB, Fallows DA, Geissler E, Glaser T, Emanuel JR, Lalley PA, Levenson R, Housman DE (1987b). Genes encoding α and β subunits of Na,K-ATPase are located on three different chromosomes in the mouse. Proc Natl Acad Sci USA 84:5369-5373.

Kozak CA, Fournier REK, Leinwand LA, Ruddle FH (1979). Assignment of the gene governing cellular ouabain resistance to Mus musculus chromosome 3 using human/mouse microcell hybrids. Biochem Genet 17:23-34.

Shull GE, Greeb J, Lingrel JB (1986). Molecular cloning of three distinct forms of the Na^+,K^+-ATPase α-subunit from rat brain. Biochemistry 25:8125-8132.

Sweadner KJ (1979). Two molecular forms of $(Na^+ + K^+)$-stimulated ATPase in brain. J Biol Chem 254:6060-6067.

Young RM, Lingrel JB (1987). Tissue distribution of mRNAs encoding the α isoforms and β subunit of rat Na^+,K^+-ATPase. Biochem Biophys Res Comm 145:52-58.

MULTIPLE FORMS OF THE Na,K-ATPase: THEIR GENES AND TISSUE SPECIFIC EXPRESSION

Jerry B Lingrel, Regina M. Young, and Marcia M. Shull
University of Cincinnati, College of Medicine
Department of Microbiology and Molecular Genetics
231 Bethesda Ave., Cincinnati, Ohio 45267/0524

INTRODUCTION

Although its basic enzymatic activity in all cells is the energy-dependent transport of Na and K ions across the cell membrane, in certain tissues the Na,K-ATPase fulfills a specialized physiological function. For example, in kidney the enzyme is responsible for sodium and water reabsorption from the glomerular filtrate, in nerve and muscle it maintains the electrochemical gradients required for excitability, in brain choroid plexus and ciliary epithelium of the eye it may be involved in determining the composition of the cerebrospinal fluid and aqueous humor respectively, and in the lung, fluid reabsorption at birth may depend in part on functioning of the Na,K-ATPase. The fulfillment of these specialized roles may require enzymes with subtle differences in turnover number, affinity for Na^+ and/or K^+, and response to effector molecules regulating activity. Related Na,K-ATPases or a single enzyme with varying α or β subunits could provide these functional differences.

The existence of multiple forms of the Na,K-ATPase was suggested by Hansen based on differences in cardiac glycoside sensitivity (Hansen, 1976). These studies were extended by Sweadner (Sweadner, 1979) who identified two isoforms of the α subunit in brain based on differences in electrophoretic mobility, reactivity with N-ethylmaleimide and affinity for cardiac glycosides. The two subunits were designated α and α+. Additional studies by Lytton and coworkers (Lytton et al., 1985) demonstrated that both α and α+ are present in adipose tissue and skeletal muscle and that these two isoforms differ in their response to insulin. The isolation of cDNA clones corresponding to mRNAs for the α and α+ subunits clearly established that these isoforms were the products of separate genes (Shull et al., 1986) rather than the result of modifications to a single protein. Cloning studies also allowed identification of a third isoform αIII (Shull et al., 1986).

Differences in mobility of the β subunit on SDS-polyacrylamide gels had been observed (Jørgensen, 1982) but whether this was the result of differences in the primary structure of the protein or in the extent of glycosylation (Sweadner et al., 1985) was unknown. The isolation of cDNA clones corresponding to the β subunit from both brain and kidney yielded cDNAs encoding a single protein, suggesting that isoforms of this subunit may not exist (Young et al., 1987).

The finding of multiple forms of the α subunit raises the question of whether other, as yet undescribed, forms of this subunit may exist. While the characterization of cDNA clones from additional tissues represents a viable approach to discovering new α and β subunit isoforms, the isolation of genomic clones for these subunits offers an alternative approach. A detailed investigation of the tissue specific distribution of each of the subunit isoforms may aid in elucidating the specific physiological function of their Na,K-ATPase activity.

RESULTS AND DISCUSSION

Characterization of Genes for the α Subunit

The construction, isolation and characterization of cDNAs corresponding to the α and α+ subunits of the Na,K-ATPase revealed the genetic basis for these isoforms. A new α isoform, αIII, was also identified in these studies. The identification of this novel isoform raised the possibility that additional α isoforms may exist. Characterization of the genes encoding this subunit represents one method for defining the number of α subunit isoforms. To pursue this approach, a human genomic library was constructed and screened using sheep α and rat α+ cDNA probes. Four sets of overlapping clones were isolated (Shull and Lingrel, 1987). The basic restriction maps of these clones are shown in Figure 1. The genes were designated αA, αB, αC and αD. Based on sequence analysis, αA corresponds to the α isoform while αB corresponds to α+. Although the other two genes show distinct homology with the α subunit, neither corresponds to the αIII cDNA sequence. It is not known whether αC and αD represent functional genes, pseudogenes or encode related transport ATPases. In accordance with standard human gene nomenclature the four genes have been renamed NAKAA1 (αA), NAKAA2 (αB), and NAKAAL1 and NAKAAL2 for Na,K-ATPase α subunit-like genes αC and αD. Ovchinnikov and coworkers (Ovchinnikov et al., 1987) have described and partially sequenced a human gene which does not correspond to any of the ones we have described to date. This gene exhibits sequence homology to the rat αIII cDNA and probably represents the human equivalent of the rat αIII gene. Thus, at least five α subunit genes are present in the human genome, three of which encode functional α subunits and two which may represent yet unidentified α isoforms. Further characterization of the sequence organization of the NAKAAL1 and L2 genes and a search for their corresponding mRNAs will be required to determine whether these α like genes are functional.

Tissue Distribution of the α Subunit mRNAs

The availability of cDNA clones corresponding to the rat α, α+, αIII and β subunits allows determination of the tissue distribution of the α isoform mRNAs. RNA was isolated from various tissues and analyzed for the presence of each α subunit mRNA using a combination of Northern blot and slot blot analysis. A necessary prerequisite of such studies was the development of probes specific for each of the subunits. Examination of the sequences of the cDNA clones revealed a region at the 5' end of each mRNA which exhibited maximum sequence differences. Restriction fragments containing these regions were subcloned and used as probes (Young and Lingrel, 1987). They were shown to hybridize specifically with their corresponding mRNAs.

Table 1 gives the relative abundance of each of the three α subunits and the β subunit in various tissues of the rat. Approximately equal amounts of the α, α+, and αIII subunits are present in the brain and are arbitrarily

assigned a value of 1. All other levels are related to
these values. As expected the α and β subunit mRNAs are
abundant in kidney. While the α isoform predominates, some
α+ mRNA is also present in kidney. Doucet and Bartlet
(Doucet and Bartlet 1986) have shown that the Na,K-ATPase
activity in the collecting tubules of the kidney is more
sensitive to ouabain than it is in the proximal tubules.
Because the α+ form is more sensitive to cardiac glycosides
than is the α form, it is conceivable that α+ is present in
the Na,K-ATPase of collecting tubules.

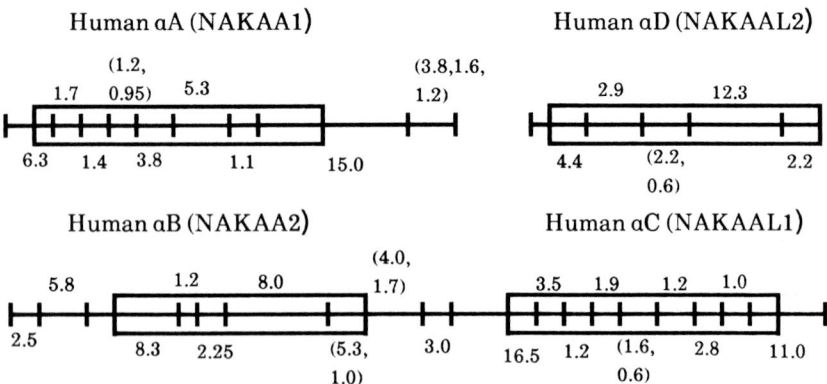

Figure 1. Human α subunit genes. EcoR1 restriction sites
are shown by the vertical lines and the boxes indicate
the approximate location of the transcription unit.

Rat heart contains predominantly the α isoform, but a
significant amount of α+ also occurs. Again, this is in
agreement with the report of two cardiac glycoside
receptors in this tissue (Grupp et al., 1981). In muscle
the predominant form is α+, with some α being present. The
stomach contains mainly α but some α+ also occurs.
Interestingly, αIII appears to be present in this tissue as
well. This is one of the few tissues where the yet
uncharacterized αIII subunit may occur. Lung and liver
contain mainly α but some αIII is also present in the lung.

TABLE 1. Tissue Distribution of the Na,K-ATPase, α, α+, αIII and β subunit mRNAs

	α	α(+)	αIII	β
Kidney	7.0	<0.05		2.0
Brain	1.0	1.0	1.0	1.0
Heart	0.7	0.1		0.2
Muscle	0.1	1.5		0.1
Stomach	0.3	<0.05	<0.05	0.5
Lung	0.3	<0.05	<0.05	<0.05
Liver	<0.05			<0.05

Amounts are in arbitrary units compared to brain which is assigned a value of 1 for each subunit.

The β subunit mRNA is present in each of the tissues examined, but in varying amounts relative to the α subunit mRNA. For example, significantly more β than α subunit mRNA is found in stomach while the reverse is true in muscle.

We have also examined the distribution of the α, α+ and β subunits in various regions of the heart. Figure 2 shows a Northern blot analysis of the RNA isolated from left and right ventricles and atria. A higher concentration of the α subunit mRNA is found in the atria as compared to the left and right ventricle. On the other hand, the α+ subunit mRNA is found in a much higher concentration in the ventricle compared to the atria. The β subunit is equally distributed among these regions of the heart and serves as an internal control. These results are in apparent agreement with pharmacological studies of the differences of cardiac glycoside sensitivities in rat atrium and ventricle (Grupp et al., 1981). The contractile response to glycosides, mediated by Na,K-ATPase, exhibits two sensitivities in the ventricle: a smaller high-affinity response (probably α+ related) and a larger low-affinity response (probably α related). The atrium, on the other hand has an even larger low-affinity response but no or very little high affinity response.

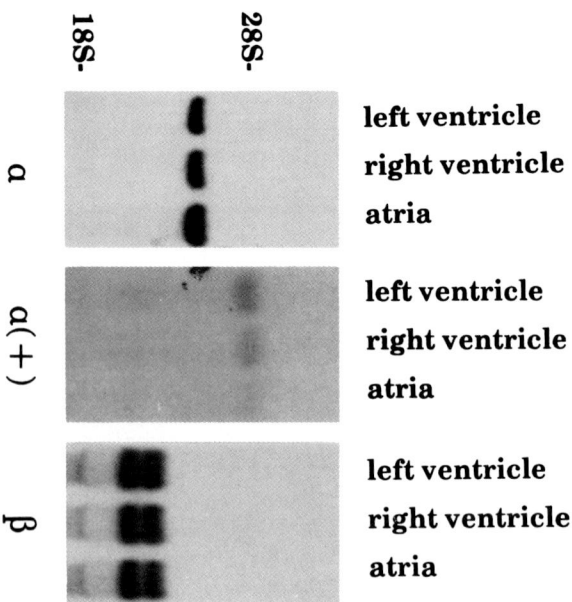

Figure 2. Distribution of α, α+ and β subunits of the Na,K-ATPase in heart ventricles and atria. Reprinted with permission from Young and Lingrel, 1987.

SUMMARY

The use of genetic tools has been invaluable for examining the potential number of α and β subunits of the Na,K-ATPase. To date at least five genes corresponding to the α subunit have been described. Two of these encode the α and α+ isoforms while the third encodes the novel αIII subunit. The products of the other two genes have not yet been described; these genes may be pseudogenes or may encode new α isoforms or related transport ATPases.

The determination of the tissue distribution of the various isoforms of the α subunit using Northern blot and slot blot analyses has yielded several interesting findings. Each tissue differs in the abundance and combination of isoforms. While α is the major isoform mRNA in kidney, α+ mRNA predominates in skeletal muscle. All three α subunit mRNAs are present in brain in roughly equal amounts. The newly discovered αIII isoform mRNA is found not only in brain but in low amounts in stomach and lung.

It is possible that individual cell types may contain a defined combination of isoforms which allow the Na,K-ATPase to fulfill its specific physiological role in those cells. It will be interesting to compare the functional properties, i.e., affinity for Na^+ and K^+, turnover number, response to external changes in environment and response to effector molecules for the enzymes containing each of the α subunit isoforms.

ACKNOWLEDGMENTS

This work was supported by NIH grants HL22619 and HL28573. R.M.Y and M.M.S. were supported by fellowships from the Albert J. Ryan Foundation.

REFERENCES

Doucet A, Bartlet C (1986). Evidence for differences in the sensitivity to ouabain of Na,K-ATPase along the nephrons of rabbit kidney. J Biol Chem 261: 993-995.

Grupp JL, Grupp G, Schwartz A (1981). Digitalis receptor desensitization in rat ventricle: ouabain produces two inotropic effects. Life Sci 29: 2789-2794.

Hansen O (1976). Non-uniform populations of g-strophanthin binding sites of $(Na^+ + K^+)$-activated ATPase. Biochim Biophys Acta 433: 383-392.

Jørgensen PL (1982). Mechanism of the Na^+, K^+ pump. Protein structure and conformations of the pure $(Na^+ + K^+)$-ATPase. Biochim Biophys Acta 694: 27-68.

Lytton J, Lin JC, Guidotti G (1985). Identification of two molecular forms of (Na^+,K^+)-ATPase in rat adipocytes. J Biol Chem 260: 1177-1184.

Ovchinnikov YuA, Monastyrskaya GS, Broude NE, Allikmets RL, Ushkaryov YuA, Melkov AM, Smirnov YuV, Malyshev IV, Dulubova IE, Petrukhin KE, Gryshin AV, Sverdlov VE, Kiyatkin NI, Kostina MB, Modyanov NN, Sverdlov ED (1987). The family of human Na,K-ATPase genes - a partial nucleotide sequence related to the α-subunit. FEBS Lett 213: 73-80.

Shull GE, Greeb J, Lingrel JB (1986). Molecular cloning of three distinct forms of the Na^+,K^+-ATPase α-subunit from rat brain. Biochemistry 25: 8125-8132.

Shull MM, Lingrel JB (1987). Multiple genes encode the human Na^+,K^+-ATPase catalytic subunit. Proc Natl Acad Sci 84: 4039-4043.

Sweadner KJ (1979). Two molecular forms of $(Na^+ + K^+)$-stimulated ATPase in brain. J Biol Chem 254: 6060-6067.

Sweadner KJ, Gilkeson R (1985). Two isozymes of the Na,K-ATPase have distinct antigenic determinants. J Biol Chem 260: 9016-9022.

Young RM, Lingrel JB (1987). Tissue distribution of mRNAs encoding the α isoforms and β subunit of rat Na^+,K^+-ATPase. Biochem Biophys Res Commun 145: 52-58.

Young RM, Shull GE, Lingrel JB (1987). Multiple mRNAs from rat kidney and brain encode a single Na^+,K^+-ATPase β subunit protein. J Biol Chem 262: 4905-4910.

IDENTIFICATION OF A THIRD FORM OF Na,K-ATPase CATALYTIC
SUBUNIT IN RAT BRAIN BY PHOTOAFFINITY LABELING

Joseph M. Lowndes, Arnold E. Ruoho and
Mabel Hokin-Neaverson
Departments of Physiological Chemistry, (J.M.L.,
M.H.), Psychiatry, (J.M.L., M.H.), and Pharmacology, (A.E.R.), University of Wisconsin, Madison,
Wisconsin 53706, USA.

INTRODUCTION

Our laboratory has been using photoaffinity labeling to study the Na,K-ATPase catalytic, alpha subunits which contain the cardiac glycoside binding site. In the course of this work we have labeled NaK-ATPase in rat brain microsomes with cardiac glycoside photolabels. We have identified a radiolabeled polypeptide which represents a saturable, high affinity cardiac glycoside binding site which is distinct from the previously reported alpha and alpha(+) sites (Sweadner, 1979). This polypeptide is refered to as alpha(-) since it has a slightly lower apparent molecular weight than alpha by SDS-PAGE.

MATERIALS AND METHODS

Brain microsomal membranes were prepared from female Sprague-Dawley rats. Whole brains were homogenized with a teflon-glass homogenizer in a solution containing 25 mM Tris-HCl, pH 7.4, 1 mM EDTA and 320 mM sucrose. The homogenate was centrifuged at 16,000 x g to pellet the mitochondria and nuclei. The supernatant was then centrifuged at 125,000 x g for 60 minutes to pellet the brain microsomal membranes. Purified canine kidney Na,K-ATP-ase was prepared as described (Jørgensen, 1974).

The synthesis of $[^{125}I]$AISC will be described elsewhere. $[^{125}I]$AIPP-GluD was prepared as described (Lowndes et al., 1987). The photoaffinity labeling protocol was essentially as previously described (Lowndes et al., 1984).

Rat brain microsomes (280 ug) and purified canine kidney NaK-ATPase (120 ug) were incubated with 1.0 nM photolabel for 45 minutes at 30°C in 0.2 ml of 50 mM imidazole-HCl, pH 7.2, which contained 4 mM $MgCl_2$ and 3 mM Tris-phosphate. Control incubations contained either 0.2 mM or 1.0 mM ouabain, 0.2 mM cymarin or 0.2 mM AISC. The incubations were diluted with ice cold buffer to 10 ml and centrifuged to collect the membranes. The membranes were diluted to 5 ml with ice cold buffer and photolyzed for 10 seconds with a 1 KW high pressure mercury lamp. The membranes were then collected by centrifugation and analyzed by SDS-PAGE (5% or 10%). Radiolabeled peptides were identified by autoradiography.

Na,K-ATPase activity was determined by measuring the liberation of $[^{32}P]P_i$ from $[gamma-^{32}P]ATP$ as previously described (Lowndes et al. 1984). Rat brain microsomes (2.5 ug protein) were incubated with ouabain, cymarin or AISC (0.1 nM to 1.0 uM) for 15 minutes at 37°C in 0.2 ml of 25 mM Tris-HCl, pH 7.4, which contained 50 mM NaCl, 10 mM KCl, 4 mM $MgCl_2$ and 1.2 mM Tris-$[gamma-^{32}P]ATP$.

To determine competitive displacement of $[^3H]$ouabain binding, rat brain microsomes (10 ug protein) were incubated in 0.5 ml of 50 mM imidazole-HCl, pH 7.2, which contained 4 mM $MgCl_2$ and 3 mM Tris-phosphate for 15 minutes, at 37°C, then $[^3H]$ouabain (3.5 nM) was added and the incubation was allowed to continue for a further 30 minutes. The membranes were then collected and washed by filtration at room temperature and the $[^3H]$ouabain bound to the membranes was determined by scintilation counting.

RESULTS

The photoaffinity label AISC (azido-iodophenethylamido-succinyl-cymarin) was characterized as a probe for the cardiac glycoside binding site of Na,K-ATPase in rat brain microsomes by showing that the compound inhibited enzyme activity and $[^3H]$ouabain binding activity. AISC inhibited Na,K-ATPase activity in approximately the manner as the parent cardiac glycoside, cymarin. AISC, cymarin and ouabain each exhibited biphasic inhibition of rat brain Na,K-ATPase activity over a broad concentration range (0.1 nM to 1.0 uM). The two K_is for AISC were K_i = 0.2 nM and 80 nM. AISC and cymarin also inhibited the binding of $[^3H]$ouabain (3.5 nM) to rat brain microsomes in a biphasic manner. The derivatization of cymarin to form AISC had no apparent effect on the IC_{50} for the high affinity site while the IC_{50} for the low affinity site was shifted from 13 nM (cymarin) to 45 nM (AISC).

Photoaffinity labeling of rat brain microsomes with 1.0 nM [^{125}I]AISC resulted in a single heavily radiolabeled polypeptide band of approximately 100 kDa when analyzed by 10% SDS-PAGE. When 1 mM ouabain was included in the pre-photolysis incubation to inhibit the binding of the photolabel to the microsomes, the radiolabeling of the polypeptide band was only partially reduced. However, the radiolabeling was completely inhibited when 0.2 mM cymarin or non-radioactive AISC was included in the pre-photolysis incubation.

An analysis of [^{125}I]AISC photolabeled rat brain microsomes by 5% SDS-PAGE resolved the single radiolabeled band observed on 10% SDS-PAGE gels into two distinct polypeptide bands. By comparison with labeled alpha from pure canine kidney Na,K-ATPase, the radiolabeled protein that was associated with the band of higher apparent molecular weight in the gel from brain microsomes appeared to be a mixture of the alpha and the alpha(+) isozymic forms of the catalytic subunit of Na,K-ATPase. The other radiolabeled protein showed an apparent molecular weight that was 2-5 kDa lower than that of alpha from pure canine kidney enzyme; it is refered to here as alpha(-). Ouabain (0.2 mM), cymarin (0.2 mM) and AISC (0.2 mM) all inhibited [^{125}I]AISC photoaffinity labeling of the alpha and alpha(+) subunits of Na,K-ATPase in rat brain microsomes. In contrast, photoaffinity labeling of alpha(-) by [^{125}I]AISC was inhibited with cymarin or AISC but not with ouabain. This difference in affinity for ouabain between the alpha(-) and the alpha and alpha(+) peptides argues against the possibility that alpha(-) is simply a proteolytic degradation product of alpha or alpha(+).

Specific radiolabeling of alpha(-) was also observed with a second cardiac glycoside photolabel, [^{125}I]AIPP-GluD (a derivative of 4-amino-4,6-dideoxy-glucosyl digitoxigenin), at a concentration of 1 nM. The amount of radioactivity in alpha(-) after photoaffinity labeling with [^{125}I]AIPP-GluD was much less than that which was seen after photoaffinity labeling with [^{125}I]AISC at the same concentration and specific activity. These results indicate that photoaffinity labeling of alpha(-) polypeptide is not a special characteristic of [^{125}I]AISC.

DISCUSSION

Photoaffinity labeling has been used successfully to identify and study specific ligand binding sites for a wide variety of biologically active molecules (Ruoho et al., 1984). In our laboratory, we have synthesized a series of radioiodinated photoaffinity labels to study the structure of the cardiac glycoside binding site of the Na,K-ATPase. The photoaffinity label AISC is the latest in our series. It has been characterized as a specific probe for the Na,K-ATPase in three ways: (i) AISC inhibits Na,K-ATPase activity with the same efficacy as that of known cardiac glycosides, (ii) AISC competitively displaces [^{3}H]ouabain binding to Na,K-ATPase, and (iii) [^{125}I]AISC specifically photoaffinity labels the alpha subunit of purified canine kidney Na,K-ATPase.

With [^{125}I]AISC, we have photoaffinity labeled a polypeptide in rat brain microsomes that represents a new cardiac glycoside binding site, alpha(-). The results of these photoaffinity labeling experiments describe several characteristics of the alpha(-) binding site in rat brain: (i) The alpha(-) site appears to bind AISC with relatively high affinity since the polypeptide was labeled at the low concentration of 1.0 nM [^{125}I]AISC. (ii) The [^{125}I]AISC-photoaffinity labeling of alpha(-) can be inhibited with the addition of non-radioactive cymarin or AISC to the prephotolysis incubation. From this , alpha(-) binding sites appear to be saturable or limited in number, as opposed to unsaturable, non-specific binding sites. The number of alpha(-) sites relative to the number of alpha and alpha(+) sites cannot be estimated by photoaffinity labeling because the efficiency of the reaction with each of the various subunits is not known. (iii) The apparent molecular weight of the alpha(-) polypeptide band is close to those of the alpha and alpha(+) polypeptide bands and this is evidence that all three represent isozymes of the catalytic subunit of Na,K-ATPase. Further support for this would be a demonstration that alpha(-) has Na,K-ATPase activity, and that it has amino acid sequence homology with the established alpha and alpha(+) isozymes.

The main characteristics which distinguish the isozymic

forms alpha and alpha(+) are the differences in apparent molecular weight, in affinity for cardiac glycosides, and in distribution in different tissues (Sweadner, 1979). The alpha(-) polypeptide also shows variation in these characteristics. It has a lower apparent molecular weight than alpha; it has a lower apparent affinity for ouabain than both alpha(+) and alpha; and it is found in rat brain microsomes, but we have not observed it in microsomes prepared from torpedo electroplax and from canine kidney and heart.

It is interesting to speculate that the alpha(-) binding site may be the translation product of the alpha(III) mRNA described by Shull et al. (1986). Other possibilities are suggested by reports that at least four different genes may encode the human Na,K-ATPase catalytic subunit (Shull and Lingrel, 1987; Sverdlov et al., 1987).

This work was supported by the National Institute of Mental Health grant MH-26494. The authors would like to thank Noemi M. Millan for her expert technical assistance.

REFERENCES

Jørgensen PL (1974). Purification and characterization of $(Na^{+}+K^{+})$-ATPase: III. Purification from the outer medulla of mammalian kidney after selective removal of membrane components by sodium dodecylsulphate. Biochem Biophys Acta 356: 36-52.
Lowndes JM, Hokin-Neaverson M, Ruoho, AE (1984). Photoaffinity labeling of $(Na^{+}K^{+})$-ATPase with [^{125}I]iodoazidocymarin. J Biol Chem 259: 10533-10538.
Lowndes JM, Hokin-Neaverson M, Ruoho AE (1987). [^{125}I]AIPPS, a heterobifunctional reagent for the synthesis of radioactive photoaffinity ligands: Synthesis of a carrier-free ^{125}I-cardiac glycoside photoaffinity label. Anal Biochem: in press.
Ruoho AE, Rashidbaigi A, Roeder PE (1984). Approaches to the identification of receptors utilizing photoaffinity labeling. In Venter JC, Harrison LC (eds): "Membranes, Detergents, and Receptor Solubilization," New York: Alan R. Liss, p 119-160.
Shull GE, Greeb J, Lingrel JB (1986). Molecular cloning of three distinct forms of the Na^{+},K^{+}-ATPase alpha subunit from rat brain. Biochem 25: 8125-8132.
Shull MM, Lingrel JB (1987). Multiple genes encode the human Na^{+},K^{+}-ATPase catalytic subunit. Proc Natl Acad Sci

USA 84: 4039-4043.

Sverdlov ED, Monastyrskaya GS, Broude NE, Ushkaryov YA, Allikmets RL, Melkov AM, Smirnov YV, Malyshev IV, Dulobova IE, Petrukhin KE, Grishin AV, Kijatkin NI, Kostina MB, Sverdlov VE, Modyanov NN, Ovchnikov YA (1987). The family of human Na^+,K^+-ATPase genes: No less than five genes and/or pseudogenes related to the alpha-subunit. FEBS Lett 217: 275-278.

Sweadner KJ (1979). Two molecular forms of (Na^++K^+)-stimulated ATPase in brain: Separation and difference in affinity for strophanthidin. J Biol Chem 254: 6060-6067.

MOLECULAR CLONING AND CHARACTERIZATION OF α-SUBUNIT ISOFORMS OF THE NA,K-ATPASE

Robert W. Mercer, Jay W. Schneider and Edward J. Benz, Jr.

Department of Cell Biology and Physiology, Washington University School of Medicine, St. Louis, MO 63110 (R.W.M) and Department of Human Genetics, Yale University School of Medicine, New Haven, CT 06310 (J.W.S., E.J.B.)

INTRODUCTION

We have previously reported the isolation of rat brain and human kidney cDNAs encoding the α and ß subunits of the Na,K-ATPase. Recently, our group (Schneider et al., 1987) and Shull et al. (1986) have independently isolated three forms of distinct but highly homologous α-subunit cDNAs. These cDNA clones represent the α isoform which predominates in the kidney, the previously described α(+) isoform originally identified in the brain, and an unidentified α isoform, termed αIII. The αIII cDNA encodes a novel isoform not previously appreciated by analysis of membrane proteins or transport activities. Comparison of the deduced primary structures of α, α(+), and αIII proteins revealed ≈85% amino acid sequence homology, and predicts no major differences in the secondary structures or membrane topologies of the isoform polypeptides (Shull et al., 1986).

To determine the tissue specificity of the α-subunit isoforms, the cDNA clones were used to characterize the RNA products of α-subunit gene expression in different rat tissues. Hybridization analysis of total cellular RNA and *in situ* hybridization histochemistry indicate that the mRNA encoding each isoform is heterogeneously distributed among various rat tissues. Interestingly, synthesis of αIII mRNA was restricted to neural tissues and fetal heart muscle. In adult brain and spinal cord, αIII mRNA represented a major transcript, similar in abundance to α and α(+) mRNA. These results suggest that the α-subunit isoforms have different patterns of regulation and expression which may be important in Na,K-ATPase function.

METHODS

Total RNA was extracted from adult and 18 day old fetal rat tissues by the guanidinium isothiocyanate/cesium chloride method of Chirgwin et al.(1979). RNA electrophoresis, blotting and hybridization was performed as previously described (Schneider et al. 1985). Cross-isoform hybridization was eliminated by high stringency washes in 0.1X SSC, 0.1% SDS for >3 hours at 68°C. *In situ* hybridization of fetal whole body or adult brain frozen sections (10-20 μm) was performed as described by Awgulewitsch et al. (1986).

RESULTS

To determine the patterns of expression of the α-subunit isoforms, the cDNA clones were used as probes for hybridization analysis. To eliminate cross-hybridization among partially homologous cDNAs and mRNAs, all blots were hybridized and washed under highly stringent conditions. Characterization of α-subunit isoform expression in adult brain, diaphragm, intestine, heart, kidney and spinal cord identified informative differences in mRNA expression. As shown in Figure 1, α cDNA annealed to a single major mRNA, ≈3.7 kilobases (kb) long, which was found in varying abundance in all tissues examined. In contrast, α(+) transcripts were confined to brain, diaphragm, heart and spinal cord. In these

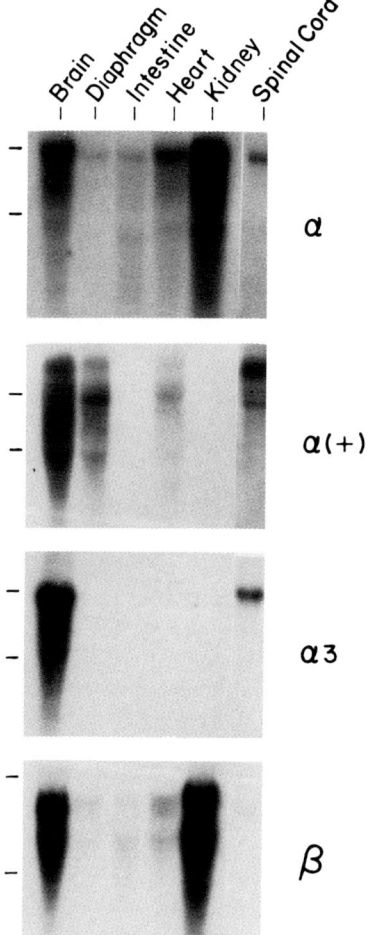

Figure 1. RNA blot analysis of Na,K-ATPase mRNAs in adult rat tissues.
   Total RNAs isolated from the identified tissues were denatured, electrophoresed in a 1% agarose-2.2 M formaldehyde gel, transferred to nitrocellulose and hybridized to α, α(+), αIII or ß cDNA probes. Twenty-five μg of RNA was added per lane. The positions of 28S and 18S ribosomal RNAs are indicated at the left.

tissues, α(+) cDNA identified two discrete mRNAs, one ≈3.4 kb and the other ≈5.3 kb long. The relative abundance of these mRNA species appears to be tissue-specific: the ≈3.4 kb transcript is predominate in muscle tissues (diaphragm and heart) while the ≈5.3 kb mRNA predominates in neural tissues (brain and spinal cord). The expression of the αIII isoform is also regulated in a tissue-specific fashion. The αIII cDNA annealed to a ≈3.7 kb transcript that could be found only in adult brain and spinal cord. Judging by the relative levels of hybridization seen in Fig. 1, αIII mRNA is present in total RNA from adult rat brain in similar abundance to α and α(+). Figure 1 also shows the tissue distribution of ß-subunit mRNA expression (Mercer et al.,1986).

Hybridization analysis of total RNA isolated from brain, heart and kidney of 18 day old fetal rats revealed both tissue and developmental specificity in mRNA expression. As shown in Fig. 2, α mRNA is present in fetal kidney and heart, and at very low levels in brain. α(+) mRNA is also present at very low levels in the fetal brain, however, α(+) mRNA could not be identified in the heart or muscle. In contrast, αIII mRNA is abundant in fetal brain, present in low abundance in the heart, and absent from kidney. Thus, both α and α(+) isoforms are minor mRNAs in fetal brain, whereas αIII is the major Na,K-ATPase transcript. In addition, αIII appears to be a fetal-specific form of rat heart Na,K-ATPase mRNA.

To further investigate the distribution of Na,K-ATPase mRNAs in fetal tissues, sagittal sections of 18 day old fetuses were probed with the α-subunit cDNAs. The stained fetal sections (A-C) and their corresponding autoradiographs (A'-C') obtained after hybridization are shown in Fig. 3. As expected the α cDNA identified abundant mRNA in ion and water transporting epithelia such as kidney (ki), gastrointestinal tract (gi), and naso-tracheal respiratory epi-

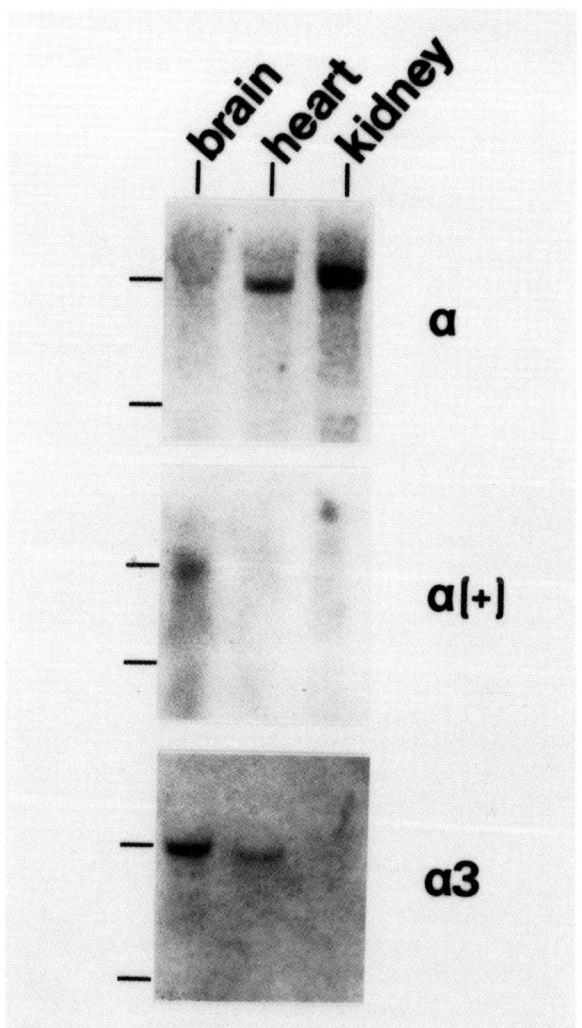

Figure 2. RNA blot analysis of Na,K-ATPase mRNAs in 18 day old fetal rat tissues.
Procedures were as described in Figure 1. The positions of 28S and 18S ribosomal RNAs are indicated at the left.

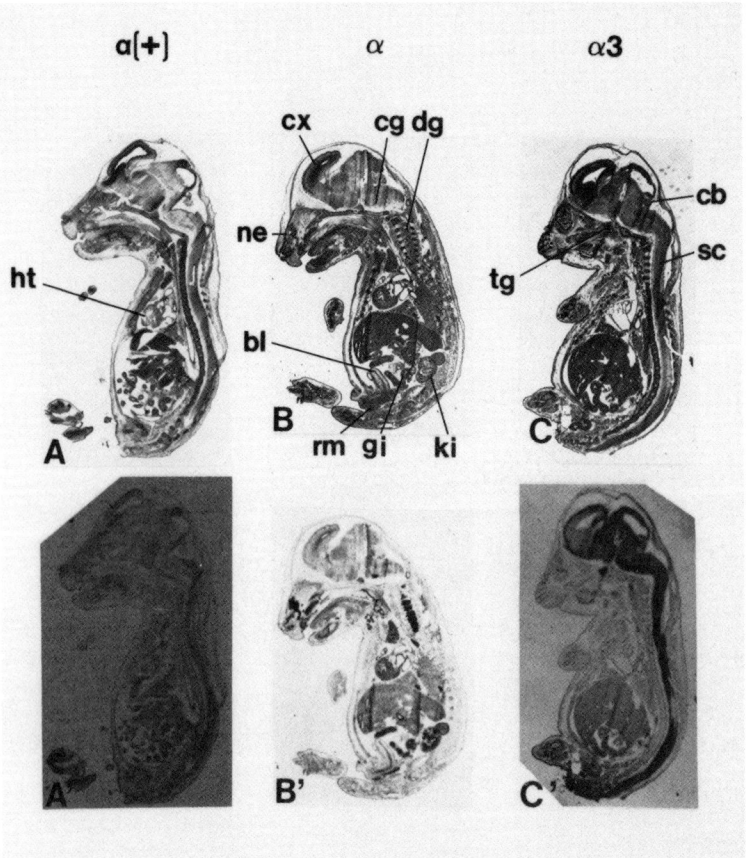

Figure 3. Localization of Na,K-ATPase mRNAs in fetal rat tissues by hybridization histochemistry.

A-C are photographs of Giemsa stained saggital sections of 18 day old embryonic rats. These sections were hybridized with either α (A), α(+)(B), or αIII (C), [^{35}S]-labeled RNA probes complementary to the Na,K-ATPase mRNAs; A', B' and C' are the corresponding autoradiographs. Abreviations: bl, bladder; cb, cerebellum; cg, cranial nerve ganglion; ne, neocortex; dg, dorsal root ganglia; gi, gastrointestinal tract; ht, heart; ki, kidney; ne, naso-tracheal epithelium; rm, rectum; sc, spinal cord; tg, trigeminal ganglion.

thelium (ne) (Fig. 3B'). Abundant α mRNA was also present in embryonic neural structures such as the dorsal root ganglia (dg) and cranial nerve ganglia (cg).

Analysis of fetal sections with the α(+) probe failed to demonstrate labeling of fetal brain or peripheral tissues (Fig. A'). This result confirms the results obtained from hybridization analysis of RNA isolated from fetal rat tissues (Fig. 2).

Expression in fetal rat tissues of mRNA encoding the αIII isoform of the Na,K-ATPase contrasted markedly with the pattern seen for the other isoforms. As shown in Fig. 3C', αIII mRNA is present in the brain, along the entire length of the spinal cord (sc), and in the trigeminal ganglion (tg). Hybridization of the αIII probe to adult rat brain sections indicate that αIII expression appears to be neuron specific. High levels of αIII expression were found in the large neurons of the cerebellum, cortex and hippocampus, the Purkinje cells of the cerebellum, and the pyramidal cells of the cortex and hippocampus (data not shown). Thus, the mRNA encoding the αIII isoform of the Na,K-ATPase is specifically expressed in the major neurons of the adult rat brain.

SUMMARY

The identification of the different α-subunit isoforms of the Na,K-ATPase has added an unexpected complexity to the understanding of the function and regulation of this important transport protein. Our results indicate that the α, α(+) and αIII isoform mRNAs have distinct, but partially overlapping, distributions in brain and peripheral tissues. Also, characterization of mRNA distribution in fetal tissues suggests that the Na,K-ATPase α-subunit isoforms are regulated during development and differentiation. The developmental regulation and strikingly different tissue specificities of the three α-subunit isoforms suggests that the different Na,K-ATPase α-subunits may be adapted to perform different roles in maintaining sodium and potassium homeostasis.

REFERENCES

Awgulewitsch A, Utset MF, Hart CP, McGinnis W, Ruddle FH, 1986. Spacial restriction in expression of a mouse homeo box locus within the central nervous system. Nature 320:328-335.

Chirgwin JM, Przybyla AE, MacDonald RJ, Rutter WJ, (1979). Isolation of biochemically active ribonucleic acid from sources enriched in ribonuclease. Biochemistry 18:5294-5299.

Mercer RW, Schneider JW, Savitz A, Emanuel J, Benz EJ, Jr, Levenson R (1986). Rat brain Na,K-ATPase ß-subunit gene: Primary structure, tissue-specific expression and amplification in ouabain-resistant HeLa cells. Mol. Cell. Biol. 6:3884-3890.

Schneider JW, Mercer RW, Benz EJ, Jr (1987). Identification by molecular cloning of isoforms of the alpha subunit of the Na,K-ATPase expressed predominantly in the brain and heart. Clin. Res. 35:595A.

Schneider JW, Mercer RW, Caplan M, Emanuel JR, Sweadner KJ, Benz EJ Jr, Levenson R (1985). Molecular cloning of rat brain Na,K-ATPase α-subunit cDNA. Proc. Natl. Acad. Sci. USA 82:6357-6361.

Shull GE, Greeb J, Lingrel JB (1986). Molecular cloning of three distinct forms of the Na^+,K^+-ATPase α-subunit from rat brain. Biochemistry 25:8125.

CHARACTERIZATION AND EXPRESSION OF A cDNA CLONE FOR THE BETA SUBUNIT OF RAT BRAIN Na+,K+-ATPase

Koichiro Omori, Kyoko Omori, Mary Flanagan, Vandana A Desai, Nuran Nabi, Yuji Sugita, Dipak Haldar, Jane M Sherman, David D Sabatini and Takashi Morimoto

Department of Cell Biology, New York University School of Medicine, New York, 10016, USA, Department of Physiology, Kansai Medical University, Osaka, Japan (K.O., K.O), Colgate Palmolive Company, N.J., 08854, USA (N.N), Showa University, Pharmaceutical Science, Tokyo, Japan (Y.S), Biological Science, St. John's University, N.Y., 11439, USA (D.H)

INTRODUCTION

The Na^+,K^+-ATPase of animal cells is an integral plasma membrane protein that plays an important role in maintaining the characteristic ionic composition of the cytoplasm and in generating electro-chemical gradients necessary for normal cell function. The enzyme consists of a nonglycosylated 100 Kd protein, the alpha-subunit, and a smaller glycoprotein, the ß-subunit, which are present in an equimolar ratio (Lane et al, 1973). The ATP hydrolysing activity resides in the alpha subunit, but the function of the ß-subunit has not yet been elucidated.

We have isolated a cDNA clone for the ß-subunit from a rat brain cDNA library and used it to study various aspects of the relationship of the ß-subunit to the membrane in which it is incorporated. In this report we present experiments designed to identify the segment of the ß-subunit that directs its cotranslational insertion into the ER. We also demonstrate the existence of several mRNA species for the ß-subunit and show that the levels of liver ß-subunit mRNA are increased by treatment with thyroid hormone.

ISOLATION AND CHARACTERIZATION OF ß-SUBUNIT cDNA CLONE

The dog kidney ß-subunit was purified by the procedure of Takemura et al. (1984) and a partial amino terminal sequence (ARGKAKEEGSWKKFIWNSEK) was determined by Edman degradation. A mixed oligonucleotide, complementary to the sequence TGG AAG(A) AAG(A) TTC(T) ATC(TA) TGG AAC(T), that encodes residues 11 to 17, was synthesized, labelled with ^{32}P and used for screening a rat brain cDNA library in the vector lambda gt11 (Gubler and Hoffman, 1983).

A clone isolated from about 250,000 recombinants contained a cDNA insert whose nucleotide sequence showed that it encoded a polypeptide of 304 amino acid residues, with the amino terminal sequence determined for the ß-subunit polypeptide.

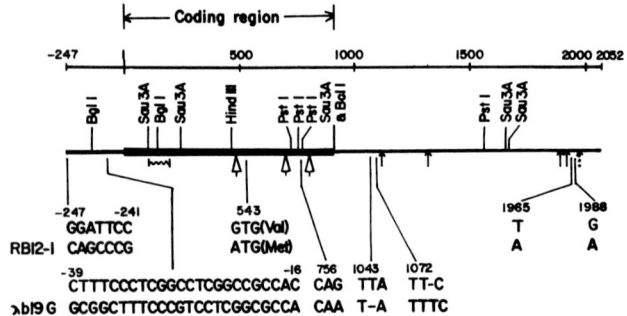

Fig.1. The restriction endonuclease map of the cDNA clone for the ß subunit of the rat brain Na$^+$,K$^+$-ATPase, and nucleotide sequences which are different from those of λb 19G(Mercer et al, 1986) and RBI2-1(Young et al, 1987). The numbers indicate nucleotide number, which starts at the A of the ATG for the initiator methionine as "1"; negative numbers refer to the 5'-untranslated region.

The cDNA insert contains 2299 bp, of which 912 bp correspond to the coding region and 247 bp and 1140 bp to the 5' and 3' untranslated regions respectively. The 3' untranslated region contains four typical (AATAAA), (Proudfoot and Brownlee, 1976) and one possible (TATAAA), (Yu-Cheng et al, 1986) polyadenylation signals (Fig. 1). The initiator methionine in the encoded polypeptide is immediately followed by the N-terminal amino acid (Alanine) of the mature ß-subunit, confirming that the ß-subunit polypeptide does not contain a cleavable amino terminal insertion signal. The presence of a single hydrophobic domain in the ß-subunit, extending from residues 35 to 62, suggests that this ß-subunit is a class II polypeptide, i.e., a polypeptide which traverses the membrane only once and has its aminoterminus exposed on the cytoplasmic side and its C-terminus on the luminal side of the membrane. The putative luminal domain (240 residues) contains three potential N-linked glycosylation sites and the putative cytoplasmic portion 34 residues.

During the completion of this work, the nucleotide sequences for two other rat brain Na^+,K^+-ATPase ß-subunit cDNA clones (λb 19G, 1186nt from Mercer et al, 1986 and RB 12-1, 2528 nt from Young et al 1987) were reported. A comparison of the sequence of those clones with the sequence derived from us shows several nucleotide differences in both the coding and non-coding regions (Fig. 1). A single substitution exists between the coding

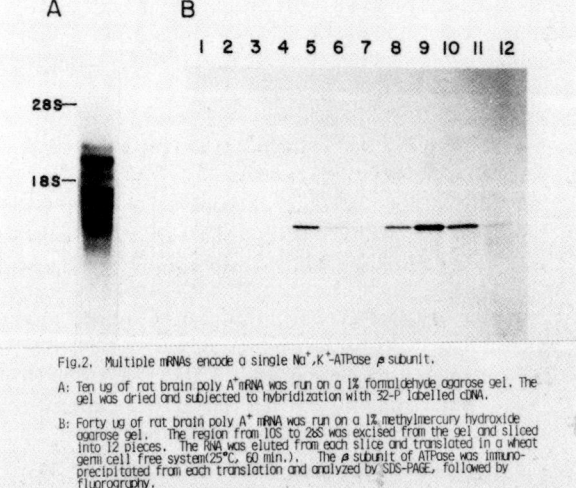

Fig.2. Multiple mRNAs encode a single Na^+,K^+-ATPase ß subunit.

A: Ten ug of rat brain poly A^+ mRNA was run on a 1% formaldehyde agarose gel. The gel was dried and subjected to hybridization with 32-P labelled cDNA.

B: Forty ug of rat brain poly A^+ mRNA was run on a 1% methylmercury hydroxide agarose gel. The region from 10S to 26S was excised from the gel and sliced into 12 pieces. The RNA was eluted from each slice and translated in a wheat germ cell free system(25°C, 60 min.). The ß subunit of ATPase was immuno-precipitated from each translation and analyzed by SDS-PAGE, followed by fluorography.

region of RB 12-1 (Young et al., 1987), and the ß-subunit cDNA clone isolated by us at nt 1006 of RB 12-1 which is nt 543 in Fig. 1, and this results in a conservative amino acid replacement (Met in RB12-1 and Val in ours). The same difference was observed by Young et al. (1987) between RB 12-1 and a rat kidney cDNA clone (RK 2-1). Our cDNA clone and λb19G (Mercer et al., 1986) have the same sequence (Val) at the position that corresponds to Met in RB12-1. It thus seems likely that, as proposed by Young et al (1987), different alleles for the ß-subunit are found in the rat population which differ at nt residue 543.

MULTIPLE mRNA SPECIES FOR THE ß-SUBUNIT POLYPEPTIDE

In Northern blots of rat brain Poly A^+ mRNA, a labelled cDNA probe that contained only the coding region hybridized to several mRNAs 2.9, 2.6, 2.25, 1.75, and 1.5 kb in length, of which the 2.6 kb species was predominant (Fig. 2A). To determine if these mRNAs encode different ß-subunits poly A^+ mRNAs were extracted from different regions of the agarose gel ranging from 10S to 28S and translated in a wheat germ system. Immunoprecipitation with anti-ß-subunit antibodies showed that a signle polypeptide of 38 kD was obtained, regardless of the size of the mRNA (Fig. 2B). This result suggested that the various mRNAs differed in the lengths of the non-coding regions. Recently Young et al (1987) have shown directly that this is the case.

Multiple mRNA species encoding a single ß-subunit are also found in rat kidney, liver and heart (Mercer et al, 1986; Young et al, 1987 and on our unpublished observations). These organs have similar sets of hybridizable mRNA, but the size of the predominant species varies in each case.

IN VITRO INSERTION INTO MICROSOMAL MEMBRANES OF THE ß-SUBUNIT ENCODED IN THE cDNA

A cDNA insert containing the complete coding region for the ß-subunit was cloned into the plasmid vector pTZ18R (Pharmacia) downstream from the T_7 polymerase promoter. The ß-subunit cDNA in the resulting plasmid (pTZ18Rß) was expressed in a coupled transcription and translation system with or without the addition of microsomal membranes. In the absence of membranes a single primary translation product of M_r of 38 kD was obtained (Fig. 3A lane a), but when translation was carried out in the presence of dog pancreas microsomes, a polypeptide of 45 kD was also synthesized (Fig. 3A lane b). The larger size suggests that the latter polypeptide represents a microsomal form of the ß-subunit which during cotranslational insertion into the membranes undergoes core glycosylation. The acquisition of high mannose oligosaccharide chains was demonstrated directly by treatment of the 45 Kd polypeptide with endoglycosidase H which shifted the apparent molecular weight to 38 kD (Fig. 3A lane c). Incubation of the translation mixture with trypsin and chymotrypsin showed that whereas the 38 kD polypeptide was completely digested to undetectable small peptides, the apparent molecular weight of the polypeptide incorporated into membranes (45 Kd) was slightly decreased by 1 or 3 Kd, first to 44 kd and then to 42 Kd, as expected from the digestion of only a short region of about 3 kD (Fig. 3A lane d) exposed on the cytoplasmic face of the membrane. On prolonged incubations the 45 kD polypeptide did not undergo further cleavage to the smaller fragments and treatment with the protease V8 caused an even smaller reduction in apparent molecular weight (2 Kd) than trypsin and chymotrypsin. These results suggest that one or only two very small segments of the ß-subunit (N-terminal, C-terminal or both) amounting to a total of 3 Kd could be exposed on the cytoplasmic side of the membrane. Because the sequence of the ß-subunit shows that potential cleavage sites for the enzymes employed are present within the first 30 aminoterminal residues of the polypeptide and of the three glycosylation sites one is within the last (C-terminal) 40 amino acid residues which are hydrophilic it seemed very likely that the N-terminal end of the ß-subunit is the only one accessible to the proteases.

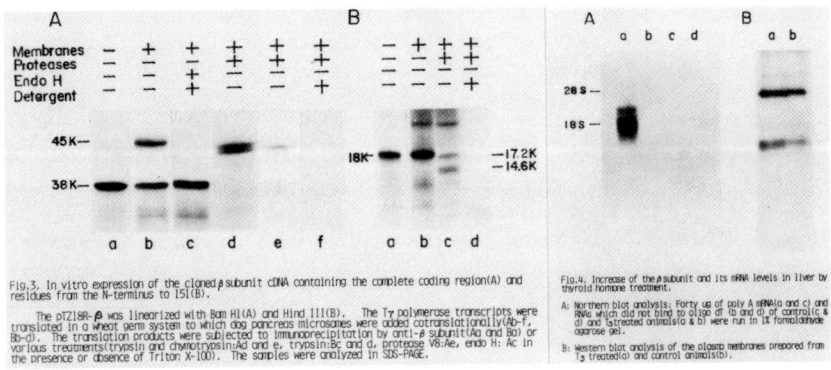

Fig.3. In vitro expression of the cloned ß subunit cDNA containing the complete coding region(A) and residues from the N-terminus to 151(B).

The pTZ18R-ß was linearized with Bam HI(A) and Hind III(B). The T7 polymerase transcripts were translated in a wheat germ system to which dog pancreas microsomes were added cotranslationally(Ab-f, Bb-d). The translation products were subjected to immunoprecipitation by anti-ß subunit(Aa and Ba) or various treatments(trypsin and chymotrypsin:Ad and e, trypsin:Bc and d, protease V8:Ae, endo H: Ac in the presence or absence of Triton X-100). The samples were analyzed in SDS-PAGE.

Fig.4. Increase of the ß subunit and its mRNA levels in liver by thyroid hormone treatment.
A: Northern blot analysis: Forty ug of poly A mRNA(a and c) and RNAs which did not bind to oligo dT (b and d) of control(c & d) and T3 treated animals(a & b) were run in 1% formaldehyde agarose gel.
B: Western blot analysis of the plasma membranes prepared from T3 treated(a) and control animals(b).

ROLE OF THE N-TERMINAL HALF OF THE ß-SUBUNIT POLYPEPTIDE IN MEMBRANE INSERTION

Studies in which the synchronized translation of the natural ß-subunit mRNA was carried out in the presence of microsomal membranes have indicated that insertion into the membrane takes place before one-third of the polypeptide is completed (Nabi et al, 1983 and Geering et al., 1985). Since in nascent polypeptides about 40 amino acid residues are thought to be enclosed within the large ribosomal subunit (Blobel and Sabatini, 1970) and the ß-subunit contains 304 amino acid residues, these kinetics of membrane insertion on the ß-subunit suggest that the domain which triggers the insertion is located approximately between residues 40 and 60. This estimate is in good agreement with the location of the single hydrophobic segment in the polypeptide (residues 35-62). To determine directly if this segment functions as an internal insertion signal and a permanent membrane anchor (Sabatini et al., 1982), the vector pTZ18R-ß was cleaved with Hind III and the open plasmid transcribed to produce an mRNA that encodes a polypeptide extending from the N-terminus to residue 151 (Fig. 1). The translation product found in the absence of membranes had a M_r of 18 kD and was immunoprecipitable with anti-ß-subunit antibodies (Fig. 3B lane a). As expected from the distribution of cleavage sites when treated with trypsin this polypeptide was extensively cleaved to fragments of about 2 kD and 4 kD. When the translation was carried out in the presence of microsomal membranes a polypeptide of higher molecular weight did not appear (Fig. 3B lane b), but in this case digestion with trypsin reduced the apparent molecular weight by only 1 or 3 Kd, first to about 17 kD (Fig. 3A lane c) and upon prolonged digestion to 15 kD, but no further cleavage was observed.

The fact that a 3 kD portion could be removed by proteolysis from both the membrne inserted ß-subunit and the amino terminal segment (151 amino acid) indicates that the exposed portion corresponds to the N-terminal hydrophilic domain of the polypeptide. It can therefore be concluded that the ß-subunit is indeed a type II membrane protein that contains a permanent insertion signal, that in the mature protein serves also as a membrane anchor that bears a 34 amino acid segment on the cytoplasmic side of the membrane.

THYROID HORMONE CAUSES AN INCREASE IN HEPATIC LEVELS OF ß-SUBUNIT mRNA

Hepatic levels of Na^+,K^+-ATPase measured by enzyme activity have been shown to be very low (Bonting et al, 1961) and to increase after thyroid hormone treatment

(Ismail-Beigi and Edelman, 1971). In normal rat liver levels of ß-subunit mRNA are almost undetectable by hybridization on Northern blots to the labelled cDNA. However, several hybridizable mRNA species (2.9, 1.7 and 1.5 kb) were easily detectable in poly A$^+$ mRNA samples obtained from the livers of rats that received five daily intraperitoneal injections (200 ug/100g body weight) of the thyroid hormone T_3 (Fig. 4A lane a).

In contrast to the large increase in ß-subunit mRNA, the levels of ß and alpha subunit polypeptides measured in liver plasma membrane fractions (Ray, 1970) by Western blotting with specific antibodies followed by ^{125}I-protein A, increased only twice over the levels of controls (Fig. 4B lanes a and b).

The effect of T_3 treatment on ß-subunit mRNA levels in brain, kidney and heart was also examined by Northern and dot blot analyses. In brain and kidney these levels did not increase significantly, but in heart T_3 caused a notable increase, although not as high as in liver. For all these organs similar sets of ß-subunit mRNAs, with the relative amounts characteristic of each organ, were found in control and T_3 treated samples. Studies on the mechanisms by which the ß-subunit mRNA levels are regulated in response to thyroid hormone treatment are now in progress.

ACKNOWLEDGEMENTS

We are grateful to Brian Zeitlow and Jody Culkin for expert preparation of the figures, and Ms. Myrna Cort and Ms. Bernice Rosen for typing the manuscript. This work was supported by NIH grant AG01461.

REFERENCES

Blobel G, Sabatini DD (1970). Controlled proteolysis of nascent polypeptides in rat liver cell fractions. I. Location of the polypeptide within ribosomes. J Cell Biol 45:130.

Bonting SL, Simon KA, Hawkins NM (1961). Studies on sodium potassium-activated adenosine triphosphatase. I. Quantitative distribution in several tissues of the cat. Arch Biochem Biophys 95:416.

Geering, K, Meyer DI, Paccolat MP, Kraehenbuhl JP, Rossier BC (1985). Membrane insertion of alpha and beta-subunits of Na^+,K^+-ATPase. J Biol Chem 260:5154.

Gubler U, Hoffman RJ (1983). A simple and very efficient method for generating cDNA libraries. Gene 25:263.

Ismail-Beigi F, Edelman IS (1971). The mechanism of the calorigenic action of thyroid hormone. Stimulation of Na^+,K^+ activated adenosine triphosphatase activity. J Gen Physiol 57:710.

Lane LK, Copenhaver JHJr, Lindenmayer GE, Schwaltz A (1973). Purification and characterization of and [^{3}H] Quabain binding to the transport adenosine triphosphatase from outer medulla of canine kidney. J Biol Chem 248:7197.

Mercer RW, Schneider JW, Savitz A, Emanuel J, Benz EJJr., Levenson R (1986). Rat brain Na,K-ATPase ß-chain gene: primary structure, tissue specific expression, and amplification in ouabain-resistant HeLa C^+ cells. Mol and Cellular Biol 6:3884.

Nabi N, Sherman J, Sabatini DD, Morimoto T (1983). Biosynthesis of rat brain Na^+,K^+-ATPase. J Cell Biol 97:117a.

Proudfoot NJ, Brownlee GG (1976). 3' non-coding region sequences in eukaryotic messenger RNA. Nature 263:211.

Ray TK (1970). A modified method for the isolation of the plasma membrane from rat liver. Biochim Biophys Acta 196:1.

Sabatini DD, Kreibich G, Morimoto T, Adesnik M (1982). Mechanisms for the incorporation of proteins in membranes and organelles. J Cell Biol 92:1.

Takemura S, Omori K, Tanaka K, Omori K, Matsuura S, Tashiro Y (1984). Quantitative immunoferritin localization of Na^+,K^+-ATPase on canine hepatocyte cell surface. J Cell Biol 99:1502.

Young RM, Shull GE, Lingrel JB (1987). Multiple mRNAs from rat kidney and brain encode a single Na^+,K^+-ATPase ß-subunit protein. J Biol Chem 262:4905.

Yu-Cheng J, Lebo RV, Clawson GA, Smuckler EA (1986). Human prion protein cDNA:molecular cloning, chromosomal mapping and biological imprecations. Science 233:364.

INTRAINDIVIDUAL TISSUE-SPECIFIC VARIATIONS IN ORGANIZATION
OF GENES CODING FOR THE Na,K-ATPase SUBUNITS

E.D. Sverdlov, N.E. Broude, G.S. Monastyrskaya,
V.E. Sverdlov, A.V. Grishin, N.S. Akopyantz,
K.E. Petrukhin, N.N. Modyanov.

M.M. Shemyakin Institute of Bioorganic Chemistry,
USSR Academy of Sciences, Moscow, USSR.

INTRODUCTION

Restriction endonuclease mapping and direct sequencing of clones isolated from human genomic libraries revealed five different genes related to the alpha subunit of Na,K-ATPase (Sverdlov et al., 1987). At least two of them transcribed in human brain (Ovchinnikov et al., 1987). Using cloned cDNA for alpha and beta subunits of pig Na,K-ATPase as hybridization probes we have unexpectedly found intraindividual restriction fragments length polymorphism (RFLP) in genomic DNA isolated from different mammalian (mouse, rabbit, human) tissues involving sites for EcoR I, Pst I and Bgl II. These tissue-specific RFLPs could be a result of rearrangements in the gene loci for alpha and beta subunits of Na,K-ATPase.

RESULTS AND DISCUSSION

Southern blot analysis of EcoR I-digested human genomic DNA isolated from different tissues (some of them from one individual) and probed with 32P-labeled alpha subunit specific cDNA fragment clearly shows the length polymorphism and variations in copy number of DNA fragments (Fig. 1A). The expected band pattern obtained after hybridization of the same blot with gamma interferon-specific cDNA probe (Fig. 1B) excludes the incomplete digestion or other artefacts as the explanations of the effect observed. We have found similar tissue-specific RFLPs in the alpha subunit genes for the Pst I and Bgl II (data not shown). Southern blot analysis of DNAs isolated from different tissues of mouse and rabbit also showed tissue-specific RFLPs (data not shown).

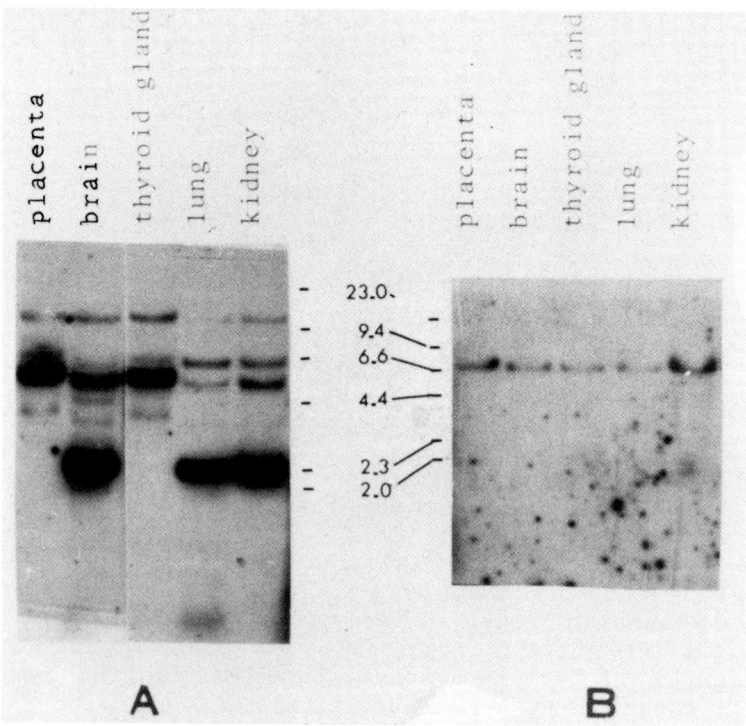

Figure 1. Southern blot analysis of human chromosomal DNA from indicated tissues. DNA was digested by EcoR I and probed with (A) 1300-bp alpha subunit cDNA fragment from the clone pB 2801 (Ovchinnikov et al., 1985) and (B) gamma interferon cDNA fragment.

Tissue-specific EcoR I and Pst I RFLPs have also been observed in the case of hybridization with beta subunit-specific cDNA probe (Fig. 2).

Analysis of EcoR I digests of DNA isolated from mouse embryos of different age and from adult mouse (Fig. 3) has demonstrated nonidentical hybridization patterns of DNA from embryos on different stages of prenatal development as well as differences between embryonic and adult mouse DNA.

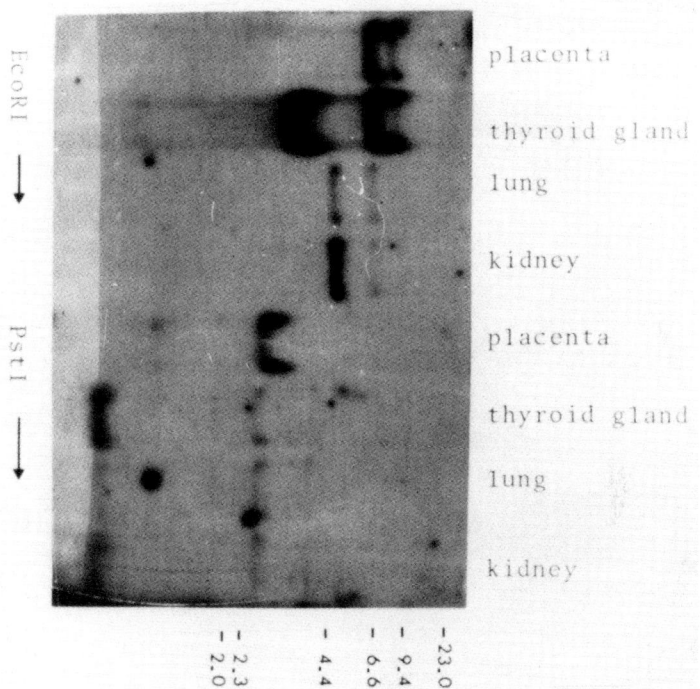

Figure 2. Southern blot analysis of human chromosomal DNA digested by EcoR I and Pst I using as probe pig kidney cDNA for beta subunit from the clone pNb31 (Ovchinnikov et al., 1986).

Comparison of alpha subunit cDNA hybridization with genomic cDNA fragments from normal human lung and two different types of lung tumor also revealed RFLP (Fig. 4). Therefore we conclude that alterations in the alpha subunit genes are likely to be a general feature of differentiation.

The question arises what are the structural reasons and functional significance of tissue-specific RFLPs in Na,K-ATPase genes. One possible explanation to RFLP could be connected with different copy number of genes coding for different isoforms of alpha and beta subunits in different tissues. It cannot be excluded that tissue-specific RFLP

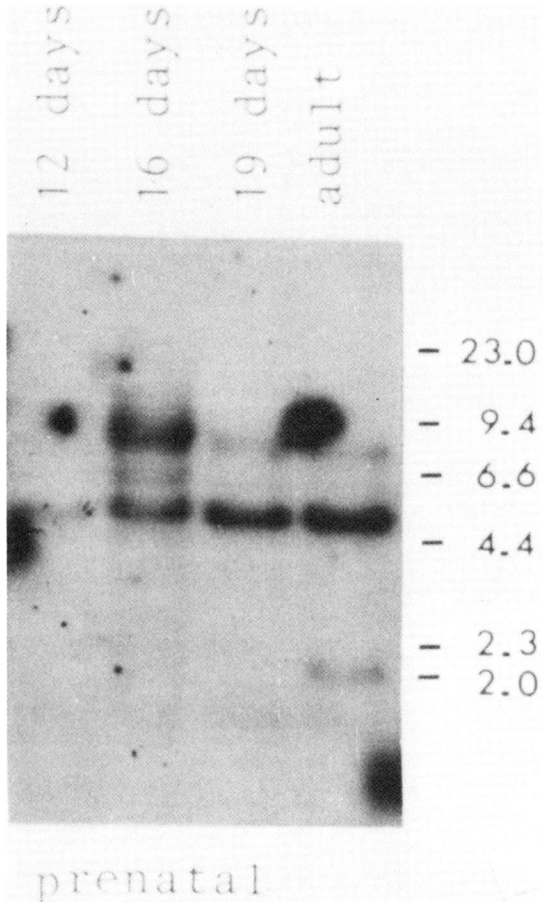

Figure 3. Southern blot analysis of mouse chromosomal DNA digested by EcoR I. DNA was isolated on different stages of embryonic and postnatal development. Hybridization was performed with pig kidney alpha subunit cDNA probe (see legend to Fig. 1).

could be associated with tissue-specific modification (e.g. methylation) of genes for the Na,K-ATPase subunits. However, the reccurence of the RFLPs when various restriction endonucleases are used is not consistent with this proposal. The third and from our point of view most probable explanation

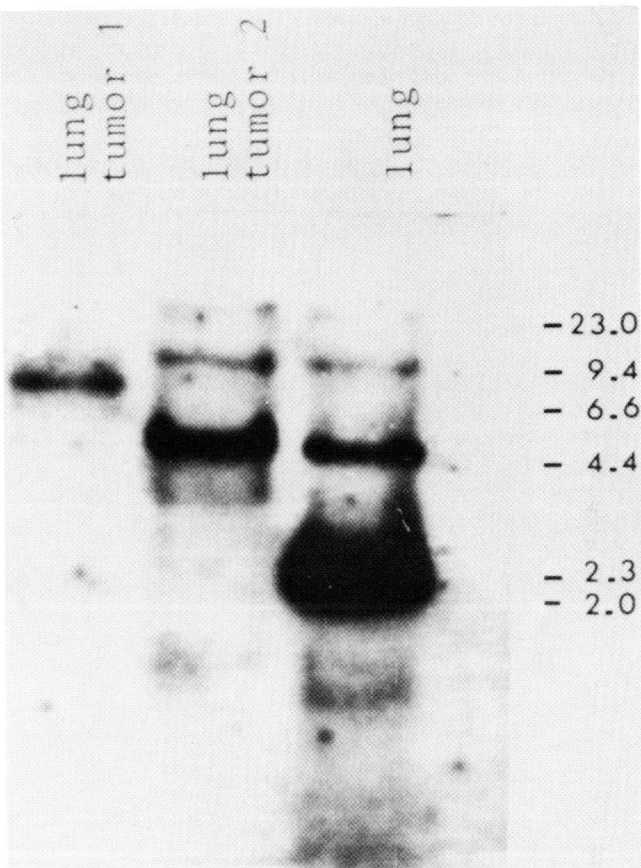

Figure 4. Southern blot analysis of human chromosomal DNA isolated from normal lung and from two types of lung tumor digested by EcoR I and probed with pig kidney alpha subunit cDNA fragment (see legend to Fig. 1).

of the RFLPs in Na,K-ATPase genes seems to be the somatic rearrangements, which at present are known in mammals only for genes of immune system.

As to functional consequences of the somatic rearrangements of Na,K-ATPase genes it is tempting to propose that there is a connection between the gene structure and the differential expression of ATPase genes in different tissues. Our experiments demonstrated the tissue-specific expression of at least two genes from the alpha subunit gene family (Fig. 5). Moreover their expression is strictly diminished in tumor tissues (Fig. 6).

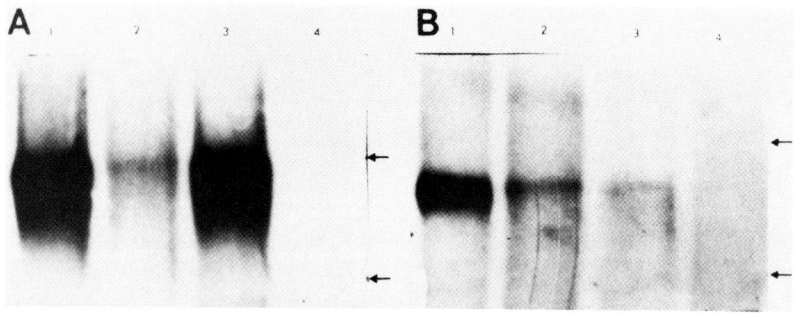

Figure 5. Northern blot hybridization of poly(A+)RNA from human kidney, brain, thyroid gland and liver (lanes 1-4, respectively) with (A) NKalphaTW-4-specific oligoprobe and (B) NKalphaR3-2-specific oligoprobe. NKalphaTW-4 and NKalphaR3-2 are the members of alpha subunit gene family (Sverdlov et al., 1987) and code for the different molecular forms of catalytic subunits. Both gene-specific oligoprobes correspond to the regions of mRNA coding for the peptides with coordinates 3-11. Arrows on the figure indicate the position of 28S and 18S ribosomal RNA.

We propose therefore that there is a correlation between the Na,K-ATPase genes organisation and the specificity and/or level of their expression. Data presented in this paper poses the question if the role of the Na,K-ATPase is limited only to ion transport or there are some additional regulatory function of the enzyme in the cell which require somatic rearrangement of Na,K-ATPase genes.

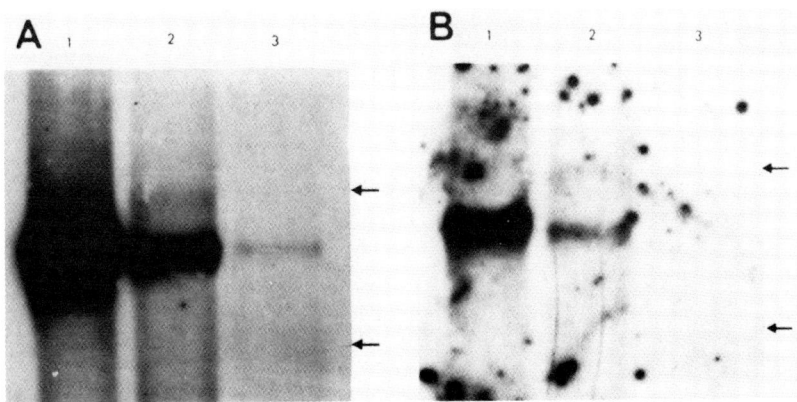

Figure 6. Northern blot hybridization of poly(A+)RNA from normal human kidney, renal carcinoma and Wilms' tumor (lanes 1-3, respectively). For the explanation (A) and (B) see legend to Fig. 5. Positions 28S and 18S ribosomal RNA are indicated by arrows.

REFERENCES

Ovchinnikov Yua, Arsenyan SG, Broude NE, Petrukhin KE, Grishin AV, Aldanova NA, Arzamazova NM, Aristarkhova EA, Melkov AM, Smirnov YuV, Guryev SO, Monastyrskaya GS, Modyanov NN (1985). Nucleotide sequence of cDNA and deduced primary structure of the Na,K-ATPase alpha subunit from pig kidney. Dokl Acad Nauk SSSR 285: 1490-1495.

Ovchinnikov YuA, Broude NE, Petrukhin KE, Grishin AV, Kiyatkin NI, Arzamazova NM, Gevondyan NM, Chertova EN, Melkov AM, Smirnov YuV, Malyshev IV, Monastyrskaya GS, Modyanov NN (1986). Nucleotide sequence of cDNA and deduced primary structure of the Na,K-ATPase beta subunit from pig kidney. Dokl Acad Nauk SSSR 286: 1491-1495.

Ovchinnikov YuA, Monastyrskaya GS, Broude NE, Allikmets RL, Ushkaryov YuA, Melkov AM, Smirnov YuV, Malyshev IV, Dulubova IE, Petrukhin KE, Grishin AV, Sverdlov VE, Kiyatkin NI, Kostina MB, Modyanov NN, Sverdlov ED (1987). The family of human Na,K-ATPase genes. A partial nucleotide sequence related to the alpha subunit. FEBS Lett 213: 73-80.

Sverdlov ED, Monastyrskaya GS, Broude NE, Ushkaryov YuA, Allikmets RL, Melkov AM, Smirnov YuV, Malyshev IV, Dulubova IE, Petrukhin KE, Grishin AV, Kiyatkin NI, Kostina MB, Sverdlov VE, Modyanov NN, Ovchinnikov YuA (1987). Family of human Na,K-ATPase genes. II. No less than five genes and/or pseudogenes related to the alpha subunit. FEBS Lett 217: 275-278.

INCREASED IN VIVO SYNTHESIS OF RETINAL NA PUMP ISOFORMS AFTER PRE-TREATMENT WITH CITRATE BUFFER

Susan C. Specht, Enid Del Valle and Sonia Castro Hernández
Department of Pharmacology and Laboratory of Neurobiology, University of Puerto Rico School of Medicine, San Juan, Puerto Rico, USA 00936

INTRODUCTION

Na,K-ATPase isolated from the rat retina can be separated by SDS-PAGE into two distinct bands between 95 and 100 kD, representing different molecular isoforms of the catalytic subunit alpha (Sweadner, 1979). The in vivo incorporation of radioactive amino acids into Na pump isoforms can be quantified by scintillation counting of the solubilized gel bands. Whether the various Na pump isoforms (Schull et al., 1986) are independently regulated is an important issue. During the course of studies with intra-ocular tetrodotoxin, a Na^+ channel blocker that is commercially supplied in citrate buffer at pH 2.8, we found evidence that citrate buffer alone provokes a parallel increase in the synthesis of Na pump isoforms present in the two gel bands.

METHODS

Adult male Sprague Dawley rats were anesthesized with ether and the sclera was punctured with a sterile 25 gauge needle under microscopic observation. A fine glass pipet containing citrate buffer (3 mM, pH 2.8) was inserted into the vitreal chamber and 10 µL were injected by application of hydrostatic pressure ; controls received a puncture wound into the vitreal chamber but no intraocular solution. At intervals of 2.6 and 24 hours following scleral puncture. 250 µCi (25 µL) of ^{35}S-methionine were injected intra-vitreally. For incorporation studies, the rats were killed at 2 hours after intraocular labeling. For axonal transport studies, retinal proteins were labeled in vivo at 6 hours after citrate treatment and rats were killed at either material or at 24 hours for material that remained after the peak of fast transport. Na,K-ATPase was purified from radioactively-labeled retinae, post chiasmatic optic tract and optic terminal nuclei (lateral geniculate and superior colliculus) as previously described (Specht and Sweadner, 1984). The Na pump isoforms were separated by SDS-PAGE, cut apart with a razor blade and dissolved (Wilson et al., 1980) and the incorporated radioactivity was counted.

RESULTS AND DISCUSSION

The alpha subunit isoforms of Na,K-ATPase isolated from rat retina migrate at about 100 kD on a 5% SDS-polyacrylamide gel.

In preliminary studies of the latent period required for citrate stimulation, rat retinae were labeled in vivo by intraocular injection of ^{35}S-methionine after varying periods of citrate pretreatment (2,6,24 and 48 hours) and killed at 24 hours after labeling. We found the greatest amount of radioactive label incorporated into Na pump alpha isoforms after 6 hours of citrate pretreatment.

In order to determine if intraocular citrate increased synthesis or decreased degradation, we examined incorporation by killing the rats at 2 hours after intraocular labeling. Both citrate-treated retinae and puncture-only controls were examined (Table 1). Incorporation of ^{35}S-methionine into the alpha subunit isoforms (Fig. 1) was identical in control and treated retinae after 2 hours, but increased 60-100% at 6 hours after citrate treatment. The increase was somewhat greater in the more rapidly migrating gel band. Radioactivity was also increased in the proteins running at the gel front, MW. $\leq$ 58 kD.

Figure 1. SDS-PAGE of Na, K-ATPase isolated from rat retina.

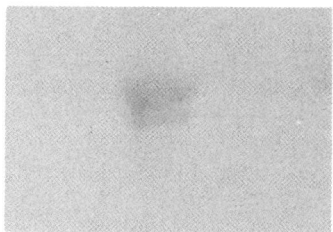

TABLE 1. The Effect of Intraocular citrate buffer on Incorporation of ^{35}S-methionine into Na Pump Isoforms of the Rat Retina.

Treatment	CPM/Gel Band / Retina		
	Alpha(+)	Alpha	Front
A. 2 hour treatment			
Citrate	1058 ± 322	812 ± 230	24519 ± 10219
Puncture	1153 ± 262	845 ± 245	18808 ± 3888
B. 6 hour treatment			
Citrate	1721 ± 78	1643 ± 183	36171 ± 2840
Puncture	1145 ± 341	1009 ± 299	19869 ± 5663

The values are mean ± SEM for three animals.

By 24 hours after retinal labeling, the incorporated radioactivity was lower in citrate-treated retinae than in puncture controls. These data suggest that citrate initially stimulated synthesis with a latent period of 2-6 hours and enhanced degradation somewhat later. More newly-synthesized Na pumps remained after 24 hours in control retinae, neither synthesis nor degradation having been stimulated.

In order to study axonally-transported Na pumps, rats were killed at 6 or 24 hours after labeling with 6 hours of citrate pretreatment. Axonally-transported Na,K-ATPase recovered from the optic tract and optic terminal nuclei of citrate-treated rats had increased radioactivity at 6 hours (122-148% of control), and decreased radioactivity after 24 hours (48-55% of control). In the same experiment, labeling of citrate-treated retinal Na pump alpha isoforms was 115% of control after 6 hours and 48% after 24 hours. The correspondence with retinal labeling data indicates that the ganglion cells had participated in the retinal response to citrate injection.

Up-regulation of retinal Na,K-ATPase by citrate may be associated with increased $\{Na\}_i$. The Na content of whole retinae was determined by flame photometry (Fernandez-Repollet et al., 1980) at 2 hours after intravitreal injection of citrate or puncture. Retinal Na content was increased 20-25% in retinae treated <u>in vivo</u> with citrate buffer for two hours as compared to puncture-only controls.

TABLE 2. Content of Na^+ in Citrate-Treated and Control Retinae

Treatment	mEq/g (wet weight)	mEq/g (tissue water)
Puncture Na^+	.120	.138
Citrate Na^+	.145	.173

Values for each group are the average of duplicate determinations (four rat retina each) by flame photometry.

These data support the hypothesis that intraocular injection of citrate buffer, pH 2.8, provokes an increase in intracellular Na^+ which, in turn, causes a parallel increase in the synthesis rate of Na pump isoforms. Although probably not all retinal Na pump isoforms are separated by this method. There was no evidence of isoform-specific regulation.

ACKNOWLEDGMENTS

Supported in part by NS-07464 and RR-08102 from NIH.

REFERENCES

Fernandez-Repollet E, Martinez-Maldonado M, Opava-Stitzer S (1980) Role of water balance in the enhanced potassium excretion and hypokalemia of rats with diabetes insipidus. J. Physiol. (London) 305:97-108.

Schull GE, Greeb J, Lingrel JB (1986) Molecular cloning of three distinct forms of the Na^+,K^+-ATPase a-subunit from rat brain. Biochemistry 25:8-25.

Specht, SC, Sweadner KJ (1984) Two different Na,K-ATPase in the optic nerve cells of origin and axonal transport. Proc Nat Acad Sci. USA, 81:1234-1238.

Sweadner KJ (1978) Purification from brain of an intrinsic membrane protein fraction enriched in $(Na^+ + K^+)$-ATPase. Biochim Biophys Acta 508:486-499.

Sweadner KJ (1979) Two molecular forms of $(Na^+ + K^+)$-stimulated ATPase in brain. J. Biol Chem 254:6060-6067.

Wilson DL, Hall ME, Stone CG, Rubin RW (1977) Some improvements in two-dimensional gel electrophoresis of proteins. Anal. Biochem. 83: 33-44.

TWO FORMS OF THE Na,K-ATPase CATALYTIC SUBUNIT DETECTED BY THEIR IMMUNOLOGICAL AND ELECTROPHORETIC DIFFERENCES. ALPHA AND ALPHA+ IN MAMMALIAN HEART.

Kathleen J. Sweadner and Sima K. Farshi

Pappas Laboratories, Neurosurgical Service
Massachusetts General Hospital, Boston, MA 02114
and Dept. of Physiology, Harvard Medical School.

INTRODUCTION

The Na,K-ATPase is the only known receptor for the cardiac glycosides, and yet the role of inhibition of the Na,K-ATPase in the genesis of cardiac inotropy has been controversial for many years. A principal problem has been that inotropic effects have frequently been obtained at doses of cardiac glycosides much lower than those thought to be required to inhibit the cardiac Na,K-ATPase. In parallel, several laboratories have reported the existence of high- affinity binding sites for cardiac glycosides that do not correspond to the relatively lower-affinity Na,K-ATPase. This has raised the question of whether there is another, unrelated cardiac glycoside receptor in the heart, not the Na,K-ATPase.

In the interim, isozymes of the Na,K-ATPase with different affinities for the cardiac glycosides were discovered in the brain (Sweadner, 1979, 1985). The isozymes fortuitously could be distinguished by their electrophoretic mobility in SDS, and they were separated and purified in active form. These isozymes (alpha+ and alpha)* were shown to have distinct antigenic determinants (Sweadner & Gilkeson, 1985). It was an attractive hypothesis that the same or similar isozymes might account for the puzzling inconsistencies in the action of cardiac glycosides on the heart.

This report summarizes our recent evidence for two isozymes in the rat heart (Sweadner & Farshi, in press), and provides further evidence that there are striking differences in the expression of Na,K-ATPase isozymes in the hearts of different mammalian and avian species.

*At the present writing, a highly homologous gene, possibly a third isozyme of the Na,K-ATPase (alphaIII) (see Lingrel et al., this volume) has not yet been identified as a protein. In this manuscript the identified isozymes are called alpha and alpha+, terms which are used as operational definitions based on both gel electrophoretic mobility and reaction with antisera raised against the Na,K-ATPases from kidney and axolemma. The protein that we identify as alpha+ is likely to correspond to the gene product that Lingrel also calls alpha+, since Young and Lingrel (1987) have detected alpha+ mRNA in the rat heart.

RESULTS

Two different forms of the Na,K-ATPase can be detected in the rat heart ventricle by their reactivity on Western blots with antisera (K3 and Ax2, respectively) raised against the purified kidney (alpha) and axolemma (alpha+) Na,K-ATPases (Figure 1). The K3 antiserum detects only an alpha form in brain, newborn rat ventricle, or adult rat ventricle. The Ax2 antiserum (formerly called A2) detects alpha+ in both the brain and heart, but the abundance of alpha+ is lower in adult ventricle than in newborn ventricle. The Ax2 antiserum also crossreacts weakly with the alpha subunit in the heart preparations;

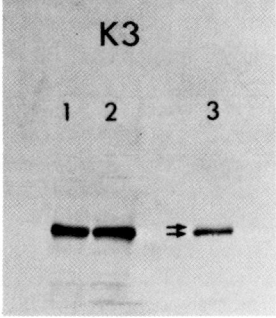 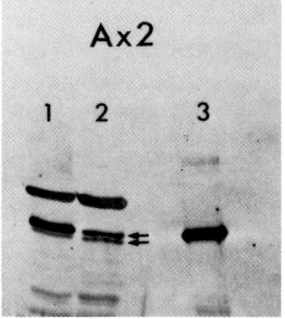

Figure 1. Immunoblot detection of alpha and alpha+ in crude particulate fractions of rat ventricle. 1) newborn ventricle; 2) adult ventricle; 3) adult rat brain. The upper arrow marks alpha+ and the lower arrow alpha.

alpha is 3-4 times as abundant as alpha+ even in the newborns (data not shown). The Ax2 antiserum also crossreacts with the Ca-ATPase (higher apparent molecular weight) and weakly with a number of other cardiac proteins.

In another publication, we have presented evidence that the band that reacts with the Ax2 antiserum is indeed an alpha+ form of the Na,K-ATPase. Besides having the same electrophoretic mobility as the alpha+ form from brain and crossreacting with an alpha+-specific antibody, it has the following characteristics. It is enriched with the alpha form in sarcolemma fractions. It copurifies with the alpha form during extraction with SDS by the Jorgensen procedure. It exhibits Na-stimulated, K-inhibited phosphorylation with [γ-32P]ATP. In addition, the presence of the alpha+ form in newborn ventricles at a measurable level (approximately 20%) correlates with the inhibition of about 20% of the Na,K-ATPase activity at low concentrations of ouabain (apparent Ki close to 10-7 M). This suggests that the alpha+ of the rat heart is very similar to the alpha+ of the rat brain in its relatively high sensitivity to cardiac glycosides.

We also investigated the time course of the change in the level of expression of the alpha+ form of Na,K-ATPase during development in the rat (Sweadner & Farshi, in press). Alpha+ levels remained high from 18 days of fetal life (the earliest time point examined) through 10 days of postnatal life. Alpha+ levels then declined between 10 and 17 days, and approached the low adult levels by 21 days of life. This closely parallels both the known changes in cardiac glycoside sensitivity of the rat heart, and the shortening of the cardiac action potential (Langer et al., 1975).

We have used immunoblot techniques to investigate the generality of the expression of two forms of the Na,K-ATPase in the hearts of several species that are popular subjects for investigations of cardiac physiology. The results are seen in Figure 2: The primate heart, like the canine heart, has roughly equal amounts of immunoreactivity for alpha+ and alpha. In marked contrast, the sheep, guinea pig, chicken, and rat hearts display only one band when stained with a mixture of the K3 and Ax2 antisera. We have previously shown that the Ax2 antiserum is capable of detecting the alpha+ Na,K-ATPase in brain preparations from

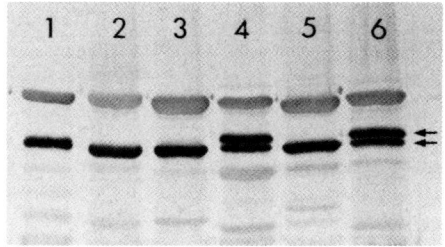

Figure 2. Immunodetection of alpha and alpha+ in crude particulate preparations from ventricles of several species. Alpha+: upper arrow, alpha: lower arrow. 1) rat; 2) chicken; 3) guinea pig; 4) dog; 5) sheep; 6) monkey (Macaca fascicularis).

sheep, guinea pig, and rat (Sweadner & Gilkeson, 1985), so the failure to detect an alpha+ in the heart of these species is not due to a lack of species crossreactivity of the antiserum. There is only one electrophoretic type of Na,K-ATPase catalytic subunit in the chicken brain (Sweadner & Gilkeson, 1985), as in the chicken heart. The small amount of alpha+ in the adult rat heart is always detected when stained with Ax2 alone (Figure 1), but that stain is often lost in the background when the blot is stained with K3 at the same time, as in Figure 2. The same may be true here for the other species, and small amounts of alpha+ may actually be present.

DISCUSSION

Low dose inotropic effects and high-affinity ouabain binding sites have been extensively described in preparations from the rat heart (Erdmann et al., 1980; Adams et al., 1982; Herzig & Mohr, 1984; Noel & Godfraind, 1984; Lelievre, Charlemagne et al., 1986). The rat heart has also been shown to change its cardiac glycoside sensitivity during postnatal maturation (Inturrisi & Papaconstantinou, 1974; Langer et al., 1975). Charlemagne et al. (1986) have made the intriguing observation that hypertrophied rat heart has more of a high-affinity, slowly-dissociating binding site than normal heart, and more closely resembles the neonatal heart in this respect.

Published reports on the presence or absence of alpha+ in the rat heart have been controversial. We found that there is alpha+ in rat ventricle preparations, but that it is a very minor component in adult heart, and makes up only about 20% of the Na,K-ATPase even in newborn heart. We routinely detect it when immunoblots are stained with Ax2 antiserum alone (Figure 1), but when an antiserum is used that stains the alpha isozyme well, the stain of alpha+ is frequently lost in the background (Figure 2). Charlemagne et al. (1986) failed to detect alpha+ by immunoblot techniques in either newborn or adult rat heart, while in the same paper demonstrating that newborn (and hypertrophied adult) heart expressed a higher-affinity ouabain receptor. Ng & Akera (1987) found no alpha+ in adult rat heart by active site phosphorylation, much as they found for guinea pig heart. These reports are approximately in agreement with our observation that alpha+ in adult rat heart is a very minor component, and that antisera raised against kidney Na,K-ATPase may fail to detect it.

Not all published observations are so consistent with our results, however. Mansier & Lelievre (1982) reported finding equal amounts of high and low-affinity ouabain binding sites in adult rat hearts that had been pretreated with a calcium chelator, and Charlemagne et al. (1987) reported that they could see roughly equal amounts of faster- and slower-migrating alpha bands in preparations from rat heart that had been reduced with dithiothreitol in 8M urea and alkylated with iodoacetamide. Reduction and alkylation also affected the electrophoretic mobility of both alpha and alpha+ in a rat brain preparation. How these observations relate to the conventionally-defined alpha and alpha+ is not clear. Lelievre, Maixent et al. (1986), reported that both high- and low-affinity sites showed reduced dissociation rates in membrane preparations from hypertrophied hearts, an observation that is not easily explained by the premise that changes in isozyme expression account for ouabain affinity differences.

Matsuda et al. (1984) were the first to detect two isozymes of Na,K-ATPase in canine heart, as phosphorylated intermediates which separated during electrophoresis in SDS. The slower (alpha+) form, like the alpha+ of the brain, was found to be more sensitive to preinhibition with pyrithiamin, a reagent they later showed to be reacting with sulfhydryl groups. More recently, Maixent et al. (1987) have reproduced the separation of isozymes from the canine heart, and have demonstrated that there are also two different ouabain binding sites. Their data does not, however, permit any conclusion about whether alpha or alpha+ is the high-affinity form in the dog, contrary to what they assume. In fact, the relative abundance of the bands on their Coomassie-stained gel suggests that the less abundant alpha+ form may correlate with the less abundant lower-affinity binding site. These authors concluded that only the high-affinity site is involved in inotropy, based on an analysis of the literature on cardiac glycoside effects in the canine heart.

Ng and Akera (1987) examined the ferret heart, and also report both the presence of two ouabain binding sites and two electrophoretically-separated catalytic subunits. In their case, preinhibition of the enzyme preparation with ouabain in Mg + Pi prior to phosphorylation with [γ-23P]ATP indicated that the alpha+ is the higher affinity form in this species, as it is in the rat brain (Sweadner, 1979). These authors found a biphasic inotropic response to ouabain. McDonough & Schmitt (1985) failed to detect any alpha+ in membranes from guinea pig heart. That this is a species difference was confirmed by Ng and Akera (1987) and again by our Figure 2.

SUMMARY

Differences in the expression of Na,K-ATPase isozymes in the heart make a significant contribution to the understanding of cardiac glycoside effects in this tissue. More explicit information is needed about the cardiac glycoside affinities of the different isozymes in various species, however, before a unified picture can emerge. Much more information is also needed on what controls the level of expression of the different isozymes, and why mammalian species differ in this respect. For the rat, at least, it appears that the relative abundance of two Na,K-ATPase isozymes is controlled by changes in gene expression during development. These changes correlate with large changes in other contractile and excitable membrane properties. Isoforms of myosin and other important cardiac proteins are known to change in hypertrophy. It may be of critical importance to understand how the isoforms of the cardiac glycoside receptor change as well.

ACKNOWLEDGMENTS

Supported by the NIH, HL36271, and by a Grant-in-Aid from the American Heart Association. KJS was an Established Investigator of the American Heart Association.

REFERENCES

Adams RJ, Schwartz A, Grupp G, Grupp I, Lee S-W, & Wallick ET (1982). High-affinity ouabain binding site and low-dose positive inotropic effect in rat myocardium. Nature 296: 167-169.

Charlemagne D, Maixent J-M, Preteseille M, & Lelievre LG (1986). Ouabain binding sites and (Na,K)-ATPase activity in rat cardiac hypertrophy. Expression of the neonatal forms. J Biol Chem 261: 185-189.

Charlemagne D, Mayoux E, Poyard M, Oliviero P, & Geering K (1987). Identification of two isoforms of the catalytic subunit of Na,K-ATPase in myocytes from adult rat heart. J Biol Chem 262: 8941-8943.

Erdmann E, Philipp G, & Scholz H (1980). Cardiac glycoside receptor, (Na + K)-ATPase activity and force of contraction in rat heart. Biochem Pharmacol 29: 3219-3229.

Herzig S & Mohr K (1984). Action of ouabain on rat heart: comparison with its effect on guinea-pig heart. Br J Pharmacol 82: 135-142.

Inturrisi CE & Papaconstantinou MC (1974). Ouabain sensitivity of the Na,K-ATPase from rat neonatal and human fetal and adult heart. Annals NY Acad Sci 242: 710-716.

Langer GA, Brady AJ, Tan ST, & Serena SD (1975). Correlation of the glycoside response, the force staircase, and the action potential configuration in the neonatal rat heart. Circ Res 36: 744-752.

Lelievre LG, Charlemagne D, Mouas C, & Swynghedauw B (1986). Respective involvements of high- and low-affinity digitalis receptors in the inotropic response of isolated rat heart to ouabain. Biochem Pharmacol 35: 3449-3455.

Lelievre LG, Maixent JM, Lorente P, Mouas C, Charlemagne D, & Swynghedauw B (1986). Prolonged responsiveness to ouabain in hypertrophied rat heart: physiological and biochemical evidence. Am J Physiol 250: H923-H931.

Mansier P, & Lelievre LG (1982). Ca-free perfusion of rat heart reveals a (Na + K)-ATPase form highly sensitive to ouabain. Nature 300: 535-537.
Matsuda T, Iwata H, & Cooper JR (1984). Specific inactivation of $\alpha(+)$ molecular form of (Na + K)-ATPase by pyrithiamin. J Biol Chem 259: 3858-3863.
Maixent JM, Charlemagne D, de la Chapelle B, & Lelievre LG (1987). Two Na,K-ATPase isoenzymes in canine cardiac myocytes. Molecular basis of inotropic and toxic effects of digitalis. J Biol Chem 262: 6842-6848.
McDonough A & Schmitt C (1985). Comparison of subunits of cardiac, brain, and kidney Na,K-ATPase. Am J Physiol 248: C247-C251.
Ng Y-C, & Akera T (1987). Two classes of ouabain binding sites in ferret heart and two forms of Na,K-ATPase. Am J Physiol 252: H1016-H1022.
Noel F & Godfraind T (1984). Heterogeneity of ouabain specific binding sites and (Na,K)-ATPase inhibition in microsomes from rat heart. Biochem Pharmacol 33: 47-54.
Sweadner KJ (1979). Two molecular forms of (Na,K)-stimulated ATPase in brain. Separation, and difference in affinity for strophanthidin. J Biol Chem 254: 6060-6067.
Sweadner KJ (1985). Enzymatic properties of separated isozymes of the Na,K-ATPase. Substrate affinities, kinetic cooperativity, and ion transport stoichiometry. J Biol Chem 260: 11,508-11,513.
Sweadner KJ, & Farshi SK. Rat cardiac ventricle has two Na,K-ATPases with different ouabain affinities. Developmental changes in immunologically different catalytic subunits. Proc Natl Acad Sci USA, in press.
Sweadner KJ & Gilkeson RG (1985). Two isozymes of the Na,K-ATPase have distinct antigenic determinants. J Biol Chem 260: 9016-9022.
Young RM & JB Lingrel (1987). Tissue distribution of mRNAs encoding the α isoforms and β subunit of rat Na,K-ATPase. Biochem Biophys Res Comm 145: 52-58.

THE EFFECT OF GROWTH IN MONENSIN OR LOW POTASSIUM ON INTERNAL SODIUM, ALPHA SUB-UNIT mRNA AND SODIUM PUMP DENSITY IN HUMAN CULTURED CELLS

Shiela Unkles, Trudi McDevitt, Carol Voy, James R. Kinghorn, Gordon Cramb and Joseph F. Lamb.

Department of Physiology and Pharmacology, and Molecular Genetics Unit, University of St.Andrews, Fife, U.K. KY16 9TS.

INTRODUCTION

It is now well established that chronic elevation of the intracellular sodium concentration ($[Na]_i$) results in an up-regulation of sodium pump density in many cell types (Algharably et al., 1985). More recently it has been shown that a down-regulation of sodium pump numbers also occurs if the $[Na]_i$ is reduced from normal values (Kim and Smith, 1986). This suggests that cells can continually adjust the density of sodium pumps on their surface to the prevailing mean concentration of sodium within them. Our current experiments are aimed to 1) examine the form of the relationship between $[Na]_i$ and pump numbers, 2) investigate whether the rate of pump activity was a factor in up-regulation and 3) determine if there was any relationship between $[Na]_i$ and [mRNA] for the pump.

METHODS

Initial experiments were carried out using the human cultured cell lines MRC5 (SV-40 transformed foetal lung fibroblasts) and HeLa (epithelial carcinoma). Cells were seeded at 4×10^4 cells/cm^2, either on 65 cm^2 plastic petri dishes for the extraction of total mRNA or in 1 cm^2 multiwell plates (Nunc) for the determination of ion contents or estimation of sodium pump numbers. Cells were grown for three days in normal growth medium and then for a further 0.5 to 24 hours in medium containing low potassium or monensin. Sodium pump numbers were determined by [^{3}H]-ouabain binding and pump activity assessed by measurement of ouabain-sensitive influx (using ^{86}Rb as the tracer) (Boardman et al., 1974). Intracellular sodium ($[Na]_i$) and potassium ($[K]_i$) were measured by flame photometry. Total RNA, isolated by a modification of the lithium precipitation method described by Cathala et al., 1983, was estimated by absorbance at 260/280 nm, serially diluted and either blotted directly onto nitrocellulose by vacuum filtration or components separated by Northern Blot analysis (Patel et al., 1981). Blots were probed with ^{32}P-labelled cDNA's for both actin and the sodium pump alpha subunit (donated by Prof. R. Levenson; see Emanuel et al., 1986) and then exposed to X-ray film. The autoradiographs were scanned by densitometer and densities of the spots recorded. The extent of hybridisation of probes to blotted RNA samples isolated from control cells and cells grown in the modified medias was compared by analysis of the relative autoradiograph densities.

RESULTS

Both decreasing the extracellular potassium concentration ($[K]_o$) or the addition of the sodium/proton antiporter monensin to the growth medium resulted in a dose dependent increase in sodium pump density (fig. 1). The experimental data was fitted by Michaelis Menten kinetics yielding a K_m of around 28 nmol/10^6 cells (corresponding to an $[Na]_i$ of 12 mM) and a V_{max} of 1.3×10^6 sites/cell.

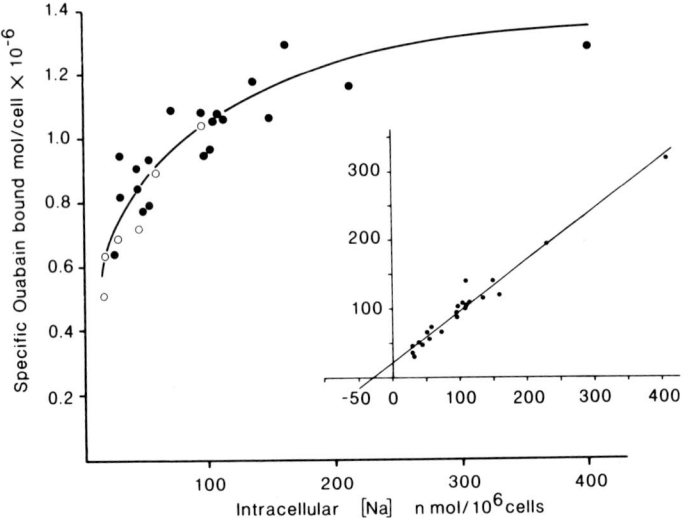

Figure 1. Relationship between $[Na]_i$ and sodium pump number in HeLa cells. $[Na]_i$ was altered by growth in low potassium (○;0.25-5.4 mM) or monensin (●;1-10 uM) for 24 hours. Inset shows a s/v vs. S plot of data.

Growth of cells in monensin or low potassium both increase $[Na]_i$. However, unlike growth in low potassium medium (Algharably et al., 1985), monensin (concentrations up to 10^{-5} M) does not result in any significant reduction in $[K]_i$ (fig. 2).

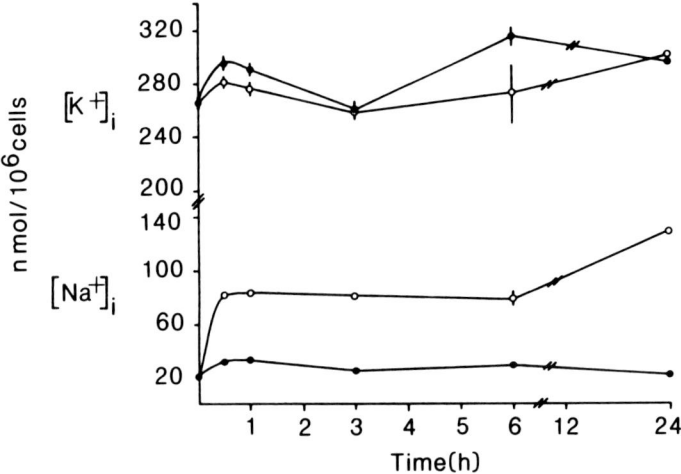

Figure 2. Time course of the effect of monensin (10^{-5}M) on $[Na]_i$ and $[K]_i$ in HeLa cells (○). Control values (●).

Northern Blot analysis of total RNA isolated from cultured cells indicates the presence of only one mRNA species hybridising to the cDNA probe for the alpha subunit protein (fig. 3). A linear relationship was found between pump alpha subunit mRNA and monensin concentration up to 10^{-5}M (fig. 4a). Likewise blots of total RNA isolated from cells grown in low $[K]_o$ or partial sodium replacement with lithium (40 mM) also showed increases in hybridisation with the cDNA probes for both actin and the sodium pump alpha subunit (fig. 4b). The increase in sodium pump mRNA but not that of actin mRNA appeared to be sensitive to changes in the serum concentration of the growth medium with low serum concentrations potentiating the increase in sodium pump mRNA response to low $[K]_i$ (fig. 4b).

CONCLUSIONS

The relationship between $[Na]_i$ and sodium pump density can be described by Michaelis Menten type kinetics, with a V_{max} occurring at $[Na]_i$ in excess of 150 nmol/10^6cells ($[Na]_i$ of 50-60 mM) and a K_m around 30 nmol/10^6cells (approximately equal to normal $[Na]_i$ around 10-15 mM). In normal cells we calculate that a change of 1mM in $[Na]_i$ alters pump density by 3%. Although growth in monensin or low potassium had similar actions on $[Na_i]$ and pump numbers the former at concentrations up to 10^{-5}M had very little effect on $[K]_i$ whereas the latter is known to reduce $[K]_i$ by up to 30% (Algharably et al., 1985). In the presence of monensin,

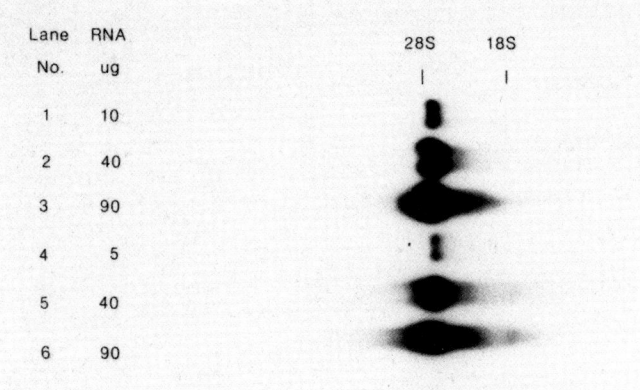

Figure 3. Northern Blot of total RNA probed with ^{32}P-labelled sodium pump alpha subunit cDNA. RNA (5-90 ug) was extracted from HeLa cells grown for four days in normal medium and electrophoresed in formaldehyde/agarose gels at 25 volts for 12 hours and blotted onto nitrocellulose. Filters were probed with ^{32}P-labelled cDNA for the alpha subunit of the pump.

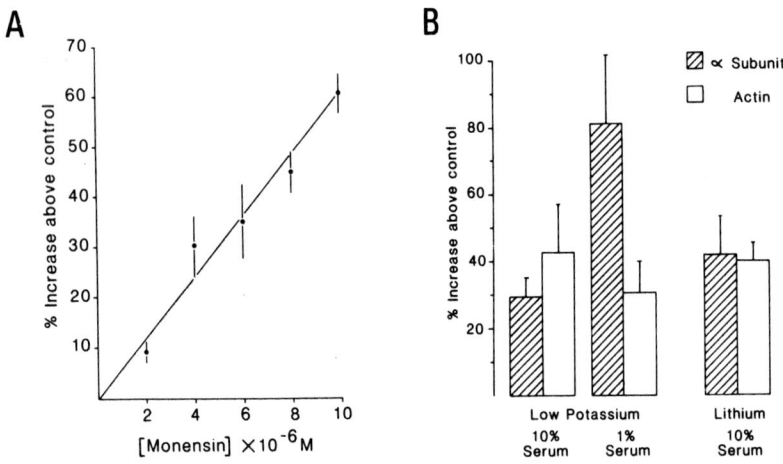

Figure 4. a) effect of growth for 24 hours in various monensin concentrations on sodium pump alpha subunit mRNA levels in HeLa cells. b) effect of growth in lithium or low potassium on mRNA levels for sodium pump alpha subunit and actin. HeLa cells were grown for 24 hours in the presence of lithium (40 mM) or low potassium (0.5 mM) in normal (10%) or reduced (1%) serum concentrations. Messenger RNA levels are measured as increases over mRNA extracted from control cells.

sodium pump activity would be maintained at an artificially high level as a result of the increased sodium leak whereas growth in low $[K]_o$ reduces pump activity and hence decreases $[K]_i$. This suggests that the signal for an increase in pump numbers is the raised $[Na]_i$ and not low $[K]_i$ or a reduction in enzyme activity.

Growth of cells in the presence of monensin or in low potassium increased the concentration of mRNA for the alpha subunit of the sodium pump by 20-30%. In the case of monensin there was a linear increase in mRNA over the concentration range used. This is in contrast to the increase in pump numbers which appear to saturate at the higher concentrations of the antiporter. Clearly, increasing the concentration of sodium pump alpha subunit mRNA above a certain level does not necessarily reflect a similar increase in expression of the functional protein.

Raised $[Na]_i$ or lithium increase the mRNA concentrations for both actin and the alpha subunit of the pump to similar extents. Both therefore may have a mitogenic effect on the cells, a finding also described by Bowen & McDonough (1987) and by Pressley et al., at this meeting. However when $[Na]_i$ is raised in cells grown in low serum concentrations (1%), the increase in mRNA for the alpha subunit is specifically potentiated (fig. 4b). This reinforces our earlier observations that increasing the serum concentration of the growth medium attenuates the intracellular sodium-mediated up-regulation (Aiton and Lamb, 1984). These results indicate that complex interactions exist between stimulatory and inhibitory factors in serum and the intracellular sodium concentration in the transcriptional control of alpha subunit sodium pump message in these cell lines. A further interesting point which needs exploring is whether inhibitory plasma factors are able to modulate the transcription of mRNA for the pump proteins, in addition to any direct effects they may have on the activity of pumps once they are in the membrane.

ACKNOWLEDGEMENT

The authors would like to thank the British Heart Foundation for financial support.

REFERENCES

Aiton JF, Lamb, JF (1984). Effect of the serum concentration of the growth medium on the sodium pump site density of cultured HeLa cells. Q J exp Physiol 69: 97-115.
Algharably N, Lamb JF, Ogden P, Owler D, Tenang, EM (1985). The turnover of sodium pumps in Human cultured cells, and its significance for the actions of cardiac glycosides. in The Sodium Pump (ed I.M. Glynn and J.C. Ellory) Company of Biologists 197-207.
Boardman LJ, Huett M, Lamb JF, Newton JP, Polson JM (1974). Evidence for the genetic control of the sodium pump density in HeLa cells. J Physiol 241: 771-794.

Bowen JW, McDonough A (1987). Pretranslational regulation of Na-K-ATPase in cultured canine kidney cells by low K$^+$. Am J Physiol 252: c179-c189.

Cathala G, Savouret JF, Mendez B, Best BL, Karin M, Martial JA, Baxter JD (1983). A method for isolation of intact, translationally active ribonucleic acid. DNA 2: 329-335.

Emanuel JR, Garetz S, Schneider J, Ash J, Benz EJ Jnr, Levenson R (1986). Amplification of DNA sequences coding for the Na,K-ATPase alpha subunit in ouabain-resistant C$^+$ cells. Mol Cell Biol 6: 2476-2481.

Kim D, Smith TW (1986). Effect of growth in low-Na medium on transport sites in cultured cells. Am J Physiol 250: C32-C39.

Patel VB, Schniezer M, Dykstra CC, Kushner SR, Giles NM (1981). Genetic organisation of the qu cluster. Proc Natl Acad Sci USA 78: 5783-5787.

Pressley TA, Ismail-Beigi F, Gick GG, Edelman IS (1987). Differences in the relative abundance of messenger RNAs encoding the alpha and beta subunits of Na,K-ATPase in a rat liver cell line. Poster No 34. this meeting.

EFFECTS OF TUNICAMYCIN ON THE CELLULAR EXPRESSION AND STRUCTURAL ARRANGEMENT OF Na K-ATPase.

D. Zamofing, B.C. Rossier and K. Geering

Institut de Pharmacologie, Université de Lausanne,
Rue du Bugnon 27,
CH-1005 Lausanne, Switzerland

INTRODUCTION

The minimal functional enzyme unit of Na,K-ATPase consists of an α/β-subunit protomer. The catalytic properties can all be assigned to the α-subunit (for review, see Jørgensen, 1986) while the precise role for the β-subunit has not yet been established. We have examined in this study whether the β-subunit assists the newly synthesized α-subunit to adopt a correct and stable membrane arrangement.

The biosynthetic pathway of the two subunits has been studied in cell free systems and in the intact cell. Recently, we have shown by in vitro translation of mRNA fractions coding for α- or β-subunits respectively, that membrane insertion of both subunits occurs cotranslationally but independently of each other and is accompanied by coreglycosylation of the β-subunit alone (Geering et al, 1985). In the intact cell, it was shown that α- and β-subunits assemble soon after synthesis (Tamkun and Fambrough, 1986) and that the β-subunit undergoes complex-type glycosylation during its intracellular transport (Geering et al, 1985). In addition, the α-subunit is subjected to a structural rearrangement (Geering et al, in press) which appears to mark the functional maturation of the enzyme, namely its ability to perform cation-dependent conformational changes (Geering et al, in press) or to bind ouabain in an ATP-dependent fashion (Caplan et al, 1985).

In view of these findings, the question arises whether the association to a structurally intact β-subunit might be important for the observed maturation process of the α-subunit. In this study, we have impaired the structural integrity of the β-subunit by inhibiting its coreglycosylation with tunicamycin (Elbein, 1984). In cultured amphibian cells labelled for short periods, we assessed the effects of such treatment on the expression and the structural arrangement of both the β- and the α-subunits in an early post-synthetic cellular transport compartment. As an experimental approach to study posttranslational processing of the two subunits we used controlled trypsinolysis of cell homogenates.

MATERIAL AND METHODS

TBM cells (derived from the urinary bladder of the toad Bufo marinus) have been obtained from J.S. Handler, National Institutes of Health, Bethesda, MD, and were cultivated on petri dishes at confluency as described by Handler (1983). Cells were pretreated or not with tunicamycin (TM, Sigma) at 5 µg/ml for 18 h before labeling with L-^{35}S-methionine (Amersham, specific activity > 1000 µCi/mmol) for 15 min. with 400 µCi/ml or for 30 min. with 250 µCi/ml at 28°C. Cells were scraped, washed, resuspended in homogenization buffer (30 mM DL-histidine 5mM EDTA, 18 mM Tris/HCl pH 7.4, 0.1 mg DNase type I, Sigma) and sonicated on ice 3 times for 3 seconds using a Branson sonifier (position 4). Controlled trypsinolysis was performed for 1 h on ice in the presence of 140 mM K-acetate, 0.1% deoxycholate and at a trypsin (Sigma type XI) to protein ratio as indicated in Fig. 1. Trypsinolysis was stopped with a five fold excess (w/w) of soybean trypsin inhibitor (Sigma) for 10 min. on ice. Samples were adjusted to 3.7% sodium dodecyl sulphate (SDS) and heated for 5 min. at 95°C before protein content (Lowry et al, 1951) and radioactivity were measured and immunoprecipitation with antisera against α- and β-subunits, 30 KD protein and citrate synthase was performed as described (Geering et al, 1985). Quantitation of the immunoprecipitated material was performed by laser densitometry (α-subunit) or by light scattering of eluted silver grains (β-subunit) (Suisa, 1983) of the bands revealed on fluorograms.

RESULTS AND DISCUSSION

To assure that tunicamycin did not have non-specific toxic effects on the cell metabolism, we tested whether pretreatment of TBM cells with this drug had any influence on the protein synthesis as assessed by a 15 min labeling with ^{35}S-methionine.

Table 1. Effect of tunicamycin (18 hours, 5 μg/ml)) on the synthesis (15 min. pulse with ^{35}S-methionine) of total proteins, and of Na,K-ATPase subunits, citrate synthase and a 30 KD membrane protein. Cell homogenization, immunoprecipitation and quantitation were as described in Material and Methods.

	Effects of tunicamycin (% of control mean ± SE)	
Total synthesis		
^{35}S-methionine incorporated/ mg protein	95 ± 10	n = 3
Immunoprecipitated antigens		
citrate synthase	100	n = 1
30 KD protein	93 ± 6	n = 2
α-subunit	*30 ± 8	n = 3
β-subunit	*23 ± 4	n = 3

* $p < 0.05$ compared to control, t test.

As can be seen in Table 1, exposure of cells to 5 μg/mg of tunicamycin for 18 hours had no effect on the incorporation of ^{35}S-methionine into total proteins or into two specific immunoprecipitated proteins, a membrane bound non-glycosylated 30 KD protein (Geering and Rossier, 1979) and the mitochondrial citrate synthase. These data indicate that under our experimental conditions, tunicamycin treatment did not affect the protein synthesis machinery in a non-specific way.

On the other hand, tunicamycin efficiently inhibited coreglycosylation of the newly synthesized β-subunit resulting in the production of a 32 KD β-peptide (Fig 1, lanes 1 and 3).

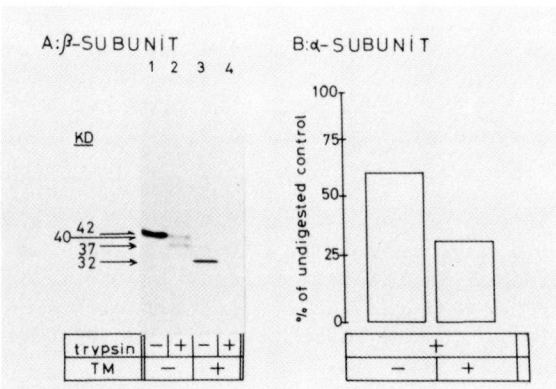

Figure 1. Effect of tunicamycin (TM) on the structural arrangement of newly synthesized Na,K-ATPase. A) Effect of TM on glycosylation and trypsin-sensitivity of the β-subunit. Cell homogenates from TBM cells incubated without (lanes 1 + 2) or with TM at 5 µg/ml (lanes 3 + 4) for 18 hours and labelled for 15 min. were digested with trypsin at a trypsin to protein ratio of 0.2 (lanes 2 + 4)) and immunoprecipitated with anti β-serum as described in Material and Methods. B) Effect of TM on trypsin sensitivity of the α-subunit. Cell homogenates from TBM cells incubated without or with TM for 18 hours and labelled for 30 minutes with ^{35}S-methionine were digested with trypsin at a trypsin to protein ratio of 0.06 and immunoprecipitated with anti α-serum as described in Material and Methods. Results are expressed as resistant α-subunit material remaining after trypsinolysis (% of undigested control)

In addition, the amount of β-subunit recovered by immunoprecipitation from tunicamycin-treated cells was reduced by about 75% compared to non-treated controls (Table 1). This result suggests either that the newly synthesized, non-glycosylated β-subunit is very unstable and rapidly degraded or that a regulatory link exists between coreglycosylation and β-subunit synthesis as has been suggested for other glycoproteins (for review, see Elbein, 1984). Interestingly, tunicamycin also decreased the amount of newly synthesized α-subunit to a similar extent as the β-subunit (Table 1). Since non-glycosylated membrane proteins other than the α-subunit (e.g. the 30 KD protein) were not affected by tunicamycin

(Table 1) and since it is likely, that the fully denatured α- and β-subunits from tunicamycin-treated and control cells are equally well recognized by our antisera, these data suggest that the decreased α-subunit expression in linked to inhibition of glycosylation and decreased expression of a glycoprotein. Therefore, we might conclude that, in the intact cell, a glycoprotein, perhaps even the β-subunit itself is responsible for an efficient and coordinate expression of newly synthesized Na,K-ATPase subunits.

A second issue concerns the role of the core sugars of the β-subunit in the structural arrangement of the Na,K-ATPase in an early posttranslational cellular compartment. To assess this question, we compared the trypsin-sensitivity of α- and β-subunits from tunicamycin-treated and control cells which we labelled for 15 or 30 min. with ^{35}S-methionine. As can be seen in Fig 1A, (lanes 1 and 2), the coreglycosylated β-subunit was partially trypsin-resistant and gave rise to two tryptic fragments. The 40 KD fragment was identical to the one obtained by trypsinolysis from the β-subunit synthesized in vitro in the presence of rough microsomes (Geering et al, 1985). Thus, this fragment is most likely produced by cleavage of the short N-terminus of the polypeptide which protrudes on the cytoplasmic side of the membrane (Shull et al, 1986). The 37 KD fragment, on the other hand, probably results from cleavage at a tryptic site in the large extracytoplasmic domains of the polypeptide. In contrast to the coreglycosylated β-subunit, the non-glycosylated β-subunit produced in the presence of tunicamycin was completely digested during controlled trypsinolysis of cell homogenates (Fig. 1, lane 3 and 4). These data suggest that coreglycosylation is associated with a structural reorganization of the newly synthesized β-chain which results in protection of most of its tryptic sites.

Tamkun and Fambrough (1986) have shown that the association of α- and β-subunits which occurs rapidly after polypeptide synthesis is not influenced by inhibition of coreglycosylation. We have tested whether the α-subunit assembled with the non-glycosylated and thus structurally perturbed β-subunit is more sensitive to trypsinolysis than when it is associated with the coreglycosylated β-subunit.

Fig. 1B illustrates that this is indeed the case. Both after a 15 min (data not shown) or a 30 min. pulse (Fig. 1B), less trypsin-resistant α-subunit related material was recovered from tunicamycin-treated than from control cells. Although structurally perturbed, the α-subunit associated to non-glycosylated β-subunit seems to reach the plasma membrane (Tamkun and Fambrough, 1986). We are presently investigating whether this non-glycosylated enzyme unit becomes functionally active.

In summary, the data of this study suggest that the expression of α- and β- subunits at the level of the endoplasmic reticulum in the intact cell is strictly synchronized. The coordinating factor could be the β-subunit itself, or else, another yet unidentified glycoprotein. In addition, acquisition of core sugars by the β-subunit is important for an efficient translation and/or the stabilization of the β-subunit as well as for the correct membrane arrangement of the newly synthesized α-subunit.

REFERENCES

Caplan MJ, Palade GE, Jamieson JD (1985). Cell surface expression and activation of newly synthesized Na,K-ATPase in MDCK cells. In Glynn I, Ellory C (eds): "The Sodium Pump", Cambridge: The Company of Biologists Ltd, pp 147-151.

Elbein AD (1984). Inhibitors of the biosynthesis and processing of N-linked oligosaccharides. CRC Crit Rev Biochem 16:21-49.

Geering K, Rossier BC (1979). Purification and characterization of ($Na^+ + K^+$)-ATPase from toad kidney. Biochim Biophys Acta 566:157-170.

Geering K, Meyer DI, Paccolat MP, Kraehenbühl JP, Rossier BC (1985). Membrane insertion of α- and β-subunits of Na^+,K^+-ATPase. J Biol Chem 260:5154-5160.

Handler JS (1983). Use of cultured epithelia to study transport and its regulation. J Exp Biol 106:55-69.

Jørgensen PL (1986). Structure, function and regulation of Na,K-ATPase in the kidney. Kidney Int 29:10-20.

Shull GE, Lane LK, Lingrel JB (1986). Amino-acid sequence of the β-subunit of the ($Na^+ + K^+$)ATPase deduced from a dDNA. Nature 321:429-431.

Tamkun MM, Fambrough DM (1986). The ($Na^+ + K^+$)-ATPase of chick sensory neurons: Studies on biosynthesis and intracellular transport. J Biol Chem 261:1009-1019.

Secondary Active Transport, Na^+, K^+-Homeostasis, and Kidney Function

Overview:
PHYSIOLOGICAL ROLE OF THE NA-K PUMP

Claude Lechene, M.D.

Laboratory of Cellular Physiology and
National Electron Probe Resource for Analysis
of Cells, Harvard Medical School and Brigham
& Women's Hospital, Boston, MA 02115

Approximately 30 years ago Jens Chr. Skou discovered an ATPase that was Na and K dependent. He described its main biochemical characteristics in two seminal papers (Skou, 1957 and 1960). Skou immediately surmised that the enzyme might be an essential part of an Na-K pump responsible for maintaining the high potassium and low sodium intracellular concentrations. Reams of information have since been accumulated demonstrating that the (Na,K)-ATPase is a transmembrane protein biochemical substrate of the Na-K pump.

The essential importance of the physiological role of the Na-K pump transcends any particular domain of physiology. The Na-K pump is a universal energy transducer allowing eukaryotic cells to perform any of their cellular functions. The central role of (Na,K)-ATPase in cellular life, however, is not yet fully appreciated as illustrated by the following quote from a recent textbook (Darnell et al., 1986):

> "The Km, or binding constant for the binding of Na+ to the ATPase is 0.2 mM. This value is well below the intracellular Na+ concentration, so Na+ ions are pumped out of the cell at the maximum rate."

that of course is wrong both factually and conceptually. The Km for sodium is 20-40 mM and the pump in normal cells is far from pumping at its maximum rate.

Any normal living mammalian cell contains

approximately the same high concentration of potassium and low concentration of sodium. These cells, however, may greatly vary in Na-K pump rates. For example, the initial rate of sodium leak after Na-K pump inhibition, an estimate of the steady-state rate of sodium pumping, varies by two orders of magnitude from human red blood cells to rat renal proximal tubular cells (Fig. 1).

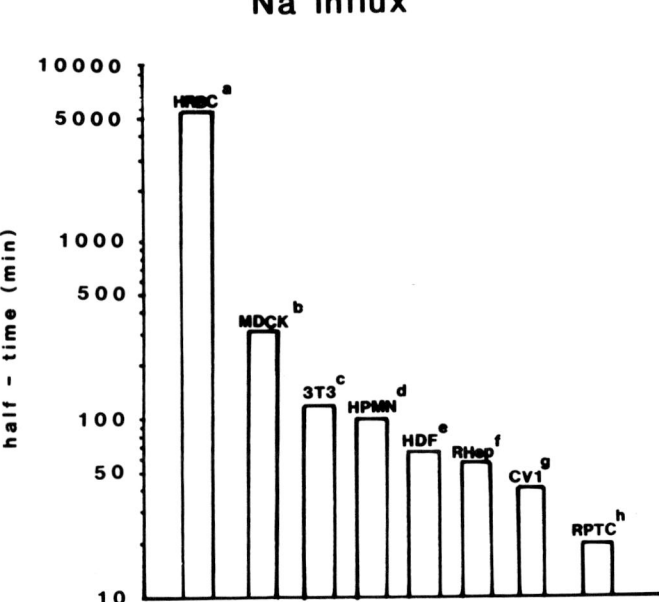

Fig. 1: a: Human red blood cells. From Garrahan and Glynn (1967); b: Canine kidney cell line. From Bolivar et al. (1987) and V. Pena-Cruz and C. Lechene, unpublished results; c: NIH 3T3 fibroblasts. From Chang-Sing and Lechene (1983); d: Human polymorphous nuclear neutrophiles. From Simchowitz et al. (1982); e: Human diploid fibroblasts. From Abraham et al. (1985); f: 2-4 days primary culture of rat hepatocytes. From B.J. Cohen and C. Lechene, (1988); g: Green monkey kidney cell line. From Epstein and Lechene (in press); h: 3-day primary culture of rat renal proximal tubular cells. From S. Larsson et al. (1986).

One can wonder why all cells have approximately the same high potassium and low sodium concentration, why these concentration are so different from the surrounding extracellular fluid and why, in spite of equivalent intracellular composition the rate of leaking and of pumping of Na and K may vary by two orders of magnitude among cell types.

Besides ionic channels, many co- and counter-transport systems have been found in the plasma membrane. They use the potential energy of the ionic chemical gradients created by the Na-K pump to move inside the cell phosphate, amino acid, glucose, to remove from the cell protons, calcium, bicarbonate, and to cotransport K and Cl with Na (Aronson, 1985; Baker, 1986; DiPolo and Beauge, 1987; Ganapathy and Leibach, 1986; Grinstein and Rothstein, 1986; Hamilton and Nilsen, 1978;Hoffman, 1986; O'Grady et al., 1987; Quamme and Shapiro, 1987; Roomans et al., 1977;Schultz et al., 1985; Wright, 1984). The dissipation through conductive pathways, symporters and antiporters of the potential energy of the ionic gradients created by the Na-K pump, allows cells to perform their general and their terminally differentiated functions and is at the core of all organs activity: nerve impulse, muscle contraction, intestinal absorption, plexus choroid secretion, tear formation, inner ear endolymph formation, tear secretion, gland secretion, urine elaboration, transport of nutrients, (phosphate, amino acids, glucose), regulation of intracellular pH, and regulation of cellular volume.

A common mechanism allows eukaryotic cells to avoid exhausting the ionic gradients driving their activities and to maintain at appropriate levels their potential energy: the activity of the Na-K pump (Fig. 2). Thus, among cell types, more metabolically active cells will leak more, and therefore will have a higher Na-K pumping rate. Within a cell type, leak and pumping activity will vary in parallel with the need for the cell to modulate its exchanges with the outside world.

The Na-K pump rate, which varies by two orders of magnitude among cell types, may vary by 3-4 fold within a cell type. Variation of Na-K pump activity may be immediate, in response to changes in substrate, mainly intracellular sodium concentration; short term, in response to extracellular (i.e. insulin) or intracellular (i.e. dopamine ?) factors; long term, with the synthesis of new pumps.

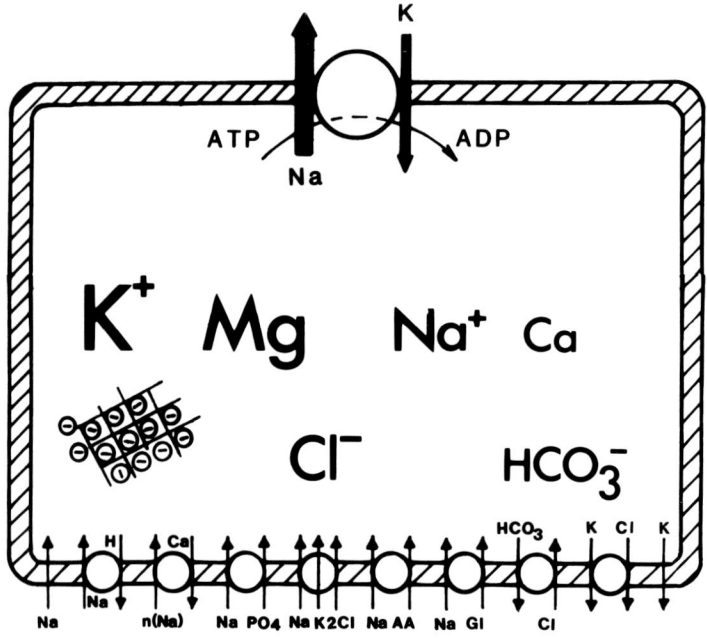

Fig. 2: One may figure cells performing their functions through the dissipation of ionic gradients using a variety of specialized leak pathways, and the Na-K pump as a general reactor that creates and maintains the potential energy of the cells and adjusts its activity in response to modulation of the intensity of the leaks.

The Na gradient is driving a majority of co- and counter-transporters. As we will see, the Na-K pump activity appears to balance exquisitely changes in Na influx. If sodium influx increases, the intracellular sodium concentration immediately increases, which in turn immediately accelerates the Na-K pump rate so that the cell will reach a new steady-state. If sodium influx returns to normal, intracellular sodium concentration and pump rate will also decrease. If the increase in sodium influx is sustained, a new type of regulation occurs: the synthesis of new pumps. The synthesis of new pumps allows the intracellular sodium concentration to return to normal; an increased sodium influx is then balanced by an increased number of pumps.

The Na-K pump rate is measured in our laboratory using electron probe analysis of a variety of cell types in culture. For measuring the maximum pump rate, cells are pre-incubated in a medium lacking potassium until they have exchanged normal intracellular content of potassium with sodium. At the time of the experiment, the Na-K cellular pump is activated by switching the cells in a medium containing 5 mM potassium. The increase in intracellular potassium and the decrease in intracellular sodium content are simultaneously measured as a function of time and the ouabain-sensitive initial rates of sodium efflux and of potassium influx are taken as a measure of the maximum Na-K pump activity. For measuring the steady-state Na-K pump rate at low, normal intracellular sodium, cells are pre-incubated in a medium in which potassium is replaced by rubidium so that all intracellular potassium is exchanged for rubidium. At the time of the experiment, the cells are switched from the rubidium-containing medium to a potassium-containing medium. The ouabain-sensitive initial rate of potassium influx is taken as a measure of the steady-state potassium pump rate. We find that the Na-K pump rate increased by a factor of approximately 3 between normal (low) and high intracellular sodium (Harris et al., 1986). Under high intracellular Na conditions, the coupling ratio of sodium efflux to potassium influx was not different from 1.5 among a variety of cells, over a range of pump rates. Surprisingly, however, when the ionic coupling ratio was estimated at low intracellular sodium, in normal steady-state conditions of cells, the values found were not significantly different from 1 in a variety of cell types (Table 1). This is in agreement with the findings of Gill and Solomon (1959) and Cereijido et al. (1981) but is at odds with the creed (Glynn, 1984).

COUPLING RATIO Na/K
(mean ± SE) (# exp)

	Low, normal Na_i	High Na_i
Fibroblasts	0.95 ± 0.17 (6) i, a	1.52 ± 0.10 (13) iii, c
	0.85 ± 0.13 (7) ii, a	
RPTC #	1.08 ± 0.04 (15) a	1.85 ± 0.26 (5) •, e
		1.47 ± 0.13 (5) ••, e
CV1 ##	0.95 ± 0.21 (3) f	1.51 ± 0.05 (8) f
MDCK	1.11 ± 0.14 (3) g	
Hepatocytes &	0.97 ± 0.11 (12) h	

Table 1: #: 2-4 days primary cultures of rat renal proximal tubular cells; ##: green monkey kidney cells; &: 2-4 days primary cultures of rat hepatocytes; i: NIH 3T3; ii: NIH 3T3 (other strain); iii: human diploid fibroblasts; •: immature kidneys (12-15 day-old rats); ••: mature kidneys (40 day-old rats);

a: NIH 3T3 fibroblasts - R. Bassin and C. Lechene, unpublished data; b: NIH 3T3 fibroblasts - G. Boseck and C. Lechene, unpublished data; c: E.H. Abraham et al. (1985); d: R.C. Harris et al. (1986); e: S. Larsson et al. (1988); f: J.A. Epstein and C. Lechene (in press); g: from Bolivar et al. (1987); h: B. Cohen and C. Lechene, unpublished data

RAPID VARIATION OF NA-K PUMP RATE

Na-K pump rate can vary very rapidly in conditions increasing sodium influx. This is illustrated in short-term primary culture of renal proximal tubular cells (Harris et al., 1986). These cells have the highest effective permeability to sodium that we have measured among a variety of cell types. The main pathways for sodium entry is through a Na-H exchanger which makes up approximately 80% of the total sodium entry flux. In control conditions amiloride inhibits the Na-H exchanger; the cell acidifies, as measured by microfluorescence, and intracellular sodium content decreases, as measured by electron probe analysis. After the cells are acidified, they recover from acidification through an increased activity of the Na-H exchanger. Recovery is inhibited in a medium lacking sodium or with sodium but in the presence of amiloride. Activation of the Na-H exchanger is immediate after acidification and results in an immediate increase in intracellular sodium content (Fig. 3).

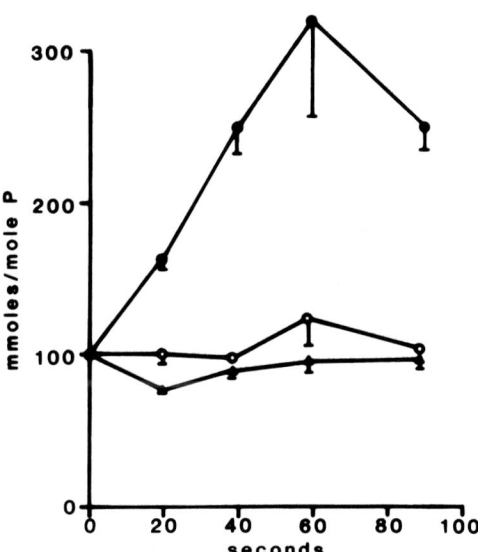

Cell Na after an acid load

Fig. 3: 3 days primary cultures of rat renal proximal tubular cells were preincubated for 20 min with 15 mM NH4Cl. At time 0 cells were placed in cultured media without NH4Cl and without (•) or with (o) 0.5 mM amiloride. Control cells (▲) were not preincubated with NH4Cl (mean ± SE, n=3) (from Harris et al., 1986).

Immediately after acidification, the pump rate was significantly decreased for about 20-40 seconds but then it increased and more than doubled relative to its control value within 2 minutes. So that, in Fig. 4., after intracellular acidification the following sequence of events occurred: immediate increase of intracellular sodium concentration followed, after 40-60 seconds, by an increase in Na-K pumping rate.

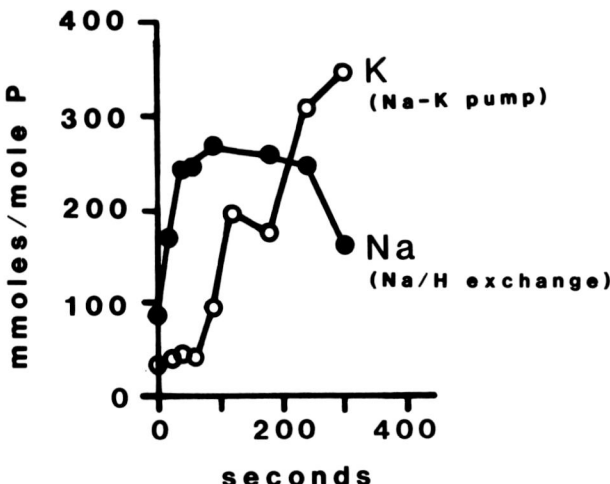

Fig. 4: Primary cultures of rat renal proximal tubular cells were preincubated for 180 min in culture medium with KCl replaced by 5 mM RbCl, and 20 minutes with 15 mM NH4Cl. AT time 0 cells were placed in culture media without NH4Cl, without RbCl and with 5 mM KCl (from Harris et al., 1986).

The immediate activation of the Na-H exchanger repairs the acid load and results in an increase in intracellular sodium concentration. The increase in Na-K pump activity in turn repairs the increase in intracellular sodium concentration. The pump rate inhibition observed immediately after acidification was likely due to an inhibition of the Na-K pump by the decreased intracellular pH.

By loading proximal tubular cells with varied contents of sodium, (the cells are first pre-incubated in rubidium-containing medium and lacking potassium, and then sodium-loaded by pre-incubating for varied amount of time in medium lacking both potassium and rubidium), one can measure the ouabain-sensitive potassium influx at different intracellular sodium concentration. This provides an estimate of the apparent Km of the pump for intracellular sodium, approximately 40 mM in primary cultures of Wistar rat renal proximal tubular cells (M. Crabos and C. Lechene, unpublished data).

SHORT TERM REGULATION OF THE PUMP

That intracellular factors may regulate the activity of the Na-K pump has not yet been demonstrated in the realm of cellular physiological functions. The recent observation that dopamine inhibits (Na,K)-ATPase activity in microdissected proximal renal tubules (Bertorello and Aperia, 1987a) may have an important physiological relevancy. Renal proximal tubular cells contain amino acid decarboxylase (AADC), so that L-dopa reabsorbed across proximal tubular cells may be transformed to dopamine inside the cells. We confirmed, in suspensions of proximal tubules, that dopamine inhibited (Na,K)-ATPase activity (Crabos et al., 1988). In short-term primary culture of renal proximal tubular cells, however, we were unable to observe any effect of L-dopa on Na-K pumping activity measured at different intracellular sodium concentrations.

We attempted to increase L-dopa uptake by incubating the cells in medium lacking amino acids that may be competing for Na cotransport. In order to attempt to increase the intracellular AADC, we cultured the cells with small doses of L-dopa or we maintained some rats on a high salt diet. In none of these conditions did we find any effect of L-dopa on the pump rate although it was increasing with intracellular sodium concentration (Fig. 5). Thus, in cultured cells the dopamine effect was not observed because either L-dopa did not enter the cells, or was not transformed in dopamine. Conversely, in tubular suspensions, or in microdissected tubules, the inhibition of (Na,K)-ATPase observed with dopamine may be due to a chemical oxidative effect of dopamine.

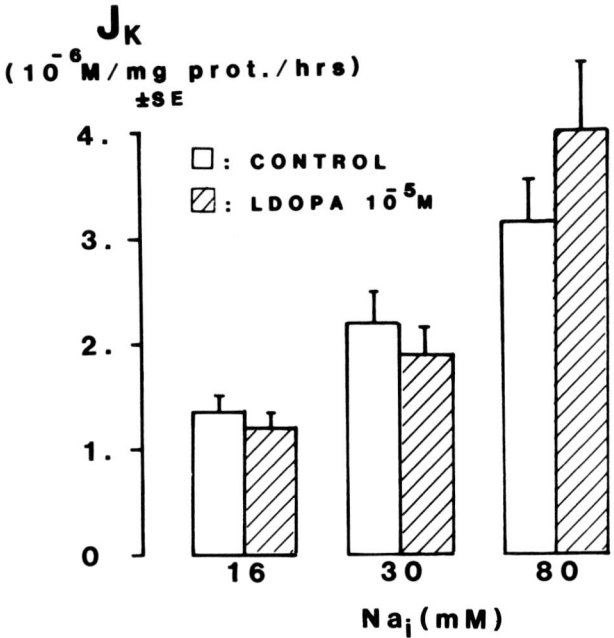

Fig. 5: Initial rate of ouabain sensitive K influx in rat renal proximal tubular cells. Cells were preincubated in medium containing 5 mM Rb instead of K for 4 hrs and then in medium lacking both K and Rb for variable times so that intracellular Na content could be varied. At time 0 cells were returned in medium containing 5 mM K. Na_i: intracellular Na content. J_K: initial rate of ouabain sensitive K influx (from Crabos et al., 1988).

The dopamine inhibition of (Na,K)-ATPase activity observed in microdissected proximal tubules may be mediated by a G-protein system. G-proteins may also be involved in modifying (Na,K)-ATPase activity in the nephron thick ascending limb (Bertorello and Aperia, 1987b)

ONTOGENIC AND HORMONAL REGULATION

The Na-K pump activity varies both during ontogenic development and with duration of culture. The Na-K pumping activity varies in parallel to changes in the magnitude of the sodium and potassium leak. In proximal tubular cells obtained from young rats before terminal differentiation, the rates of Na and K leaks do not change with time in culture. On the contrary, cells obtained from adult rats, after terminal differentiation, start with a leak twice that of immature cells and decrease after four days of culture to a leak rate equivalent to that of immature cells (Fig. 6).

Effects of duration in culture on K and Na leak pathways of cells from mature (adult) and immature (young) rat kidneys

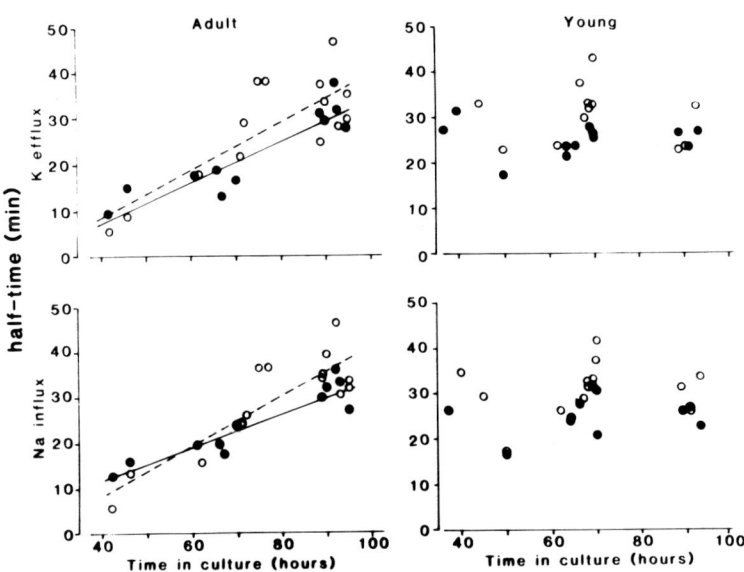

Fig. 6: Half time of K efflux and Na influx in adult and young (immature kidney) rat renal proximal tubular cells. Cells were cultured for 2 to 4 days, and fluxes were measured after exposing the cells to 1 mM ouabain for variable period of time. •: cells continuously grown with 10% fetal bovine serum. o: cells serum deprived for 24 hrs before an experiment (from Larsson et al., 1986).

The magnitude of Na and K leak was studied as a function of the degree of terminal differentiation, by preparing renal cells obtained from 10 to 15 day old rats, during terminal differentiation, and by culturing the cells for a fixed amount of time (3 days). Na and K leaks increased with increasing degree of maturation (Fig. 7).

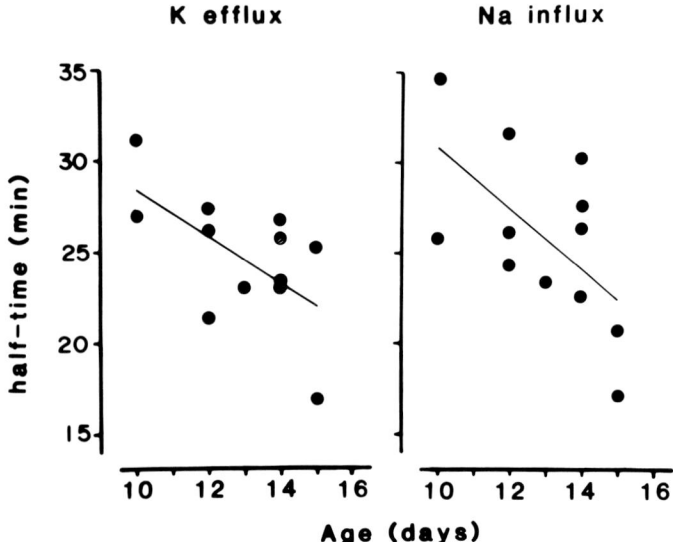

Fig. 7: Fluxes were measured as a function of the age of the animal during the period of renal terminal differentiation. Cells were cultured for 3 days. The correlation between age of animal and half-time of K efflux and Na influx are significant (from Larsson et al., 1986).

Similar observations were made when measuring Na-K pump activity as a function of time in culture, and as a function of the degree of terminal differentiation. After 2 days of culture, adult rat cells had a much higher rate of pumping activity than immature rat cells. But after 4 days of culture, adult rat cells' Na-K pump rate had decreased such that it reached the low level of Na-K pump rate of immature rat cells (Fig. 8).

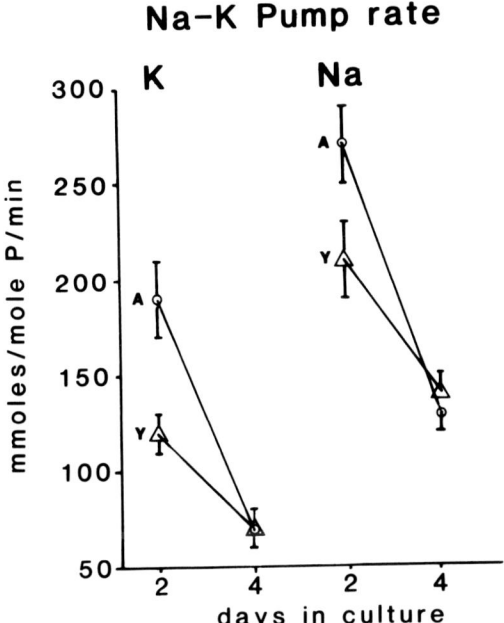

Fig. 8: 2-4 days primary culture of rat renal proximal tubular cells were preincubated in medium lacking K for 5 hrs. At time 0 the cells were returned in media containing 5 mM K and the initial rates of ouabain sensitive K influx and Na efflux were measured. A: cells from adult rat kidney. Y: cells from young rat kidney (not terminally differentiated) (from Larsson et al., 1988).

In the renal proximal tubular cells, ontogeny of Na and K leak pathways precedes that of (Na,K)-ATPase. The component of the Na leak pathways which increases during ontogenic development is essentially the Na-H exchanger. Chronic administration of amiloride during renal terminal differentiation blocked (Na,K)-ATPase increase in the proximal tubules but not in the thick ascending limb of the Loop of Henle (without any measurable effects on either other protein synthesis or DNA synthesis) (Rane et al., 1987; Aperia et al., 1988). Thus the ontogenic increase in sodium entry into the cells may be a determinant for the increased synthesis and insertion of (Na,K)-ATPase in the cellular membrane. That intracellular sodium concentration serves as the primary signal for regulating cell surface levels of (Na,K)-ATPase had been suggested in a variety of cells in tissue culture (Boardman et al., 1972; Fambrough et al., 1987; Ismail-Beigi et al., 1986; Pollack et al., 1981; Pressley et al., 1986).

Developmental and hormonal regulation of the Na-K pump may be related to the existence of isoforms of the enzyme. In cardiac ventricles, the ratios of the two isoforms changed with post-natal maturation. So that the ratio of $\alpha(+)$ to α is higher in newborns than in adult. Besides difference in tissue distribution, the two isoforms α and $\alpha(+)$ (Sweadner, 1979) exhibit differential expression during development and in response to hormones (Lytton, 1985; Schmitt and McDonough, 1986; Sweadner and Farshi, 1987).

Blanco and Beauge (this book) report a developmental increase in (Na,K)-ATPase activity in the rat hippocampal formation, with no changes in the properties of the enzyme with development. Different forms of (Na,K)-ATPase could be identified on the basis of their sensitivity to strophantidin with the high affinity component increasing with maturation.

Regulatory inhibition of Na-K pumping activity with a parallel decrease of ouabain binding (both being reduced virtually to 0) were observed after exposure of isolated xenopus oocytes to progesterone. Pump inhibition did not seem a consequence of a degradation of (Na,K)-ATPase (Richter and Passow, 1985).

RELATIONSHIPS BETWEEN MEASUREMENTS OF (Na,K)-ATPase ACTIVITY AND Na-K PUMPING ACTIVITY

Several works point to the difference there may be between studies in whole cells and studies in isolated membranes.

In studies by Mandel's group, it was concluded that the affinity of the Na-K pump for ATP, measured in suspension for proximal tubules, was much lower than that of (Na,K)-ATPase measured in lysed membrane preparations (Mandel, 1986).

The nerve growth factor dependency of the Na-K pump of chick dorsal rat ganglia neurons decreased with increasing ganglionic age although cell-free membrane preparations from NGF-deprived or NGF-supported cells displayed equal activity and characteristics of (Na,K)-ATPase. These observations raise the possibility that the pump or its membrane domain may be affected in the absence of NGF via a reversible process which involve some intracellular constituent not residing in the plasma membrane itself (Varon and Skaper, 1983).

Insulin appears to act directly on the Na-K pump of rat adipocytes (Resh et al., 1980). Insulin may increase the affinity of both forms α and $\alpha(+)$ of the enzyme for Na. Despite the large difference between Na affinity of α and $\alpha(+)$ in the whole cell, once the membranes have been isolated, there is no apparent low affinity component. Insulin may release a partial inhibition which could exist in cells, but the release of inhibition could be completed during the course of membrane isolation (Lytton, 1985).

CELLULAR COOPERATION AND SHARING OF Na-K PUMPS

That the pump responds to an increased load is further demonstrated in the following experiments (Bolivar et al., 1987). When the mutant ouabain-resistant MDCK cell line and the normal wild-type MDCK, ouabain-sensitive, are mixed together in equal proportion and plated at confluent density, the resistant-type of MDCK protect the wild-type from detaching from the petri dish after exposure to ouabain. Protection may be due to histological attachment between wild-type and resistant-type, or through maintenance of some metabolic activities in the wild-type despite ouabain. In the former case, the cells of the wild-type in the mixed population would be expected to have lost potassium and gained sodium, while the resistant-type would have maintained a high potassium and a low sodium concentration. In the latter case, all the cells in the mixture would be able to maintain a relatively high potassium and low sodium. After ouabain treatment of the mixture of resistant and wild-type cells, we found that the mean intracellular sodium concentration moderately increased and the mean intracellular potassium concentration slightly decreased, but all the cells still maintained high potassium and low sodium concentrations so that the potassium over sodium ratio was greater than 4.

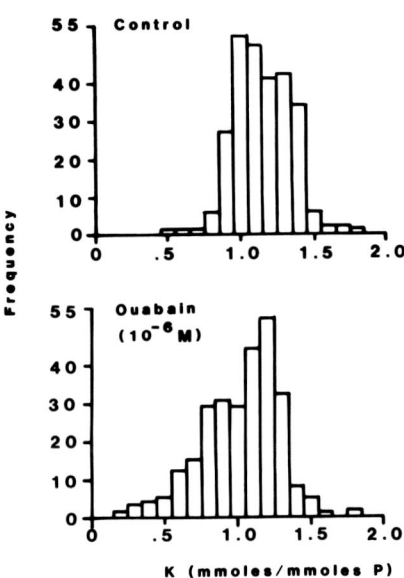

Fig. 9: A mixture of 1:1 wild type (ouabain sensitive) and mutant type (ouabain resistant) MDCK cells were plated at confluency. K content was measured in individual cells of the monolayers either in control conditions, or after 6 hrs of exposure to 10-6M ouabain (from Bolivar et al., 1987).

The distribution of intracellular potassium content in a population of cells from the mixture exposed to ouabain, although wider than in the mixed monolayer under control conditions, remained unimodal, showing that the cells behaved like a single population resistant to ouabain (Fig. 9).

DISSOCIATION BETWEEN Na-K PUMP ACTIVITY AND LEAK

There is in general tight coupling between leak and Na-K pump activity, of course to maintain intracellular steady-state of ionic and water content. One can uncouple leak from pump, however, when the cells have to accumulate or reject a net amount of water as it may be observed in the so-called volume regulatory situations. When renal proximal tubular cells are switched from a 300 mOsm medium to a hypotonic medium of 140 mOsm, they lose potassium after a few minutes of exposure to hypotonicity. In this situation, there is no change in the rate of pumping activity, but there is a marked increase in the rate of potassium leak (Fig. 10).

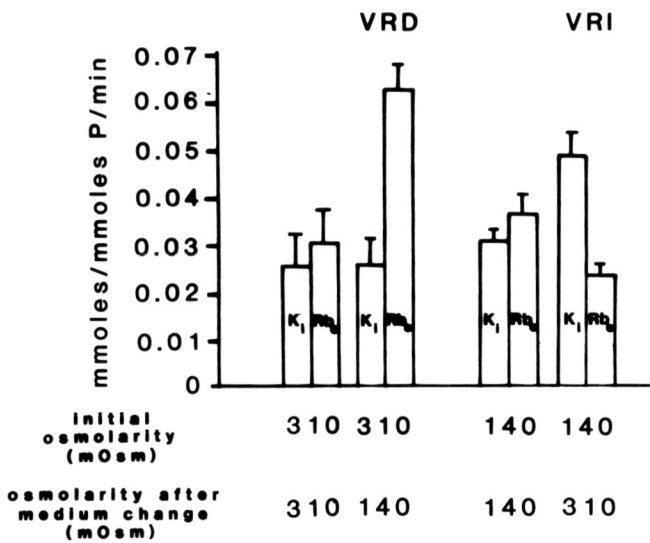

Fig. 10: Initial rates of Rb efflux and K influx in rat renal proximal tubular cells preincubated for 180 minutes in medium with Rb substituted for K and then preincubated for 30 minutes in medium of indicated osmolality. RPTC were then returned to medium of indicated osmolality containing 5 mM K (from Savin et al., 1986 and Harris et al., 1987).

If the cells are pre-incubated in 140 mOsm hypotonic medium, and are returned to 300 mOsm medium, they quickly re-accumulate potassium. This potassium re-accumulation is due to both a decrease in potassium leak rate and an increase in Na-K pump activity (Fig. 10). The pathway for increased Na-K pump activity after shrinkage, appears to be increased sodium entry by activation of the Na-H exchanger, leading to secondary activation of the pump.

HETEROGENEITY OF THE Na-K PUMP

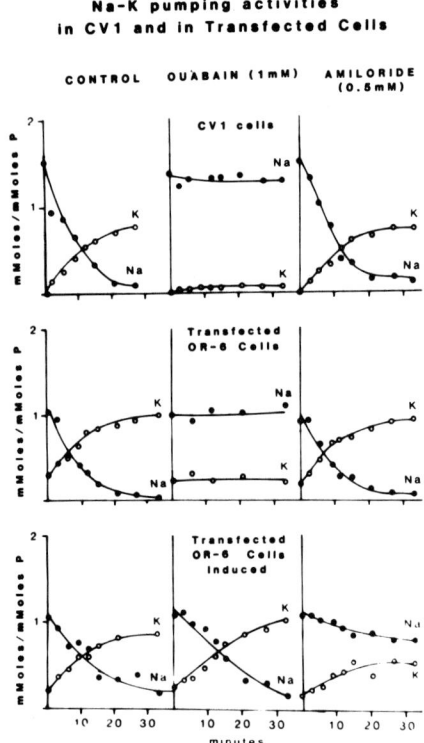

Fig. 11: Cells were preloaded with Na by incubating for 4 hours in medium lacking K. At time 0 cells were returned to medium containing 5.5 mM K and either no drug or ouabain or amiloride (from Epstein and Lechene, in press).

Ouabain sensitivity is not necessarily a fixed phenotypic characteristic. This is demonstrated by the following observations. Levenson et al. (1984) transfected a ouabain-resistant gene in CV-1 green monkey kidney cell line. We found that after culturing the transfected cells in the presence of 10 µM of ouabain, their Na-K pump activity became resistant to 1 mM of ouabain (Epstein and Lechene, in press). Surprisingly, the Na-K pumping activity of the cells which had been induced to be resistant to ouabain had become amiloride-sensitive, contrary to the non-induced transfectant (Fig. 11).

If induced cells are cultured in the absence of ouabain, they revert and become sensitive to the drug. Thus, in this transfected cell line, ouabain resistance is an inducible and reversible phenomenon. Related to the ouabain resistance gene, we found that the transfected cells, induced or not induced, undergo extreme morphological changes when they are challenged with low serum, low potassium, or c-AMP.

HETEROGENEITY OF PUMPING ACTIVITY IN A COLONY OF CELLS

There are some indirect indications that in a population of growing cells, Na-K pumping activity is different depending upon the location of the cells in the colony (Lipman et al., 1986). In primary cultures of renal proximal tubular cells only the most peripherally located cells synthesized DNA, as was shown by autoradiography of tritiated thymidine. The peripheral cells were more alkaline than the centrally located cells, as was shown by microfluorescence. The peripheral cells had a potassium over sodium ratio which was significantly higher than the K/Na ratio of the more centrally located cells as was shown using electron probe analysis. This difference in potassium over sodium ratio was likely to reflect a difference in pumping activity. These observations may be related to the general question of the relations between alkaline pH, ionic transport, DNA synthesis and cellular division.

CONCLUSION

The Na-K pump has a central role in eukaryotic animal cells. The Na-K pump is the motor creating and maintaining the cellular potential energy in the form of the K and Na gradients between cellular and extracellular world. This potential energy is used by the cells to perform their general and specialized functions.

Na-K pump rate varies by orders of magnitude amongst cell types, and may vary rapidly by several fold within a cell type.

Immediate and long-term Na-K pump rate variations may be secondary to changes in leak rate. The response to an increase in cellular influx of Na may be an immediate increase in pump rate secondary to an increase in intracellular Na concentration followed by a latter increase in number of active pump sites.

During ontogeny, an increase in number of Na-K pumps may be the response to the developmental expression of an increase in Na permeability pathways.

When the metabolic activity of a cell increases, the cell has to use its potential energy faster, dissipating faster K and Na gradients. In order to maintain these gradients, cells have to increase their rate of Na-K pumping. In consequence, more metabolically active cells will have higher overall leak rates and higher Na-K pump activity.

REFERENCES

Abraham EH, Breslow JL, Epstein J, Chang-Sing P Lechene C (1985). Preparation of individual human diploid fibroblasts and study of ion transport. Am J Physiol (Cell Physiol 17) 248: C154-C164.

Aperia A, Fukada Y, Lechene C (1988). Ontogenic increase of Na+-H+ exchange induces increase Na-K ATPase in rat proximal convoluted tubule (RPTC). Kidney Intl 33: 415.

Aronson PS (1985). Properties of the renal Na-H+exchanger. Ann NY Acad Sci 456: 220-228.

Baker PF (1986). Exchange of sodium for calcium across the plasma membrane. J Cardiovasc Pharm 8 (suppl 8): S25-S32.

Bertorello A, Aperia A (1987a). Effect of L-dopa, dopamine, dihydroxyphenyl acetic acid and homovanillic acid on Na,K-ATPase activity in rat proximal tubule segments. Acta Physiol Scand 130: 571-574,

Bertorello A, Aperia A (1987b). Pertussis toxin modulates dopamine inhibition of Na-K ATPase activity in rat proximal convoluted tubule segments. Am J Physiol 252: F39-F45.

Blanco G, Beauge L (1988). Pools of (Na^+, K^+)-ATPase isoforms in unborn and adult rats can be detected by their sensitivity to strophanthidin, but have similar reactivity to Na, K and ATP. The Proceedings of the Fifth International Conference on Na,K-ATPase: 1-8.

Boardman LJ, Lamb JF, McCall D (1972). Uptake of [$3H$] ouabain and Na pump turnover rates in cells cultured in ouabain. J Physiol (London) 225: 619-635.

Bolivar JJ, Lazaro A, Fernandez S, Stefani E, Pena-Cruz V. Lechene C, Cereijido M (1987). Rescue of a wild-type MDCK cell by a ouabain-resistant mutant. Am. J. Physiol (Cell Physiol 22) 253: C151-C161.

Bowen J. McDonough A (1987). Pretranslational regulation of Na-K-ATPase in cultured canine kidney cells by low K^+. Am J Physiol (Cell Physiol) 21: 179-189.

Cereijido M, Ehrenfeld J, Fernandez-Castelo S, Meza I (1981). Fluxes, junctions and blisters in cultured monolayers of epitheloid cells (MDCK) Ann NY Acad Sci 372: 422-441.

Chang-Sing P and Lechene (1983). Electron probe microanalysis of mouse Swiss albino 3T3 fibroblasts; Effects of ADH and serum on Na permeability. Fed Proc 42 (no 3): 321.

Cohen BJ, Lechene C (1988). Alanine increases Na and K leaks in cultured rat hepatocytes independently of Na-K pump activity. 72nd Annual Meeting of FASEB.

Crabos M, Aperia A, Lechene C (1988). Lack of L-dopa effect on Na-K pumping activity of cultured rat renal proximal tubular cells. Kidney Intl 31 (1): 173.

Darnell J, Lodish H, Baltimore D (1986). Molecular Cell Biology. New York: Scientific American Books p 626.

DiPolo R, Beauge L (1987). The squid axon as a model for studying plasma membrane mechanisms for calcium

regulation. Hypertension 10: I15-I19.
Epstein JA, Lechene C (1988). Ouabain resistant amiloride sensitive Na-K pumping activity, and morphological changes are inducible in a cultured cell line. Am J Physiol (in press).
Fambrough DM, Wolitzky BA, Tamkun MM, Takeyasu K (1987). Regulation of the sodium pump in excitable cells. Kidney Intl (Suppl. 23) 32: S97-S11.
Ganapathy V, Leibach FH (1986). Carrier-mediated reabsorption of small peptides in renal proximal tubule. Am J Physiol 251: 945-953.
Garrahan PJ, Glynn IM (1967). Factors affecting the relative magnitudes of the sodium-potassium and sodium-sodium exchanges catalyzed by the sodium pump. J Physiol 192: 189-216.
Gill TJ, Solomon AK (1959). Effect of ouabain on sodium flux in human red cells. Nature 183: 1127-1128.
Glynn IM (1984). "The electrogenic sodium pump." In: Electrogenic Transport: Fundamental Principles and Physiological Implications, edited by MP Blaustein and M Lieberman, New York: Raven Press.
Goldshlegger RS, Karlish JD, Raphaeli A, Stein WD (1987). The effect of membrane potential on the mammalian sodium-potassium pump reconstituted into phospholipid vesicles. J Physiol 387: 331-355.
Grinstein S, Rothstein A (1986). Mechanisms of regulation of the Na+/H+ exchanger. J Membr Biol 90: 1-12.
Hamilton RT, Nilsen M (1978). Transport of phosphate in membrane vesicles from mouse fibroblasts transformed by SV40 virus (SV3T3). J Biol Chem 253: 8247-8256.
Harris RC, Seifter JL, Lechene C (1986). Coupling of Na-H exchange and Na-K pump activity in cultured rat proximal tubule cells. Am J Physiol (Cell Physiol 20) 251: C815-C824.
Harris RC, Savin VJ, Lechene, C (1987). Rat renal proximal tubular cells (RPTC) increase net ionic content following return to isotonicity from hypotonicity. Kidney Intl 31: 435.
Hoffman EK (1986). Anion transport systems in the plasma membrane of vertebrate cells. Biochim Biophys Acta 864: 1-31.
Ismail-Beigi F, Haber RS, Loeb JN (1986). Stimulation of active Na+ and K+ transport by thyroid hormone in a rat liver cell line: Role of enhanced Na+ entry. Endocrinology 119 (6): 2527-2536.

Larsson S, Aperia A, Lechene C (1986). Studies on final differentiation of rat renal proximal tubular cells in culture. Am J Physiol (Cell Physiol 20) 251: C455-C464.

Larsson S, Aperia A, Lechene C (1988). Studies on final differentiation of rat renal proximal tubular cells in culture: II. Ouabain sensitive Na and K transport. Acta Physiol Scand 132: 129-134.

Levenson R, Rancaniello V, Albritton L, Housman D (1984). Molecular cloning of the mouse ouabain resistance gene. Proc Natl Acad Sci 81: 1489-1493.

Lipman RD, Harris RC, Seifter JL, Brenner BM, Lechene C (1986). Growth of rat proximal tubular cells (RPTC) occurs at periphery of a colony where cells are more alkaline and have a higher K/Na ratio. Kidney Intl 29: 397.

Lytton, J. (1985). Insulin affects the sodium affinity of the rat adipocyte (Na^+,K^+)-ATPase. J. Biol. Chem. 260: 10075-10080.

Mandel LJ (1986). Primary active sodium transport, oxygen consumption, and ATP: Coupling and regulation. Kidney Int 29: 3-9.

O'Grady SM, Palfrey HC, Field M (1987). Characteristics and functions of Na-K-Cl cotransport in epithelial tissues. Am J Physiol 253 (Cell Physiol 22): C177-C192.

Pollack LR, Tate EH, Cook JS (1981). Turnover and regulation of the Na-K-ATPase in HeLa cells. Am J Physiol 241: C173-C183.

Pressley TA, Haber RS, Loeb JN, Edelman IS, Ismail-Beigi F (1986). Stimulation of Na,K-activated adenosine triphosphate and active transport by low external K^+ in a rat liver cell line. J Gen Physiol 87: 591-606.

Quamme GA, Shapiro RJ (1987). Membrane controls of epithelial phosphate transport. Can J Physiol Pharmacol 65: 275-286.

Rane S, Aperia A, Lechene C (1987). Serum stimulated amiloride inhibitable Na influx is expressed at terminal differentiation of renal proximal tubular cells (RPTC). Kidney Intl 31: 178.

Resh MD, Nemonoff RA, Guidotti G (1980). Insulin stimulation of (Na^+,K^+)-adenosine triphosphatase-dependent $^{86}Rb^+$ uptake in rat adipocytes. J Biol Chem 255: 10938-10945.

Richter HP, Passow H (1985). Regulatory changes of the operation of the sodium-potassium pump during progesterone-induced maturation of xenopus oocytes. The Comp of Biologists Ltd: 171-178.

Savin VJ, Harris RC, Seifter JL, Brenner BM, Lechene C (1986). Hypotonicity causes net efflux of K and Cl in normal rat renal proximal tubular cells (RPTC) and net efflux of Na and Cl in Na loaded RPTC. Kidney Intl 29: 406.

Schmitt CA, McDonough AA (1986). Developmental and thyroid hormone regulation of two molecular forms of Na^+-K^+-ATPase in brain. J Biol Chem 261: 10349-10444.

Schultz SG, Hudson RL, Lapointe JY (1985). Electrophysiological studies of sodium cotransport in epitheliae: toward a cellular model. Ann NY Acad Sci 456: 127-135.

Simchowitz L, Spillberg I, De Weer P (1982). Sodium and potassium fluxes and membrane potential of human neutrophils. J Gen Physiol 79: 453-479.

Skou JC (1957). The influence of some cations on an adenosine triphosphate from peripheral nerves. Biochim Biophys Acta 23: 394-401.

Skou JC (1960). Further investigations of a Mg^{++} and Na^+-activated adenosintriphosphate possibly related to the active, linked transport of Na^+ and K^+ across the nerve membrane. Biochim Biophys Acta 42: 6-23.

Sweadner, KJ (1979). Two molecular forms of (Na^+ and K^+)-stimulated ATPase in brain. Separation and difference in affinity for strophantidine. J Biol Chem 256: 6060-6067.

Sweadner K, Farshi SK (1987). Rat cardiac ventricle has two Na^+,K^+-ATPases with different affinities for ouabain: Developmental changes in immunologically different catalytic subunits. Proc Natl Acad Sci, USA 84: 8408-8407.

Varon S, Skaper SD (1983). The Na^+,K^+ pump may mediate the control of nerve cells by nerve growth factor. Trends in Biochemical Sciences.

Wright EM (1984). Electrophysiology of plasma membranes vesicles. Am J Physiol 247: F363-F372.

Overview:
MAINTENANCE OF Na,K-HOMEOSTASIS BY Na,K-PUMPS IN STRIATED MUSCLE

Ole M. Sejersted

Department of Physiology, National Institute of Occupational Health, P.O. Box 8149 DEP, 0033 Oslo 1, Norway

INTRODUCTION

Electrical activity in excitable tissues is associated with a dissipation of the Na^+ and K^+ gradients across the plasma membrane. The maximum rate at which these two ions can be transported in the opposite directions by the Na,K-pump, might therefore represent a tolerance limit with respect to the stimulation frequency the tissue can sustain over a prolonged period. At lower frequencies adjustment of pump rate allow the muscle cells to attain steady state, possibly as a result of increased intracellular Na^+ concentration ($[Na^+]_i$).

It is difficult to predict the behaviour of the intact cell from known characteristics of the isolated Na,K-ATPase. This paper is intended to present selected information on characteristics of the Na,K-pump in situ in its normal environment. The isolated enzyme has been described in great detail (for review see e.g. Robinson and Flashner, 1979), and there is a growing literature on pump behavior in reconstituted vesicles where the environment of intracellular and extracellular sites of the enzyme can be varied independently. Still, the intact cell, especially in excitable tissues and epithelia, represents a unique regulatory setting. The Na,K-pump has been studied extensively in the red cell where it in fact is absent in some species, and hence connection to cell function is at best difficult to judge. Studies on Na,K-pump properties in intact excitable cells are fewer and available methods restrict the information that can be gained. In recent years information about pump rate has been obtained from measurements of pump current and temporal changes in intracellular concentrations of Na^+ and K^+ ($[K^+]_i$) in addition to older standard methods utilizing isotopes.

As recently reviewed by Clausen (1986) the Na,K-pump is subject to acute and long-term regulation. Slow changes of the amount of enzyme or number of pump sites have important consequences for tissue function. However, I will restrict this review to three factors which alone or in concert may change pump rate rapidly: extracellular K^+ concentration ($[K^+]_o$), reciprocal changes of $[Na^+]_i$ and $[K^+]_i$ and catecholamines.

Na,K-PUMP RESPONSE TO $[K^+]_o$.

It is generally agreed that K^+ binds to the K^+-site of the isolated Na,K-ATPase with an affinity that gives half maximal ATP hydrolysis rate at less than 1 mM (Robinson and Flashner, 1979). It is therefore a notable discrepancy when studies on intact cell preparations report $K_{0.5}$ values for the Na,K-pump ranging from 0.8 to 33 mM (Table 1). Two possible sources of artifacts have been pointed out.

First, Cohen et al. (1984) have argued that much of the observed difference in estimates of $k_{0.5}$ for $[K^+]_o$ could be reduced when variable saturation of the internal sodium site was taken into account. Second, Eisner et al. (1981) suggested an extracellular cleft depletion of K^+. During net uptake of K^+ by the cells in multicellular preparations a diffusion gradient from the surface towards the cell membrane is created. Hence, the K^+ concentration of the

TABLE 1. $K_{0.5}$ values for activation by K^+ or Rb^+.

	Measured variable	$K_{0.5}$, mM	
Gadsby (1980)	Pump current	1	(K^+)
Eisner & Lederer (1980)	Pump current	6.3	(Rb^+)
Eisner et al. (1981)	Decline of $[Na^+]_i$	1 or 4	(Rb^+)
Deitmer & Ellis (1978)	Decline of $[Na^+]_i$	10	(K^+)
Clausen et al. (1987)	Uptake of K^+	33	(K^+)
Sejersted et al (1988)	Decline of $[Na^+]_i$	9	(Rb^+)
Cohen et al. (1987)	Pump current	0.8	(K^+)

superfusate is much higher than the concentration at the extracellular pump site. This problem can be circumvented by performing experiments on isolated cells as recently done by Cohen et al. (1987). Pump current was measured during voltage clamp conditions in Purkinje fiber cells at various $[K^+]_o$. Pump current was measured during sudden lowering of $[K^+]_o$ at normal $[Na^+]_i$. Changes in $[Na^+]_i$ were slow and did not affect the result. $K_{0.5}$ for pump stimulation by extracellular K^+ equalled 0.8 mM.

Na^+ transport can also be measured during variable $[K^+]_o$ in organs with intact circulation. In the intact kidney Na^+ transport in cells with high $[Na^+]_i$ was reduced only when plasma K^+ concentration fell below 4 mM (Monclair et al., 1980). It is therefore reasonable to conclude that the affinity to K^+ at the extracellular site is the same in intact cells and isolated enzyme preparations.

NORMAL VARIATION OF $[K^+]_o$

Is $[K^+]_o$ an important determinant of pump rate? Assuming Michaelis-Menten kinetics a K_m of 0.8 mM will cause 85 % activation at a concentration of 4.5 mM which is close to the normal extracellular concentration in many species. Hence, increments in $[K^+]_o$ can only cause modest activation, whereas hypokalemia can lower pump rate significantly.

The most extreme changes in the distribution of K^+ between the extracellular and intracellular compartments is seen with high intensity exercise. During 1 min maximal running on the treadmill at an inclination of 6 ° the arterial plasma K^+ concentration reached close to 8 mM and fell significantly below control level a few min later (Medbø et al., 1982; Hermansen et al., 1984). Half time for the decline of arterial plasma K^+ was less than 30 s. Provided these measurements reflect the changes in the interstitium, variation of $[K^+]_o$ can only account for a change of pump rate of 20 % at most.

Ilebekk et al. (1986) measured the K^+ concentration in coronary sinus plasma in intact porcine hearts during sudden increments and subsequent decrements of heart rate by 50 beats·min^{-1}. At the onset of pacing there was a transient increase in plasma K^+ amounting to at most 0.2 mM.

The opposite change was observed when heart rate was abruptly normalized. This means that during activation of a smaller muscle mass which loses insufficient K$^+$ to cause a general rise in extracellular K$^+$ concentration, only transient elevations of [K$^+$]$_o$ will occur locally, since K$^+$ is washed out by the circulation. Clearly, transient responses of this magnitude cannot cause sustained and significant pump activation.

In conclusion, physiological changes of plasma K$^+$ concentration are often transient during variation of stimulation frequency and occur within a range that can cause at most 20 % change in Na,K pump rate. There is an interesting discrepancy between activation of a small and a large skeletal muscle mass in the intact organism. In the former case a continuous loss of K$^+$ from the muscle is observed (Sjøgaard et al 1986), whereas with cycling or running the a-v difference for K$^+$ across the leg disappears after a few min (Vøllestad and Sejersted, 1985). The explanation could be the lack of rise of extracelllar K$^+$ concentration with activation of a small as opposed to a large muscle mass. Thus, a larger outward concentration gradient during the repolarization phase of the action potential is maintained and a small, but perhaps important increase in pump rate is prevented since K$^+$ is washed away by the blood.

Na,K PUMP RESPONSE TO [Na$^+$]$_i$ AND [K$^+$]$_i$.

It is generally agreed that the Na,K-ATPase in isolated membrane fractions is half maximally stimulated at a Na$^+$ concentration of about 20 mM. Activation curves obtained from vesicle studies on intact cell preparations tend to give significantly lower $k_{0.5}$ values for [Na$^+$]$_i$. In fig. 1 the two upper curves from the studies of Philipson and Nishimoto (1983) using cardiac enzyme and Karlish and Stein (1985) using kidney enzyme were both obtained from vesicles with zero [K$^+$]$_i$. They both give $k_{0.5}$ values less than 10 mM and exhibit slightly different sigmoidicity. Since in the normal cell Na$^+$ and K$^+$ vary inversely, these curves are not representative for Na$^+$ activation of the pump in intact cells. However, the classical curve obtained on red cells by Garay and Garrahan (1973) takes into account reciprocal variation of [Na$^+$]$_i$ and [K$^+$]$_i$, and $k_{0.5}$ for [Na$^+$]$_i$ is close to 10 mM.

There have been few studies on muscle describing the pump kinetics over a wide range of $[Na^+]_i$. In a recent investigation $k_{0.5}$ in terms of concentration was estimated to about 14 mM in intact sheep Purkinje fibers (Fozzard et al., 1987; Sejersted et al., 1988). In these fibers $[Na^+]_i$ was measured continuously with a Na^+-selective microelectrode. Na^+-loading of the fibers was accomplished by removing extracellular K^+ and Mg^{++} and lowering Ca^{++} to 10^{-8} M. By these means the Na,K-pump was inhibited and the Ca^{++}-channels opened up to Na^+ so that $[Na^+]_i$ rose to desired level. After loading, Mg^{++} was added again to close the Ca^{++}-channels, and the pump was reactivated by adding K^+ or Rb^+. Extracellular Na^+ was lowered to 2.4 mM to prevent back-leak of Na^+ as $[Na^+]_i$ fell. The decline of $[Na^+]_i$ from very high (80-130 mM) to non-measurable levels occurred over a period of 10-20 min and was completely inhibited by acetylstrophantidin. The experimental model thus provided a functional isolation of the Na,K-pump in intact Purkinje fibers.

$[Na^+]_i$ and $[K^+]_i$ varied almost reciprocally during reactivation of the pump. The pump activation curve in

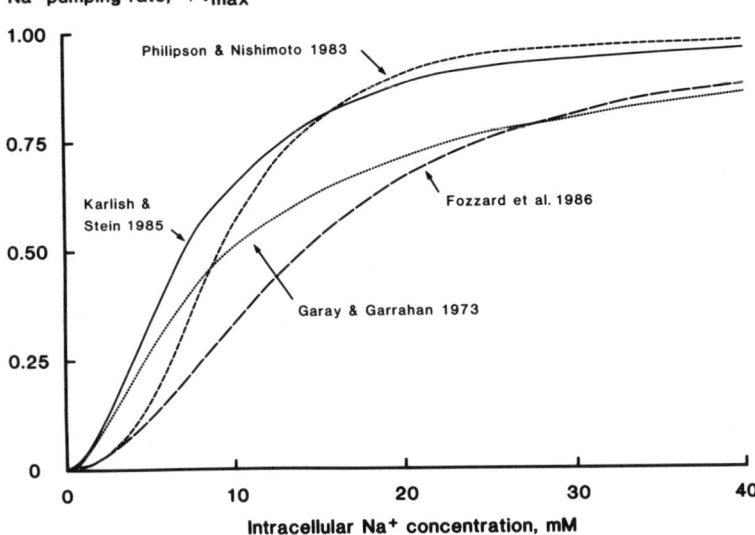

Figure 1. Relationship between Na,K-pump rate and $[Na^+]_i$ in four different investigations.

Fig. 1 (Fozzard et al., 1986) is therefore representative for the intact cell. The sigmodicity is described by a Hill coefficient close to 2. In normal, resting Purkinje fibers $[Na^+]_i$ is close to 10 mM which means that relative pump activation is close to 35 %.

Two factors which have not been taken into account in these experiments could influence the pump response to intracellular Na^+. Karlish and Stein (1985) showed in vesicle studies that extracellular Na^+ has an allosteric effect which causes a decrease of the Hill coefficient and slightly higher $k_{0.5}$. Gadsby et al. (1985) observed that in in isolated myocytes clamped at different potentials, pump current was maximal in the range of -50 to +50 mV, whereas it approached zero at -150 mV. At physiological membrane potentials in these cells pump rate could therefore be 10 to 20 % lower than maximum.

EXCITATION RELATED CHANGES OF $[Na^+]_i$ AND $[K^+]_i$.

The following is an attempt at applying the characteristics of the pump in Purkinje fibers in an analysis of changes of the Na^+ and K^+ balance in intact muscle during activation. The influence of external Na^+ and membrane potential will not be taken into account. The additional information needed is twofold. What is maximal pump rate in skeletal and cardiac muscle, and what are the Na^+ and K^+ fluxes associated with each action potential.

In Table 2 are collected some reported data on maximum Na^+-efflux rates from studies carried out with high enough $[Na^+]_i$ to ensure almost complete saturation of the Na^+-site. Flux rates have been related either to membrane surface or to tissue weight. Converted figures are given in brackets. The highest pumping rates have recently been observed in rat soleus and Purkinje fibers and they fall in the same range as observed in squid axon (Table 2). Hence, at 37°C and high $[Na^+]_i$, pump rates in skeletal muscle are at least 100 $nmol \cdot g^{-1} \cdot s^{-1}$.

Maximum pump capacity can also be determined as the number of ouabain binding sites multiplied by the turnover number which at 37°C is close to 8.000 min^{-1} (Clausen 1986). Ouabain binding sites have been determined in intact cardiac tissue and skeletal muscle as exemplified in

TABLE 2. Estimates of maximum Na^+ pumping rates.

	Na-transport pmol/cm²·s	Na-transport nmol/g·s	Tp °C	$[Na^+]_i$ mM
Glitsch et al. (1976) Guinea pig auricles	(30)	30	35	10-88
DeWeer et al. (1986) Squid axon	33	-		50
Mullins & Frumento (1963) Frog sartorius	17	8	20	40-60
Clausen et al. (1987) Rat soleus	(83)	108	30	126
Sejersted et al. (1988) Sheep Purkinje	(25)	115	37	0-100
	K-transport			
Clausen et al. (1987) Rat soleus		97	30	126
	Current uA/cm²			
Gadsby et al. (1985) Isolated guinea pig heart cells	(42)	3	36	34-41

Table 3. Conversion to Na^+-pumping capacities gives values above 110 nmoles·g^{-1}·s^{-1} and in Fig. 2, left hand panel, a V_{max} of 150 nmoles·g^{-1}·s^{-1} has been implied.

The broken lines show the anticipated relationships between Na^+ influx and $[Na^+]_i$ at different stimulation frequencies, based on an average influx of Na^+ of 3-5 nmoles·g^{-1} per action potential. Zero influx has been assumed at an intracellular Na^+ concentration of 150 mM. The points of intercept between pump rate and Na^+-influx predict steady state $[Na^+]_i$ at different stimulation frequencies and are plotted in the right hand panel (Fig. 2).

An obvious problem with the model is the fact that pump rate in resting rat soleus muscle is probably considerably lower than 50 nmoles·g^{-1}·s^{-1} which Fig. 2 predicts (Clausen and Flatman, 1977). This could be due to different Na^+ sensitivity of the pump in skeletal muscle and cardiac Purkinje fibers. Alternatively, electrical stimulation of the tissues could alter this relationship.

Some general conclusions can be drawn:

TABLE 3. Ouabain binding sites in intact tissue

		Max binding pmol·g^{-1}	Estimated maximum pump rate nmol·g^{-1}·s^{-1}
Nørgaard et al. (1986)	Human heart	413	160
Sejersted et al. (1985)	Pig heart	826	330
Sejersted et al. (1988)	Sheep Purkinje	401	160
Nørgaard et al. (1984)	Human vastus	280	110
Clausen & Hansen (1982)	Mouse soleus	670	270
Kjeldsen et al. (1985)	Rat soleus	310	125

1) At high, but still physiological stimulation frequencies there seems to be a limit above which the cells can no longer maintain a stable $[Na^+]_i$. Breakdown of transmembrane ionic gradients could contribute to exhaustion possibly by means of the rise of $[K^+]_o$.

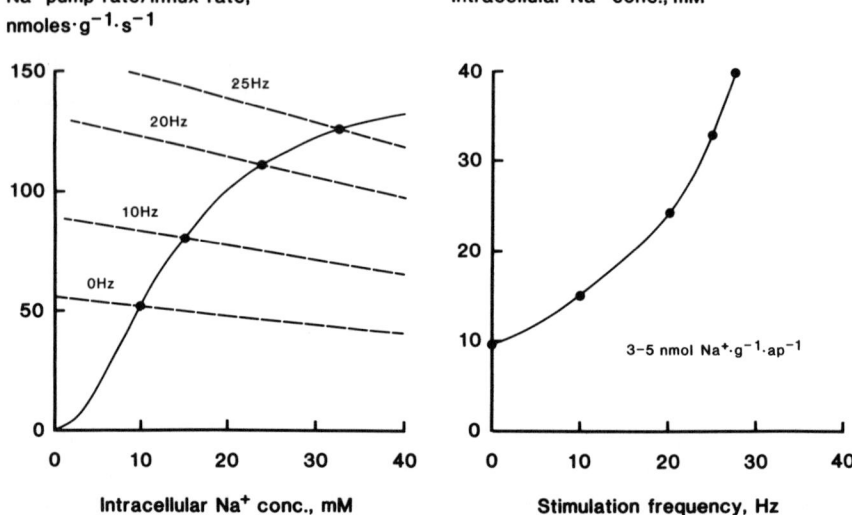

Figure 2. Na,K-pump rate (drawn line) and Na$^+$-influx for different stimulation frequencies (broken lines) at various $[Na^+]_i$ (left panel). Estimated steady state $[Na^+]_i$ as a function of stimulation frequency in skeletal muscle (right panel).

2) In cardiac ventricular tissue where the number of Na,K-pumps tend to be larger than in skeletal muscle and stimulation frequencies do not exceed 3-4 Hz, only small changes in $[Na^+]_i$ would be expected since pump capacity is even greater than in skeletal muscle.

NORMAL VARIATION OF $[Na^+]_i$ and $[K^+]_i$.

$[Na^+]_i$ has been measured both in isolated skeletal and cardiac tissue after changes of stimulation frequency. In mouse soleus Juel (1986) noted a rise in $[Na^+]_i$ of 12 mM during stimulation at a mean frequency of 16 Hz. This is slightly more than predicted from fig. 2 taking into account the higher number of ouabain binding sites in these muscles. His results showed that restoration of $[Na^+]_i$ occurs rapidly with a $t_{1/2}$ of less than one min.

In Purkinje fibers Cohen et al. (1982) observed a rise of 1.4 mM when initially quiescent fibers were stimulated at 1 Hz. This is a far larger increase than presently predicted. However, with further increments in stimulation frequency the rise of $[Na^+]_i$ was only 0.3-0.4 mM per Hz, which is in good agreement with fig. 2

The observed changes in $[K^+]_i$ are more difficult to interpret, especially in skeletal muscle. Juel (1986) observed a decline of $[K^+]_i$ of 33 mM in the mouse soleus which was three times more than the rise of $[Na^+]_i$. Since cell volume increased during stimulation, changes of $[Na^+]_i$ and $[K^+]_i$ do not directly reflect the net exchange across the sarcolemma. Even so it seems that loss of K^+ exceeded gain of Na^+. It is also an intriguing question how $[K^+]_i$ is restored considering the 3Na:2K ratio of the Na,K-pump. In intact porcine hearts loss of K^+ during an increment in beating frequency of 1 Hz corresponds to a $[K^+]_i$ fall of 0.3 mM in agreement with the observed changes of $[Na^+]_i$ in Purkinje fibers (Ilebekk et al., 1986).

In conclusion some reports support the predictions of fig. 2, but whether the rise of $[Na^+]_i$ during increased stimulation frequency is sufficient to explain the increase in pump rate especially when starting from the quiescent state is still unresolved. The changes of $[K^+]_i$ can be out of proportion to the change of $[Na^+]_i$, but is probably of little regulatory consequence as regards pump rate.

Na,K-PUMP STIMULATION BY CATECHOLAMINES.

Clausen and Flatman (1977) originally showed that upon exposure to adrenaline the rat soleus muscle loses Na^+ at the same time as ouabain-sensitive Na^+-efflux is doubled. This strongly indicates a direct stimulation of the Na,K-pump. In support of this Wasserstrom et al. (1982) observed that isoproterenol caused $[Na^+]_i$ to fall by about 2 mM in the course of a few min in isolated sheep and canine ventricular tissue. At the same time there is a significant uptake of K^+ of the same order of magnitude (Ellingsen et al. 1987). As opposed to skeletal muscle the effect in cardiac tissue is mediated by beta-1 and not beta-2-adrenoceptors. It is not know whether the Na^+-activation curve is changed significantly by catcholamines. The effect of catecholamines might significantly amplify the pump response to a rise of $[Na^+]_i$ and would be of major importance e.g. during exercise.

REFERENCES

Clausen T, (1986). Regulation of active Na^+-K^+ transport in skeletal muscle. Physiol Rev 66: 542-580.

Clausen T, Everts ME, Kjeldsen K (1987). Quantification of the maximum capacity for active sodium-potassium transport in rat skeletal muscle. J Physiol 388: 163-181.

Clausen T, Flatman JA (1977). The effect of catecholamines on Na-K transport and membrane potential in rat soleus muscle. J Physiol 270: 383-414.

Clausen T, Hansen O (1982). The Na^+-K^+-pump, energy metabolism and obesity. Biochem Biophys Res Commun 104: 357-362.

Cohen CJ, Fozzard HA, Sheu S-S (1982). Increase in intracellular sodium activity during stimulation in mammalian cardiac muscle. Circ Res 50: 651-662.

Cohen I, Falk R, Gintant G (1984). Saturation of the internal sodium site of the sodium pump can distort estimates of potassium affinity. Biophys J 46: 719-727.

Cohen IS, Datyner NB, Gintant GA, Mulrine NK, Pennefather P (1987). Properties of an electrogenic sodium-potassium pump in isolated canine Purkinje myocytes. J Physiol 383: 251-167.

Deitmer JW, Ellis D (1978). The intracellular sodium activity of cardiac Purkinje fibres during inhibition and reactivation of the Na K pump. J Physiol 284: 241-259.

DeWeer P, Gadsby DC, Rakowski RF (1986). Voltage dependence of Na/K pump-mediated ^{22}Na efflux and current in squid giant axon. J Physiol 371: 144P.
Eisner DA, Lederer WJ (1980). Characterization of the electrogenic sodium pump in cardiac Purkinje fibres. J Physiol 303: 441-474.
Eisner DA, Lederer WJ, Vaughan-Jones RD (1981). The effects of rubidium ions and membrane potential on the intracellular sodium activity of sheep Purkinje fibres. J Physiol 317: 189-205.
Ellingsen Ø, Sejersted OM, Leraand S, Ilebekk A (1987). Catecholamine-induced myocardial potassium uptake mediated by b_1-adrenoceptors and adenylate cyclase activation in the pig. Circ Res 60: 540-550.
Fozzard HA, Sejersted OM, Wasserstrom JA (1986) Sodium activation of the Na,K-pump in isolated sheep cardiac Purkinje strands. J Physiol 381: 91P
Gadsby DC (1980). Activation of electrogenic Na^+/K^+ exchange by extracellular K^+ in canine cardiac Purkinje fibers. Proc Natl Acad Sci 77: 4035-4039.
Gadsby DC, Kimura J, Noma A (1985). Voltage dependence of Na/K pump in isolated heart cells. Nature 315: 63-65.
Garay RP Garrahan PJ (1973). The interaction of sodium and potassium with the sodium pump in red cells. J Physiol 231: 297-325.
Glitsch HG, Pusch H, Venetz K (1976). Effects of Na and K ions on the active Na transport in guinea-pig auricles. Pflügers Arch 365: 29-36.
Hermansen L, Orheim A, Sejersted OM (1984). Metabolic acidosis and changes in water and electrolyte balance after maximal exercise of short duration. Int J Sports Med 5(Suppl.): 110-115
Ilebekk A, Andersen FR, Sejersted OM (1986). Magnitude of myocardial potassium changes during acute alterations in pacing frequency of the in situ pig heart. Cardiovasc Res 20: 176-181.
Juel C. (1986). Potassium and sodium shifts during in vitro isometric muscle contraction, and the time course of the ion-gradient recovery. Pflügers Arch 406: 458-463.
Karlish SJD, Stein WD (1985). Cation activation of the pig kidney sodium pump: transmembrane allosteric effects of sodium. J Physiol 359: 119-149.
Kjeldsen K, Norgard A, Clausen T (1985). Effects of ouabain, age and K-depletion on K-uptake in rat soleus muscle. Pflügers Arch 404: 365-373.

Medbø JI, Sejersted OM, Hermansen L, (1982). Changes in blood lactate and plasma electrolytes after maximal exercise. Acta Physiol Scand 116 (Suppl 508): 52.

Monclair T, Sejersted OM, Kiil F (1980). Influence of plasma potassium concentration on the capacity for sodium reabsorption in the diluting segment of the kidney. Scand J Clin Lab Invest 40: 27-36.

Mullins LJ, Frumento AS (1963). The concentration dependence of sodium efflux from muscle. J Gen Physiol 46: 629-654.

Nørgaard A, Kjeldsen K, Clausen T (1984). A method for the determination of the total number of ^{3}H-ouabain binding sites in biopsies of human skeletal muscle. Scand J Clin Lab Invest 44: 509-518.

Nørgaard A, Kjeldsen K, Hansen D, Clausen T, Larsen CG, Larsen FG (1986). Quantification of the ^{3}H-ouabain binding site concentration in human myocardium: a postmortem study. Cardiovasc Res 20: 428-435.

Philipson KD, Nishimoto AY (1983). ATP-dependent Na^+ transport in cardiac sarcolemmal vesicles. Biochim Biophys Acta 733: 133-141.

Robinson JD, Flashner MS (1979). The $(Na^+ + K^+)$-activated ATPase, Enzymatic and transport properties. Biochim Biophys Acta 549: 145-176.

Sejersted OM, Andersen FR, Ilebekk A (1985). Potassium balance and ouabain binding sites in intact porcine hearts during isoproterenol infusion. In Dhalla NS, Hearse DJ (eds): Adv Myocardiol, vol 6, New York: Plenum Publ Co, pp 83-95.

Sejersted OM, Wasserstrom JA, Fozzard HA (1988). Na,K-pump stimulation by intracellular sodium in isolated, intact sheep cardiac Purkinje fibers. J Gen Physiol (In press)

Sjøgaard G (1986). Water and electrolyte fluxes during exercise and their relation to muscle fatigue. Acta Physiol Scand 128 (Suppl 556): 129-136.

Vøllestad NK, Sejersted OM (1985). Plasma K^+ during exercise of various intensity in normal humans. Clin Physiol 5 (Suppl 4): 151.

Wasserstrom JA, Schwartz DJ, Fozzard HA (1982). Catecholamine effects on intracellular sodium activity and tension in dog heart. Am J Physiol 243: H670-H675.

OVERVIEW:

ROLE OF Na-K-ATPase IN KIDNEY FUNCTION

Adrian I. Katz

Department of Medicine, The University of Chicago
Pritzker School of Medicine, Chicago, Illinois 60637

INTRODUCTION

While Na-K-ATPase* is ubiquitous and regulates cell composition and volume in many tissues, it plays a particularly important role in epithelial cells, whose function is to perform vectorial active transport. The kidney is a prime example of such an organ, being composed of thousands - or in man, millions - of nephrons involved in the transport of numerous solutes. To appreciate the uniqueness of the kidneys it is of interest to recall their contribution to overall energy consumption: These organs constitute only 0.5% of the body mass in man yet they receive one fifth of the total cardiac output, and consume 8% of the total O_2 utilized by the body at rest. Because of its evolutionary heritage the human kidney filters 25,000 mEq Na daily in order to excrete 100. It is chiefly to recover the remaining 24,900 mEq of Na, much of it by active transport, that the metabolic cost of kidney function is so high. Thus, while tissues at rest spend 20-40% of their energy consumption to maintain transmembrane Na and K gradients, 80% or more of the energy utilized by the kidneys is invested in active transport.

Considering the statistics just mentioned, it comes as no surprise that the kidney is one of the richest sources of Na-K-ATPase, and has served prominently as object for the study of its properties and regulation for more than 2 decades. Nevertheless, by comparison with the spectacular recent progress in our understanding of the structure and

---

* The terms Na-K-ATPase and Na:K pump are used interchangeably in this paper.

reaction mechanism of the Na:K pump, elucidation of its physiologic role in complex organs like the kidney lags behind. Still lacking is a clear understanding of the factors that modulate the biosynthesis and activity of the renal enzyme, and – given its slow turnover – of its role in rapid adjustments in tubular Na and K transport.

The limited scope of this article permits only a selective survey of current knowledge of the localization and function of Na-K-ATPase in the kidney; interested readers are referred to several recent sources where this topic is treated in greater detail (Jorgensen, 1980; Katz, 1982; Doucet, 1984; Katz 1986). The present review describes the distribution profile of Na-K-ATPase along the nephron, discusses its less familiar role in potassium secretion and reabsorption, and summarizes some of the still unsettled issues regarding the regulation of renal Na-K-ATPase activity, – especially the nature of the signals involved in this process.

LOCALIZATION OF Na-K-ATPase IN THE KIDNEY

Na-K-ATPase is not distributed evenly in the kidney, its activity being highest in the outer medulla, intermediate in the cortex, and lowest in the inner medulla-papilla (Hendler et al., 1971). Because the concentration of Na:K pumps varies along the nephron, these differences reflect the relative abundance of the different nephron segments in each of the 3 zones of the kidney. For example, the renal outer medulla is rich in Na-K-ATPase because it contains chiefly thick ascending limbs of Henle's loops, one of the segments with the highest density of Na:K pumps in the nephron.

Distribution Profile of Na-K-ATPase Along the Nephron

The kidney is a remarkably heterogeneous organ, due to the structural and functional diversity of consecutive segments of the nephron. This fact underscores the limitations inherent in examination of Na-K-ATPase in whole tissue homogenates or subcellular fractions thereof: Such preparations, derived not only from a variety of tubular structures but also from glomeruli, blood vessels, and interstitial cells, are ill-suited to quantitative correlations between pump activity and transport events measured with micropuncture or isolated tubule microperfusion techniques. In addition, alterations in tubular transport produced by hormones or dietary manipulations take place in certain segments of the nephron and not in others, and therefore corresponding changes in pump activity are likely to remain undetected when measured in preparations from the entire cortex or medulla. Clearly, the important question is not where is Na-K-ATPase located in the kidney, but where it is in the nephron.

To determine enzyme activity in individual nephron segments it was necessary to perfect the technique of tubule microdissection, and at the same time to scale down the assays to the minute amount of tissue (usually < 1 µg protein) in single pieces of tubule. The first measurements of Na-K-ATPase in discrete nephron segments were reported by Schmidt and Dubach (1969), who examined lyophilized tubules and used ultramicrochemical techniques to determine the inorganic phosphate liberated during ATP hydrolysis; the latter technique was substituted in subsequent papers by an enzymatic cycling method based on the fluorometric measurement of generated NADPH (Schmidt and Dubach 1971, 1974). Because the use of freeze-dried tissue has certain limitations and the method in general was too cumbersome for routine application, my colleagues and I developed an alternative technique that utilizes freshly microdissected tubules and measures directly radiolabeled P_i liberated from the hydrolysis of $[\gamma^{32}-P]ATP$ (Doucet et al., 1979; Katz et al., 1979). This method is simple enough to allow simultaneous measurement of Na-K-ATPase in large numbers of tubules, and sufficiently sensitive to determine the enzyme activity in tubule segments 100-200 um in length. Alternative techniques that combine certain features of those listed above have been introduced more recently (Garg et al., 1981; O'Neil and Dubinsky, 1984).

Na-K-ATPase activity has been determined in individual nephron segments of several species, including the rabbit, rat, mouse, and human. Not surprising, the absolute levels of enzyme activity measured by different investigators using diverse methods corresponded only for some segments and differed for others. In contrast, however, the enzyme distribution profile along the nephron was remarkably similar regardless of species studied or methods used, the greatest activity being always measured in the distal convoluted tubule and thick ascending limb. Considerable activity is also found in proximal convoluted tubules, but the enzyme is present in lesser amounts in the collecting tubule and pars recta, and is barely detectable in the thin limbs of Henle's loops (Schmidt and Dubach, 1969; Katz et al., 1979; Garg et al., 1981; El Mernissi and Doucet, 1984). When measured with the same technique, Na-K-ATPase activity was in general comparable in the three species studied, being greater in the rat, lesser in the rabbit, and intermediate in the mouse (Katz et al., 1979) (Fig. 1).

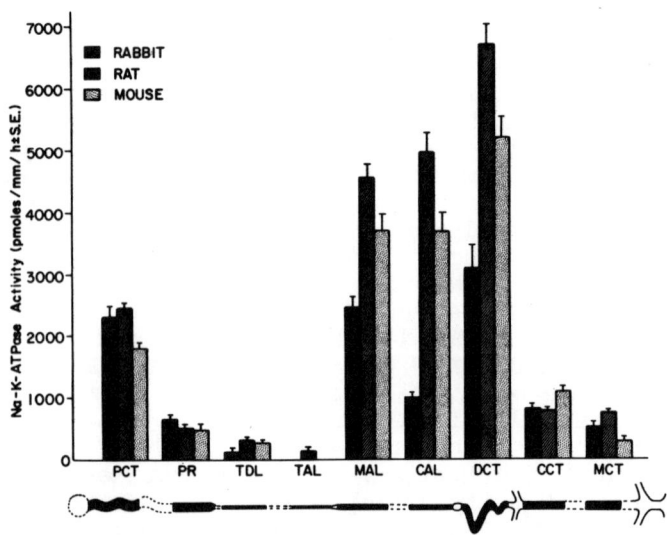

Figure 1. Distribution profile of Na-K-ATPase activity along the rabbit, rat, and mouse nephron (From Katz et al., 1979).

Study of renal Na-K-ATPase in microdissected tubules offers several advantages over alternative methods besides that of working with well-defined nephron populations of unequivocal provenance. The tubules have preserved their linear architecture and cell orientation, obviating the need for detergents required to treat "inside-out" membrane vesicles. Enzyme activity can be expressed conventionally per μg tissue protein or, more relevant to correlations with ion transport, per unit tubule length. Most important, recent refinements of this methodology allow a more comprehensive examination of the various aspects of Na-K-ATPase expression. These include measurements of specific ouabain binding, which reflects the number of pumps in each nephron segment (El Mernissi and Doucet, 1984), and of ouabain-sensitive ^{86}Rb uptake, an indicator of the functional capacity of the pumps (Fujii et al., 1987). The capability to measure pump density in a given nephron segment permits, of course, determination of its turnover rate, and thus the distinction between alterations in pump synthesis (or catabolism) and changes in its kinetics in cases where Na-K-ATPase activity is varied experimentally. Equally useful is the assessment of the pump's potassium (^{86}Rb) transporting capacity, a more direct measurement of Na-K-ATPase function than its ATP hydrolytic activity (see below). Further improvement of the isolated tubule micromethodology should extend its capacity to quantitative measurements of the various pump parameters in individual tubule cells (e.g. principal and intercalated cells of the collecting tubule), as evaluation of Na-K-ATPase activity in various cell types is now possible only with qualitative, albeit elegant, immunocytochemical techniques (Kashgarian et al., 1985).

Intracellular Localization

In renal tubular cells, as in other absorbing epithelia, Na-K-ATPase is located predominantly (or exclusively) in the basolateral membrane, i.e. the cell surface facing the "blood side", where its presence is predicated by the direction of the electrochemical potential. This location has been demonstrated consistently by numerous investigators using a wide variety of techniques, including manual dissection of the luminal and basal surface of proximal tubular cells (Schmidt and Dubach, 1971), isolation of luminal and basolateral membrane vesicles (Kinne et al., 1971), immunostaining with ferritin (Kyte, 1976), ouabain binding and autoradiography (Shaver and Stirling, 1978) and monoclonal antibodies (Kashgarian et al., 1985). Whether the renal enzyme may be present under special circumstances on other aspects of the plasma membrane (Hayashi and Katz, 1987a), or intracellularly during biogenesis and preceding its insertion in the membrane, as it does in other tissues (Tamkun and Fambrough, 1986), remains to be established.

Characteristics of Renal Na-K-ATPase - Is it the Same Everywhere in the Nephron?

The marked morphologic and functional heterogeneity of the nephron mentioned earlier raises the question whether the Na:K pump has the same characteristics in all its various subdivisions. This issue was first examined in tissue homogenates by Hendler et al. (1971), who found that the cortical and outer medullary Na-K-ATPase had the same affinity (K_m) for ATP and sodium, although the hydrolytic activity (V_{max}) of the cortical enzyme was substantially lower. Similarly, studies with microdissected tubules revealed comparable K_m for potassium despite the variable V_{max} of the enzyme from thick ascending limbs, proximal convoluted, and cortical collecting tubules (Doucet and Barlet, 1986). Furthermore, pump density measured by [^{3}H]ouabain binding varies along the nephron exactly in parallel with Na-K-ATPase hydrolytic activity, so that the turnover rate of the pump ($\sim$ 2000 ATP/ouabain binding site/ minute) is the same in all nephron segments (El Mernissi and Doucet, 1984).

The data reviewed above support, at least in a general way, the concept that renal Na-K-ATPase is of a single type, and the variations in its activity along the nephron reflect merely the density of active enzyme units in each segment, which varies between 3×10^6/cell in the outer medullary collecting tubule and 5×10^7/cell in the distal convoluted tubule (Doucet, 1984). The possibility that this may not be the case, however, was raised by observations that the sensitivity to ouabain appears to vary by substantial margins in several segments of the rabbit tubule, increasing progressively from the early to the late nephron (Doucet and Barlet, 1986). As pointed out by the authors, this variability probably reflects the local influence of factors extrinsic to the pump rather than the presence of different pump forms for various nephron segments. The latter hypothesis, besides being counterintuitive, is not supported (at the time of this writing) by any evidence for multiple isoforms of renal Na-K-ATPase, such as have been demonstrated in the mammalian brain and heart.

FUNCTION OF RENAL Na-K-ATPase

By effecting the coupled countertransport of Na and K against their respective electrochemical gradients at the basolateral surface of renal tubular cells, Na-K-ATPase mediates the active reabsorption of sodium from lumen to blood and secretion of potassium in the opposite direction. The central place of Na-K-ATPase in renal function is further underscored by its indirect role in the translocation of other solutes ("secondary active" transport) via the sodium gradient that it generates, as well as by its probable participation in active tubular K reabsorption by potassium-depleted animals.

Sodium Reabsorption and Secondary Active Transport

Reabsorption of sodium from the glomerular filtrate against an unfavorable gradient is a cardinal function of the renal tubule, both in quantitative terms and because it engenders the secondary-active reabsorption of several other key solutes. The involvement of renal Na- K-ATPase in active sodium reabsorption was suspected by renal physiologists in the early years after the enzyme was discovered, and compelling evidence supporting this association was amassed by the end of the 1960's. Katz and Epstein (1967) sought to demonstrate the physiologic role of renal Na-K-ATPase by correlating alterations in enzyme activity with experimentally-induced changes in tubular Na reabsorption, based on the analogy to increasing or decreasing traffic along the pathway catalyzed by a rate-limiting enzyme. It was found that Na-K-ATPase activity of kidney microsomes paralleled changes in net sodium reabsorption both when the latter was enhanced and reduced, i.e. changed in an adaptive way in response to the requirements for active Na transport. These studies and subsequent ones by the same authors and others (reviewed in Katz, 1982) identified a number of hormones and physiologic or pathophysiologic conditions (e.g. salt loading or depletion, partial renal ablation, high-protein feeding, pregnancy, adrenal- and thyroid hormone excess or deficiency) that modify renal Na-K-ATPase activity. The one characteristic shared in common by all these factors (apart from any direct effect on the pump the hormones listed may have) is that each alters the reabsorptive sodium load and/or sodium entry into tubular cells, and thus its access to the basolateral membrane enzyme (Katz and Epstein, 1967; Lindheimer and Katz, 1971; Katz and Lindheimer, 1973; Petty et al., 1981; O'Neil and Hayhurst, 1985; Scherzer et al., 1985). Indeed, availability of sodium for tubular transport seems to be an overriding determinant of the pump's activity, even when such potent regulators as the mineralocorticoids are considered (Westenfelder et al., 1977).

What fraction of the total sodium reabsorption of the kidney is active, and therefore mediated by Na-K-ATPase ? One way of answering this question is to inhibit the enzyme and observe the effect on sodium excretion. Use of ouabain, the specific inhibitor of the pump, for this purpose is limited by its cardiotoxicity which precludes attainment of fully inhibitory concentrations of the glycoside in vivo. To circumvent this problem, Besarab et al (1976) utilized the isolated perfused rat kidney, and found that very high concentrations of ouabain, no doubt sufficient to inhibit Na-K-ATPase completely, reduced net sodium reabsorption by about 50 %. This value is in good agreement with the decrement in Na reabsorption observed in vivo at maximal infusion rates of vanadate, a non-specific inhibitor of Na-K-ATPase (Day et al., 1980), and is consistent with observations that in proximal convoluted tubules only one third of the reabsorption is active (the rest being accomplished by solvent drag and electrical transference).

Quantitative Correlations. An obvious question in assessing the function of renal Na-K-ATPase is whether there is enough enzyme to account for the transport processes attributed to it. (Because the energy requirements for sodium transport are by far larger than those for potassium, only the former are considered here). For the kidney as a whole, Jorgensen (1980) and Doucet (1984) calculated that the Na-K-ATPase present, even if operating at limited (40 %) efficiency, is sufficient to drive the reabsorption of up to 50 % of the filtered sodium load - which corresponds to the fraction determined in the inhibitor studies just cited. These estimates are reinforced by calculation of the rate of synthesis based on oxygen consumption by the kidney, which show that generation of the nucleotide actually exceeds the needs for active Na reabsorption, i.e. that some of it is available for other energy-consuming processes. Similar calculations by Maude (1974) also show that the energy provided by renal oxidative metabolism slightly exceeds the estimated requirements for active tubular sodium transport.

Measurements of Na-K-ATPase activity in individual nephron segments offer the advantage of more direct comparisons with rates of active sodium transport determined with micropuncture or isolated tubule microperfusion techniques. It is emphasized, however, that such correlations are only approximate, principally because Na transport is measured under physiological conditions (that at least mimic those present in vivo), whereas enzyme activity is measured under optimal (V_{max}) conditions in vitro. Furthermore, the exact fraction of sodium reabsorption dependent on the pump in each nephron segment is uncertain, and the relevant measurement, unidirectional Na flux, also includes exchange diffusion. Given these limitations it is remarkable that there is a reasonably good concordance (assuming the stoichiometry of 3 Na/ATP) between Na-K-ATPase activity and sodium reabsorption measured by different investigators using diverse methods (Katz et al., 1979; Jacobson, 1981; Garg et al., 1981; O'Neil and Dubinsky, 1984). Significantly, this analysis indicates that, in general, the Na-K-ATPase present is theoretically sufficient for the active component of Na transport in those nephron segments where data for such comparisons are available (detailed in Doucet, 1984; Katz, 1986).

Secondary Active Transport. The low intracellular sodium concentrations generated by the Na:K pump promote sodium movement into the cell down its concentration gradient, and the energy thus dissipated is coupled with the uphill movement of other solutes via specific membrane carriers. That this type of transport, aptly termed "secondary-active", is indirectly mediated by Na-K-ATPase is proved by observations that it is inhibited by ouabain or omission of potassium from the extracellular milieu (reviewed in Doucet, 1984).

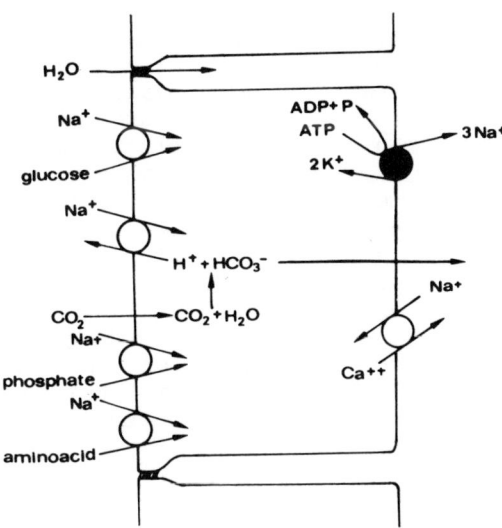

Figure 2. Secondary-active transport driven by Na-K-ATPase in proximal convoluted tubule cell (From Doucet, 1984).

The nature of the solutes transported via this mechanism and of the coupling with Na varies in different regions of the nephron. In proximal convoluted tubule cells the inward movement of glucose, amino acids and phosphate is in the form of cotransport with sodium, whereas a portion of proton secretion and calcium extrusion from the cells (Na:H and Na:Ca exchange) are accomplished by their uphill countertransport with Na moving inward along its concentration gradient (Fig. 2). (The latter are also transported actively by processes independent of Na-K-ATPase, i.e. by specific proton- or calcium pumps in the plasma membrane). It is obvious that for the secondary-active type of transport to operate it requires both adequate Na:K pump activity and the simultaneous presence, on the appropriate side of the cell membrane, of both Na and the solute to be transported.

As the reabsorption of organic solutes like glucose and amino acids is largely complete before the filtrate leaves the proximal tubule, secondary active transport involves different species in more distal parts of the nephron. In the thick ascending limb of Henle's loop, for example, the downhill movement of Na at the luminal membrane is coupled with the uphill entry of Cl, together with Na and K in the ratio of $2Cl^-:1Na^+:1K^+$ via a furosemide-sensitive carrier. It is now widely accepted that luminal chloride reabsorption is driven by the contraluminal Na-K-ATPase rather than being primarily active as previously proposed, and that in this segment of the nephron as in all others it is sodium that is the actively transported member of the NaCl pair (Knepper and Burg, 1983; Greger and Schlatter, 1983). The physiological importance of secondary-active transport, responsible as it is for the reclamation of filtered organic solutes, phosphate and bicarbonate, proton secretion, and urine concentration or dilution, cannot be overemphasized; the role of Na-K-ATPase in mediating this type of transport is no less crucial to the function of the kidney than the primary one played by the enzyme in the translocation of sodium and potassium across the renal tubule.

Potassium Secretion

Most of the filtered potassium is reabsorbed before the end of the distal convoluted tubule, so that urinary K is derived primarily from its secretion beyond this point, i.e. in the collecting tubule (Wright and Giebisch, 1985). The first (and rate-limiting) step in this process, the active uptake of K into the collecting tubule cell, is mediated by Na-K-ATPase present in the basolateral membrane; potassium then diffuses across the luminal or basolateral aspect of the cell, depending on the relative K permeability of the two membranes and the prevailing electrochemical gradients. K secretion into the lumen is usually passive, but if an unfavorable gradient is present, apical K exit may be coupled with the downhill entry of sodium - another example of secondary-active transport mediated indirectly by the pump. Highlighting the importance of Na-K-ATPase in K secretion, ouabain was found to inhibit this process in kidneys of amphibia, birds and mammals, as well as by the isolated cortical collecting tubule (reviewed in Katz, 1982; Doucet, 1984).

Animals or humans chronically subjected to increased dietary potassium become conditioned to it, being able to excrete all the ingested potassium so as to maintain external K balance ("potassium adaptation"), and to dispose rapidly of an acute K load ("potassium tolerance"). Silva and colleagues (1973) were the first to propose a role for Na-K-ATPase in renal potassium adaptation when they found an increase in enzyme activity of kidney homogenates from chronically K-loaded rats. Subsequent studies with isolated perfused kidneys from potassium-adapted rats indicated that their enhanced K excretory capacity is maintained independently of systemic influences and is completely inhibited by ouabain (Silva et al., 1975), i.e., it is contingent on the enhanced Na-K-ATPase activity.

Further insight into the process of renal K adaptation has been gained in recent years from experiments using microdissected tubules and morphometric techniques. These studies unequivocally located the nephron sites of the increment in Na:K pump activity (and by extension of increased K secretion) in the collecting duct, and not in the distal convoluted tubule as previously thought (Doucet and Katz, 1980; Stanton et al., 1981; Le Hir et al., 1982). The early portions of this nephron segment (initial and cortical collecting tubules) are the main sites of K adaptation; although the outer medullary collecting tubule may also be involved (Doucet and Katz, 1980; Rastegar et al., 1980), it seems that this process does not extend into the inner stripe of this segment (Garg and Narang, 1985; Mujais et al., 1986). The increased Na-K-ATPase activity that characterizes chronic renal K adaptation is due to a higher V_{max} of the enzyme, as its affinity for K is unaltered (Doucet and Katz, 1980). The increased pump density occurs in the principal cells of the collecting tubule which alone show a marked amplification of their basolateral membrane surface area after chronic potassium loading, whereas the appearance of intercalated cells remains unchanged (Rastegar et al., 1980; Stanton et al., 1981; Kaissling and Le Hir, 1982). Importantly, the stimulation of Na-K-ATPase activity in cortical collecting tubules (CCT) is dose-dependent and can be elicited by modest increments in K intake, such as may occur spontaneously in nature (Mujais et al., 1986).

One unresolved issue regarding renal potassium adaptation relates to the role of aldosterone in this process. Such a role is strongly suggested by the fact that aldosterone concentrations increase in K-loaded animals, and that increased plasma levels of mineralocorticoids lead to structural and biochemical changes in the CCT similar to those elicited by a high potassium intake (reviewed in Mujais et al., 1986). A number of studies using adrenalectomized animals or specific aldosterone antagonists (spironolactone or canrenoate) have yielded seemingly conflicting results, yet on closer scrutiny it appears that such discrepancies may be more apparent than real. Demonstration of Na-K-ATPase stimulation in CCT of adrenalectomized rabbits fed high K diets (Garg and Narang, 1985) and of dietary modulation of active K secretion by the CCT of such animals (Wingo et al., 1982) suggest that enhanced production of aldosterone is not a mandatory requirement for renal K adaptation. At the same time it is clear that this process is seriously impaired when the increment in aldosterone levels typically produced by potassium surfeit is prevented by adrenalectomy or blocked by antagonists: Not only are the induction of Na-K-ATPase and the basolateral membrane amplification blunted in such animals, but the enhanced K excretion is achieved at the price of dangerous hyperkalemia (Kaissling and Le Hir, 1982; Le Hir et al., 1982; Young and Paulsen, 1983; Stanton et al., 1985a; Mujais et al., 1986). Furthermore, besides its effect on K uptake by the pump, aldosterone also promotes its secretion by increasing apical K conductance (Sansom and O'Neil, 1985). It can be concluded that various components of the renal response to increased K load are facilitated by increased availability of aldosterone, and therefore that this hormone plays an important permissive role in potassium adaptation, even though it is not essential for its occurrence.

Critical to understanding the role of Na-K-ATPase in potassium adaptation is the temporal relationship between the increments in K excretion and pump activity. We have recently reported a large increase in enzyme activity in CCT of animals adapted to a high potassium diet for 7 days. Urinary potassium increased however to levels matching intake already 2 days after K loading began (the first time interval studied), whereas Na-K-ATPase activity did not change significantly until day 4 (Mujais et al., 1986).

This apparent dissociation could be explained either if the early response to K loading is not dependent on stimulation of the pump (e.g. takes place via an increase in apical K conductance or paracellular potassium flux), or if a change in pump function has occurred in vivo but was not detectable by measurement of its hydrolytic activity under V_{max} conditions in vitro. To explore the latter possibility we examined other indices of Na-K-ATPase function in rat CCT, including the directly relevant one of K-transporting capacity (measured as the ouabain-sensitive ^{86}Rb uptake; Fujii et al., 1987).

Results of the short-term experiments outlined above confirmed that Na-K-ATPase activity (ATP hydrolysis) remained unchanged after 2 days of high K intake. Likewise, the number of Na:K pumps (assessed from the specific binding of [^{3}H]ouabain) was also unaltered at this time. In contrast, ^{86}Rb uptake increased significantly both after 2 days and one day of dietary K loading, when urinary excretion of potassium was already comparable to the amount ingested. Furthermore, animals fed the K-enriched diet for such short intervals displayed an enhanced capacity to excrete an acute i.v. potassium load ("potassium tolerance"). These observations indicate that Na-K-ATPase of CCT is instrumental in the development and maintenance of potassium adaptation and tolerance, and that its participation assumes different forms in the early and late phases of these events. In the early stages there is an enhanced transporting capacity (increased turnover rate) by existing pump units to accommodate a homeostatic need too immediate to be met by an increase in pump number; later, if the requirement for increased K secretion is sustained, a different pattern emerges which is characterized by an increase in the number of membrane pumps (Fujii et al., 1987).

Potassium Reabsorption

In potassium-depleted animals the reabsorption of this cation across the luminal membrane of certain nephron segments must proceed against an unfavorable gradient (Strieder et al., 1974), and micropuncture studies have located this type of transport in the medullary collecting tubule (Linas et al., 1979; Backman and Hayslett, 1983; Wright and Giebisch, 1985). Curiously, active luminal potassium uptake in the course of K reabsorption has not been hitherto linked to Na-K-ATPase, although this step may be regarded as analogous to that occurring at the basolateral membrane during K secretion and was shown some time ago to be inhibited by ouabain (Strieder et al., 1974). The likely reason for this omission is that the concept of a luminal Na:K pump seems, at first sight, to be at odds with the prevailing view that Na-K-ATPase is present in the basolateral membrane. However, while this is certainly true under normal conditions, a different situation may obtain in other circumstances such as potassium depletion.

Intrigued by several reports that Na-K-ATPase activity increases in nonrenal cells exposed to low ambient K in culture, we examined the effect of a low potassium diet on Na-K-ATPase from various segments of the rat nephron (Hayashi and Katz, 1987a). K-depletion produced a striking, time-dependent increase in Na-K-ATPase activity selectively in the inner stripe of outer medullary collecting tubules ($MCT_{i.s.}$). After 3 weeks on a K-free diet, at a time when plasma aldosterone was significantly reduced, enzyme activity in $MCT_{i.s.}$ was over 4-fold higher than in control animals (!). Because the $MCT_{i.s.}$ plays an important role in K reabsorption and the maximum increment in pump activity coincided with the virtual elimination of potassium from the urine, we postulated that the enhanced Na-K-ATPase activity may be involved in K reabsorption, which predicates its location in the luminal membrane. Further supporting this hypothesis we found that the binding of [^{3}H]ouabain to intact tubules, which reflects the number of enzyme units on the basolateral membrane, was similar in K-depleted and control animals. On the other hand, when tubules were permeabilized by hypotonic lysis and freeze-thawing, we observed an increase in ouabain binding in K-depleted tubules that roughly paralleled that in Na-K-ATPase activity. As this procedure exposes pump units otherwise inaccessible in intact cells, the results suggest that the additional pumps were located either intracellularly or on the luminal membrane, where they would be involved in potassium reabsorption (Hayashi and Katz, 1987a).

The hypothesis that the increased number of Na:K pumps in K-depleted animals may be involved in potassium reabsorption was further tested in renal function experiments. We compared the response to ouabain of isolated perfused kidneys from 3-weeks K-depleted and control rats, based on the assumption that if the increased Na-K-ATPase activity in the former group represents a K-reabsorptive pump, ouabain ought to increase K excretion. This prediction was indeed confirmed, as ouabain increased potassium excretion by K-depleted kidneys at each concentration of potassium in the perfusate. In contrast, the response of control kidneys was variable, often showing a reduction in K excretion due presumably to inhibition of the peritubular (K-secretory) pump (Hayashi and Katz, 1987b). Finally, recent experiments using microdissected $MCT_{i.s.}$ reveal a substantially higher ouabain-inhibitable ^{86}Rb uptake in K-depleted tubules with open lumen, in which rubidium gains access to the luminal surface as well, compared to those with collapsed lumen, where ^{86}Rb uptake occurs only at the basolateral membrane (Fujii and Katz, unpublished observations).

The combined results of the studies summarized above provide persuasive, albeit indirect, evidence for the operation of a ouabain-sensitive, K-reabsorptive pump in the medullary collecting tubule of K-depleted rats. Confirmation of this hypothesis would be of great theoretical interest for the dual reason that a) it would extend the known functions of renal Na-K-ATPase to include its participation in tubular potassium reabsorption, and b) it would indicate that under certain conditions Na:K pump can be directed to either aspect of the plasma membrane, depending on the prevailing stimulus and the homeostatic needs of the organism.

REGULATION OF RENAL Na-K-ATPase

The principal factors known to modulate renal Na-K-ATPase activity are presented in Table 1. This list, which is not intended to be all-inclusive, is loosely based on the outline of Trachtenberg et al. (1981), who classified as "intrinsic" the immediate (seconds to minutes) modulation of enzyme activity by its direct interaction with substrates, modifiers, or inhibitors, and as "extrinsic" the long-term (hours to days) regulation brought about by the sustained influence of endocrine or developmental events. Factors regulating renal Na-K-ATPase are grouped according to these 2 broad categories above and below the solid line of Table 1, respectively; the broken line in the first group further divides substrates/activators from inhibitors of the enzyme.

Table 1. Regulation of Renal Na-K-ATPase

Intracellular Sodium

? Extracellular Potassium

? ATP Concentration

Endogenous Circulating Inhibitors	Hypothalamic Digoxin-like
? Vanadate	

Mineralocorticoids	Aldosterone (DOCA)
Glucocorticoids	Corticosterone (Dexamethasone)
Thyroid Hormones	T_3
? Insulin	
Cell Maturation	

A detailed discussion of all the factors listed in the table is beyond the scope or available space for this review, especially if one considers the abundance of question marks denoting uncertainty or controversy. Instead, I shall focus on the "intrinsic" regulators, emphasizing the role of intracellular sodium as possible mediator of the hormonal message and proximate signal for the regulation of pump activity.

To evaluate factors potentially involved in the short-term regulation of renal Na-K-ATPase it is helpful to determine which of its major reactants is likely to be rate-limiting under the conditions that prevail at the two sides of the tubular cell membrane. This type of analysis, based on results with a pure Na-K-ATPase preparation, suggests that the enzyme's activity is determined chiefly by cytoplasmic Na (Na_i), this being the only reactant normally present in nonsaturating concentrations (Jørgensen, 1980; Figure 3). While this premise is certainly correct, the situation in the intact cell may be less straightforward, at least in regards to ATP.

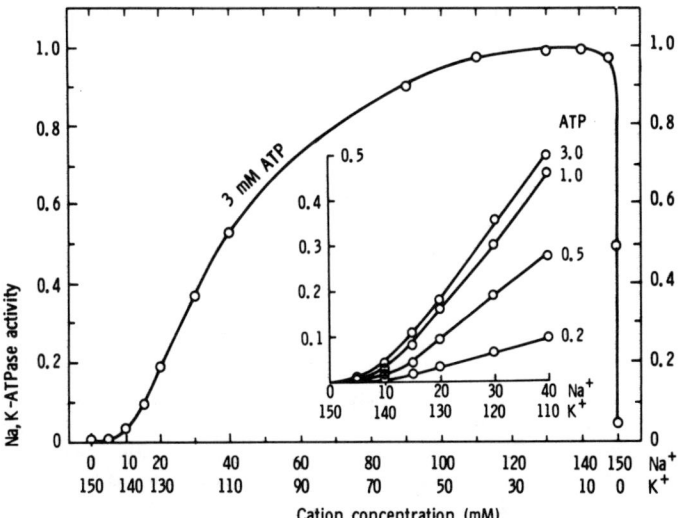

Figure 3. Dependence of pure renal Na-K-ATPase activity on Na, K, and ATP concentrations. The left and right parts of the graph reflect conditions in the intra - and extra-cellular domains of the tubule cell, respectively (From Jørgensen, 1980).

Because cell ATP concentrations, normally $\geq$ 2 mM, are close to those required for maximal ATPase⁻ hydrolytic activity, variations about this value were not considered likely to affect the enzyme. Recent reports demonstrate, however, a discrepancy between the ATP dependence of pump activity measured in intact cells and of hydrolytic activity measured in broken cell preparations. Thus both in a suspension of proximal tubules and in isolated proximal cells the uptake of K (or ^{86}Rb) had a linear, nonsaturating dependence on ATP concentration, whereas the hydrolytic activity of the enzyme in lysed cells showed the familiar curvilinear relationship, with saturation at $\sim$ 2 mM ATP (Soltoff and Mandel, 1984; Tessitore et al., 1986). These findings suggest that in vivo ATP concentrations may be rate-limiting, in which case factors affecting its synthesis or consumption could modulate the transporting capacity of the pump.

The relation of extracellular potassium concentration to renal Na-K-ATPase is also somewhat uncertain. Many studies using variably purified preparations show a high affinity of the enzyme for external K, so that half-maximal activation is achieved at ≤ 0.5 mM, and saturation at ~ 1 mM (e.g., Katz and Lindheimer, 1975; Jorgensen, 1980). Such figures indicate that K concentrations prevailing in the extracellular compartment of normal or even K-depleted animals are saturating, and would appear to exclude alterations in ambient K as the mechanism whereby the pump is activated during K-loading. However, using isolated intact tubules we found a linear dependence of ^{86}Rb uptake on K concentration up to 5 mM (Fujii and Katz, unpublished observations), and even with permeabilized (but otherwise structurally intact) tubules, Na-K-ATPase hydrolytic activity reached V_{max} only at 5 mM K (Doucet et al., 1979). Comparable values have been noted in other tissues as well (Trachtenberg, 1981) and therefore, as in the case of ATP, it appears that activation kinetics of the pump by K may differ in situ and in isolated membrane preparations.

In contrast to the ambiguity regarding ATP and potassium, a large body of evidence unequivocally points to the intracellular Na concentration (or Na:K ratio) as the major determinant of Na:K pump activity. The Na concentration required for half-maximal activation of the pure enzyme is 37 mM (Figure 3) when assayed at K concentrations that mimic those present in the cytosol.* As Na_i concentrations (determined by electron probe microanalysis) are 19 and 10 mmoles/kg wet weight in proximal and distal tubule cells, respectively (Beck et al., 1982), the enzyme likely operates in both locations at a fraction of its maximal capacity. An increase in Na_i will accelerate the turnover rate of the pump, calculated by Jorgensen (1980) to double when Na_i rises from 20 to 30 mM. Numerous investigators have demonstrated this relationship between intracellular sodium and Na-K-ATPase in various tissues or in cultured cells, using ionophores like monensin or nystatin to increase, or amiloride to block, sodium entry. We have also confirmed these observations in intact cortical collecting tubules treated with monensin, in which

---

* In kidney microsomes we found the $K_{0.5}$ for Na to be 17 mM (Katz and Lindheimer, 1975). This value was obtained, however, at a constant K concentration of 20 mM, which shifts the Na activation curve to the left compared to the higher K concentrations present in the intact cell.

ouabain-sensitive ^{86}Rb uptake nearly doubled as sodium concentration increased from 10 to 25 mM (Fujii and Katz, unpublished observations) (Fig. 4).

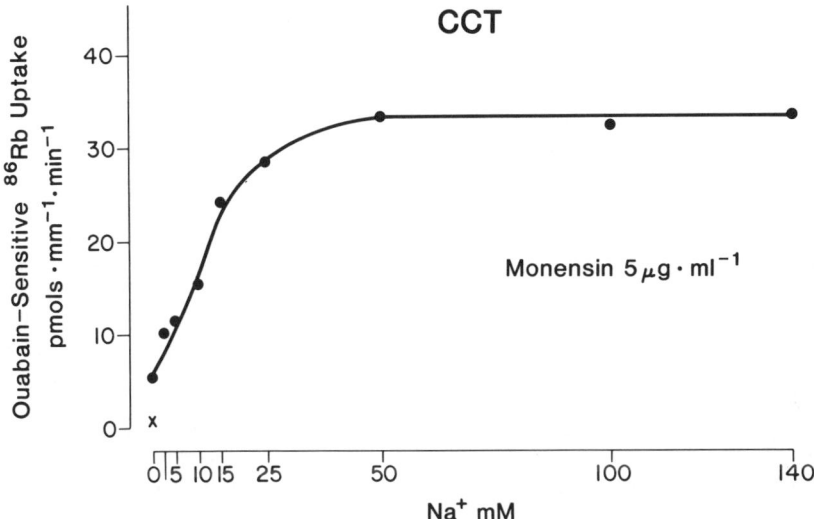

Figure 4. Relationship between intracellular Na concentration and Na-K-ATPase function in the cortical collecting tubule.

Although the effect of cell sodium concentration on the pump is not peculiar to the kidney, it assumes special importance in this organ as the potential mediator for the action of several hormones and other factors on renal Na-K-ATPase. This is perhaps best illustrated in the case of aldosterone, which promotes sodium reabsorption and potassium secretion by the collecting tubule. That the Na:K pump in the kidney is under partial mineralocorticoid control is attested to by oft-repeated observations that its activity drops after adrenalectomy and is restored by hormone replacement (reviewed in Marver and Kokko, 1983; Mujais et al., 1985). Furthermore, dietary manipulations of endogenous aldosterone production or chronic administration of exogenous mineralocorticoid modulate Na-K-ATPase activity in the CCT, the main target site of aldosterone in the nephron, in a manner concordant with its requirements for active transport (Schwartz and Burg, 1978; Garg et al.,

1981; Le Hir et al., 1982; O'Neil and Hayhurst, 1985). It is now generally accepted that the long-term action of aldosterone in the CCT is associated with an increase in the number of Na:K pump units, an effect evident morphologically in the selective amplification of the principal cells' basolateral membrane surface area (Mujais et al., 1985; Stanton et al., 1985b).

In contrast to the consensus regarding the chronic effects of the hormone, the timing and mechanism of the early events in the stimulation of Na-K-ATPase by aldosterone remain matters of debate. Much of the controversy centers on the nature and role of the "aldosterone-induced proteins", and on discrepancies between the length of the latent period preceding the aldosterone effect on Na-K-ATPase biosynthesis and on electrolyte transport. Even so, the preponderance of current opinion holds that among the earliest actions of aldosterone in kidney cells is an alteration in the permeability of the apical membrane (synthesis or activation of amiloride-sensitive Na channels) that promotes sodium entry (Marver and Kokko, 1983; Sariban-Sohraby et al., 1983; 1984; Sansom and O'Neil, 1985). This effect on luminal Na conductance and the resultant increase in Na_i are considered prerequisite for the subsequent induction of Na-K-ATPase, because the latter can be blocked by amiloride (Petty et al., 1981; Handler et al., 1981; Sariban-Sohraby et al., 1983; 1984; O'Neil and Hayhurst, 1985). Said otherwise, the late stimulation of enzyme synthesis by aldosterone is dependent on the acute elevation of intracellular Na (and presumably on its effect on the turnover rate of existing pumps) produced by the hormone.

The remarkable aspect of the relationship between intracellular sodium and stimulation of the pump is its widespread occurrence in various species and tissues, including nonepithelial ones. Enhanced sodium entry and the attendant increase in Na_i have been found to precede (and probably be the required signal for) Na-K-ATPase induction not only by aldosterone but also by glucocorticoids (Rayson and Gupta, 1985), triiodothyronine (Ismail-Beigi et al., 1986), and insulin (Fehlman and Freychet, 1981). Increased Na_i is also responsible for the stimulation of the pump during K depletion , a mechanism clearly demonstrated in cultured kidney cells (Bowen and McDonough, 1987) but probably underlying this process in vivo as well (Beck et al., 1982; Hayashi and Katz, 1987a). Finally, this same mechanism explains the stimulating effect of serum on

Na-K-ATPase in a variety of cultured cells, which is seemingly required for DNA synthesis and cell growth (Smith and Rozengurt, 1978; Mendoza et al., 1980). The general nature of this phenomenon establishes Na_i as a primary modulator of Na-K-ATPase activity and as the intracellular "messenger" of the effect of certain hormones and other factors on expression of the pump.

CONCLUSION

This review emphasized the role of Na-K-ATPase in kidney function and dealt briefly with its regulation. Much progress has been made in understanding these subjects, and a good deal remains to be made. Among the unresolved issues regarding the renal enzyme several come to mind that are pertinent to the topics discussed. First, regarding the role of Na_i in pump regulation, it should be noted that this thesis is not accepted by all (Geering et al., 1982; Doucet and Barlet-Bas, this symposium). Are there then situations in which direct genomic activation (e.g. via binding of hormone to nuclear receptors) occurs without the intermediate step of increased sodium entry (Verrey et al., 1987)? If, on the other hand, the late stimulation of pump biosynthesis indeed depends on the short-term increase in the turnover rate of existing pumps by increased Na_i, how are these signals related? In other words, how is the requirement for increased transport transmitted to the genome to produce the coordinated sequence of increased synthesis of the α and β subunits of the enzyme and their assembly and insertion into the membrane? How long does this stimulus (increased pumping rate by a fixed number of pumps) have to be sustained in order to trigger Na-K-ATPase gene transcription? And, finally, how does potassium loading increase the turnover rate of the enzyme in the short run and its biosynthesis chronically if baseline K concentrations are already saturating and aldosterone is not essential for this phenomenon? It is hoped - and predicted - that answers to some of these questions will be forthcoming before the next International Conference on Na-K-ATPase.

ACKNOWLEDGMENTS

My work cited in this article was supported by grants DK 13601 and DK 19250 from the National Institutes of Health. The expert secretarial assistance of Ms. Christina Foster is gratefully acknowledged.

REFERENCES

Backman KA, Hayslett JP (1983). Role of the medullary collecting duct in potassium conservation. Pflügers Arch 396:297–300.

Beck F, Dörge A, Mason J, Rick R, Thurau K (1982). Element concentrations of renal and hepatic cells under potassium depletion. Kidney Int 22:250–256.

Besarab A, Silva P, Epstein FH (1976). Multiple pumps for sodium reabsorption by the perfused kidney. Kidney Int 10:147–153.

Bowen JW, McDonough A (1987). Pretranslational regulation of Na-K-ATPase in cultured canine kidney cells by low K^+. Am J Physiol 252:C174–C189.

Day H, Middendorf D, Lukert B, Heinz A, Grantham J (1980). The renal response to intravenous vanadate in rats. J Lab Clin Med 96:382–395.

Doucet A (1984). Na-K-ATPase: General considerations, role and regulation in the kidney. Adv Nephrol 14:87–159.

Doucet A, Barlet C (1986). Evidence for differences in the sensitivity to ouabain of Na-K-ATPase along the nephron of rabbit kidney. J Biol Chem 261:993–995.

Doucet A, Katz AI (1980). Renal potassium adaptation: Na-K-ATPase activity along the nephron after chronic potassium loading. Am J Physiol 238:F380–F386.

Doucet A, Katz AI, Morel F (1979). Determination of Na,K-ATPase activity in single segments of the mammalian nephron. Am J Physiol 237:F105–F113.

El Mernissi G, Doucet A (1984). Quantitation of [^{3}H]ouabain binding and turnover of Na,K-ATPase along the rabbit nephron. Am J Physiol 247:F158–F167.

Fehlman M, Freychet P (1981). Insulin and glucagon stimulation of (Na^+-K^+)-ATPase transport activity in isolated rat hepatocytes. J Biol Chem 256:7449–7453.

Fujii Y, Mujais SK, Katz AI (1987). Response patterns of the cortical collecting tubule Na:K pump to potassium loading. Trans Assoc Am Physcns, in press.

Garg LC, Knepper MA, Burg MB (1981). Mineralocorticoid effects on Na,K-ATPase in individual nephron segments. Am J Physiol 240:F536–F544.

Garg LC, Narang N (1985). Renal adaptation to potassium in the adrenalectomized rabbit. Role of distal tubular sodium-potassium adenosine triphosphatase. J Clin Invest 76:1065–1070.

Geering K, Girardet M, Bron C, Kraehenbuhl JP, Rossier BC (1982). Hormonal regulation of (Na^+,K^+)-ATPase biosynthesis in the toad bladder. Effect of aldosterone and 3,5,3'-triiodo-l-thyronine. J Biol Chem 257:10338–10343.

Greger R, Schlatter E (1983). Properties of the basolateral membrane of the cortical thick ascending limb of Henle's loop of rabbit kidney. A model for secondary active chloride transport. Pflügers Arch 396:325–334.

Handler JS, Preston AS, Perkins FM, Matsumura M, Johnson JP, Watlington CO (1981). The effect of adrenal steroid hormones on epithelia formed in culture by A6 cells. Ann NY Acad Sci 372:442–454.

Hayashi M, Katz AI (1987a). The kidney in potassium depletion. I. Na^+-K^+-ATPase activity and [^{3}H]ouabain binding in MCT. Am J Physiol 252:F437–F446.

Hayashi M, Katz AI (1987b). The kidney in potassium depletion. II. K^+ handling by the isolated perfused rat kidney. Am J Physiol 252:F447–F452.

Hendler ED, Toretti J, Epstein FH (1971). The distribution of sodium–potassium–activated adenosine triphosphatase in medulla and cortex of the kidney. J Clin Invest 50:1329–1337.

Ismail-Beigi F, Haber RS, Loeb JN (1986). Stimulation of acute Na^+ and K^+ transport by thyroid hormone in a rat liver cell line: Role of enhanced Na^+ entry. Endocrinology 119:2527–2536.

Jacobson HR (1981). Functional segmentation of the mammalian nephron. Am J Physiol 241:F203–F218.

Jørgensen PL (1980). Sodium and potassium ion pump in kidney tubules. Physiol Rev 60:864–917.

Kaissling B, Le Hir M (1982). Distal tubular segments of the rabbit kidney after adaptation to altered Na- and K-intake. I. Structural changes. Cell Tissue Res 224:469–492.

Kashgarian M, Biemesderfer D, Caplan M, Forbush B (1985). Monoclonal antibody to Na,K-ATPase: Immunocytochemical localization along nephron segments. Kidney Int 28:899–913.

Katz AI (1982). Renal Na-K-ATPase: its role in tubular sodium and potassium transport. Am J Physiol 242:F207–F219.

Katz AI (1986). Distribution and function of classes of ATPases along the nephron. Kidney Int 29:21–31.

Katz AI, Doucet A, Morel F (1979). Na,K-ATPase activity along the rabbit, rat and mouse nephron. Am J Physiol 237:F114–F120.

Katz AI, Epstein FH (1967). The role of sodium–potassium–activated adenosine triphosphatase in the reabsorption of sodium by the kidney. J Clin Invest 46:1999–2011.

Katz AI, Lindheimer MD (1973). Renal sodium– and potassium–activated adenosine triphosphatase and sodium reabsorption in the hypothyroid rat. J Clin Invest 52:796–804.

Katz AI, Lindheimer MD (1975). Relation of Na-K-ATPase to acute changes in renal tubular sodium and potassium transport. J Gen Physiol 66:209-222.

Kinne R, Schmitz JE, Kinee-Saffran E (1971). The localization of the Na^+-K^+-ATPase in the cells of the rat kidney cortex. A study on isolated plasma membranes. Pflügers Arch 329:191-206.

Knepper M, Burg M (1983). Organization of nephron function. Am J Physiol 244:F579-F589.

Kyte J (1976). Immunoferritin determination of (Na^++K^+) ATPase over the plasma membranes of renal convoluted tubules. I. Distal segment. J Cell Biol 68:287-303.

Le Hir M, Kaissling B, Dubach UC (1982). Distal tubular segments of the rabbit kidney after adaptation to altered Na- and K-intake. II. Changes in Na-K-ATPase activity. Cell Tissue Res 224:493-504.

Linas SL, Peterson LN, Anderson RJ, Aisenbrey GA, Simon FR, Berl T (1979). Mechanism of renal potassium conservation in the rat. Kidney Int 15:601-611.

Lindheimer MD, Katz AI (1971). Kidney function in the pregnant rat. J Lab Clin Med 78:633-641.

Marver D, Kokko JP (1983). Renal target sites and the mechanism of action of aldosterone. Miner Electrolyte Metab 9:1-18.

Maude DL (1974). Mechanism of tubular transport of salt and water. In Thurau, K. (ed.), "Kidney and Urinary Tract Physiology", Baltimore: University Park, Vol. 6, pp. 39-78.

Mendoza SA, Wigglesworth NM, Pohjanpelto P, Rozengurt E (1980). Na entry and Na-K pump activity in murine, hamster, and human cells - Effect of monensin, serum platelet extract, and viral transformation. J Cell Physiol 103:17-27.

Mujais SK, Chekal MA, Hayslett JP, Katz AI (1986). Regulation of renal Na^+-K^+-ATPase in the rat: role of increased potassium transport. Am J Physiol 251:F199-F207.

Mujais SK, Chekal MA, Jones WJ, Hayslett JP, Katz AI (1985). Modulation of renal sodium-potassium-adenosine triphosphatase by aldosterone. Effect of high physiologic levels on enzyme activity in isolated rat and rabbit tubules. J Clin Invest 76:170-176.

O'Neil RG, Dubinsky WP (1984). Micromethodology for measuring ATPase activity in renal tubules: Mineralocorticoid influence. Am J Physiol 247:C314-C320.

O'Neil RG, Hayhurst RA (1985). Sodium-dependent modulation of the renal Na-K-ATPase: Influence of mineralocorticoids on the cortical collecting duct. J Memb Biol 85:169-179.

Petty KJ, Kokko JP, Marver D (1981). Secondary effect of aldosterone on Na,K-ATPase activity in the rabbit cortical collecting tubule. J Clin Invest 68:1514–1521.

Rastegar A, Biemesderfer D, Kashgarian M, Hayslett JP (1980). Changes in membrane surfaces of collecting duct cells in potassium adaptation. Kidney Int 18:293–301.

Rayson BM, Gupta RK (1985). Steroids, intracellular sodium levels, and Na^+/K^+-ATPase regulation. J Biol Chem 260:12740–12743.

Sansom SC, O'Neil RG (1985). Mineralocorticoid regulation of apical cell membrane Na^+ and K^+ transport of the cortical collecting duct. Am J Physiol 248:F858–F868.

Sariban-Sohraby S, Burg MB, Turner RJ (1983). Apical sodium uptake in toad kidney epithelial cell line A6. Am J Physiol 245:C167–C171.

Sariban-Sohraby S, Burg MB, Turner RJ (1984). Aldosterone-stimulated sodium uptake by apical membrane vesicles from A6 cells. J Biol Chem 259:11221–11225.

Scherzer P, Wald H, Czaczkes JW (1985). Na-K-ATPase in isolated rabbit tubules after unilateral nephrectomy and Na^+ loading. Am J Physiol 248:F565–F573.

Schmidt U, Dubach U (1969). Activity of (Na^+K^+)-stimulated adenosintriphosphatase in the rat nephron. Pflügers Arch 306:219–226.

Schmidt U, Dubach U (1971). Na K stimulated adenosinetriphosphatase: Intracellular localization within the proximal tubule of the rat nephron. Pflügers Arch 330:265–270.

Schmidt U, Dubach U (1974). Induction of Na K ATPase in the proximal and distal convolution of the rat nephron after uninephrectomy. Pflügers Arch 346:39–48.

Schwartz GJ, Burg MB (1978). Mineralocorticoid effects on cation transport by cortical collecting tubules in vitro. Am J Physiol 235:F576–F585.

Shaver JLF, Stirling C (1978). Ouabain binding to renal tubules of the rabbit. J Cell Biol 76:278–292.

Silva P, Hayslett JP, Epstein FH (1973). The role of Na-K-activated adenosine triphosphatase in potassium adaptation: stimulation of enzymatic acitivity by potassium loading. J Clin Invest 52:2665–2671.

Silva P, Ross BD, Charney AN, Besarab A, Epstein FH (1975). Potassium transport by the isolated perfused kidney. J Clin Invest 56:862–869.

Smith JB, Rozengurt E (1978). Serum stimulates the Na^+,K^+ pump in quiescent fibroblasts by increasing Na^+ entry. Proc Natl Acad Sci USA 75:5560–5564.

Soltoff SP, Mandel LJ (1984). Active ion transport in the renal proximal tubule. III. The ATP dependence of the Na pump. J Gen Physiol 84:643–662.

Stanton BA, Biemesderfer D, Wade JB, Giebisch G (1981). Structural and functional study of the rat distal nephron: Effects of potassium adaptation and depletion. Kidney Int 19:36-48.

Stanton B, Giebisch G, Klein-Robbenhaar G, Wade J, De Fronzo RA (1985a). Effects of adrenalectomy and chronic adrenal corticosteroid replacement on potassium transport in rat kidney. J Clin Invest 75:1317-1326.

Stanton B, Janzen A, Klein-Robbenhaar G, De Franzo R, Giebisch G, Wade J (1985b). Ultrastructure of rat initial collecting tubule. Effect of adrenal corticosteroid treatment. J Clin Invest 75:1327-1334.

Strieder N, Khuri R, Wiederholt M, Giebisch G (1974). Studies on the renal action of ouabain in the rat. Effects in the non-diuretic state. Pflügers Arch 349:91-107.

Tamkun MM, Fambrough DM (1986). The (Na^++K^+)-ATPase of chick sensory neurons. Studies on biosynthesis and intracellular transport. J Biol Chem 25:1009-1019.

Tessitore N, Sakhrani LM, Massry SG (1986). Quantitative requirement for ATP for active transport in isolated renal cells. Am J Physiol 251:C120-C127.

Trachtenberg MC, Packey DJ, Sweeney T (1981). In vivo functioning of the Na^+-K^+-activated ATPase. Curr Topics Cell Reg 19:159-217.

Verrey F, Schaerer E, Zoerkler P, Paccolat MP, Geering K, Kraehenbuhl JP, Rossier BC (1987). Regulation by aldosterone of Na^+, K^+-ATPase mRNAs, protein synthesis, and sodium transport in cultured kidney cells. J Cell Biol 104:1231-1237.

Westenfelder C, Arevalo GJ, Baranowski RL, Kurtzman NA, Katz AI (1977). Relationship between mineralocorticoids and renal Na^+,K^+-ATPase: sodium reabsorption. Am J Physiol 233:F593-F599.

Wingo CS, Seldin DW, Kokko JP, Jacobson HR (1982). Dietary modulation of active potassium secretion in the cortical collecting tubule of adrenalectomized rabbits. J Clin Invest 70:579-586.

Wright FS, Giebisch G (1985). Regulation of potassium excretion. In Seldin DW, Giebisch G (eds.), "The Kidney. Physiology and Pathophysiology", New York: Raven Press, pp. 1223-1249.

Young DB, Paulsen AW (1983). Interrelated effects of aldosterone and plasma potassium on potassium excretion. Am J Physiol 244:F28-F34.

COMPARISON OF THE MAXIMUM CAPACITY FOR ACTIVE SODIUM-POTASSIUM TRANSPORT IN THE LEFT AND RIGHT VENTRICLE OF MAMMALIAN HEART.

Tamás Bányász, Tibor Kovács, János Somogyi

Department of Physiology, Medical University of Debrecen, (T.B., T.K.), I.Department of Chemistry-Biochemistry, Semmelweis Medical University, Budapest (J.S.), Hungary

INTRODUCTION

A growing body of evidence suggests that that there are considerable differences in the concentrations of ions, enzymes and metabolites in the different regions of the mammalian heart (Aomine et al.,1982; 1983; Palfi et al., 1978; Poole-Wilson et al., 1975; Stanley et al., 1978). Yamamoto et al. (1979) observed a marked difference in the ouabain-suppressible 86-Rb uptake between ventricular and atrial muscle of guinea pig heart. Previously it has been shown, that the resting ouabain-suppressible 42-K and 86-Rb influx into the right ventricle of the rat was considerable higher compared to the left one (Kovács, 1984). Since, Na,K-ATPase activity of the myocardium from laboratory animals has been found to show marked regional variation (Yamamoto et al., 1981), therefore it cannot be ruled out that a regional difference in the density of pumping sites may influence the Rb uptake in the left and right ventricle.

The aim of this study was to compare the vanadate facilitated binding of 3-/H/-ouabain and the maximum rates of ouabain-suppressible Rb uptake in specimens of left and right ventricle from rat and guinea pig.

MATERIALS AND METHODS

All experiments were carried out on forty weeks old male Wistar rats (180-220g body weight) and twenty weeks old male guinea pigs (400-500 g). Ventricular ring- and strip preparations were isolated from the left and right ventricle of the hearts.

3-/H/-ouabain binding in vitro to ventricular preparations was performed by vanadate-facilitated method as it was described by Norgard et al. (1983), and Kjeldsen et al. (1985). To maintain the low concentration of sodium and potassium, that would not interfere with the vanadate-facilitated binding of 3-/H/-ouabain, samples were prewashed for 4x15 min at 0 °C in a vanadate-Tris-sucrose buffer. Incubation took place in a buffer containing 3-/H/-ouabain (40 kBq/ml; Amersham), in a final ouabain concentration of 10 - 5000 nmol/l (37°C for 180 min). Unspecific uptake of 3-/H/-ouabain was measured in the presence of excess unlabeled ouabain. After incubation the samples were washed for 4x30 min at 0º C in the Tris-sucrose buffer to reduce the fraction of labelled ouabain not bound to the receptors. After blotting, the samples (20-30 mg) were extracted in 5% trichloroacetic acid overnight, and after counted in liquid scintillation fluid and the total amount of 3-/H/-ouabain taken up by the samples calculated based on their specific activity. Correction for unspecific uptake, and the results expressed as picomoles per gramm wet weight of muscle tissue. Additional corrections were made for radiopurity of the isotope (98.9% and loss of specifically bound 3-/H/-ouabain during the washout.

86-Rb uptake was measured using isolated ring preparations. To increase the intracellular sodium concentration, preparations were preincubated in K-free Tyrode solution containing 0.1 mmol/l Ca at 30° C for 60 min. Ouabain (1 mmol/l) was added to half of the preparations during the last 15 min of preincubation. Ventricular preparations were incubated for 5 min in Tyrode solution containing 86-Rb (80 kBq/ml) and varying concentration of K (5.5-150 mmol/l). When the K cocentration of Tyrode was increased, the Na concentration was decreased by an equimolar amount. At the end of the exposition, the extracellular space was washed out by soaking samples in an inactive Tyrode solution at 0 °C for 5 min. Samples were then digested by concentrated nitric acid and taken for counting. On the basis of specific activity of 86-Rb in the incubation medium, Rb uptake was then calculated and expressed as nmol/g wet weight. The ouabain-suppressible 86-Rb uptake was calculated by substracting the ouabain-resistant Rb uptake from the total.

RESULTS

In the isolated resting myocardium preparations of rats or guinea pigs around 45 and 60 per cent of the total K influx was ouabain-suppressible measured at 30 C in standard incubation medium containing 5.5 mM K. It seems reasonable to assume that these components reflect the transport across sarcolemma via Na-K pump. When the uptake of K was assessed using 86-Rb as tracer under similar conditions as used for the experiments presented in Fig 1A, the ouabain-suppressible Rb uptake at low external K concentration was about 1.6 fold higher in the right ventricles than in the left ventricle from both rats and guinea pigs. Comparing the ouabain-suppressible Rb uptake between the two ventricle over the concentration range of 5.5-150 mM K, it was found that 86-Rb uptake remained 1.7-2.5 fold higher in the right ventricles of both types of experimental animals as compared to the left ventricle.

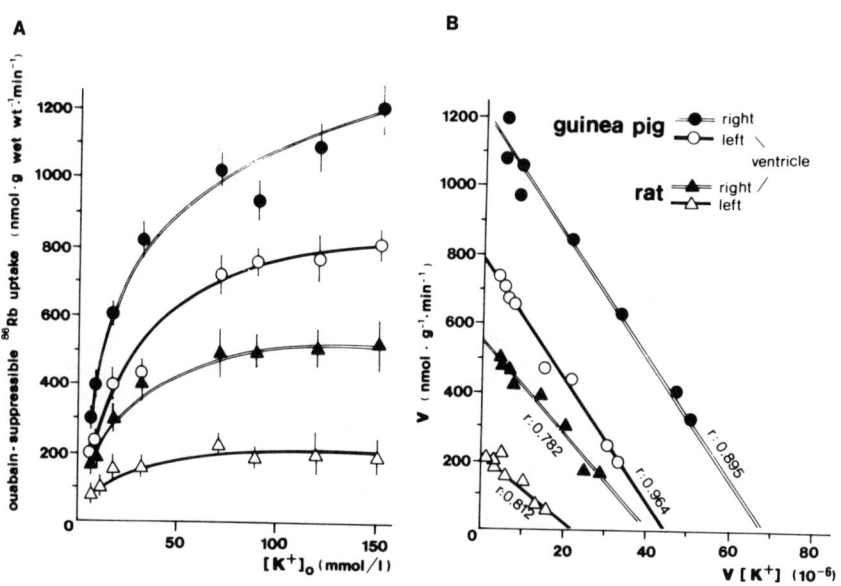

Figure 1. Ouabain-suppresible 86-Rb uptake in the left and right ventricle obtained from rats and guinea pigs as a function of the K concentration (A).
Eadie-Hofstee plot of the data gave maxima for ouabain-suppressible 86-Rb uptake (B).

From the Eadie-Hofstee plot of the ouabain-suppressible component of 86-Rb uptake it could be estimated that the maximum rates were higher in both cardiac chamber of the guinea pig than that of the rat (Fig 1B, Table 1).

Results suggest regional differences in the K influx which in turn may be a consequence of a regional variation in the number of pumping sites. For this purpose the 3-/H/-ouabain binding site concentrations were determined for each ventricles of animals. The basic characteristics of 3-/H/-ouabain binding to the samples of left and right ventricles from rats and guinea pigs were asssed in a serial of experiments. The effect of the concentration of free 3-/H/-ouabain was assesed by measurements in a concentration range of 100-500 nmol/l for rat and 50-1000 nmol/l for guinea pig. In Fig 2 the specific binding of 3-/H/-ouabain was plotted as a function of specifically

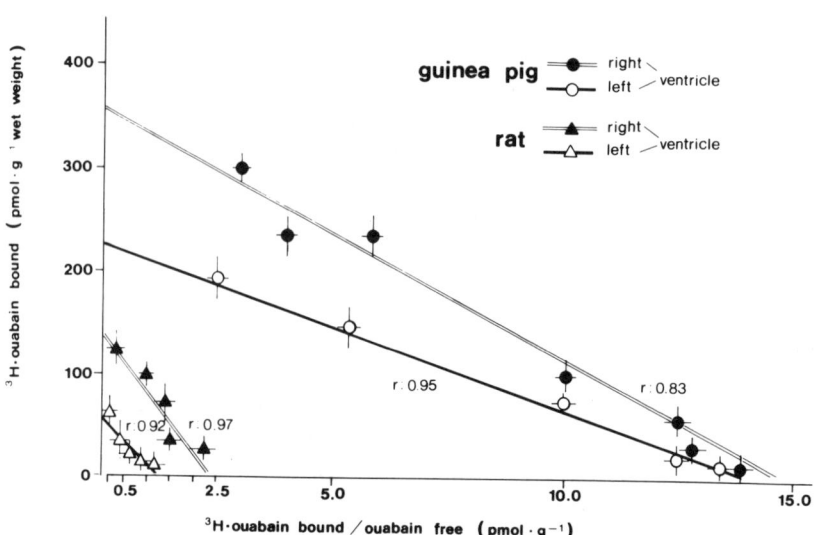

Figure 2. Bound ouabain as a function of bound/free ouabain ratio. The lines of this Scatchard type plot have been constructed using linear regression analysis. Each point represents the mean value of observation on eight to ten muscle speciments obtained from four to five different animals. The bars denote S.E.

bound 3-/H/-ouabain relative to the amount of free 3-/H/-ouabain in the incubation medium. This Scatchard-type plot gives no evidence for more than one population of binding sites. The ouabain binding site concentrations were significantly higher in the right ventricles compared to that of the left ventricle from each experimental species.

Table I. Comparison of maximum ouabain-suppressible ^{86}Rb-uptake and ^{3}H-ouabain binding site concentration in right and left ventricle of rat and guinea pig

animal	ventricle	ouabain-suppressible ^{86}Rb-uptake (nmol/g wet wt./min.)	^{3}H-ouabain binding site concentration (pmol/g wet weight)	number of K^+ ions transported per ^{3}H-ouabain binding site per min. (min^{-1}.)
rat	left	228±31 (24)	55 ± 7 (14)	4145
rat	right	566±42 (24)	144±29 (14)	3930
guinea pig	left	810±91 (15)	222±31 (9)	3648
guinea pig	right	1256±148 (15)	349±64 (9)	3599

The number of animals is given in parentheses

These data are in good agreement with the ouabain-suppresible Rb uptake data, and indicate that the number of pumping sites is lower in the left than in the right ventricle (Table 1). In both cardiac chambers obtained either from rats or guinea pigs, the maximum ouabain-suppresible 86-Rb uptake was found to vary in proportion to the concentration of 3-/H/-ouabain binding sites. It should be noted that the number of K ions transported per ouabain binding site per minute was around 3800 (mean values in the last column of Table 1) only 45 per cent of theoretical maximum, presumably due to the incomplete Na loading.

Although the physiological role of the heterogenous distribution of pumping sites in the left and right ventricle is an open question yet, but it may have important influence on myocardium performance, especially in falling heart when cardiac glycosides are applied.

REFERENCES

Aoamni M, Arita M, Imanishi S, Kiyosuet T (1982). Isotachophoretic analyses of metabolites of cardiac and skeletal muscle in four species. Jap J Physiol 32:741-760.

Aomine M, Arita M, Kiyosute T, Imanishi S (1983). Na,K-ATPase activity and Pi concentrations in various regions of monkey heart and their relation to post-overdrive hyperpolarization. J Jap Physiol 33:351-365.

Kjeldsen K, Norgard A, Hansen O, Clausen T (1985) Significance of skeletal muscle digitalis receptors for H-ouabain distribution in the guinea pig. J Pharm Exp Ther 234:720-727

Kovács T, (1984). Subcellular location and distribution of alkali cation pathways in various types of muscles. Symp Biol Hung 26:103-116.

Norgard A, Kjeldsen K, Hansen O, Clausen T (1983). A simple and rapid method for the determination of the number of H-ouabain binding site in biopsies of skeletal muscle. Biochim Biophys Res Comm 111:319-325

Palfi FJ, Besch HR, Watanabe AM (1978). Ouabain sensitivity of the Na,K-ATPase activity from single bovine cardiac Purkinje fibre and adjacent papillary muscle. J Mol Cell Card 10:1149-1155.

Poole-Wilson PA, Cameron IR (1975). ECS, intracellular pH, and electrolytes of cardiac and skeletal muscles. Am J Physiol 229:1229-1304.

Stanley RL, Conatser J, Dettbarn WD, (1978). Acetylcholine, choline acetyltransferase and cholinesterase in the rat heart. Bioch Pharm 27:2409-2411.

Yamamoto S, Akera T, Brody TM (1979). Sodium influx rate and ouabain sensitive rubidium uptake in isolated guinea pig atria. Biochim Biophys Acta 555:270-284.

Yamamoto S, Akera T, Kim DH, Brody TM (1981). Tissue concentration of Na,K-adenosine triphosphatase and the positive inotropic action of ouabain in guinea pig heart. J Pharm Exp Ther 217:701-707

IS THE Na,K-PUMP CAPACITY IN SKELETAL MUSCLE INADEQUATE DURING SUSTAINED WORK?

Torben Clausen and Maria Elisabeth Everts

Institute of Physiology
University of Aarhus
8000 Århus C, DENMARK

THE PROBLEM

During excitation of muscle cells, each action potential is associated with an influx of Na$^+$ and an efflux of K$^+$. Several studies have demonstrated that muscle work is associated with an appreciable net loss of cellular K$^+$, both in vivo and in vitro (Clausen, 1986). In the isolated rat diaphragm, Creese et al. (1958) found that at 38°C and a stimulation frequency of 2 Hz, the net loss of K$^+$ per contraction was 7.4 nmol/g wet wt. Isotopic flux measurements showed that with correction for diffusional delay, the unidirectional efflux of K$^+$ per contraction was considerably larger (16.5 nmol/g wet wt.). In isolated single frog muscle fibers, measurements of the K$^+$-efflux per contraction at 21°C gave a value of 3.9 nmol/g wet wt. (Hodgkin and Horowicz, 1959). Finally, electrical stimulation in vivo at a frequency of 5 Hz caused a net K$^+$-loss from rat soleus corresponding to 4.6 nmol/g wet wt. per contraction (Sreter, 1963).

All of these observations indicate that during muscle work, the amount of K$^+$ lost by excitation may exceed the capacity to reaccumulate K$^+$ via the Na,K-pump. The ensuing rise in extracellular K$^+$ is likely to interfere with muscle fiber excitation and cause fatigue. The present study was undertaken in an attempt to quantify the passive leaks of Na$^+$ and K$^+$ and to compare them with the rate of active Na,K-transport.

K$^+$-EFFLUX AND Na$^+$-INFLUX PER CONTRACTION

The net loss of K$^+$ induced by electrical stimulation was determined by flame-photometric measurement of the K$^+$ content in rat soleus muscle after various periods of incubation without or with electrical stimulation at 1 Hz. Unstimulated muscles maintained an almost constant K^+ content for 120 min of incubation. Electrical stimulation at 1 Hz for 30 min induced a net loss of K^+ corresponding to 3.9 nmol/g wet wt. per contraction, which is in good agreement with the value obtained in the rat soleus in vivo (Sreter, 1963).

Since part of the K^+ released during excitation will return to the muscle fibers via the Na,K-pump, this route was blocked in order to get a better estimate of the K^+-efflux per contraction. Stimulation in the presence of ouabain gave a net K^+-loss of 6.1 nmol/g wet wt. per contraction. However, since ouabain reduces the resting membrane potential (Clausen, 1986), the loss of K^+ per action potential may be reduced, and therefore, even this somewhat larger value represents an underestimate.

The efflux of K^+ per contraction was also estimated in isotope flux experiments. Following loading for 60 min with ^{86}Rb, the fractional loss of ^{86}Rb was measured in non-radioactive buffer. Electrical stimulation at 1 Hz gave a large rise in the fractional loss. By multiplying this increase by the K^+-content at the onset of stimulation and correcting for the use of ^{86}Rb as a tracer for K^+, a K^+-efflux per contraction of 7.0 nmol/g wet wt. was obtained (Table 1). Similar measurements performed in the presence of ouabain (1 mM) showed a K^+-efflux per contraction of 9.4 nmol/g wet wt.

Table 1. K^+-efflux and Na^+-influx during excitation

Preparation and Reference	K^+-efflux	Na^+-influx
	(nmol/g wet wt. per contraction)	
Rat diaphragm (38°C) (Creese et al., 1958)	16.5	-
Single fibers of frog sartorius (21°C) (Hodgkin and Horowicz, 1959)	3.9	6.3
Rat soleus (30°) (present study) without ouabain	7.0	6.8
with ouabain (1 mM)	9.4	8.5

The influx of Na$^+$ per contraction was determined by performing electrical stimulation during 2 min incubations in the presence of ^{22}Na. The amount of ^{22}Na taken up into the cytoplasm was measured following a washout for 4x15 min at 0°C during which extracellular isotope was removed (Clausen and Kohn, 1977). Using this procedure, the influx of Na$^+$ amounted to 6.8 nmol per contraction which is almost the same as the efflux of K$^+$ per contraction. It should be noted that this value is likely to represent an underestimate, partly because the isotopic Na$^+$ will not reach the central fibers of the muscles, partly because some of the ^{22}Na will be extruded from the cells via the Na,K-pump. When the experiment was repeated with 10 min incubation in the presence of 1 mM ouabain, the Na$^+$-influx per contraction was somewhat larger (8.5 nmol/g wet wt.).

BASAL AND MAXIMAL K$^+$-UPTAKE VIA THE Na,K-PUMP

Measurements performed using a variety of mammalian skeletal muscle preparations have shown that under resting conditions, the ouabain-suppressible K$^+$-influx is between 115 and 700 nmol per g wet wt. per min (Clausen, 1986). When compared with the estimated K$^+$-efflux per contraction, it is obvious that already at a stimulation frequency as low as 1 Hz, the efflux of K$^+$ (which is at least 60x7 = 420 nmol/g wet wt. /min) may easily exceed the reaccumulation of K$^+$ via the Na,K-pump. This may account for the net loss of K$^+$ incurred during continuous stimulation at 1 Hz.

It is usually assumed that the net loss of K$^+$ is counteracted by the simultaneous stimulation of active Na,K-transport. However, it can be calculated that in the isolated rat soleus muscle, 1 min of electrical stimulation at 1 Hz will produce a Na$^+$-gain of 60x6.8 = 408 nmol/g wet wt. Since the intracellular water space in this preparation is around 50%, the intracelular Na$^+$ concentration will only increase by 0.8 mM. This corresponds to a rise of only 4% above the intracellular Na$^+$ concentration in unstimulated muscles (Clausen and Kohn, 1977). Although this is likely to represent an underestimate of the rise actually taking place, it is insufficient to produce the required activation of the Na,K-pump.

At higher frequencies of stimulation, the situation rapidly gets worse (Fig. 1). Assuming that the K$^+$-efflux induced by excitation increases as a linear function of the frequency of stimulation and using the value of 7 nmol/g wet wt. per contraction

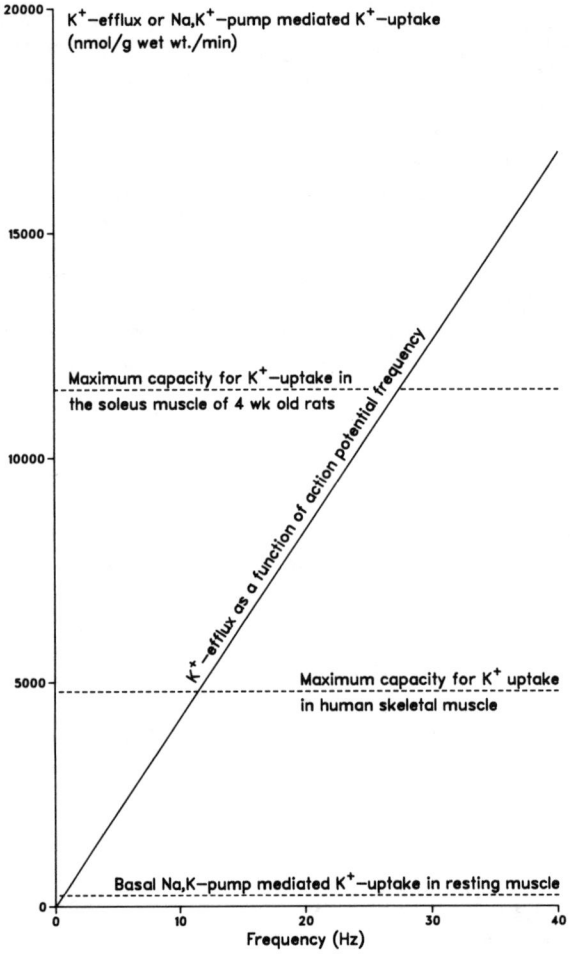

Fig. 1. K⁺-efflux associated with excitation and K⁺-uptake via the Na,K-pump in muscle. On the basis of the undivectional K⁺-efflux during stimulation at 1 Hz (Table 1) the K⁺-efflux was calculated as a function of stimulation frequency. The values are compared to the ouabain suppressible component of ^{42}K-influx as determined at 30°C under resting conditions (dashed line at the bottom) or the theoretical maximum capacity for K⁺-uptake at 37°C via the Na,K-pump as calculated for soleus muscles obtained from 4-wk-old Wistar rats (uppermost dashed line) or for the vastus lateralis muscle from normal adult human subjects (dashed line in the middle).

it can be calculated that at 30 Hz, K^+-efflux will reach 12,600 nmol/g wet wt./min. This exceeds the theoretical maximum rate of active K^+-influx via the Na,K-pump. Assuming an ATP turnover number of 8,000/min and two K^+ ions transported per site, the measured concentration of Na,K-pumps in the soleus muscle of 4-wk-old rats (0.72 nmol/g wet wt. - Clausen, 1986), will allow a maximum rate of K^+-influx at 37°C of 16,000x0.72 = 11,520 nmol/g wet wt./min (Fig. 1).

It should be noted that since the K^+-efflux induced by electrical stimulation was determined at 30°C, even larger losses could be expected at 37°C. Therefore, the maximum Na,K-pump capacity may be exceeded already at lower frequencies. The theoretical maximal capacity for K^+-influx that can be calculated for human skeletal muscle amounts to 4,800 nmol/g wet wt. per min (0.25x16,000; Nørgaard et al., 1984). Assuming a K^+-efflux per contraction of 7 nmol/g wet wt. it can be seen that already at 12 Hz, which is a moderate muscle activity level, the capacity of the Na,K-pump for reaccumulation of K^+ may be exceeded. These values should be compared to the observations that during short-lasting intense muscle contractions, the action potential frequency exceeds 50 Hz and that work on a standard bicycle ergometer corresponds to a frequency of around 20 Hz (Sjøgaard et al., 1985).

Obviously, such comparisons are only of significance if it can be assumed that measurements of the [^{3}H]ouabain-binding site concentration quantify functional Na,K-pumps. Experiments with isolated rat soleus muscle showed that following Na^+ loading so as to reach an intracellular Na^+ concentration above 120 mM, the ouabain-suppressible components of K^+-influx and Na^+-efflux were close to the theoretical maximum calculated at the temperature of measurement (Clausen et al., 1987).

CONCLUSIONS AND IMPLICATIONS

1. Even at moderate work intensities (or low frequencies of stimulation), the loss of K^+ from skeletal muscle exceeds the capacity of the Na,K-pump for reaccumulation of K^+.
2. This is the most likely explanation for the net loss of K^+ from working muscles and the hyperkalemia seen during exercise (Sjøgaard et al., 1985).
3. It may also account for the observation that in rabbit muscle, continuous electrostimulation at 10 Hz in vivo caused a marked net loss of K^+ and gain of Na^+ which showed rapid recovery after the cessation of stimulation (Sreter et al., 1980).

4. Since the Na,K-pump concentration is rate-limiting for K^+ reaccumulation during continuous activity, it can be expected that this will generate a demand for increased density of Na,K-pumps in the plasma membrane.
5. Indeed, endurance training (6 hours of swimming per day) was found to increase the concentration of [^{3}H]ouabain-binding sites in the working muscles by around 45% (Kjeldsen et al., 1985).

ACKNOWLEDGMENT

This study was supported by a grant from The Danish Medical Research Council.

REFERENCES

Clausen T (1986). Regulation of active Na^+-K^+ transport in skeletal muscle. Physiol Rev 66: 542-580.

Clausen T, Everts ME, Kjeldsen K (1987). Quantification of the maximum capacity for active sodium-potassium transport in rat skeletal muscle. J Physiol 388: 163-181.

Clausen T, Kohn PG (1977). The effect of insulin on the transport of sodium and potassium in rat soleus muscle. J Physiol Lond 265: 19-42.

Creese R, Hashish SEE, Scholes NW (1958). Potassium movements in contracting diaphragm muscle. J Physiol 143: 307-324.

Hodgkin AL, Horowicz P (1959). Movements of Na and K in single muscle fibers. J Physiol 145: 405-432.

Kjeldsen K, Richter EA, Galbo H, Lortie G, Clausen T (1986). Training increases the concentration of [^{3}H]ouabain-binding sites in rat skeletal muscle. Biochim Biophys Acta 860: 708-712.

Nørgaard A, Kjeldsen K, Clausen T (1984). A method for the determination of the total number of ^{3}H-ouabain binding sites in biopsies of human skeletal muscle. Scand J Clin Lab Invest 44: 509-518.

Sjøgaard G, Adams RP, Saltin B (1985). Water and ion shifts in skeletal muscle of humans with intense dynamic knee extension. Am J Physiol 248: R190-R196.

Sreter FA (1963). Cell water, sodium, and potassium in stimulated red and white mammalian muscles. Am J Physiol 205: 1295-1298.

Sreter FA, Mabuchi K, Köver A, Gesztelyi I, Nagy Z, Furka I (1980). Effect of chronic stimulation on cation distribution and membrane potential in fast-twitch muscles of rabbit. In Pette D (ed.): "Plasticity of Muscle," Walter de Gruyter & Co., Berlin - New York, pp 441-451.

Na,K-ATPase CO-DISTRIBUTES WITH ANKYRIN AND SPECTRIN IN RENAL TUBULAR EPITHELIAL CELLS

Michael Kashgarian, Jon S. Morrow, Harald G. Foellmer, Andrea S. Mann, Carol Cianci, and Thomas Ardito

Departments of Pathology, Physiology and Medicine, Yale University, New Haven, CT

INTRODUCTION

Na,K-ATPase has been shown to be confined to the lateral and infolded surfaces of the cell membranes of renal tubular epithelial cells (Kashgarian et al., 1985). This localization of Na,K-ATPase is undoubtedly responsible for generating the driving force for sodium and fluid reabsorption and indirectly for the co-transport of amino acids, glucose and phosphate and the counter transport of protons. The mechanisms responsible for establishing and maintaining this high degree of order of polar distribution of transport proteins are poorly understood (Simons and Fuller, 1985). Several components of the cytoskeleton have been implicated in the establishment and maintenance of this distribution, including microtubules, actin and the cortical cytoskeletal protein fodrin (Rodriquez-Boulan, 1983). In the erythrocyte, ankyrin and protein 4.1 have been demonstrated to mediate the attachment of the spectrin-actin based cortical cytoskeleton to the anion exchange protein band 3 (Anderson and Marchesi, 1985). To investigate whether similar mechanisms are operative in the polar distribution of Na,K-ATPase, the distribution of the cytoskeletal proteins was studied in renal epithelial cells and in Madin Darby Canine Kidney (MDCK) cells in culture.

METHODS

Purification of Proteins

Na,K-ATPase was prepared from membrane vesicles obtained from fresh frozen canine kidneys by homogenizing the outer medullas in 0.25 M sucrose, 0.03 M histidine, pH 7.2 followed by isolation of the membrane fraction by sedimentation using the method of Jorgensen (1974). Ankyrin was prepared from fresh human erythrocyte membranes by extraction of spectrin depleted erythrocyte vesicles with 1 M KCl followed by ion exchange chromatography using DE-52 cellulose. The details of the procedure have been previously described (Horne et al., 1985). The cytoplasmic domain of protein 3 was solubilized by α-chymotryptic digestion of the ankyrin depleted erythrocyte membrane vesicles which had been stripped of any residual peripheral proteins by washing. The preparation of this fragment has been previously described (Fukuda et al., 1978).

Antibodies

The antibody to the α-subunit of Na,K-ATPase MAB C62.4 (Kashgarian, 1985) and the anti-fodrin antisera (Harris et al., 1986) have been previously characterized and described.

Immunoelectron Microscopy

Rat kidneys were fixed by retrograde perfusion with periodate lysine paraformaldehyde (McLean and Nakane, 1974). MDCK cells were fixed by immersion with the same fixative. Samples were then cryoprotected with 10% DMSO, frozen and tissue samples sectioned at 16 microns with a cryostat. The immunolabeling procedure utilized was identical to that previously described (Kashgarian et al., 1985). Samples were examined without counterstaining using a Zeiss EM10B electron microscope at 60 kv.

Tissue Culture Procedures

Stock MDCK1 cells are maintained in 100 mm Corning dishes in MEM supplemented with 10% fetal bovine serum and 1% penicillin streptamycin. Millicell membranes were coated with gelatin, purified type I collagen (bovine), type IV, type V or laminin (EHS tumor). Plain and coated Millipore filters were seeded with trypsinized stock MDCK1 cells and grown to confluence. Na,K-ATPase was assayed by the method of Forbush (1983) and transmonolayer was measured in an Ussing chamber.

Binding Assays
The binding of ^{125}I labeled ankyrin to the dog kidney membranes was measured by co-sedimentation of the bound ankyrin with membranes through a 5% sucrose barrier in isotonic buffer. Assays were done as previously described for the binding of spectrin to erythrocyte membrane vesicles (Harris et al., 1986). Binding inhibition studies were performed by incorporating increasing amounts of the 43 KD cytoplasmic domain of protein 3 into the incubation medium of the assay. The direct binding of ^{125}I labeled ankyrin to Na,K-ATPase was detected after the transfer of Na,K-ATPase to nitrocellulose sheets following electrophoresis in 7% polyacrylimide gels.

RESULTS

Immunolocalization Studies
Na,K-ATPase, fodrin and ankyrin were shown to co-distribute and be confined to the lateral and basal infoldings of renal tubular epithelial cells. There was a relative exclusion of all of these proteins from the flat basal regions where the plasmalemma was directly opposed to the basal lamina in a pattern which has been previously described (Kashgarian et al., 1985). A similar pattern was seen when MDCK cells were grown to confluence on Millipore filters coated with laminin, type IV collagen or type V collagen. When MDCK cells were grown on uncoated filters or grown on filters coated with type I collagen or gelatin the proteins distributed on all portions of the basal lateral membrane without exclusion where plasma membrane and matrix were opposed. A loss of polarity with a generalized distribution of all cell membrane surfaces was seen in MDCK cells grown in suspension.

Binding Studies
Dog kidney membranes were shown to bind red blood cell ankyrin with a KD of approximately 1.3 µM and an estimated saturation value of 77 mg. of ankyrin per mg. of kidney membrane protein. The binding was competively inhibited by the 43 KD fragment of band 3. In a solid phase binding assay, the α-subunit of Na,K-ATPase was

shown to specifically bind ^{125}I labeled ankyrin in a manner similar to the binding of ankyrin to band 3 obtained from red cells.

Transmonolayer Resistance

The time course of the development of confluence as indicated by a high transmonolayer resistance for MDCK cells grown on different matrix coatings is presented in Table 1.

TABLE 1

TRANSEPITHELIAL RESISTANCE (Ohms·cm^2) OF MDCK1 CELLS GROWN ON DIFFERENT MATRIX COATINGS

	4 DAYS	5 DAYS	7 DAYS
FILTER	300 ± (100)	2100 ± (200)	1500 ± (400)
GELATIN	700 ± (200)	1300 ± (500)	1200 ± (400)
TYPE I	800 ± (500)	900 ± (500)	1400 ± (300)
TYPE IV	500 ± (200)	2500 ± (500)	4400 ± (800)
TYPE V	1300 ± (600)	3100 ± (300)	4200 ± (1000)
LAMININ	300 ± (200)	2500 ± (400)	4800 ± (500)

Confluence appeared to be achieved between 5-7 days and cells grown on uncoated filters reached a maximum transmonolayer resistance of approximately 1,500 Ohms·cm^2. Coating of the filters with gelatin or type I collagen showed no difference. In contrast, cells grown on filters coated with type IV collagen, type V collagen or laminin demonstrated a four fold increase with transmonolayer resistances in excess of 4,000 Ohms·cm^2 by seven days.

DISCUSSION

These experiments demonstrate that Na,K-ATPase co-distributes with ankyrin and spectrin both in intact renal tubular epithelial cells and in MDCK cells. This co-distribution suggests that there is more than a casual association between these proteins. In the erythrocyte, it has been demonstrated that spectrin and ankyrin have a similar association with the band 3 protein through interaction of its cytoplasmic domain (Anderson and Marchesi, 1985) and a similar co-distribution of ankyrin, spectrin and band 3 has been demonstrated in the intercalated cell of the collecting duct of the kidney (Drenckhahn et al., 1985). Further evidence that this is a specific association is seen in the experiments with MDCK cells. In nonconfluent MDCK cells Na,K-ATPase and fodrin lose their polar distribution with Na,K-ATPase present on all membrane surfaces and fodrin approximately evenly divided between cytoplasmic and plasma membrane associated sites in a random fashion. When confluence occurs, a coincidence segregation of Na,K-ATPase and fodrin to the basolateral portions of the membrane is seen. A further restriction to the in vivo distribution is seen when MDCK cells are grown on basement membrane matrix components. Coincident with the redistribution of these proteins to an in vivo pattern is an increase in the transmonolayer resistance. This constellation of findings suggests that there are specific interactions of Na,K-ATPase and cytoskeletal proteins which regulate its distribution in the cell and which may be related to other cytoskeletal membrane interactions which determine tight junctional integrity.

A specific interaction is further supported by the binding studies. These demonstrated that ankyrin binds specifically to the α-subunit of Na,K-ATPase and that this binding is inhibited by the cytoplasmic domain of band 3. These studies suggest that the interaction of the α-subunit with the cytoskeletal proteins is similar to that seen in the erythrocyte where ankyrin and protein 4.1 mediate the attachment of the spectrin actin base cortical cytoskeleton to the integral membrane proteins. Taken together, these studies indicate that the renal analogs of ankyrin mediate the organization of the cortical cytoskeleton via a direct interaction with the α-subunit of Na,K-ATPase. Furthermore, this interaction is highly specific and appears to be identical to that involved with protein 3 in the erythrocyte. The cortical cytoskeleton may thus play an integral role in the establishment an maintenance of polarity in renal epithelial cells.

REFERENCES

Anderson JP Marchesi VT (1985). Regulation of the association of membrane skeletal protein 4.1 with glycophorin by a phosphoinositide. Nature 318:295-298.

Drenckhahn D, Schluter K, Allen DP, Bennett V (1985). Co-localization of band 3 with ankyrin and spectrin at the basal membrane of intercalated cells in the rat kidney. Science 230:1287-1289.

Forbush B III (1983). Assay of Na,K-ATPase in membrane preparations. Anal Biochem 128:157-163.

Fukuda M, Eshdat Y, Tarone G, Marchesi VT (1978). Isolation and characterization of peptides derived from the cytoplasmic segments of band 3. J Biol Chem 253:2419-2428.

Harris AS, Anderson JP, Yurchenco PD, Green D, Ainger KJ, Morrow JS (1986). Mechanism of cytoskeletal regulation: Functional and antigenic diversity in human erythrocyte and brain beta spectrin. J Cell Biochem 30:51-69.

Horne WB, Leto TS, Marchesi VT (1985). Differential phosphorylation of multiple sites in protein 4.1 and protein 4.9 by phorbol ester activated and cyclic AMP dependent protein kinases. J Biol Chem 260:9073-9076.

Jorgensen PL (1974). Purification and characterization of Na,K-ATPase III. Biochim Biophys Acta 356:36-52.

Kashgarian M, Biemesderfer D, Caplan M, Forbush B (1985). Monoclonal antibody to Na,K-ATPase: Immunocytochemical localization along nephron segments. Kid Int 28:899-913.

McLean IW, Nakane PK (1974). Periodate-lysine paraformaldehyde fixative. J. Histochem Cytochem 22:1077-1083.

Rodriquez-Boulan E (1983). Membrane biogenesis enveloped RNA viruses and epithelial polarity. Mol Cell Biol 1:119-170.

Simons K, Fuller SD (1985). Cell surface polarity in epithelia. Ann Rev Cell Biol 1:243-288.

Na,K-ATPase CONCENTRATION IN SKELETAL MUSCLE: QUANTIFICATION, REGULATION, AND SIGNIFICANCE

Keld Kjeldsen, Maria E. Everts and Aage Nørgaard

Institute of Physiology, University of Aarhus, DK-8000 Århus C (K.K., M.E.E., A.N.) and Medical Department B 2142, State University Hospital, Blegdamsvej 9, DK-2100 Copenhagen Ø, Denmark (K.K.)

QUANTIFICATION

Various methods are available for studies of skeletal muscle Na,K-ATPase. For quantitative studies, however, it is a prerequisite that measurements take place with complete recovery and activation of Na,K-pumps. Alternatively, it should be documented that the recovery and activation of the Na,K-ATPase are identical in tissue from control as well as experimental groups. Table 1 gives the recovery/activation of various methods.

Table 1. Approximate recovery/activation of skeletal muscle Na,K-ATPase by various methods:

Method	Recovery/activation
Na,K-ATPase activity	2-18%
K^+-dependent 3-O-MFPase activity	100%
^{3}H-ouabain binding	100%
Standard ouabain suppressible K^+-uptake	1-7%
Maximal stimulated ouabain suppressible K^+-uptake	90-100%

In studies where Na,K-ATPase activity in skeletal muscle is measured the recovery of Na,K-ATPase is only a few per cent. This is due to the extensive purification that has to be carried out to reduce the large amount of unspecific ATPases in muscle tissue. It is difficult to ensure that the recovery of membrane fractions and thus Na,K-ATPase is identical in tissue from control and experimental groups after purification (Jones and Besch, 1984;

Kjeldsen, 1986).

The activity of the K^+-dependent 3-O-methylfluorescein phosphatase (3-O-MFPase), an enzyme closely related to the Na,K-ATPase can, however, be determined at optimum conditions. This is due to the fact that it can be carried out on crude muscle homogenates (Askari and Koyal, 1968; Nørgaard et al., 1984).

The recovery problem can also be avoided by measuring ^3H-ouabain binding to small intact muscle samples (around 5 mg wet weight) (Kjeldsen, 1986). Like K^+-dependent 3-O-MFPase activity measurements, ^3H-ouabain binding assays quantify muscle Na,K-ATPase with high accuracy. The major advantages of ^3H-ouabain binding studies are that they are simple and rapid and that errors of the method can be easily identified and corrected for (Kjeldsen, 1986; 1987).

In standard measurements of ouabain suppressible K^+-uptake where muscle fibers are incubated in Krebs-Ringer bicarbonate buffer at 30°C (Kjeldsen et al., 1985a), the Na,K-pump is only activated to 1-7% of its maximal activity (Clausen et al., 1987). This is due to the relatively low activation of the Na,K-pump by the intracellular Na^+ and extracellular K^+ concentrations under physiological conditions. Thus, in order to obtain comparable results reflecting the Na,K-pump concentration by this method the Na^+ and K^+ concentrations should be identical in samples from control and experimental animals or differences should be taken into account. Hence, it is necessary to determine these concentrations and to correct for differences. It is, however, a major problem to determine the intracellular concentration of Na^+ with sufficient accuracy.

By measuring the ouabain-suppressible K^+-uptake in Na^+-loaded muscles exposed to high extracellular K^+ (130 mM) it is possible to obtain an activation of the Na,K-ATPase which is close to the theoretical maximum (Clausen et al., 1987). Under such maximally stimulated conditions the ouabain suppressible K^+-uptake can be used to quantify variations in muscle Na,K-ATPase concentration. The method has, however, as compared to ^3H-ouabain binding the disadvantage that it has to be carried out on intact muscle fibers.

REGULATION

Table 2 summarizes the changes in Na,K-ATPase concentration under various physiological and pathological conditions as measured by ^{3}H-ouabain binding capacity by our group. Where possible, changes obtained by ^{3}H-ouabain binding studies have been confirmed by measurements of K$^+$-dependent 3-O-MFPase activity (Kjeldsen et al., 1984a; 1984b; Nørgaard et al., 1984) and maximally stimulated ouabain suppressible K$^+$-uptake (Clausen et al., 1987).

Table 2. Up- and down-regulation of skeletal muscle Na,K-ATPase concentration. Up- and down-regulation is indicated by + and -, respectively. No regulatory change by 0. A dash indicate that no study has yet been done. Values give changes in per cent:

Condition	Per cent change	
	Animals	Humans
Age	+433/-69,+100/+64	0
Potassium-depletion	-78	-18
Magnesium-depletion	-21	-18
Semi-starvation	-25	-
Potassium-loading	+36	-
Hypothyroidism	-50	-50
Hyperthyroidism	+195	+68
Diabetes mellitus	-49	-
Hypertension	0	-
Training	+46	0
Immobilisation	-30	-

Major developmental changes are observed in rodent skeletal muscle Na,K-ATPase concentration (Sperelakis, 1972). In rats an increase within the first 4 weeks of life (+433%) is followed by a decrease to a plateau 69% below the maximum value. In guinea pigs only the decrease is seen after birth (-62%). An upregulation is observed in utero (Kjeldsen et al, 1984a; 1985b). In human subjects no age-dependent changes have been observed.

Dietary abnormalities cause major changes in skeletal muscle Na,K-ATPase concentration. During K$^+$-depletion a reduction is seen in animals as well as in humans (Kjeldsen et al., 1984b; Brown et al., 1986; Kjeldsen 1987; Dørup et al., 1987). This is also the case for Mg$^+$-depletion (Kjeldsen and Nørgaard, 1987; Dørup et al., 1987). Reduction of food supply causes a minor reduction in rat skeletal muscle Na,K-ATPase concentration (Kjeldsen et al.,

1986b). In contrast K+-loading causes an upregulation (Blachley et al., 1986).

Endocrine changes influence the Na,K-ATPase concentration. A reduction is seen with hypothyroidism and an upregulation with hyperthyroidism in animals as well as humans (Asano et al., 1976; Kjeldsen et al., 1984c, 1986a). Streptozotocin induced diabetes mellitus caused a downregulation of up to 48% in rat skeletal muscle Na,K-pump concentration (Kjeldsen et al., 1987).

It should be noted that studies of ^{3}H-ouabain binding to skeletal muscles and resistance vessels from spontaneously hypertensive rats gave no evidence for any involvement of the Na,K-ATPase concentration in the spontaneous hypertensive process (Aalkjær et al., 1985).

The state of training seems to influence muscle Na,K-ATPase concentration (Knochel et al., 1985). In rats an upregulation is seen with training and a down-regulation with immobilization (Kjeldsen et al., 1986c). A study on vastus lateralis muscle biopsies from healthy young volunteers before and after a light training program showed, however, no increase.

SIGNIFICANCE

Changes in skeletal muscle Na,K-ATPase concentration are of importance for basic as well as specific cellular functions (intracellular electrolyte homeostasis, contractility, excitability). Since skeletal muscle constitutes a major distribution volume for digitalis glycosides, changes in Na,K-ATPase concentration (digitalis receptor concentration) may change the amount of digitalis glycoside available in plasma for binding to the heart during digitalization (Kjeldsen et al., 1985b). Changes are furthermore of importance for body potassium-homeostasis. During muscle activity K+ is lost from the cells and extracellular K+ concentration increases up to 15 mmol/l. Optimum function of the Na,K-ATPase is important under muscle work. The Na,K-pump must together with the K+-permabilities allow the extracellular K+ concentration to increase to cause vasodilation so that blood flow to the working muscles can increase. Furthermore, the Na,K-pump must rapidly pump K+ back to the cells so that muscle excitation and contraction can occur again and so that a major rise in plasma-potassium with subsequent risk for cardiac arrytmics is prevented.

CONCLUSION

^{3}H-ouabain binding to intact muscle samples (down to a few mg wet weight) is a simple and accurate method for quantifying regulatory changes in skeletal muscle Na,K-ATPase concentration. Major changes occur with development, dietary abnormalities, endocrinological disturbances and physical activity. The changes may be of importance for cellular functions, digitalis distribution and potassium-homeostasis.

REFERENCES

Aalkjær C, Kjeldsen K, Nørgaard A, Clausen T, Mulvany MJ (1985). Ouabain binding and Na$^+$ content in resistance vessels and skeletal muscles of spontaneously hypertensive rats and K$^+$-depleted rats. Hypertension 7: 277-286.

Asano Y, Liberman UA, Edelmann IS (1976). Relationship between Na$^+$-dependent respiration and Na$^+$-K$^+$-adenosine triphosphatase activity in rat skeletal muscle. J Clin Invest 57: 368-379.

Askari A, Koyal D (1968). Different oligomycin sensitivities of the Na$^+$+K$^+$-activated adenosine triphosphatase and its partial reactions. Biochim Biophys Res Comm 32: 227-232.

Blachley JD, Crider BP, Johnson JH (1986). Extrarenal potassium adaptation: role of skeletal muscle. Am J Physiol 251: F313-F318.

Brown L, Wagner G, Hug E, Erdmann E (1986). Ouabain binding and inotropy in acute potassium depletion in guinea pigs. Cardiovasc Res 20: 286-293.

Clausen T, Everts ME, Kjeldsen K (1987). Quantification of the maximum capacity for active sodium-potassium transport in rat skeletal muscle. J Physiol 388: 163-181.

Dørup I, Skajaa K, Clausen T, Kjeldsen K (1987). Decreased concentration of K, Mg and Na,K-pumps in human skeletal muscle during diuretic treatment. Br med J, submitted.

Jones LR, Besch HR (1984). Isolation of canine cardiac sarcolemmal vesicles. Meth Pharmacol 5: 1-12.

Kjeldsen K (1986). Complete quantification of the total concentration of rat skeletal-muscle Na$^+$+K$^+$-dependent ATPase by measurements of [^{3}H]ouabain binding. Biochem J 240: 725-730.

Kjeldsen K (1987). Regulation of the concentration of ^{3}H-ouabain binding sites in mammalian skeletal muscle. Effects of age, K-depletion, thyroid status and hypertension. Dan Med Bull 34: 15-46.

Kjeldsen K, Brændgaard H, Sidenius P, Larsen JS, Nørgaard N (1987). Diabetes decreases Na$^+$-K$^+$ pump concentration in skeletal muscles, heart ventricular muscle, and peripheral nerves of rat. Diabetes 36: 842-848.

Kjeldsen K, Everts ME, Clausen T. (1986a). The effects of thyroid hormones on ^{3}H-ouabain binding site concentration, Na,K-contents and ^{86}Rb-efflux in rat skeletal muscle. Pflügers Arch 406: 529-535.

Kjeldsen K, Everts ME, Clausen T (1986b). Effects of semi-starvation and potassium deficiency on the concentration of [^{3}H]ouabain-binding sites and sodium and potassium contents in rat skeletal muscle. Br J Nutr 56: 519-532.

Kjeldsen K, Nørgaard A (1987). Effect of magnesium depletion on ^{3}H-ouabain binding site concentration in rat skeletal muscle. Magnesium 6: 55-60.

Kjeldsen K, Nørgaard A, Clausen T (1984a). The age-dependent changes in the number of ^{3}H-ouabain binding sites in mammalian skeletal muscle. Pflügers Arch 402: 100-108.

Kjeldsen K, Nørgaard A, Clausen T (1984b). Effects of K-depletion on ^{3}H-ouabain binding and Na-K-contents in mammalian skeletal muscle. Acta Physiol Scand 122: 103-117.

Kjeldsen K, Nørgaard A, Clausen T (1985a). Effects of ouabain, age and K-depletion on K-uptake in rat soleus muscle. Pflügers Arch 404: 365-373.

Kjeldsen K, Nørgaard A, Gøtzsche CO, Thomassen A, Clausen T (1984c). Effects of thyroid function on number of Na-K pumps in human skeletal muscle. Lancet ii: 8-10.

Kjeldsen K, Nørgaard A, Hansen O, Clausen T (1985b). Significance of skeletal muscle digitalis receptors for [^{3}H]ouabain distribution in the guinea pig. J Pharmacol Exp Therap 234: 720-727.

Kjeldsen K, Richter EA, Galbo H, Lortie G, Clausen T (1986c). Training increases the concentration of [^{3}H]ouabain-binding sites in rat skeletal muscle. Biochim Biophys Acta 860: 708-712.

Knochel JP, Blachley JD, Johnson JH, Carter NW (1985). Muscle cell electrical hyperpolarization and reduced exercise hyperkalemia in physically conditioned dogs. J Clin Invest 75: 740-745.

Nørgaard A, Kjeldsen K, Hansen O (1984). (Na$^+$+K$^+$)-ATPase activity of crude homogenates of rat skeletal muscle as estimated from their K$^+$-dependent 3-O-methylfluorescein phosphatase activity. Biochim Biophys Acta 770: 203-209.

Sperelakis N (1972). (Na$^+$,K$^+$)-ATPase activity of embryonic chick heart and skeletal muscles as a function of age. Biochim Biophys Acta 266: 230-237.

QUANTIFICATION OF THE Na,K-PUMP IN HEART MUSCLE BY MEASUREMENT OF 3-O-METHYLFLUORESCEIN PHOSPHATASE ACTIVITY

Jim S. Larsen, Keld Kjeldsen and Aage Nørgaard

Institute of Physiology, University of Aarhus, DK-8000 Århus C (J.S.L., K.K., A.N.) and Medical Department B 2142, State University Hospital, Blegdamsvej 9, DK-2100 Copenhagen Ø, Denmark (K.K.)

INTRODUCTION

The Na,K-ATPase catalyses a K^+-activated hydrolysis of 3-O-methylfluorescein phosphate (3-O-MFP). This K^+-dependent phosphatase activity is assumed to be closely related to the Na,K-activated adenosine triphosphate (ATP) hydrolysis (Albers and Koval, 1966; Askari and Koyal, 1968). Since the K^+-dependent 3-O-MFPase activity represents up to 30-50% of the total 3-O-MFPase activity in crude homogenates of skeletal and heart ventricular muscle (Nørgaard et al., 1984; 1985) it can be determined without purification and ensuing loss of enzyme activity. Thus, in contrast to direct measurements of Na,K-ATPase activity the K^+-dependent 3-O-MFPase activity can be quantified in crude homogenates of muscle tissue with complete recovery (Jones and Besh, 1984; Kjeldsen, 1986; Hansen and Clausen, 1987).

METHOD

Muscle tissue samples of around 1 gram wet weight were prepared. After weighing and mincing using a scalpel the samples were homogenized in histidine-sucrose buffer. 9 ml was is used per gram wet weight of tissue. The homogenization was carried out using a homogenizator with a rotating pestle fitting tightly to the glass. When the solution appeared homogeneous, 100 μl of the preparation was suspended in 900 μl imidazole-sucrose buffer. To unmask 3-O-MFPase concealed in vesicles formed during homogenization the vesicles were made leaky by adding detergent. In preparations of rat as well as guinea pig skeletal and heart muscle an incubation with 0.08% sodium deoxycholate (DOC) for 30 min at 24°C was found to give optimum activation. 10 μl of this final homogenate corresponding to 0.01% or 100 μg of the wet weight of the starting material was added to the assay medium containing 3-O-MFP. By using a spectrofluorometer the formation of 3-O-MF causing fluorescense was determined. K^+-dependent phosphatase activity was calculated by subtracting the value obtained in the absence from that obtained in the presence of K^+ (Nørgaard et al., 1984; 1985; Nørgaard, 1986).

QUANTIFICATION

The K^+-dependent 3-O-MFPase activity found in rat skeletal muscle and in rat, hamster, guinea pig and human heart left ventricular muscle is given in Table 1. On the basis of a molecular activity of the 3-O-MFPase of 550 min^{-1} (Nørgaard et al., 1985) these values were converted to Na,K-pump concentration. It should be noted that the determination of the molecular activity of 3-O-MFPase was carried out on partially purified homogenates by measurement of K^+-dependent 3-O-MFPase activity and ^{3}H-ouabain binding capacity.

Table 1. K^+-dependent 3-O-MFPase activity and Na,K-pump concentration in mammalian skeletal and left heart ventricular muscle. Tissue was obtained from 12 week old rats, 28 week old hamsters, 4 week old guinea pigs and cadavers from adult human subjects

Species and muscle	K^+-dependent 3-O-MFPase activity (μmol/g wet wt./min)	Na,K-pump concentration (pmol/g wet wt.)
Rat gastrocnemius	0.19	346
Rat left ventricle	1.16	2109
Hamster left venticle	1.93	3509
Guinea pig left ventricle	0.80	1455
Human left ventricle	0.17	309

In the gastrocnemius muscle from adult rats the K^+-dependent 3-O-MFPase activity was 0.19 μmol/g wet wt./min (Nørgaard et al., 1984) corresponding to a Na,K-pump concentration of 346 pmol/g wet wt. This agrees well with the ^{3}H-ouabain binding site concentration usually found in this muscle (Kjeldsen et al., 1984a). It can be seen that the K$^+$-dependent 3-O-MFPase activity in the heart ventricle shows major variations among species, the value being largest in the hamster (Nørgaard et al., 1987) and lowest in human subjects. It should be noted that the concentration of Na,K-pumps in guinea pig and human left heart ventricle has also been quantified by ^{3}H-ouabain binding to intact muscle samples and that these measurements correspond well with the 3-O-MFPase activity determinations (Kjeldsen et al., 1985; Nørgaard et al., 1986).

The rat is well known to be more resistant to digitalis glycosides than the guinea pig. In agreement herewith a concentration of 1×10^{-2} mol/l of ouabain was required to achieve complete inhibition of the K$^+$-dependent 3-O-MFPase activity in homogenates from the rat heart whereas a concentration of 1×10^{-4} mol/l was sufficient for the guinea pig heart (Nørgaard et al., 1985). It should be noted that the apparent dissociation constant, K_D, for ^{3}H-ouabain binding in rat and guinea pig soleus muscle was 0.6×10^{-7} and 0.5×10^{-7} mol/l, respectively (Kjeldsen et al., 1984a; 1985). Thus, the high tolerance to digitalis glycosides found in rats as compared to guinea pigs seems to be related to a high K_D for digitalis glycosides in the rat heart whereas the affinity of skeletal muscle excibits no major difference between the 2 species.

QUANTITATIVE CHANGES

Table 2 summarizes the changes observed in K$^+$-dependent 3-O-MFPase activity of left heart ventricular muscle with various physiological and pathological conditions as found by our group.

Following potassium-depletion in rats a reduction in myocardial K$^+$-dependent 3-O-MFPase activity of 14% is observed (Nørgaard et al., 1985). This decrease is much smaller than that seen in skeletal muscle (where up to 78% reduction was produced (Nørgaard et al., 1981; Kjeldsen et al., 1984b; Brown et al., 1986; Kjeldsen, 1987), probably because the loss of K$^+$ from the heart during potassium-depletion takes place more slowly than from the skeletal muscles (Kjeldsen et al., 1984b).

Endocrine changes in rat was associated with significant up- and down-regulation of the K$^+$-dependent 3-O-MFPase activity in the heart. Thus, in hypothyroidism a down-regulation of 27% and in hyperthyroidism induced by T_3-treatment an up-regulation of 13% was observed (Curfman et al., 1977; Nørgaard et al., 1985). In diabetes mellitus induced by streptozotocin administration to rats a down-regulation was observed. The changes in the heart Na,K-pump concentration with thyroid status and diabetes mellitus were in the same direction but smaller than the changes observed in skeletal muscles under these conditions (Asano et al., 1976; Kjeldsen et al., 1986a; 1987).

After training of rats an up-regulation was found in skeletal muscle Na,K-pump concentration (Knochel et al., 1985; Kjeldsen et al., 1986b). It is of interest that the reduced myocardial function in the hamster hereditary cardiomyopathy is associated with a reduced Na,K-pump concentration (Panagia et al., 1985; Makino et al., 1985; Nørgaard et al., 1987).

Table 2. Relative changes in Na,K-pump concentration of rodent left heart ventricular muscle quantified by measurements of K^+-dependent 3-O-MFPase activity. The changes are given in per cent.

Condition	Change (%)
Potassium-depletion	-14
Hypothyroidism	-27
Hyperthyroidism	+13
Diabetes Mellitus	-21
Training	+20
Cardiomyopathy	-33

CONCLUSION

Measurement of K^+-dependent 3-O-MFPase activity in crude homogenates is an accurate method for quantifying heart muscle Na,K-ATPase in rodents and human subjects. Up-regulation is found in rodents with hyperthyroidism and training, whereas down-regulation is found with potassium-depletion, hypothyroidism, diabetes mellitus, immobilisation and cardiomyopathy. The changes might be of importance for heart muscle function under these conditions.

REFERENCES

Albers RW, Koval GJ (1966). Sodium-potassium-activated adenosine triphosphatase of electrophorus electric organ. J Biol Chem 241: 1896-1898.

Asano Y, Liberman UA, Edelmann IS (1976). Relationship between Na^+-dependent respiration and Na^+-K^+-adenosine triphosphatase activity in rat skeletal muscle. J Clin Invest 57: 368-379.

Askari A, Koyal D (1968). Different oligomycin sensitivities of the Na^++K^+-activated adenosine triphosphatase and its partial reactions. Biochim Biophys Res Comm 32: 227-232.

Brown L, Wagner G, Hug E, Erdmann E (1986). Ouabain binding and inotropy in acute potassium depletion in guinea pigs. Cardiovasc Res 20: 286-293.

Curfman GD, Crowley TJ and Smith TW (1977). Thyroid-induced alterations in myocardial sodium- and potassium-activated adenosine triphosphatase, monovalent cation active transport, and cardiac glycoside binding. J Clin Invest 59: 586-590.

Jones LR, Besch HR (1984). Isolation of canine cardiac sarcolemmal vesicles. Meth Pharmacol 5: 1-12.

Hansen O, Clausen T (1987). Quantitative determination of Na,K-ATPase and other sarcolemmal components in muscle cells. Am J Physiol, in press.

Kjeldsen K (1986). Complete quantification of the total concentration of rat skeletal-muscle Na^++K^+-dependent ATPase by measurements of ^3H-ouabain binding. Biochem J 240: 725-730.

Kjeldsen K (1987). Regulation of the concentration of ^3H-ouabain binding sites in mammalian skeletal muscle - Effects of age, K-depletion, thyroid status and hypertension. Dan Med Bull 34: 15-46.

Kjeldsen K, Brændgaard H, Sidenius P, Larsen JS, Nørgaard A (1987). Diabetes decreases Na^+-K^+ pump concentration in skeletal muscles, heart ventricular muscle, and peripheral nerves of rat. Diabetes 36: 842-848.

Kjeldsen K, Everts ME, Clausen T. (1986a). The effects of thyroid hormones on ^3H-ouabain binding site concentration, Na,K-contents and ^{86}Rb-efflux in rat skeletal muscle. Pflügers Arch 406: 529-535.

Kjeldsen K, Nørgaard A, Clausen T (1984a). The age-dependent changes in the number of ^3H-ouabain binding sites in mammalian skeletal muscle. Pflügers Arch 402: 100-108.

Kjeldsen K, Nørgaard A, Clausen T (1984b). Effects of K-depletion on ^{3}H-ouabain binding and Na-K-contents in mammalian skeletal muscle. Acta Physiol Scand 122: 103-117.

Kjeldsen K, Nørgaard A, Hansen O, Clausen T (1985). Significance of skeletal muscle digitalis receptors for ^{3}H-ouabain distribution in the guinea pig. J Pharmacol Exp Ther 234: 720-727.

Kjeldsen K, Richter EA, Galbo H, Lortie G, Clausen T (1986b). Training increases the concentration of ^{3}H-ouabain binding sites in rat skeletal muscle. Biochim Biophys Acta 860: 708-712.

Knochel JP, Blachley JD, Johnson JH, Carter NW (1985). Muscle cell electrical hyperpolarization and reduced exercise hyperkalemia in physically conditioned dogs. J Clin Invest 75: 740-745.

Makino N, Jasmin G, Beamish RE, Dhalla NS (1985). Sarcolemmal Na^+-Ca^{2+} exchange during the development of genetically determined cardiomyopathy. Biochem Biophys Acta 133: 491-497.

Nørgaard A (1986). Quantification of the Na,K-pumps in mammalian skeletal muscle. Acta Pharmacol Toxicol 58 Suppl. 1: 1-34.

Nørgaard A, Baandrup U, Larsen JS, Kjeldsen K (1987). Heart Na,K-ATPase activity in cardiomyopathic hamsters as estimated from K-dependent 3-O-MFPase activity in crude homogenates. J Mol Cell Cardiol 19: 589-594.

Nørgaard A, Kjeldsen K, Clausen T (1981). Potassium depletion decreases the number of ^{3}H-ouabain binding sites and the active Na-K-transport in skeletal muscle. Nature 293: 739-741.

Nørgaard A, Kjeldsen K, Hansen O (1984). (Na^++K^+)-ATPase activity of crude homogenates of rat skeletal muscle as estimated from their K^+-dependent 3-O-methylfluorescein phosphatase activity. Biochim Biophys Acta 770: 203-209.

Nørgaard A, Kjeldsen K, Hansen O (1985). K^+-dependent 3-O-methylfluorescein phosphatase activity in crude homogenate of rodent heart ventricle: Effect of K^+-depletion and changes in thyroid status. Eur J Pharmacol 113: 373-382.

Nørgaard A, Kjeldsen K, Hansen O, Clausen T, Larsen CG, Larsen FG (1986). Quantification of the ^{3}H-ouabain binding site concentration in human myocardium: a postmortem study. Cardiovasc Res 4: 201-206.

Panagia V, Singh JN, Anand-Srivastave MB, Pierce GN, Jasmin G, Dhalla NS (1985). Sarcolemmal alterations during the development of genetically determined cardiomyopathy. Cardiovasc Res 18: 567-572.

ALPHA SUBUNITS (α AND α^+) ISOFORMS OF THE Na^+, K^+-ATPase IN DOG HEART. ALTERATION OF α^+ IN ISCHEMIA.

Jean-Michel MAIXENT, Pierre BIRKUI,
Simone FENARD and Lionel G. LELIEVRE

Laboratoire Nativelle, (J.M.M., S.F.),
1 Chemin Saulxier - 91160 Longjumeau.
INSERM U 141, (P.B.), and INSERM U 127,
(L.G.L.), Hôpital Lariboisière, Université
Paris 7, 75010 Paris, France.

INTRODUCTION

There is a compelling evidence that isoforms of the alpha subunit of the active Na, K-ATPase exist in many tissues and species. (Sweadner, 1979, Siegel et al., 1986).

At the cardiac level, the presence of two forms has been recently demonstrated in rat (Lelièvre et al., 1986 and Charlemagne et al., 1987) and in Ferret (NG and Akera, 1987) . In dog heart, kinetics studies have suggested the presence of high and low-affinity digitalis receptor sites (Wellsmith and Lindenmayer, 1980). The two receptor types has been separated by Matsuda et al., (1984) however both forms were of low sensitivity for digitalis. Thus, the structural difference was not related to the high sensitivity to digitalis of this species.

Our objective was to correlate the presence in normal dog heart, of high and low affinity Na, K-ATPase enzyme forms in vitro with inotropic and toxic effects of digitalis in vivo.

In parallel with ischemic dog heart in which digitalis has deleterious effects (Ferrier et al., 1985) we have analyzed the fate of each digitalis receptor type.

Our results show that in normal canine heart, the alpha plus form (of high affinity) represented the inotropic site whereas the alpha form was associated with toxicity. In ischemia, the inotropic site was gradually inactivated whereas the toxic site remained unaffected.

MATERIALS AND EXPERIMENTAL PROTOCOL

Male mongrel dogs, 20 to 35 Kg in weight, were anaesthetized with pentobarbital (30 mg/Kg). Positive pressure ventilation was maintained through a tracheal cannula by means of a Harvard respirator at 100 % fractional inspired oxygen. A clamp placed on the left anterior descending coronary artery (LAD) just below the first branch produced a wedge-shaped area of histologically identified infracted muscle limited to the anterior wall. The presence of ischemia was confirmed by the presence of an area of visible epicardial cyanosis. The heart was suspended in a pericardial cradle and the LAD was dissected free to be occluded with a vascular clamp just distal to the first major diagonal branch. At the end of each experiment, the heart was rapidly excised and perfused as previously described (Maixent et al. 1987). Full-thickness left ventricular samples weighing 10 g were taken from the central ischemic zone and from the nonischemic zone

Na, K-ATPase Enriched Preparations

The same isolation procedure (Maixent et al., 1987) was applied to both normal and ischemic canine ventricular muscle. The tissue (10 g of ventricle) was diced and homogenized with a Waring blendor and a Polytron PT20 in a buffer containing 20 mM Na pyrophosphate, 0.1 mM phenylmethansulfonylfluoride, 250 mM sucrose, 80 mM KCl and 20 mM imidazole/HCl pH 7.4. The 500 x g supernatant was subfractionated by two sequential differential centrifugations. The final pellet obtained after a 31, 000 x g spin for 30 min. represents the microsomes. The sarcolemmal vesicles, isolated from the microsomes by a centrifugation on a sucrose cushion, were resuspended in 100 mM NaCl, 250 mM sucrose and 30 mM imidazole/HCl pH 7.4, kept frozen at -80°C and used for the enzymatic assays within one week.

Sensitivity of Na, K-ATPase to Ouabain

It has been tested in microsomes from canine myocytes and in sarcolemmal vesicles from whole dog heart. The enzymatic activity in the absence or presence of various concentrations of ouabain from 0.3 mM to 30 uM) was determined according to Lelièvre et al., (1986) using a coupled assay method. The assays were performed either with native vesicles or with vesicles opened by sodium dodecyl sulfate treatment (0.2 mg/mg of proteins for 30 min. at 20°C)

RESULTS

As shown in Table 1, alpha and alpha plus were two biochemically distinct enzyme forms with different K_D. Their difference in ouabain affinity was equal to 150.

Table 1. Respective affinities of canine cardiac and kidney Na, K-ATPases to ouabain

	K_D (nM)	contribution %	k^{-1} ($10^{-4} \times sec^{-1}$)
Cardiac alpha plus	2 + 1*	61	1.3 - 1.8
Cardiac alpha	300 + 200*	39	14 - 23
Kidney alpha**	150 + 100	100	no tested

* The same results were obtained with membranes either from isolated cardiac myocytes or from the whole heart.
** Commercial Na,K-ATPase from outer medulla (Sigma, lot A0142).

The contribution of high to low affinity processes was 1.5 to 1 revealed by the activity associated with each form. The two forms exhibited two distinct dissociation processes. Ouabain dissociated from the high-affinity sites at a rate 10-14 fold slower than from the low affinity sites. In dog kidney preparations, the only site detected corresponded to the alpha form associated with an apparent affinity of 150 + 100 nM.

In microsomes from ischemic hearts, the K_D values and the dissociation rate constants of each receptor type were similar. However, the ratio of high to low affinity sites shifted from 1.5/1 in normal to 1/6 in 60 min.-ischemia. The reason of such a shift resided in a specific alteration of the alpha plus form whereas the change in alpha form was not significatif (Table 2).

Table 2. Differential effect of the duration of ischemia on the relative importance of the two receptor forms

	Time of ischmia (min)			
	0	15	30	60
alpha plus	39 %	41 %	46 %	43 %
alpha	61 %	41 %	32 %	7 %

From the data presented in Table 3, it is clear that the inotropic effect of ouabain (up to 10 mM) exclusively involves the inhibition of high affinity from, i.e apha plus. At ouabain doses higher than 100 mM, toxicity that occured was associated with the inhibition of the low affinity form.

Table 3. Physiological involvement of the canine Na, K-ATPase isoforms.

% of maximun enzyme inhibition	positive inotropy	toxic effect	Isoform(s) involved	
			alpha +	alpha
47(a) 59(b)	yes	no	yes	no
75(a,c) 86(d)	yes	yes	yes	yes

(a) Akera and Brody, 1987 (b) Besch et al., 1970 (c) Hougen and Smith, 1979.
(e) Bernabei and Vassale, 1984 (f) Rosen et al., 1973.

DISCUSSION

We clearly show that two functional Na, K-ATPase forms exist in canine cardiac myocytes. The discriminatory criterion was a different affinity for ouabain : a/The alpha plus isoform (high affinity for ouabain) was the "in vitro" manifestation of the inotropic site. b/The alpha form was associated with toxic effect of digitalis. c/In ischemia, the inotropic site was selectively inactivated whereas alpha seemed to remains unaffected. The depressed inotropic effect of digitalis in vivo could be explained by the inactivation of the functional high affinity sites. A larger recruitement of the low-affinity receptors would support the increased toxicity when high doses of digitalis are applied to ischemic heart. The presence of high and low affinity isoenzymes of Na, K-ATPases in dog cardiac myocytes, differentially modified by pathological conditions could reflect a physiological different role each digitalis receptor type.

REFERENCES

Akera T, Brody TM (1977). The role of Na, K-ATPase in the inotropic action of digitalis. Pharmac Rev 29: 187-220.

Bernabei R, Vassale M (1984). The inotropic effects of strophantidin in Purkinje fibers and the sodium pump. Circulation 69 : 618-631

Besh HR, Allen JC, Gick G, Schwartz A (1970). Correlation between the inotropic action of ouabain and its effects on subcellular enzyme systems from canine myocardium. J Pharmacol Exp Ther 171: 1-12.

Charlemagne D, Mayoux E, Poyard M, Oliviero P, Geering K (1987). Identification of two isoforms of the catalytic subunit of Na, K-ATPase in myocytes from adult rat heart. J. Biol. Chem. 262 : 8941-8943.

Ferrier GR, Moffat MP, Lukas A (1985). Possible mechanisms of ventricular arrhythmias elicited by ischemia followed by reperfusion. Circ Res 56: 184-194.

Hougen TJ, Smith TW (1979). Effects of inotropic and arrhythmogenic digoxin doses and digoxin specific antibody on myocardial monovalent cation transport in the dog. Circ Res 44: 23-31.

Lelievre LG, Maixent JM, Lorente P, Mouas C, Charlemagne D, Swynghedauw B (1986). Prolonged responsiveness to ouabain in hypertrophied rat heart. A physiological and biochemical study. Am J Physiol 251: H923-931.

Maixent JM, Charlemagne D, de la Chapelle B, Lelievre L (1978). Two Na, K-ATPase isoenzymes in canine cardiac myocytes. Molecular basis of inotropic and toxic effects of digitalis. J Biol Chem 262: 6842-6848.

Matsuda T, Iwata H, Cooper JR (1984). Specific inactivation of (α +) molecular form of (Na+, K+)-ATPase by Pyrithamin. J Biol Chem 259: 3858-3863.

NG YC, Akera T (1987). Two classes of ouabain binding sites in ferret heart and two forms of Na+, K+ ATPase. Am J Physiol 252: H1016-H1022.

Rosen MR, Gelband H, Merker C, Hoffman BF (1973). Mechanisms of digitalis toxicity. Circulation 47: 681-689.

Sweadner KJ (1979). Two molecular forms of Na+ K+ stimulated ATPase in rat brain. J Biol Chem 254: 6060-6067.

Siegel GJ, Desmond T, Ernst SA (1986). Immuno reactivity and ouabain-dependent phosphorylation of (Na+,K+) - Adenosinetriphosphatase catalytic subunit doublets. J Biol Chem 261: 13768-13776.

Wellsmith NV, Lindenmayer GE (1980). Two receptor forms for ouabain in sarcolemma enriched preperations from canine ventricle. Circ Res 47: 710-720.

THE Na,K-PUMP IN HUMAN MYOCARDIUM: QUANTIFICATION IN NORMAL SUBJECTS AND PATIENTS WITH SUSPECTED CARDIOMYOPATHY

Aage Nørgaard and Keld Kjeldsen

Institute of Physiology, University of Aarhus (A.N., K.K.), Department of Cardiology, Aarhus Municipal Hospital, DK-8000 Aarhus C (A.N.), and Medical Department B 2142, State University Hospital, Blegdamsvej 9, DK-2100 Copenhagen Ø, Denmark (K.K.)

INTRODUCTION

The Na,K-ATPase or Na,K-pump is of major importance for myocardial function. Thus, the active transport of Na and K across the cell membrane by the Na,K-pump is essential for excitability and contractility (Skou, 1965). The reuptake by the Na,K-pump of K lost from the cell during the action potential prevents a possible arrhytmogenic rise of interstitiel K. By maintenance of the Na gradient across the cell membrane and the associated Na-Ca exchange the Na,K-pump is of significance for Ca clearence of the cell. Finally, the Na,K-pump is accepted to be the cellular receptor for digitalis glycosides (Akera and Brody, 1978).

The concentration of Na,K-pumps in the myocardium may be estimated either from measurements of the Na,K-ATPase activity or from the ^{3}H-ouabain binding capacity of isolated membranes prepared from homogenates. However, the isolation of membrane preparations inevitably leads to incomplete recovery of the Na,K-ATPase and the results obtained may give rise to misleading interpretations. In human skeletal muscle vanadate facilitated binding of ^{3}H-ouabain to biopsy as well as to necropsy specimens has proved adequate for the determination of the total concentration of Na,K-pumps (Nørgaard et al., 1984; Nørgaard et al., 1985) and changes hereof (Kjeldsen et al., 1984).

A decrease in Na,K-ATPase activity of the cardiac plasma membrane has been shown to be one of the earliest abnormalities during the course of hereditary cardiomyopathy in the hamster (Panagia et al., 1984; Nørgaard et al., 1987) and it might be of importance for the associated intracellular Ca accumulation, mitochondrial calcification and cell necrosis (Bajusz et al., 1969). As the concentration of Na,K-pumps in the human myocardium may also undergo changes with various physiological and pathophysiological conditions there was a demand for a simple and reliable method for the quantification of the Na,K-pump in this tissue. Therefore, the binding of ^{3}H-ouabain has been assessed first to necropsy specimens of the human myocardium (Nørgaard et al., 1986) and second to biopsy specimens obtained during heart catheterisation (Kjeldsen et al., 1987).

METHODS

Patients

Necropsy specimens of around 5 mg were obtained within the first 6-24 h post mortem from 15 patients who died for reasons other than cardiovascular. In a few experiments necropsy specimens of around 1-2 mg were obtained at autopsy from the endomyocardium using a Cordis^r biotome.

A total of 15 consecutive patients with suspected dilated cardiomyopathy underwent a diagnostic left heart catheterization with ventricular cineangiography and coronary arteriography. Endomyocardial biopsy specimens were taken from the left ventricle using a Cordis^r biotome. The ventricular ejection fraction (EF) was calculated as the ratio between the angiographic stroke volume and end diastolic volume of the left ventricle. This part of the study was approved by the Local Ethics Commitee according to Helsinki Declaration II.

^{3}H-ouabain Binding

This assay was performed as previously described in detail for necropsy and biopsy specimens from skeletal muscle (Nørgaard et al., 1984; Nørgaard et al., 1985). The specimens were prewashed for 2x10 min at 0°C in a tris sucrose buffer (pH 7.3) containing 1 mmol/l vanadate (NaVO$_3$), 3 mmol/l MgSO$_4$, 10 mmol/l Tris chloride, and 250 mmol/l sucrose. Incubation took place in the tris sucrose buffer containing 1x10^{-6} mol/l ^{3}H-ouabain at 37°C for 2x60 min. This was followed by a wash at 0°C in unlabelled buffer for another 4x30 min to remove ^{3}H-ouabain from the extracellular space. After blotting, the specimens were weighed and 5% trichloroacetic acid was added. Following extraction overnight and liquid scintillation counting, the amount of ^{3}H-ouabain bound to the specimens was calculated on basis of the specific activity of the incubation medium. Following correction for unspecific uptake determined in the presence of surplus of unlabelled ouabain it was expressed as pmol/g wet weight of muscle tissue.

RESULTS

The binding of ^{3}H-ouabain to necropsy specimens of around 5 mg from the left human myocardium was specific and saturable. From a Scatchard plot in the concentration range $1 \times 10^{-8} - 5 \times 10^{-6}$ mol/l the binding appeared to take place to only one population of high affinity binding sites with an apparant K_D of 1.1×10^{-8} mol/l. Experiments carried out at higher concentrations of ouabain, i.e. in the range $1 \times 10^{-4} - 5 \times 10^{-3}$ mol/l gave no evidence for a population of low affinity binding sites.

Although digoxin is the clinically most widely used digitalis glycoside ouabain was used in the present study because it is readily water soluble. However, the concentration of ^{3}H-digoxin and ^{3}H-ouabain binding sites were identical, i.e. 347 ± 24 vs. 342 ± 38 pmol/g wet weight (SEM, n=5, P<0.95) and excess of unlabelled ouabain completely prevented the specific binding of ^{3}H-digoxin. This indicates that the binding of digoxin and ouabain takes place at the same receptor.

In 15 patients who died for reasons other than cardiovascular the concentration of ^{3}H-ouabain binding sites in necropsy specimens of around 5 mg from the left human myocardium varied from 233 to 577 pmol/g wet weight with a mean value of 413 ± 26 pmol/g wet weight. From 6 to 24 h post mortem a mean decrease in ^{3}H-ouabain binding site concentration of 11% was seen in 5 patients being only significant in 1 after 24 h indicating a slow post mortem degradation of the ^{3}H-ouabain binding capacity.

Table 1. Concentration of ^{3}H-ouabain binding sites in various regions of the human left ventricle obtained 6 h post mortem

tissue	sampling procedure	wet weight (mg)	^{3}H-ouabain binding site concentration (pmol/g wet weight)	p
myocardium	scalpel	4-6	355 ± 16 (6)	
epimyocardium	scalpel	4-6	315 ± 18 (6)	> 0.10
endomyocardium	scalpel	4-6	377 ± 13 (6)	> 0.30
endomyocardium	scalpel	1-2	339 ± 32 (6)	> 0.60
endomyocardium	Cordis^r biotome	1-2	318 ± 22 (6)	> 0.20

Since biopsy specimens obtained during heart catheterization have a wet weight around 1-2 mg, measurements were carried out on necropsy specimens in this weight range. Table 1 shows that the ^{3}H-ouabain binding site concentration was virtually the same in 1-2 and 4-6 mg specimens of endomyocardium and myocardium, respectively. This indicates that determination of ^{3}H-ouabain binding site concentration in biopsy specimens obtained during heart catheterization may be representative of the whole myocardium.

The left ventricular function in dilated cardiomyopathy is characterized by a decrease in EF. According to this the 15 patients were divided into a group with an EF below and a group with an EF above 40%. As can be seen from Fig. 1 the 9 patients with EF below 40% had a mean EF of 29% and the 6 patients with EF above 40% had a mean EF of 68% (P<0.001). In endomyocardial biopsy specimens from the left ventricle of patients in these 2 groups a ^{3}H-ouabain binding site concentration of on average 322+32 and 505+41 pmol/g wet wt was found, respectively (n=9 and 6, P<0.001).

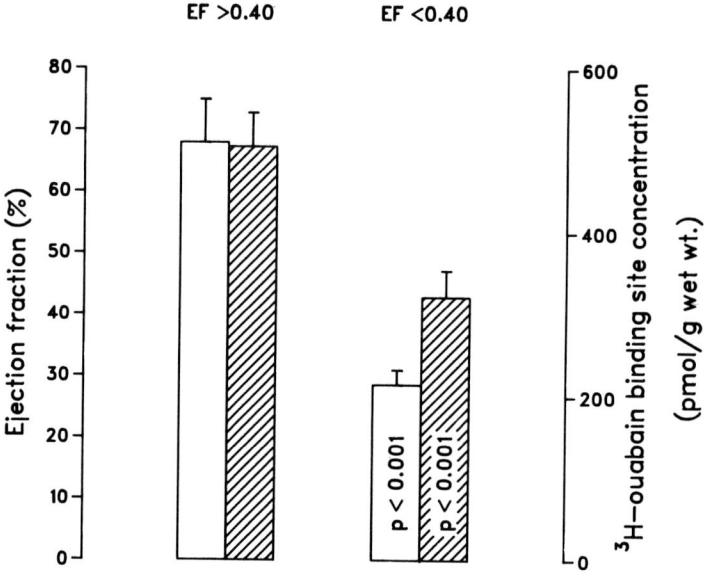

Figure 1. Left ventricular ejection fraction (EF) (□) and ^{3}H-ouabain binding site concentration (▨) in biopsy specimens of the left ventricle from 15 patients undergoing diagnostic left heart catheterization

DISCUSSION

Although the Na,K-ATPase or Na,K-pump has been demonstrated in the human myocardium it has been quantified in only one study (Erdmann and Brown, 1983). Most studies characterizing the Na,K-ATPase in the heart of laboratory animals are based upon membrane fractions containing less than a few per cent of the total enzyme activity available in the starting material. It is not known whether these fractions represent a true random sample of the plasma membrane of the myocardium or they arise from a localized or subspecialized region such as the transverse tubules (Jones and Besh, 1984). This represents a serious disadvantage in quantitative estimates of changes in the total Na,K-ATPase concentration arising under various conditions.

The binding of ^{3}H-ouabain to necropsy specimens from the myocardium was shown to be specific and saturable and might be assumed to represent the concentration of Na,K-pumps. In 15 patients the ^{3}H-ouabain binding site concentration in necropsy specimens of the left myocardium obtained 6 h post mortem was on average 413 + 26 pmol/g wet weight. This is approximately 1.7 times higher than the ^{3}H-ouabain binding site concentration of 249 pmol/g wet weight found in crude homogenates of myocardium obtained during open heart surgery (Erdmann and Brown, 1983). This discrepancy can among other factors probably be related to insufficient detergent treatment of vesicles arising during homogenization.

Of the 15 patients with suspected dilated cardiomyopathy 9 had an EF less than 40%. In endomyocardial biopsy specimens of the left ventricle from these patients a decrease in ^{3}H-ouabain binding site concentration of 36% was seen as compared to patients with an EF larger than 40%. This indicates a correlation between the concentration of Na,K-pumps and the EF in the human myocardium. The decrease in ^{3}H-ouabain binding site concentration with dilated cardiomyopathy is not the simple outcome of an increased amount of fibrosis in the myocardium since there is no correlation between EF and volume fraction of collagen tissue in the myocardium from cardiomyopathic patients (Baandrup and Olsen, 1981).

In conclusion the results indicate that it is possible to determine the concentration of ^{3}H-ouabain binding sites in the human myocardium by measurement of ^{3}H-ouabain binding to necropsy specimens obtained within the first hours post mortem or to biopsy specimens obtained during heart catheterisation. The results also suggest that a decrease in the concentration of Na,K-pumps may be of importance for myocardial dysfunction under various pathophysiological conditions such as cardiomyopathy.

REFERENCES

Akera T, Brody TM (1978). The role of Na,K-ATPase in the inotropic action of digitalis. Pharmacol Rev 29:187-220.

Baandrup U, Olsen EGJ (1981). Critical analysis of endomyocardial biopsies from patients suspected of having cardiomyopathy. II: Comparison of histological and clinical/haemodynamic information. Br Heart J 45:487-493.

Bajusz E, Baker JR, Nixon CW, Homburger F (1969). Spontaneous, hereditary myocardial degeneration and congestive heart failure in a strain of Syrian hamsters. Ann Ny Acad Sci 156:105-229.

Erdmann E, Brown L (1983). The cardiac glycoside-receptor system in the human heart. Eur Heart J 4 Suppl. A:61-65.

Jones LR, Besh HR (1984). Isolation of canine cardiac sarcolemmel vesicles. Meth Pharmacol 5:1-12.

Kjeldsen K, Nørgaard A, Gøtzsche CO, Thomassen A, Clausen T (1984). The effect of thyroid status on the number of Na-K-pumps in human skeletal muscle. Lancet 2:8-10.

Kjeldsen K, Bjerregaard P, Richter E, Bloch Thomsen PE, Nørgaard A (1987). The Na,K-ATPase concentration in rodent and human heart and skeletal muscle: Apparent relationship to muscle performance. Cardiovasc Res in press.

Nørgaard A, Kjeldsen K, Clausen T (1984). A method for the determination of the total number of ^{3}H-ouabain binding sites in biopsies of human skeletal muscle. Scand J clin Lab Invest 40:509-518.

Nørgaard A, Kjeldsen K, Larsen JS, Larsen CG, Larsen FG (1985). Estimation of stability of ^{3}H-ouabain binding site concentration in rat and human skeletal muscle post mortem. Scand J clin Lab Invest 45:139-144.

Nørgaard A, Kjeldsen K, Hansen O, Clausen T, Larsen CG, Larsen FG (1986). Quantification of the ^{3}H-ouabain binding site concentration in human myocardium: A post mortem study. Cardiovasc Res 20:428-435.

Nørgaard A, Baandrup U, Larsen JS, Kjeldsen K (1987). Heart Na,K-ATPase activity in cardiomyopathic hamsters as estimated from K-dependent 3-O-MFPase activity in crude homogenates. J Mol Cell Cardiol 19:589-594.

Panagia V, Singh JN, Anand-Srivastava MB, Perce GN, Jasmin G, Dhalla NS (1984). Sarcolemmal alterations during the development of genetically determined cardiomyopathy. Cardiovasc Res 18:567-572.

Skou JC (1965). Enzymatic basis for active transport of Na^+ and K^+ across cell membrane. Physiol Rev 45:596-617.

Regulation of Na^+,K^+-ATPase: Hormones, Inhibitors, and Inotropic Action

OVERVIEW:
HORMONAL REGULATION OF NA,K-ATPASE

G.G. Gick, F. Ismail-Beigi and I.S. Edelman

Departments of Biochemistry & Molecular Biophysics (G.G.G. & I.S.E.), and Medicine (F.I-B.), Columbia University, New York, N.Y. 10032

INTRODUCTION

The Na,K-pump (or Na,K-ATPase) is vital to all animal cells and serves a variety of crucial functions (summarized in Fig. 1). Because of the energetics of the pump and diverse coupled events, hormonal modulation can and does serve a myriad of physiological purposes. The present review will briefly summarize the actions of selected hormones on the activity of the pump, and provide a more detailed analysis of the action of thyroid hormone.

Regulators of the Na,K-pump fall into two classes, those that act with minimal latency, e.g. in minutes (vasopressin, catecholamines, insulin) and those with latent periods of 1-12 hours (steroids and thyroid hormone). The analysis of the mode of action of the former group has centered on either effects on Na^+ and K^+ permeability (i.e. rapid changes in the local transport substrate concentrations) or on direct activation of the enzyme. The latter group has been studied largely from the perspective of biogenesis of new pumps or related processes that effect pump abundance.

VASOPRESSIN

The neurohypophyseal nonapeptide, vasopressin, regulates water reabsorbtion by renal collecting ducts and transepithelial active Na^+ transport across amphibian skin and bladder by activation of the adenylate cyclase system.

In the toad bladder, antidiuretic hormones increase the

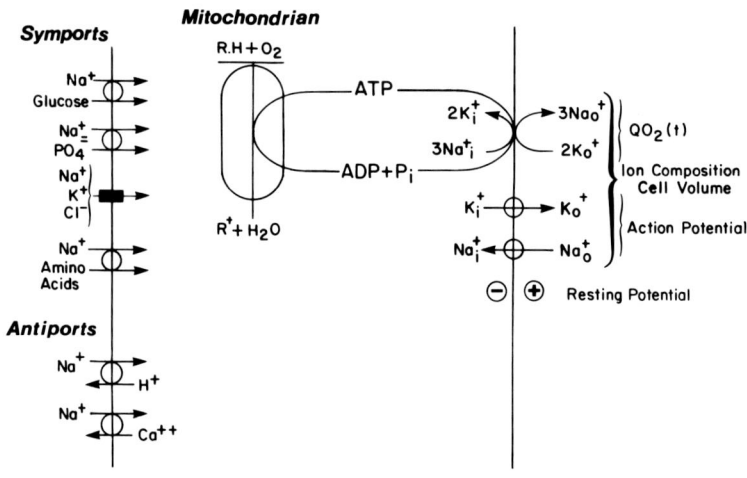

Figure 1. Functions of the Na,K-pump.

permeability of the apical membrane to Na^+ by increasing the number of functional Na^+ channels (Li et al., 1982). The resulting increased Na^+ entry can account for stimulation of transepithelial active Na^+ transport. The action of vasopressin has also been investigated in an established canine renal tubular epithelial cell line, MDCK (Reznik et al., 1985). Incubation of MDCK cells with vasopressin increases Na^+ entry with a resultant stimulation of ouabain-sensitive (86)Rb uptake. Similar effects of vasopressin on Na^+ uptake and Na^+ transport activity have been reported in mouse 3T3 cells (Mendoza et al., 1980). In addition to the short-term action, whether vasopressin increases the abundance of Na,K-ATPase in responsive tissues over a longer period of time remains to be clarified.

CATECHOLAMINES

Efforts to delineate the action of catecholamines on Na,K-ATPase activity have primarily focused on skeletal muscle, brain, liver and adipose tissue. Catecholamine stimulation of Na,K-pump was inferred from results in isolated frog sartorius muscles demonstrating strophanthidin-sensitive, adrenaline-stimulated (22)Na efflux (Hays et al., 1974). Similarly, in isolated rat soleus muscle, addition of adrenaline stimulated (22)Na efflux and (42)K influx (Clausen and Flatman, 1977). Following 90 min of incubation, intracellular Na^+ content was reduced to 1/3 of the control level, and was accompanied by hyperpolarization. The effects were blocked by the beta-adrenergic antagonist, propanolol, and not affected by alpha-adrenergic blockers. Furthermore, Clausen and Flatman (1977) reported catecholamine-induced cAMP-mediated stimulation of Na,K-ATPase activity, independent of changes in Na,K-pump number (Clausen and Hansen, 1977). Alternatively, catecholamine activation of muscle Na,K-ATPase may involve a chelation of vanadate, thereby reversing vanadate inhibition of Na,K-ATPase activity, as proposed by Cantley et al. (1978).

Cerebral Na,K-pump activity was stimulated by acute catecholamine release (Swann et al., 1981). Studies by Wu and Phillis (1980) provide additional support for catecholamine stimulation of brain Na,K-ATPase activity. In contrast to the muscle Na,K-pump, cerebral Na,K-ATPase activity does not appear to be regulated by a cAMP pathway.

Catecholamine binding to alpha-adrenergic receptors has been implicated in a rapid and transient stimulation of hepatic Na,K-ATPase activity, an effect observed exclusively in rat liver (Berthon et al., 1985). Lynch et al. (1986) proposed that catecholamines and vasopressin stimulate rat hepatic Na,K-ATPase activity via an increased synthesis of diacylglycerol and activation of the protein kinase C pathway. In summary, evidence has been presented which supports a role for catecholamines in the regulation of Na,K-pump activity, although the molecular mechanism is not clearly understood.

INSULIN

The peptide hormone, insulin, lowers serum K^+ concentration as a result of stimulation of net cellular uptake of K^+ and coupled active Na^+ transport. This effect is exerted on diverse tissues including kidney, skeletal and cardiac muscle, adipocytes and hepatocytes (Rosic et al., 1985). Tissue-specific differences appear to exist, however, in the mechanism of insulin stimulation of the Na,K-pump. Although evidence has been presented which suggests insulin recruitment of latent muscle cell Na,K-pumps (Erlij and Grinstein, 1976), in rat, mouse and guinea pig soleus muscle, ouabain-sensitive (42)K influx was stimulated by insulin treatment without an effect on the total number of (3)H-ouabain-binding sites (Clausen and Hansen, 1977). It is possible, however, that insulin might unmask cryptic pumps that are accessible to ouabain but not participants in transmembrane Na transport, prior to exposure to insulin. In an established mouse muscle line, insulin stimulated amiloride-sensitive (22)Na influx by 200%, associated with an 80% increase in intracellular Na^+ content, and a 60% increase in active Rb^+ uptake (Rosic et al., 1985). They concluded that insulin promoted $Na^+:H^+$ exchange which augmented intracellular Na^+ activity. In contrast, however, Clausen and Flatman (1987) reported that in intact rat soleus muscle the effect of insulin on Rb^+ uptake was not suppressed by amiloride.

In rat liver cell culture systems (hepatocytes and hepatoma cells), insulin binding to high affinity receptors is associated with a rapid increase in amiloride-sensitive (22)Na uptake, with no apparent change in the number of Na,K-pumps (Fehlman and Freychet, 1981; Gelehrter et al., 1984). The stimulatory effect of insulin was mimicked by monensin, suggesting that in hepatocytes insulin stimulated Na^+ uptake, thereby providing Na^+ ions to internal unsaturated transport sites of the Na,K-pump.

Resh et al. (1980) observed that in adipocytes, insulin rapidly stimulated ouabain-inhibitable uptake of (86)Rb resulting in an increase in the intracellular concentration of K^+. Ouabain-dependent-phosphorylation and (3)H-ouabain-binding indicated that the number of Na,K-pumps was invariant with insulin treatment. Analysis of the kinetics of ouabain inhibition revealed two classes of Na,K-pumps with high and low affinity for ouabain. Stimulation of (86)Rb transport was attributed to activation of the high affinity class. Lytton et al., (1985) identified the high ouabain-affinity, insulin-stimulated subset as the alpha+ isoform and the low ouabain-affinity subset, as the alpha isoform, by immunoassay. The abundance of the insulin-sensitive adipocyte alpha+ form constituted 75% of the total number of Na,K-pumps. In the absence of exogenous insulin, the Na^+ affinities of Na,K-ATPase alpha and alpha + were 17 and 52 mM, respectively. Incubation in insulin-containing medium resulted in a reduction in Na^+ affinity to 33 mM for alpha+, and to 14 mM for alpha (Lytton, 1985). Since in the absence of exogenous insulin the concentration of Na^+ in adipocytes is 15-20 mM, the activity of the Na,K-ATPase alpha+ form would likely be minimal. Lytton proposed that insulin stimulates Na,K-ATPase activity by shifting the Na^+ affinity of the alpha+ form without a concommitant increase in intracellular Na^+.

Insulin stimulation of Na,K-ATPase activity appears to be a receptor-mediated, tissue-specific process, apparently involving an increase in the concentration of intracellular Na^+ in hepatocytes and cultured myocytes, while in muscle and adipose tissue the activity of the Na,K-pumps is not mediated by changes in intracellular Na^+. The action of insulin does not seem to be mediated by recruitment of latent pumps nor by de novo synthesis.

MINERALOCORTICOIDS

Numerous studies have convincingly demonstrated a significant reduction in renal Na^+ reabsorption and Na,K-ATPase activity after adrenalectomy (Chignell and Titus, 1966; Westenfelder et al., 1977). The administration of mineralocorticoids to adrenalectomized animals results in partial or complete restoration of Na,K-ATPase activity. Although mineralocorticoids stimulate renal Na,K-ATPase activity, the precise mechanism remains to be established. Two models have been proposed for mineralocorticoid action. The first postulates a direct stimulation of Na,K-ATPase in the early phase of the response, while the second suggests an indirect affect secondary to enhanced Na^+ entry.

Aldosterone action on transepithelial Na^+ transport is dependent on the induction of both RNA and protein synthesis (Edelman et al., 1963; Wilce et al., 1977). The direct model predicts that one of the aldosterone-induced-proteins would, in fact, be Na,K-ATPase. Early evidence for a direct action of mineralocorticoids came from Knox and Sen (1974) who reported that the synthesis of Na,K-ATPase alpha subunit was increased in adrenalectomized rat kidney after 3 hr of aldosterone treatment. Schmidt et al. (1975) reported that aldosterone treatment for 1 hr was sufficient to completely restore renal Na,K-ATPase activity in adrenalectomized rats. The demonstration of a stimulatory effect of aldosterone on Na,K-ATPase activity in isolated nephron segments lends further support for a direct effect of aldosterone (Horster et al., 1980). The effect on Na,K-ATPase activity was evident after 1 hr of hormone treatment in preparations of renal tubules from adrenalectomized rabbits. A comparable concentration of dexamethasone was ineffective and spirolactone abolished the aldosterone-induced increase in Na,K-ATPase activity. Similar results on rat outer medulary collecting tubules were obtained by Rayson and Lowther (1984). It should be noted, however, that the effect of aldosterone on the number of Na,K-ATPase units has not yet been described in isolated nephron systems.

In an established cell line derived from the kidney of Xenopus laevis, A6, aldosterone given for 6 hr elicited qualitatively similar increases in Na^+ transport and biosynthesis of immunoprecipitable Na,K-ATPase alpha and beta subunits (Verrey et al., 1987). The cytoplasmic abundances of mRNA's encoding the alpha and beta subunits were augmented to the same extent as protein biosynthesis. These results are consistent with a model whereby the aldosterone-receptor complex interacts directly with DNA regulatory sites, yielding stimulation of Na,K-ATPase gene expression. It must be noted, however, that Verrey et al. (1987) do not present evidence that aldosterone induces an increase in functional Na,K-ATPase units during the early or late aldosterone response. Indeed, Johnson et al. (1986) failed to detect an increase in Na,K-ATPase activity of A6 cells at 6 hr of aldosterone treatment.

Although evidence has been presented for an aldosterone induction of the Na,K-ATPase transport enzyme, several investigators have not detected an early aldosterone-induced stimulation in Na,K-ATPase activity in either mammalian kidney or toad bladder at a time when enhanced

Na^+ transport was clearly evident (Chignell and Titus, 1966; Hill et al., 1973; Doucet and Katz, 1981; Park and Edelman, 1984).

An alternative to the direct induction model is the proposal of enhanced Na^+ entry as the mediator. Jorgensen (1972) reported an aldosterone-stimulated increase in rat renal Na,K-ATPase activity which was preceded by the increase in Na^+ reabsorption. Experimental evidence supporting this contention was provided by Petty et al. (1981). Treatment of adrenalectomized rabbits with physiological doses of aldosterone resulted at 3 hr in a restoration of Na,K-ATPase activity of cortical collecting tubules to normal. Pretreatment with amiloride, a blocker of apical Na^+ channels, abolished the aldosterone effect. Thus, augmented passive entry of Na^+ across the luminal membrane may mediate a subsequent increase in Na,K-ATPase activity. This proposal was further addressed in a recent study of Rayson and Gupta (1985). In outer rat medullary tubules, preincubation with amiloride blocked the stimulatory effect of aldosterone on Na,K-pump activity. They also noted that incubation of tubules with ouabain for 2 hr increased intracellular Na^+ without a demonstrable effect on Na,K-ATPase activity. Thus, a prior increase in intracellular Na^+ may not be the primary event responsible for the action of aldosterone on the Na,K-ATPase system. Other factors, therefore, may play a role in aldosterone action. A Na^+-independent action of aldosterone on toad bladder Na,K-ATPase activity has also been suggested by Geering et al. (1982). Incubation of toad bladders for 18 hr in the presence of 80 nM aldosterone yielded a 2.5-fold stimulation of the biosynthesis of Na,K-ATPase alpha and beta subunits. The aldosterone-induced stimulation of Na,K-pump biosynthesis was not inhibited by amiloride. Johnson et al. (1986) also found that in A6 cells aldosterone stimulation of Na,K-ATPase enzyme activity at 18 hrs was insensitive to amiloride. In summary, it is apparent that a simple model can not yet be constructed to account for the regulation of Na,K-ATPase activity by mineralocorticoids.

GLUCOCORTICOIDS

Specific and distinct glucocorticoid and mineralocorticoid receptors exist in rat kidney (Feldman et al., 1972), and daily glucocorticoid administration to adrenalectomized animals restores Na,K-ATPase activity (Chignell and Titus, 1966). In a similar study, the effect of glucocorticoids on Na,K-ATPase activity was not a generalized effect on membrane proteins, since the activities of Mg-ATPase, adenylate cyclase and 5'-nucleotidase were invariant (Hendler et al., 1972). The increase in Na,K-ATPase activity was associated with a parallel increase in (3)H-ouabain binding. Whether the observed effects of glucocorticoid were the result of ilicit occupancy of mineralocorticoid receptors was examined by Rodriquez et al. (1981). A single physiological dose of dexamethasone elicited a 40% increase in renal Na,K-ATPase activity after 2 hrs; whereas a comparable physiological dose of aldosterone had no such early effect.

Administration of dexamethasone to adrenalectomized rats for 2 hr produced a 40-50% stimulation of Na,K-ATPase activity in renal cortex and medulla (Sinha et al., 1981). This effect was absent when dexamethasone was added to isolated, broken cell membrane preparations. Renal Na,K-ATPase activity was increased 100-150% after 5 days of treatment with dexamethasone: V_{max} for Na^+, K^+ and ATP, were increased in concert, with no change in the apparent affinity for Na^+ or K^+. The increase in phospho-enzyme content and in ouabain-sensitive p-nitrophenylphosphatase activity (V_{max}) implied a glucocorticoid-induced increase in the number of Na,K-pumps. A direct effect on renal Na,K-ATPase activity was confirmed in a study of superfused distal segments of kidney tubules incubated with glucocorticoids in vitro (Rayson and Edelman, 1982). Distal tubules from adrenalectomized rats demonstrated a 27% and 32% increase in Na,K-ATPase activity in response to dexamethasone, at 6 and 24 hr, respectively. This response was glucocorticoid-specific. Additional observations on isolated renal tubules support the contention that glucocorticoids act directly rather than via effects on Na^+ content (Rayson and Lowther, 1984; Garg et al., 1985). A permissive role of intracellular Na^+, however, was suggested in studies of Rayson and Gupta (1985).

THYROID HORMONE

The thyroid hormones thyroxine (T_4) and triiodothyronine (T_3) stimulate oxygen consumption and active Na^+ and K^+ transport in many mammalian tissues. Treatment of either euthyroid or hypothyroid animals with T_3 augments Na,K-ATPase activity of liver, skeletal muscle, kidney, intestine and cardiac muscle (Ismail-Beigi and Edelman, 1970; Guernsey and Edelman, 1983). In contrast, adult rat brain Na,K-ATPase activity is unaffected by thyroid status, although the activity and abundance of Na,K-ATPase in the developing rat brain is thyroid dependent (Schmitt and McDonough, 1986). Administration of T_3 to thyroidectomized rats increases the Vmax (ATP) of skeletal muscle, renal cortex and heart with no change in K_m for ATP or $K_{1/2}$ for Na^+ and K^+ (Asano et al., 1976; Lo et al., 1976; Philipson and Edelman, 1977). Furthermore, the increase in Na,K-ATPase activity was proportionate to the increase in abundance of the enzyme in a number of studies (Lo et al., 1976; Curfman et al., 1977; Lin and Akera, 1978; Liberman et al., 1979).

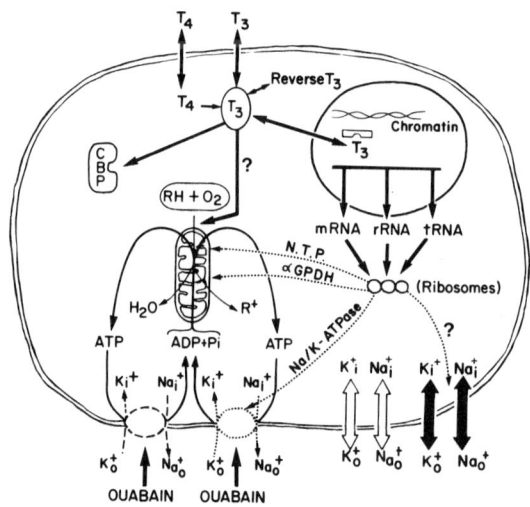

Figure 2. Thyroid hormone action.

To explore the nature of the T_3-stimulated increment in Na,K-ATPase content, the rates of synthesis and degradation of rat renal alpha and beta polypeptides were quantified (Lo and Edelman, 1976; Lo and Lo, 1980). Thyroidectomized rats were given single doses of T_3 and Na,K-ATPase was labelled with a 1 hr infusion of (3)H- or (35)S-methionine. At 20 hrs post-infusion, the synthesis of both subunits was augmented to the same extent as the peak increase in Na,K-ATPase activity. The rate constants of degradation of both subunits were approximately 0.15/day, and invariant with respect to thyroid status.

A current model of thyroid hormone action posits hormone binding to high affinity nuclear receptors with a resultant modulation of gene transcription (Figure 2).

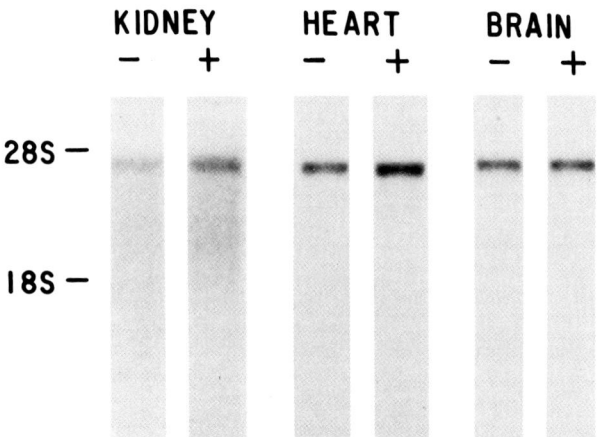

Figure 3. Kidney, heart and brain RNA blots probed with cDNA alpha.

Accordingly, renal poly(A^+)RNA was isolated from hypo- and hyperthyroid animals and translated in a rabbit reticulocyte lysate system (McDonough, 1985; Edelman et al., 1985). The translational activity of Na,K-ATPase alpha mRNA was increased 2 to 3-fold by administration of T_3. The recent cloning of alpha and beta subunits of mammalian Na,K-ATPase has enabled further exploration of the basis for T_3-induced changes in Na,K-ATPase activity (Shull et al., 1985; Schneider et al., 1985; Shull et al., 1986; Mercer et al., 1986). Northern blot analysis of total RNA from rat renal cortex, heart and brain, hybridized with cDNA alpha, indicated that alpha mRNA abundance increased in kidney and heart, but not in brain, in response to T_3 (Figure 3) (Chaudhury et al., 1987). This increase was confirmed by RNA dot blot analysis (e.g. as in Figure 4). Injection of T_3 elicited 3.9 and 2.1-fold increases in myocardial mRNA alpha abundance and in Na,K-ATPase activity (expressed per unit DNA), respectively, at 72 hrs. Similarly, renal cortex mRNA alpha abundance was stimulated 1.9-fold while Na,K-ATPase activity was increased 1.6-fold. These results are in accord with the mRNA alpha translation assays reported earlier. The lack of an effect of T_3 on brain mRNA alpha content noted by Chaudhury et al. (1987) is consistent with the absence of a thermogenic or Na,K-ATPase response in the adult brain (Ismail-Beigi and Edelman, 1971).

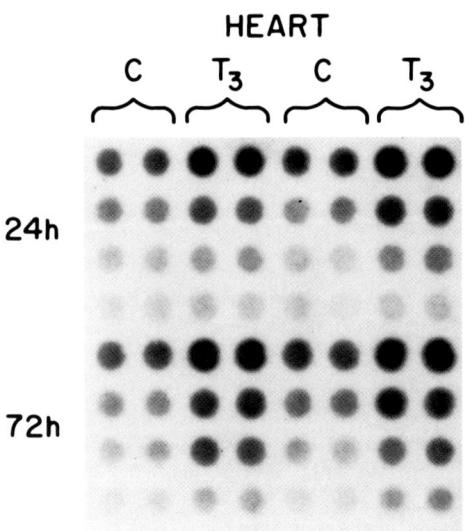

Figure 4. RNA dot blot of heart total RNA hybridized with cDNA alpha.

At the time of these studies on the effects of T_3 on mRNA alpha, little information was available on alpha isoforms, other than at the peptide level (Sweadner, 1979). Distinct Na,K-ATPase alpha isoform cDNAs have been cloned and the tissue distribution of alpha isoform mRNAs has been investigated (Shull et al., 1986; Young and Lingrel, 1987). The rat alpha isoform mRNA is predominate in kidney, heart and liver, whereas in brain the abundance of the three alpha mRNA isoforms are equivalent. Additional studies are necessary to ascertain the effect of T_3 on alpha mRNA isoforms.

T_3 has a striking and tissue-specific effect on Na,K-ATPase mRNA beta abundance (Gick et al., unpublished observations). Northern blot analysis of renal cortex total RNA indicates that T_3 administration stimulates mRNA beta content 2-fold; similar to the increase in mRNA alpha content. Accordingly, in renal cortex changes in mRNA abundance are coordinate but somewhat greater than the increase in the synthesis of the subunits (Lo and Lo, 1980). In our latest results on the rat heart, T_3 elicited a 9-fold augmentation in mRNA alpha content and a 22-fold increase in mRNA beta abundance (Gick et al., unpublished observations). The discrepancy in the magnitude of the T_3-induced increments in myocardial alpha and beta messages, and the greater accumulation of these mRNAs relative to the 2-fold increase in Na,K-ATPase activity strongly implicates translational and/or post-translational mechanisms in the action of T_3.

Seventy two hours after administration of T_3, rat hepatic Na,K-ATPase activity increased by 50%. In contrast, mRNA alpha abundance increases 6 to 8-fold (Gick et al., unpublished observations). Remarkably, hepatic mRNA beta content was unchanged in these same rats. Moreover, in euthyroid rat liver the abundance of the alpha message is significantly greater than that of the beta message. The low abundance of hepatic mRNA beta may explain the inability of Hubert et al. (1986) to detect the beta subunit in the liver by immunoprecipitation. Thus, hepatic Na,K-pumps either consist solely of alpha subunits or these subunits may be associated with variant beta subunits as compared to the beta subunits in non-hepatic tissues. If the hepatic beta subunit diverges in amino acid sequence, our results may indicate a corresponding divergence in the nucleotide sequence of the authentic hepatic mRNA beta.

The beta Northern blots revealed 4 bands migrating at 22, 20, 18 and 17 S which hybridize with both the rat brain and kidney cDNAs, which raises the possibility that these bands are derived from non-hepatocyte cellular elements that are unresponsive to T_3. Alternatively, if we are quantifying bonafide hepatic mRNA beta with the existing probes, thyroidal regulation of functional hepatic Na,K-pumps may involve both translational and post-translational levels of control.

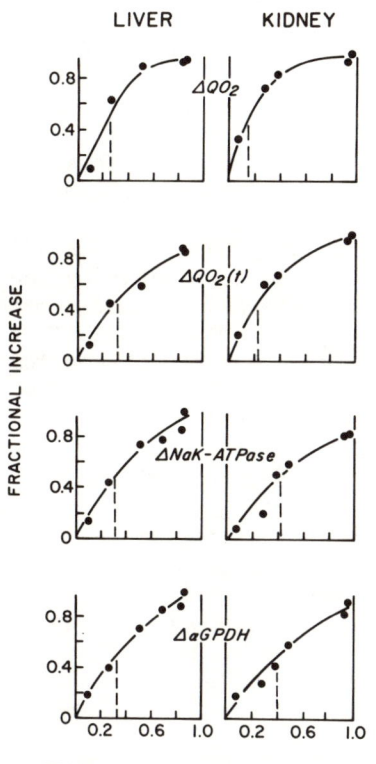

Figure 5. Nuclear T_3 receptor occupancy and physiological responses.

To determine if T_3 stimulates transcription of the alpha and beta genes, nuclei were analyzed by <u>in vitro</u> transcription assays. At 72 hrs, renal alpha and beta gene transcription was increased 80% in response to T_3 (Gick et al., unpublished observations). Analogous run-on transcription assays demonstrated a 30-40% stimulation of both alpha and beta gene transcription of liver nuclei. The discrepancies between the T_3-induced increments in abundance of renal and hepatic mRNA alpha and that of alpha gene transcription implies a predominant post-transcriptional effect of T_3. Both transcriptional and post-transcriptional events have also been implicated in thyroidal regulation of the expression of other genes (Dozin et al., 1986; Jump et al., 1986).

Although T_3 clearly increases Na,K-ATPase gene expression in responsive tissues, the mechanism remains to be clarified. The high degree of correlation between nuclear receptor occupancy by T_3, and renal and hepatic oxygen consumption, alpha GPDH activity and Na,K-ATPase activity supports a nuclear T_3-mediated mechanism (Figure 5) (Somjen et al., 1981). The nuclear T_3 receptor complexes could directly react with regulatory sequences of the Na,K-ATPase alpha and beta genes. A well-characterized example of such a direct activation is the regulation of growth hormone expression in pituitary tumor cell lines by T_3. Specific regions have been delineated in the 5'-flanking DNA of the rat growth hormone gene which mediate T_3-stimulated transcription (Flug et al., 1987). Alternatively, T_3 stimulation of Na,K-ATPase gene expression could be an indirect or secondary result of T_3 regulation of other genes, whose expression may then increase Na,K-ATPase gene transcription and Na,K-ATPase mRNA stability or processing.

REFERENCES

Asano Y, Liberman UA, Edelman IS (1976). Relationships between Na-dependent respiration and Na+K-adenosine triphosphatase activity in rat skeletal muscle. J Clin Invest 57:368-379.
Berthon B, Capiod T, Claret M (1985). Effects of noradrenaline, vasopressin and angiotensin on the Na-K pump in rat isolated liver cells. Br J Pharmac 86:151-161.
Cantley LC Jr., Ferguson JH, Kustin K (1978). Norepinephrine complexes and reduces vanadium(V) to reverse

vanadate inhibition of the (Na,K)-ATPase. J Amer Chem Soc 100:5210-5212.
Chaudhury S, Ismail-beigi F, Gick GG, Levenson R, Edelman IS (1987). Effects of thyroid hormone on the abundance of Na,K-adenosine triphosphatase alpha subunit messenger ribonucleic acid. Mol Endo 1:83-89.
Chignell CF, Titus E (1966). Effect of adrenal steroids on a Na+ and K+ requiring adenosine triphosphatase from rat kidney. J Biol Chem 241:5083-5089.
Clausen T, Flatman JA (1977). The effect of catecholamines on Na-K transport and membrane potential in rat soleus muscle. J Physiol 270:383-414.
Clausen T, Hansen O (1977). Active Na-K transport and the rate of ouabain binding. The effect of insulin and other stimuli on skeletal muscle and adipocytes. J Physiol 270:415-430.
Clausen T, Flatman JA (1987). Effects of insulin and epinephrine on Na^+-K^+ and glucose transport in soleus muscle. Am J Physiol 252: E492-E499.
Curfman GD, Crowley TJ, Smith TW (1977). Thyroid-induced alteration in myocardial sodium-and potassium-activated adenosine triphosphatase, monovalent cation active transport, and cardiac glycoside binding. J Clin Invest 59:586-590.
Doucet A, Katz AI (1981). Short-term effect of aldosterone on Na,K-ATPase in single nephron segments. Am J Physiol 241:F273-F278.
Dozin B, Magnuson MA, Nikodem VM (1986). Thyroid hormone regulation of malic enzyme synthesis. J Biol Chem 261:10290-10292.
Edelman IS, Bogoroch R, Porter GA (1963). On the mechanism of action of aldosterone on sodium transport: the role of protein synthesis. Proc Natl Acad Sci USA 50:1169-1177.
Edelman Is, Pressley TA, Hiatt A (1985). Regulation of mammalian Na,K-ATPase. In Glynn I, Ellory C (eds): "The Sodium Pump," Cambridge, UK: The Company of Biologists Limited, pp 153-159.
Erlij F, Grinstein S (1976). The number of sodium ion pumping sites in skeletal muscle and its modification by insulin. J Physiol 259:13-31.
Fehlmann M, Freychet P (1981). Insulin and glucagon stimulation of (Na^+-K^+)-ATPase transport activity in isolated rat hepatocytes. J Biol Chem 256:7449-7453.
Feldman D, Funder JW, Edelman IS (1972). Subcellular mechanisms in the action of adrenal steroids. Am J Med 53:545-560.

Flug F, Copp RP, Casanova J, Horowitz ZD, Janocko L, Plotnick M, Samuels HH (1987). Cis-acting elements of the rat growth hormone gene which mediate basal and regulated expression by thyroid hormone. J Biol Chem 262:6373-6382.

Garg LC, Narang W, Wingo CS (1985). Glucocorticoid effects on Na-K-ATPase in rabbit nephron segments. Am J Physiol 248:F487-F491.

Geering K, Girardet M, Bron C, Kraehenbuhl JP, Rossier BC (1982). Hormonal regulation of (Na^+,K^+)-ATPase biosynthesis in the toad bladder. J Biol Chem 257:10338-10343.

Gelehrter TD, Shreve PD, Dilworth VM (1984). Insulin regulation of Na/K pump activity in rat hepatoma cells. Diabetes 33:428-434.

Guernsey DL, Edelman IS (1983). Regulation of thermogenesis by thyroid hormones. In Oppenheimer JH, Samuels HH (eds): "Molecular Basis of Thyroid Hormone Action," New York: Academic Press, pp293-394.

Hays ET, Horowicz DP, Swift JG (1974). Epinephrine action on sodium fluxes in frog striated muscle. Am J Physiol 227:1340-1347.

Hendler ED, Torretti J, Kupor L, Epstein F (1972). Effects of adrenalectomy and hormone replacement on Na-K-ATPase in renal tissue. Am J Physiol 222:754-760.

Hill JH, Cortas N, Walser M (1973). Aldosterone action and sodium- and potassium-activated adenosine triphosphatase in toad bladder. J Clin Invest 52:185-189.

Horster M, Schmid H, Schmidt U (1980). Aldosterone in vitro restores nephron Na-K-ATPase of distal segments from adrenalectomized rabbits. Pflugers Arch 384:203-206.

Hubert JJ, Schenk DB, Skelly H, Leffert HL (1986). Rat hepatic (Na^+,K^+)-ATPase: alpha subunit isolation by immunoaffinity chromatography and structural analysis by peptide mapping. Biochemistry 25:4156-4163.

Ismail-Beigi F, Edelman IS (1970). Mechanism of thyroid calorigenesis: role of active sodium transport. Proc Nat Aca Sci 67:1071-1078.

Ismail-Beigi F, Edelman IS (1971). The mechanism of the calorigenic action of thyroid hormone. J Gen Physiol 57:710-722.

Johnson JP, Jones D, Wiesmann WP (1986). Hormonal regulation of Na^+-K^+-ATPase in cultured epithelial cells. Am J Physiol 251:C186-C190.

Jorgensen PL (1972). The role of aldosterone in the regulation of (Na^+-K^+)-ATPase in rat kidney. J Steroid Biochem 3:181-191.

Jump DB, Tao TY, Towle HC, Oppenheimer JH (1986). Dissociation of hepatic messenger ribonucleic acid levels and nuclear transcriptional rates in suckling rats. Endocrinology 118:1892-1986.

Knox WH, Sen AK (1974). Mechanism of action of aldosterone with particular reference to (Na+K)-ATPase. Ann NY Acad Sci 242:471-488.

Li JH, Palmer LG, Edelman IS, Lindemann B (1982). The role of sodium-channel density in the natriferic response of the toad urinary bladder to an antidiuretic hormone. J Memb Biol 64:77-89.

Liberman UA, Asano Y, Lo CS, Edelman IS (1979). Relationship between Na^+-dependent respiratory and Na^+-K^+-adenosine triphosphatase activity in the action of thyroid hormone on rat jejunal mucosa. Biophysical J 27:127-144.

Lin MH, Akera T (1978). Increased (Na^+,K^+)-ATPase concentrations in various tissues of rats caused by thyroid hormone treatment. J Biol Chem 253:723-726.

Lo CS, August TR, Liberman UA, Edelman IS (1976). Dependence of renal (Na^++K^+)-adenosine triphosphatase activity on thyroid status. J Biol Chem 251:7826-7833.

Lo CS, Edelman IS (1976). Effect of triiodothyronine on the synthesis and degradation of renal cortical (Na^++K^+)-adenosine triphosphatase. J Biol Chem 251:7834-7840.

Lo CS, Lo TN (1980). Effect of triiodothyronine on the synthesis and degradation of the small subunit of renal cortical (Na^++K^+)-adenosine triphosphatase. J Biol Chem 255:2131-2136.

Lynch CJ, Wilson PB, Blackmore PF, Exton JH (1986). The hormone-sensitive hepatic Na^+-pump. Evidence for regulation by diacylglycerol and tumor promoters. J Biol Chem 261:14551-14556.

Lytton J, Lin JC, Guidotti G (1985). Identification of two molecular forms of (Na^+,K^+)-ATPase in rat adipocytes. Relation to insulin stimulation of the enzyme. J Biol Chem 260:1177-1184.

Lytton J (1985). Insulin affects the sodium affinity of the rat adipocyte (Na^+,K^+)-ATPase. J Biol Chem 260:-10075-10080.

McDonough A (1985). Immunodetection of Na,K-ATPase in various tissues and regulation with thyroid hormone. In Glynn I, Ellory C (eds.) "The Sodium Pump", UK, The Company of Biologists Ltd. pp 161-163.

Mendoza SA, Wigglesworth NM, Rozengurt E (1980). Vasopressin rapidly stimulates Na entry and Na-K pump activity in quiescent cultures of mouse 3T3 cells. J Cell Physiol 105:153-162.

Mercer RW, Schneider JW, Savitz A, Emanuel J, Benz Jr. EJ, Levenson R (1986). Rat-brain Na,K-ATPase beta-chain gene: primary structure, tissue-specific expression, and amplication in ouabain-resistant HeLa C^+ cells. Mol Cell Biol 6:3884-3890.

Park CS, Edelman IS (1984). Effect of aldosterone on abundance and phosphorylation kinetics of Na,K-ATPase of toad urinary bladder. Am J Physiol 246:F509-F516.

Petty KJ, Kokko JP, Marver D (1981). Secondary effect of aldosterone on Na,K-ATPase activity in the rabbit cortical collecting tubule. J Clin Invest 68:1514-1521.

Philipson KD, Edelman IS (1977). Characteristics of thyroid-stimulated Na^+-K^+-ATPase of rat heart. AM J Physiol 232:C202-C206.

Rayson BM, Edelman IS (1982). Glucocorticoid stimulation of Na,K-ATPase in superfused distal segments of kidney tubules in vitro. Am J Physiol 243:F463-F470.

Rayson BM, Lowther SO (1984). Steroid regulation of Na^+-K^+-ATPase: differential sensitivities along the nephron. Am J Physiol 246:F656-F662.

Rayson BM, Gupta RK (1985). Steroids, intracellular sodium levels, and Na^+/K^+-ATPase regulation. J Biol Chem 260:12740-12743.

Resh MD, Nemenoff RA, Guidotti G (1980). Insulin stimulation of (Na^+,K^+)-Adensine triphosphatase-dependent $^{86}Rb^+$ uptake in rat adipocytes. J Biol Chem. 255:10938-10945.

Reznik VM, Shapiro RJ, Mendosa SA (1985). Vasopressin stimulates DNA synthesis and ion transport in quiescent epithelial cells. Am J Physiol 249:C267-C270.

Rodriguez HJ, Sinha SK, Starling J, Klahr S (1981). Regulation of renal Na^+-K^+-ATPase in rat by adrenal steroids. Am J Physiol 241:F186-F195.

Rosic NK, Standaert ML, Pollet RJ (1985). The mechanism of insulin stimulation of (Na^+,K^+)-ATPase transport activity in muscle. J Biol Chem 260:6206-6212.

Schmidt U, Schmid J, Schmid H, Dubach UC (1975). Sodium- and potassium-activated ATPase, a possible target of aldosterone. J Clin Invest 55:655-660.

Schmitt CA, McDonough AA (1986). Developmental and thyroid hormone regulation of two molecular forms of Na^+-K^+-ATPase in brain. J Biol Chem. 261:10439-10444.

Schneider JW, Mercer RW, Caplan M, Emanuel JR, Sweadner KJ, Benz Jr. EJ, Levenson R (1985). Molecular cloning of rat

brain Na,K-ATPase alpha-subunit cDNA. Proc Natl Acad Sci USA 82:6357-6361.
Shull GE, Schwartz A, Lingrel JB (1985). Amino acid sequence of the catalytic subunit of the (Na^+-K^+)-ATPase deduced from a complementory DNA. Nature 316:691-695.
Shull GE, Lane LK, Lingrel JB (1986). Amino-acid sequence of the beta-subunit of the (Na^+-K^+)-ATPase deduced from a cDNA. Nature 321:429-431.
Shull GE, Greeb J, Lingrel JB (1986). Molecular cloning of three distinct forms of the Na^+,K^+-ATPase alpha-subunit from rat brain. Biochemistry 25:8125-8132.
Sinha SK, Rodriguez HJ, Hogan WC, Klahr SC (1981). Mechanisms of activation of renal (Na^++K^+)-ATPase in the rat. Effects of acute and chronic administration of dexamethasone. Biochim Biophys Acta 641:20-35.
Somjen D, Ismail-Beigi F, Edelman IS (1981). Nuclear binding of T_3 and effects on QO_2, Na,K-ATPase, and alpha-GPDH in liver and kidney. Am J Physiol 240:E46-E154.
Swann AC, Crawley JN, Grant SJ, Maas JW (1981). Noradrenergic stimulation in vivo increases (Na^+,K^+)-adenosine triphosphatase activity. Life Sciences 28:251-256.
Verrey F, Schaerer E, Zoerkler P, Paccolat MP, Geering K, Kraehenbuhl JP, Rossier BC (1987). Regulation by aldosterone of Na^+,K^+-ATPase mRNAs, protein synthesis, and sodium transport in cultured kidney cells. J Cell Biol 104:1231-1237.
Westenfelder C, Arevalo GJ, Aranaowski RL, Kartzman NA, Katz AI (1977). Relationship between mineralocorticoids and renal Na^+-K^+-ATPase: sodium reabsorption. Am J. Physiol 233:F593-F599.
Wilce PA, Rossier BC, Edelman IS (1976). Actions of aldosterone on polyadenylated ribonucleic acid and Na^+ transport in the toad bladder. Biochemistry 15:4279-4285.
Wu PH, Phillis JW (1980). Characterization of receptor-mediated catecholamine activation of rat brain cortical Na^+-K^+-ATPase. Int J Biochem 12:353-359.
Young RM, Shull GE, Lingrel JB (1987). Multiple mRNAs from rat kidney and brain encode a single Na^+,K^+-ATPase beta-subunit protein. J Biol Chem 262:4905-4910.
Young RM, Lingrel JB (1987). Tissue distribution of mRNAs encoding the alpha isoforms and beta subunits of rat Na^+,K^+-ATPase. Biochem Biophys Res Comm 145:52-58.

OVERVIEW:

PHYSIOLOGICAL INHIBITORS OF Na, K-ATPase: CONCEPT AND STATUS

Garner T. Haupert, Jr., M.D.

Renal Unit, Massachusetts General Hospital and
Harvard Medical School
Boston, Massachusetts   02114

   Experimental observations which ultimately led to the concept of the existence of an endogenous regulator of the mammalian Na,K-ATPase began more than 25 years ago in laboratories of renal physiology. De Wardener and co-workers (1961) reported in 1961 that intravascular expansion using saline in dogs produced a brisk natriuresis despite the fact that renal perfusion pressure did not rise (inflation of aortic balloon above the renal arteries), that glomerular filtration rate (GFR) remained unchanged or diminished, and that decreases in mineralocorticoid activity could not occur (administration of suraphysiologic doses of mineralocorticoid during the course of volume expansion). Since perfusion pressure and filtration rate did not change, and the natriureitic effects of extracellular fluid volume expansion in one animal also occured in a second animal cross-circulated with the blood of the first, the presumption was that the natriuresis was due in part to a circulating substance which exerted its effects directly on the renal tubular Na^+ reabsorptive process, without affecting renal hemodynamics. The initial studies by de Wardener and colleagues were not fully accepted because the saline infusion used to expand the circulation changed the composition of the blood, altering physical factors such as peritubular oncotic pressure and renal vascular resistance. But more persuasive evidence came out of a second generation of cross-circulation experiments and experiments using transplanted kidneys or isolated perfused kidneys while at the same time controlling blood composition. Such experiments again showed an increase in Na^+ excretion with expansion of blood volume despite the absence of nervous system activty or measurable changes in GFR or renal blood flow (Bahlman et al., 1967; Kaloyanides and Azer, 1971; Lichardus and Nizet, 1972;)

Since crude urine or plasma samples might contain many substances which could produce the observed in vivo effects, fractionation of extracts was carried out prior to physiological testing. These early purification attempts were, however, limited largely to deproteination, desalting, and some concentration of activity using gel filtration chromatography, and the emphasis in the field remained mainly on physiological testing.

It was not clear whether this putative humoral substance acted by altering intrarenal hemodynamics or by directly inhibiting Na^+ transport in the renal tubule. Therefore, investigators turned to in vitro approaches to demonstrate biological activity consistent with the natriuretic hormone hypothesis. The bioassays involved the demonstration of changes in active Na^+ transport across anuran membranes such as the toad urinary bladder, considered a functional analogue of the human distal nephron (Bourgoignie et al, 1971; Kaplan et al, 1974; Haupert and Sancho, 1979), or in isolated, perfused segments of mammalian renal tubule (Fine et al, 1976). Results of these experiments confirmed that active extracts from plasma, urine, and tissue sources which were natriuretic in vivo exerted their effects by a direct action on transepithelial sodium transport. Hillyard et al (1976) first documented directly that the observed effects on cation transport were associated with inhibition of the Na,K-ATPase.

At the same time, in another area of circulatory physiology, evidence was beginning to accumulate linking kidney, brain, and the cardiovascular system in the genesis of certain forms of experimental animal and human essential hypertension. It had been observed that the development of volume-expanded ("low renin") hypertension in animal models followed a predictable sequence of events. Initially there is an increase in cardiac output due to an increase in cardiac contractility and an increase in venous return of blood to the heart (decreased venous capacitance). In a short time, however, cardiac output returns to normal, associated with an increase in total peripheral vascular resistance which is sustained and results in tonic elevation of arterial pressure. Na,K-ATPase activity in cardiovascular tissues in these animals with volume expanded hypertension was found to be decreased (Haddy and Overbeck, 1976). Haddy and Ovebeck (1976) postualted that the physiological sequence of events and the Na,K-ATPase measurements could be explained by the appearance of a circulating Na,K-ATPase inhibitor which might be linked to pathogenesis of the hypertension. Indeed, Overbeck et al. (1976) showed that volume expansion was associated with the elaboration of a heat-stable substance in plasma that inhibited ouabain-sensitive cell membrane Na^+ transport in vascular muscle. Brody and colleagues (1978), seeking a locus of anatomical control, demonstrated that lesions in the region of the anteroventral third ventricle of the brain prevented the hypertension of volume expansion. Pamnani et al. (1981), Songu-Mize et al. (1982), and Bealer et al. (1983) subsequently showed that these lesions prevented the secretion of the Na^+ transport inhibitor. It now seemed possible to explain a number of puzzling physiological phenomena, including both the natriuresis and hypertension of volume expansion, on the basis of a circulatory substance released by the midbrain in response to volume expansion. The postulated mechanism of action of this substance(s) was the modulation of renal tubular Na^+ reabsorption and vascular smooth muscle tone by regulation of Na,K-ATPase activity.

The overall hypothesis as recently outlined (de Wardener, 1987) proposes that in hereditary forms of hypertension there is a persistent tendancy to renal retention of Na^+. This may be due to increased Na^+-K^+ cotransport in the proximal tubule as found in the Milan hypertensive strain of rats (Bianchi et al., 1986), or to augmented Na^+-H^+ exchange in the proximal tubule occurring as a local manifestation of a generalized genetic defect in Na^+-Na^+ (Na^+-Li^+) countertransport which exists in the erythrocytes of some essential hypertensives and their first degree normotensive relatives (Canessa et al., 1980; Canessa, 1986). The renal Na^+ retention leads to a transient increase in extracellular fluid volume which serves as stimulus for the release of a Na,K-ATPase inhibitor, probably from the hypothalamus. The Na^+-pump inhibitor acts on the renal tubule to promote Na^+ excretion thus restoring extracellular fluid volume to normal (or even low) levels, but with similar inhibitory effects on the Na,K-ATPase in vascular smooth muscle cells resulting in a tonic increase in vascular tone, increased total peripheral resistance and arterial hypertension. The persistence of the Na,K-ATPase inhibitor in the circulation in the absence of increased intravascular volume (which is measured to be normal or low in established essential hypertension) appears paradoxical, but could be related to persistent increase in intrathoracic vascular pressures (London et al., 1985) acting as an afferent stimulus.

How Na,K-ATPase inhibition in vascular smooth muscle results in increased cytosolic free calcium concentrations which must occur to produce the arterial vasoconstriction remains unclear. Blaustein has proposed (1977) and defended (Blaustein et al., 1986) the hypothesis that altered Na^+-Ca^{2+} exchange resulting from partial sodium pump inhibition can account for the increased intracellular free Ca^{2+} concentration in vascular smooth muscle. Other authorities maintain that there is little evidence for intracellular Ca^{2+} regulation by Na^+-Ca^{2+} exchange in vascular smooth muscle (Somlyo, 1986), except perhaps in extreme situations of Ca^{2+} overload (Mulvaney, 1985).

The effects of Na^+-pump inhibition in excitable tissues provides nevertheless for other possibilities. Increased intracelluar free Ca^{2+} concentration could occur through membrane depolarization effects of Na,K-ATPase inhibition via the voltage-dependent Ca^{2+} channel (Mulvany et al., 1984). Pamnani et al. (1982) have measured such a change in potential using a microelectrode in rat tail arteries bathed in hypertensive rat plasma containing a Na^+ transport inhibitor. Since norepinephrine uptake into adrenergic nerve terminals is sodium dependent and inhibited by ouabain (Vanhoutte and Lorenz, 1984), another possibility is that the circulating Na,K-ATPase inhibiitor acts indirectly on smooth muscle cells through actions on sympathetic nerve endings. Normal canine plasma has been shown to block uptake of ^{3}H-norepinephrine in isolated saphenous vein, and this ability of the plasma was resistant to boiling suggesting the effects were due to a heat stable, non-protein component (Freas et al., 1982).

Although there was no reason to assume a structural identity between the posutlated endogenous Na,K-ATPase inhibitor and the cardiac glycosides, digitalis is a potent inhibitor of Na,K-ATPase, and has been shown to cause both natriuresis (Hook, 1969) and an increase in vascular resistance (Vatner et al., 1971), although these are not its major pharmacological effects. The widespread availability of the digoxin radioimmunoassay (Smith et al., 1969) stimulated the measurement of immunoreactivity in plasma in situations where the inhibitor might be elevated. Klingmuller et al (1982) found digitalis-like immunoreactivity in the urine of Na^+ loaded normal human subjects. Graves et al. (1983) made a similar observation when they examined the plasma of uremic subjects and suggested that this phenomenon was of a prevalence and magnitude sufficient to distort the interpretation of the digoxin assay in clinical situations. Cross-reactivity in digoxin radioimmunoassays was also shown in the plasma of volume expanded dogs (Gruber et al., 1980), the plasma and urine of hypertensive patients (Crabos et al., 1984), and the plasma of volume-expanded hogs and normal human subjects purified by high performance liquid chromatography (Tamura et al., 1985; Kelly et al., 1986). In the latter two cases, the responsible compounds were identified as non-esterified fatty acids and lysophospholipids, illustrating the problem of specificity of effects in the radioimmunoassays.

If antibodies specific for the digitalis glycosdes are indeed capable of binding the Na,K-ATPase inhibitor, one would expect that the administration of high affinity antibodies in vivo would inhibit the activity of the inhibitor in the same way that such antibodies inhibit the pharmacological effects of digoxin (Smith et al., 1982). In a hypertensive model, in which rats were heminephrectomized and then treated with deoxycorticosterone and salt, Kojima et al. (1982) showed that the administration of anti-digoxin antibody caused a marked decrease in blood pressure. Because whole antibody was used in this study rather than Fab fragments, the possibility that the fall in blood pressure was due to immune-complex-mediated release of vasodilators cannot be excluded. Control animals, nevertheless, did not manifest a similar hypotensive response when infused with the same antibody solution. Subsequently the same investigators showed that digoxin-like immunoreactivity as well as Na,K-ATPase binding activity was present in the plasma of these hypertensive animals (Kojima, 1984).

The background of these experimental and clinical observations has stimulated a concerted effort by a number of laboratories to isolate and identify the endogenous Na,K-ATPase inhibitor. Although considerable effort has been expended over a number of years, a purified substance of known structure is not yet in hand. Controversy still exists even about the chemical nature of the substance; some maintain it is a peptide, while others affirm that it has properties manifestly inconsistent with this class of compounds. It should be remembered, however, that vertebrates are capable of synthesizing substances closely analagous in structure and pharmacological properties to the digitalis glycosides. The bufodieneolides, present in the skin of certain toads, are, like the digitalis glycosides, steroid-lactone compounds that inhibit Na,K-ATPase (Flier et al., 1980) and are potent cardiac inotropic agents (Shimoni et al., 1984).

In evaluating the various efforts to isolate the endogenous Na,K-ATPase inhibitor, it is important to keep in mind that certain criteria must be met before a putative substance can be considered a physiological regulator. Like all enzymes, Na,K-ATPase is readily inhibited by a large number of substances. For example, the common fatty acids, linoleic and linolenic, are effective inhibitors (Bidard et al., 1984; Tamura et al., 1985), though they exhibit a rather high inhibitory constant (K_i) well beyond their physiological range of concentration. For these reasons a believable, physiologically relevant inhibitor should have a very high binding affinity for the enzyme. Ouabain, one of the digitalis glycosides, exhibits a K_i of 2 nM. One must expect the K_i of an endogenous Na,K-ATPase inhibitor to be at least within that order of magnitude.

If the function of the putative inhibitor is to regulate the Na^+-K^+ pump, such inhibition must be reversible. Na,K-ATPase regeneration by protein synthesis would seem too sluggish a mechanism to allow quickly changing homeostatic needs to be met. Specificity for membrane Na,K-ATPase is also essential, since it would be cumbersome for the inhibitor to regulate several enzymes at the same time. Finally, one would like to observe appropriate changes in inhibitor plasma concentration in response to relevant stimuli. None of the substances put forward as candidates for the endogenous Na,K-ATPase inhibitor meets all these criteria, and most have not even been convincingly tested for any of them.

As inferred above, many investigators have hypothesized that the endogenous inhibitors of Na,K-ATPase may have a chemical structure analogous to the digitalis glycosides, or at least share immunological cross-reactivity with them. Although this structural similarity would simplify measurement or purification of the inhibitor, it cannot be considered a necessary criterion for identification of the substance. Nor would it be necessary for the mechanism of enzyme inhibition to duplicate exactly that of the cardiac glycosides.

PURIFICATION EFFORTS

Various laboratories have used plasma, urine, or tissue as potential sources of the Na,K-ATPase inhibitor. Most of these efforts are at a preliminary stage. To date none has yielded molecular characterization, and few have addressed the essential tests just mentioned. Conclusions as to the chemical nature of the material are often contradictory.

Table 1 summarizes the efforts from a number of laboratories to isolate and purify the endogenous Na,K-ATPase inhibitor. Gruber et al. (1980) used high performance liquid chromatography (HPLC) to partially purify a substance from canine plasmna that showed immunological cross-reactivity with digitalis. The same fractions inhibited Na,K-ATPase, and concentration varied with volume expansion, as would be predicted by the earlier experiments of de Wardener et al (1961). Gruber postulated that the material was a peptide. An inhibitory constant could not be measured, and further information on purification and structural analysis is not available. Using an enzyme-linked assay, Hamlyn et al. (1982) found Na,K-ATPase inhibitory activity in deproteinized plasma of patients with essential hypertension, and levels of this inhibitor(s) correlated with the degree of hypertension. Biochemical and structural characterizations have not been reported. Tamura et al. (1985) and Kelly et al. (1986) fractionated hog and human plasma respectively obtaining several peaks which inhibited Na,K-ATPase activity, ouabain binding, and showed cross-reactivity with digoxin-specific

TABLE 1. Summary of efforts to isolate and characterize an endogenous Na, K-ATPase inhibitor

Source	Extraction	Assay	Affinity	Chemical Nature
Canine plasma[30]	Deproteinized Diafiltration HPLC	Digoxin RIA ATPase activity	NR	? Peptidic
Human plasma[34]	Deproteinized	ATPase activity	NR	NR
Hog plasma[71]	Acetone/methanol; HPLC	ATPase activity Ouabain binding Digoxin RIA	$K_i=100$ µM	Oleic acid Linoleic acid
Human plasma[46]	Deproteinized HPLC	ATPase activity Ouabain binding Digoxin RIA	$K_i=20$ µM	Nonesterified fatty acids Lysophospholipid
Human serum[51] (hemofiltrate)	Ion-exchange Octadecyl C18 Affinity chromatography	Sodium efflux (lymphocytes) ATPase activity Ouabain binding Digoxin RIA	NR	NR
Bovine plasma[70]	Methanol Sep-Pak (C18) HPLC	Ouabain binding ATPase activity	NR	NR
Human urine[50]	Gel filtration Ion exchange HPLC	Natriuresis (rat) Digoxin RIA	NR	Peptidic
Human urine[14]	Ion exchange HPLC	ATPase activity Ouabain binding Digoxin RIA	NR	Nonpeptidic ? steroidal
Bovine hypothalamus[35]	Methanol Ion exchange Lipophilic gel	Na transport (toad bladder, human RBCs) ATPase activity Ouabain binding	$K_i=1.4$ nM	Hydrophilic nonpeptidic zwitterionic
Brain (rat,[54] sheep[64])	Acid/acetone Methanol Gel filtration HPLC	ATPase activity Ouabain binding Cardiac muscle contractility	NR	Hydrophilic Nonpeptidic
Bovine hypothalamus[1]	Acid/acetone Gel filtration Ion exchange TLC	ATPase activity Ouabain binding	NR	Peptidic

TABLE 1. (cont.)

Source	Extraction	Assay	Affinity	Chemical Nature
Rat hypothalamus[57]	Acidified acetone Electrophoresis HPLC	Cytochemical (G6PD stimulation; Na,K-ATPase inhibition) Digoxin RIA	NR	Polar, basic Nonpeptidic
Hypothalamic cell culture supernatants[58]	Deproteinized (heat) Gel filtration HPLC	ATPase activity Ouabain binding Na transport (human RBCs)	NR	Peptidic
Guinea pig heart[18]	Water Methanol Ion-exchange Gel filtration	ATPase activity Ouabain binding Digoxin RIA	NR	NR
Human CSF[33]	None	Na transport (human RBCs) ATPase activity	NR	NR
Human CSF[53]	Methanol HPLC	ATPase activity Ouabain binding	$K_i = 0.5$ μM	Nonpeptidic

ATPase activity, inhibition of Na, K-ATPase; CSF, cerebral spinal fluid; HPLC, high performance liquid chromatography; K_i, inhibition constant; NR, not reported; RIA, radioimmunoassay; RBC, red blood cell; TLC, thin-layer chromatography. (Adapted from Haber and Haupert, 1987.)

antibodies. On further analysis these peaks were shown to contain either non-esterified fatty acids or lysophospholipid, both of which appear to be non-specific inhibitors of Na,K-ATPase. As discussed above, the relatively high affinity constants for enzyme inhibition by these compounds makes it unlikely that they play any role in the physiologic regulation of the Na,K-ATPase. A role in pathophysiology remains to be explored. Crabos et al. (1984) also used HPLC to fractionate plasma and urine in normal and hypertensive subjects. They found several peaks of Na,K-ATPase inhibitory activity, some of which cross-reacted in a digoxin radioimmunoassay and were present in greater amounts in hypertensive subjects than in normal subjects. The same group (Cloix et al., 1985) proceeded with purification of the urinary Na^+ transport inhibitor. Preliminary analysis by nuclear magnetic resonance and mass spectroscopy suggests that compound is non-peptidic, possibly steroidal with amino and sugar moieties. This inhibitor from urine showed a number of similarities with ouabain in the inhibition of Na,K-ATPase, but was not specific for the Na,K-ATPase as significant inhibition of Mg2+ and Ca2+-dependent ATPase could also be demonstrated (Crabos et al., 1987). An inhibitory constant was not reported. Klingmuller et al. (1982), who, as mentioned previously, had identified digoxin-like immunoreactivity in human urine, effected a partial purification utilizing the binding properties of digitalis antibody. A detailed structural characterization of this material is not yet available, although it is suspected to be a polypeptide (Kramer et al., 1985). Of some concern is the observation in this more recent work that the digoxin radioimmunoassay did not correlate with either Na^+ intake or total natriuretic activity by bioassay. Gault et al. (1983) have reported that radioimmunoassay results on plasma extracts with digoxin-specific antibody in untreated hypertensive subjects do correlate with Na^+ loading. Further work from the same laboratory suggests that dehydroepiandrosterone accounts for some of the digoxin immunoreactivity in plasma, and that this steroid is also a Na,K-ATPase inhibitor (Vasdev et al., 1985).

Because the midbrain has been implicated in the control of circulating inhibitors, the brain has been a favored source of tissue for study. Haupert and Sancho (1979) extracted material from the hypothalamus. Progress in this work will be briefly summarized below. Fishman (1979) prepared a fraction of guinea pig brain that inhibited ^3H-ouabain binding to brain microsomes and the uptake of $^{86}Rb^+$ into human erythrocytes. Lichtstein and Samuelov (1980) also extracted a low molecular weight "ouabain-like compound" from whole rat brain. This material inhibited ^3H-ouabain binding to rat brain synaptosomes and Na,K-ATPase activity. Material prepared from sheep brain by the same procedure and further purified by HPLC was shown to exert positive inotropic effects in frog cardiac muscle (Shimoni et al., 1984). Preliminary characterization efforts suggested that the material was of low molecular weight, polar, and non-peptidic, in agreement with the findings of Haupert and Sancho (1979). Akagawa et al. (1984) also used bovine hypothalamus to obtain a fraction that, after gel filtration and ion-exchange chromatography, inhibited Na,K-ATPase activity and specific ouabain binding to rat brain microsomes in an apparently competitive manner. However, this activity was destroyed in part by a proteolytic enzyme, indicating that at least some of the inhibition was due to a peptide. De Wardener and colleagues, using an indirect cytochemical assay (Fenton et al., 1982), found G6PD-stimulating (Na,K-ATPase inhibiting) activity in acidified-acetone extracts of rat hypothalamus purified by electrophoresis and HPLC, and the level of this activity was greater in Okamoto spontaneously hypertensive rats than in their normotensive Wistar controls (Millett et al., 1986). Preliminary characterization suggests the molecule to be polar, basic, and non-peptidic; binding affinity to Na,K-ATPase has not been reported (Millet et al., 1987).

The Na,K-ATPase inhibitor extracted in our laboratory from bovine hypothalamus has a molecular mass of approximately 500, is non-peptidic, resistant to acid hydrolysis but inactivated by base hydrolysis or ashing. It is highly polar with zwitterionic properties, due in part to a carboxylate function which is necessary for biological activity (Carilli et al., 1985). Initial studies showed this hypothalamic factor to inhibit active sodium transport in toad urinary bladder, to prevent ^{3}H-ouabain from binding to the Na,K-ATPase in frog urinary bladder, and to inhibit directly Na,K-ATPase prepared from rabbit kidney (Haupert and Sancho, 1979). Working with Carilli and Cantley, the effects of this hypothalamic factor on the Na,K-ATPase were characterized biochemically in some detail. It is a high affinity (K_i = 1.4 nM), reversible inhibitor of the purified enzyme (Haupert, 1984), whose effects are apparently specific since at concentrations which inhibit the Na,K-ATPase in the plasma membrane of human erythrocytes by 75%, the activities of the Ca^{2+}- and Mg^{2+}-dependent ATPases were unaffected (Carilli et al., 1985). It acts only from the extracellular surface of the cell by a mechanism similar to but not identical to that of the cardiac glycosides (Carilli et al., 1985). Kinetic studies in cultured mammalian renal tubular epithelial cells show binding and dissociation characteristics consistent with physiologic relevance in vivo (Haupert et al., 1986). Complete structural characterization of the molecule is in progress using mass spectroscopy.

Although endogenous Na^+ transport inhibitory activity

need not be restricted to the central nervous system, most investigators have been unable to find it in other organs. Alaghband-Zadeh et al. (1983), using the cytochemical assay for Na,K-ATPase activity mentioned above, processed numerous organ tissues from the rat (including cerebral cortex tissue) and found inhibitory activity only in extracts from the hypothalamus. However, Godfraind, De Pover and coworkers (De Pover et al., 1982) extracted and partially purified inhibitory activity from guinea pig heart. Characteristics of its interaction with the Na,K-ATPase, including high and low affinty binding sites in cardiac tissue, organ and species sensitivites, and cross reactivity with different antiglycoside antibodies have been studied (De Pover et al, 1982; De Pover, 1985; Fagoo and Godfraind, 1985b), but structural information has not been reported.

Two additional lines of evidence suggest that the brain is an important source of this inhibitory factor: two groups (Halperin et al., 1983; Lichtstein et al., 1985) have isolated "ouabain-like" compounds from human cerebrospinal fluid. Since the material has not been substantially purified by these investigators, both specificity of ATPase inhibition and estimates of concentration are in doubt. Morgan et al. (1985) have recovered and partially characterized a Na,K-ATPase inhibitor from fetal rat hypothalamic cells in culture. This latter substance is completely destroyed by several proteolytic enzymes, which indicates that it is a peptide and is thus chemically different from the nonpeptidic substances reported by most investigators who have processed whole brain or hypothalamus. Finally, Jandhyala and Ansari (1986) have recently reported that perfusion of cerebral ventricles in the dog with artificial cerebral spinal fluid containing an elevated Na^+ concentration leads to release into the circulation of a Na,K-ATPase inhibitor with resulting inhibition of Na^+-K^+ pump activity in the animal's blood vessels. To this point the bulk of evidence suggests that the brain, and more specifically, the hypothalamus, represents an enriched source of the endogenous inhibitor of Na,K-ATPase, if not the site of production.

The tabulation in Table 1 indicates the continuing uncertainty about the molecular identity of the inhibitor(s) and underscores the relative lack of critical biochemical data, such as the demonstration of high binding affinity, necessary to support the notion that a given substance may be a physiologically relevant regulator of the Na^+-K^+ pump.

Is there target organ specificty for the endogenous Na,K-ATPase inhibitor, and what could be the mechanism for particular effects in various target tissues? In a preliminary set of experiments we have addressed this issue with regard to the hypothalamic factor (HF) by measuring its inhibitory effects on Na,K-ATPase purified from three putative target organs in the same animal (Haupert et al., 1987). Guinea pig Na,K-ATPase was purified from kidney and brain using the method of Jorgensen (1974), and from heart by the method of Jones et al. (1984). Enzyme activity was measured as the hydrolysis of ATP using both a kinetic coupled-enzyme assay (Haupert et al., 1984) and the release of the gamma phosphate of ^{32}P-ATP. The amount of protein in the experiments was adjusted so that HF effects could be determined on enzyme with equal specific ATP hydrolytic activity. Results for 1 and 2 units (U) of HF (a concentration close to the K_m for inhibition of purified canine kidney Na,K-ATPase) are shown in Table 2. The enzymes from brain and kidney were clearly more susceptible to inhbibition by HF than than the enzyme from heart. Furthermore, the apparent steepness of the dose-response relationship differed for Na,K-ATPase from kidney and brain.

TABLE 2. Inhibition by hypothalamic factor (HF) of Na,K-ATPase purified from kidney, brain and heart of the guinea pig.

% Inhibition v. control enzyme activity, mean + SEM			
HF	Kidney	Brain	Heart
1U[#]	23 ± 5 (n=6)	80 ± 6* (8)	16 ± 2 (6)
2U	95 ± 2+ (5)	100 ± 0+ (4)	23 ± 3 (5)

*$p<0.0001$ v. kidney and heart
+$p<0.0001$ v. heart
[#]1 unit of hypothalamic factor is defined as the amount required for 50% inhibition of purified Na,K-ATPase from canine renal medulla under standard assay conditions at 37°C (Haupert et al., 1984).

One possibility for the differences observed in this experiment, and for the rationale for regulation by an endogenous inhibitor in general, could lie in organ-specific isoenzyme composition of the Na,K-ATPase (Sweadner, 1979). For this reason we assayed the effects of HF on purified alpha + and alpha from rat axolemma and kidney respectfully, kindly supplied by Kathleen Sweadner. Initial results show that concentrations of HF which inhibit alpha + by 82%, inhibit purified alpha by only 18%, a direction parallel to that of ouabain (1.0 mM) which inhibited alpha + and alpha by 90% and 45% respectively, using the same experimental protocol.

Why has the endogenous Na,K-ATPase inhibitor been so difficult to isolate and characterize? This question becomes all the more poignant in view of the astonishing rapidity with which another class of compounds affecting sodium and water metaboolism and vascular tone, the atrial natriuretic peptides, were isolated, purified, sequenced and synthesized in the active form (for review, see Graham and Zisfein, 1986). The answer to this question is, of course, speculative. In the author's view the reason is in part due to the fact that most groups are working with a compound which is non-peptidic, and hence all the sophistication of recent advances in protein chemistry and molecular biology are to no avail. Secondly, most investigators in this field find that they are dealing with exceedingly small amounts of bioactive inhibitor, thus requiring enormous quantities of starting material which must be processed, at least initially, on a cumbersomely large scale. An example from our own laboratory may illustrate the point. Approximately 3 units of HF activity can be recovedred per gram wet weight of bovine hypothalamus (Haupert, 1984). One unit of activity is defined as the amount of HF required to inhibit 1 ug of purified Na,K-ATPase by 50% in a 50 ul incubation at 37°C for 30 min. Our kinetic analysis of the binding of HF to purified Na,K-ATPase estimated that one unit/50 ul is 15 nM in HF (Haupert, 1984). Therefore, 1 unit of activity corresponds to about 1 pmol of HF, and 1 kg of starting tissue yields about 3 nmol of inhibitory activity. It follows that approximately 33 Kg of hytpothalamic tissue must be processed to purify 100 nmol of the inhibitor, assuming no losses during the various steps, and ignoring the inhibitor which must by consumed in the bioassays during the purification sequence!

Is the endogenous Na,K-ATPase inhibitor(s) truly "digitalis-like"? To answer this question in the affirmative in the strict sense of the word would require that at least the following criteria be met: Interaction with the Na,K-ATPase must be of high affinity, with an inhibitory constant in the nanomolar range; the inhibition must be reversible; the inhibition must be specific for the Na,K-ATPase, and competitive with K^+; and the endogenous inhibitor must show positive inotropic effects on cardiac muscle. None of the compounds reviewed above has yet demonstrated all these criteria. Even those characterized in the greatest biochemical detail (Carilli et al., 1985; Crabos et al., 1987) manifest some differences with the cardiac glycosides. In the case of the inhibitor purified from urine (Crabos et al., 1987), specificity for the Na,K-ATPase is not rigourous. The inhibitor from bovine hypothalamus has ligand requirements for maximal in vitro enzyme inhibition distinct from those of ouabain (Haupert, 1984), has opposite effects from ouabain on phosphorylation from inorganic phosphorous and magnesium of the active-site aspartate residue (Carilli, 1985), and shows dissociation kinetics from intact renal epithelial cells much more rapid than those of ouabain (Haupert et al., 1986). Immuno cross-reactivity with digitalis antibodies has been reported by some groups (Gruber et al., 1980; Klingmuller et al., 1982; Crabos et al., 1984; Fagoo and Godfraind, 1985a), but cannot be demonstrated by others (Hamlyn et al., 1982; Haupert, unpublished experiments; Millet et al., 1987; Kuske et al., 1987). In two instances, cross-reactivity from plasma was found to be due to nonspecific effects of plasma fatty acids or lysolipids (Tamura et al., 1985; Kelly et al., 1986). Even in cases where a still unidentified inhibitor shows displacement curves parallel to those of the specific ligand, the interaction does not predict nor require structural identity with the cardiac glycosides (Fagoo and Godfraind, 1985b).

In the author's view, the issue of being "digitalis-like" is mainly a semantic one. It is fair to say that most of the isolated endogenous Na,K-ATPase inhibitors share some of the biological effects of the cardiac glycosides. Identity with the cardiac glycosides in terms target organ specificity, mechanism of inhibition of the Na,K-ATPase, and molecular structure is not a requirement for relevance to normal biology or pathophysiology, provided a finally purified and structurally identified compound meets the general criteria discussed above for physiologically relevant regulators. Since, in the few instances where detailed studies have been reported, the unknown inhibitors have all been found to differ from the cardiac glycosides in at least some fundamental aspect, the term "digitalis-like" is best considered a convenient metaphor, whilst remaining something of a misnomer.

REFERENCES

1. Akagawa K, Hara N, Tsukada Y (1984). Partial purification and properties of the inhibitors of Na, K-ATPase and ouabain-binding in bovine central nervous system. J Neurochem 42:775.

2. Alaghband-Zadeh J, Fenton S, Hancock K, Millett J, De Wardener, HE (1983). Evidence that the hypothalamus may be a source of a circulating Na,K-ATPase inhibitor. J Endocrinol 98:221.

3. Bahlman NJ, McDonald SJ, Ventom MG and deWardener HE (1967). The effect on urinary sodium excretion of blood volume expansion without changing the composition of blood in the dog. Clin Sci 32:403.

4. Bealer SL, Haywood JR, Gruber KA, Buckalew VM Jr, Fink GD, Brody MJ, Johnson AK (1983). Preoptic hypothalamic periventricular lesions reduce natriuresis to volume expansion. Am J Physiol 244:R51.

5. Bianchi G, Ferrari P, Cusi D, Barber BR, Salardi S, Torielli L, Tripodi MG, Niutta E, Vezzoli G, Barlassina C (1986). Membrane abnormalities in essential hypertension. Physiologic and genetic links. Ann NY Acad Sci 488:266.

6. Bidard J-N, Rossi B, Renaud J-F, Lazdunski M (1984). A search for an "ouabain-like" substance from the electric organ of Electrophorus electricus which led to arachidonic acid and related fatty acids. Biochimn Biophys Acta 769:245.

7. Blaustein MP (1977). Sodium ions, calcium ions, blood pressure regulation, and hypertension: a reassessment

and a hypothesis. Am J Physiol 232:C165.

8. Blaustein MP, Ashida T, Goldman WF, Wier WG, Hamlyn JM (1986). Sodium/calcium exchange in vascular smooth muscle: a link between sodium metabolism and hypertension. Ann NY Acad Sci 488:199.

9. Bourgoignie JJ, Klahr S and Bricker NS (1971). Inhibition of transepithelial sodium transport in the frog skin by a low molecular weight fraction of uremic serum. J Clin Invest 50:303.

10. Brody MJ, Fink GD, Buggy J, Haywood JR, Gordon FJ, Johnson AK (1978). The role of the anterventral third ventricle (AV3V) region in experimental hypertension. Circ Res 43 (Suppl I):I-2.

11. Canessa ML (1986). Pathophysiology of the Na exchange and Na-K-Cl cotransport in essential hypertension: new findings and hypotheses. Ann NY Acad Sci 488:276.

12. Canessa M, Adragna N, Solomon HS, Connolly TM, Tosteson DC (1980). Increased sodium-lithium countertransport in red cells of patients with essential hypertension. N Engl J Med 302:772.

13. Carilli CT, Berne M, Cantley LC, Haupert GT Jr (1985). Hypothalamic factor inhibits the (Na,K)ATPase from the extracellular surface. J Biol Chem 260:1027.

14. Cloix JF, Crabos M, Wainer IW, Ruegg U, Seiler M, Meyer P (1985). High yield-purification of a urinary Na^+-pump inhibitor. Biochem Biophys Res Commun 131:1234.

15. Crabos M, Grichois M-L, Guicheney P, Wainer IW, Cloix J-F (1987). Further biochemical characterization of an Na^+ pump inhibitor purified from human urine. Eur J Biochem 162:129.

16. Crabos M, Wainer IW, Cloix JF (1984). Measurement of endogenous Na+, K+-ATPase inhibitors in human plasma and urine using high-performance liquid chromatography. FEBS Lett 176:223.

17. De Pover A (1985). Isolation and interaction with

Na,K-ATPase of an endogenous digitalis-like factor from rat heart. In Bevan JA et al (eds): "Vascular Neuroeffector Mechanisms," Elsevier Science Publishers B.V., p 23.

18. De Pover A, Castaneda-Hernadez G, Godfraind T (1982). Water versus acetone-HCl extraction of digitalis-like factor from the guinea pig heart. Biochem Pharmacol 31:267.

19. De Wardener HE (1987). Natriuretic and sodium-transport inhibitory factors associated with volume control and hypertension. In Mulrow PJ, Schrier RW (eds): "Atrial Hormones and Other Natriuretic Factors," Bethesda: American Physiological Society, p 127.

20. De Wardner HE, Mills IH, Clapham WF, Hayter CJ (1961). Studies on the efferent mechanism of the sodium diuresis which follows the administration of intravenous saline in the dog. Clin Sci 21:249.

21. Fagoo M, Godfraind T (1985a). Further characterization of cardiodigin, Na,K-ATPase inhibitor extracted from mammalian tissues. FEBS lett 184:150.

22. Fagoo M, Godfraind T (1985b). Interaction of cardiodigin, endogenous inhibitor of Na^+,K^+ ATPase, with antidigoxin and antidigitoxin antibodies. Biochem Biophys Res Comm 129:553.

23. Fenton SE, Clarkson G, MacGregor G, Alaghband-Zadeh J, De Wardener HE (1982). An assay of the capacity of biological fluids to stimulate renal glucose-6-phosphate dehydrogenase activity in vitro as a marker of their ability to inhibit sodium potassium-dependent adenosine triphosphatase activity. J Endocrinol 94:99.

24. Fine LG, Bourgoignie JJ, Hwang KH, Bricker NS (1976). On the influence of the natriuretic factor from patients with chronic uremia on the bioelectric properties and sodium transport of the isolated mammalian collecting tubule. J Clin Invest 58:590.

25. Fishman MC (1979). Endogenous digitalis-like activity

in mammalian brain. Proc Natl Acad Sci USA 76:4661.

26. Flier J, Edwards MW, Daly JW, Myers CW (1980). Widespread occurrence in frogs and toads of skin compounds interacting with the ouabain site of Na,K ATPase. Science 208:503.

27. Freas W, Muldoon SM, Haddy FJ (1982). Accumulation of ^{3}H-norepinephrine in canine saphenous vein: influence of plasma. Am J Physiol 243:H424.

28. Gault MH, Vasdev SC, Longerich LL, Fernandez P, Prabhakaran V, Dawe RT, Maillet C (1983). Plasma digitalis-like factor(s) increase with salt loading [Letter]. N Engl J Med 309:1459.

29. Graves SW, Brown B, Valdes R Jr (1983). An endogenous digoxin-like substance in patients with renal impairment. Ann Intern Med 99:604.

30. Gruber KA, Whitaker JM, Buckalew VM Jr (1980). Endogenous digitalis-like substance in plasma of volume-expanded dogs. Nature 287:743.

31. Haber E, Haupert GT Jr (1987). The search for a hypothalamic Na,K-ATPase inhibitor. Hypertension 9:315.

32. Haddy FJ, Overbeck HW (1976). The role of humoral agents in volume expanded hypertension. Life Sci 19:935.

33. Halperin J, Schaeffer R, Galvez L, Malave S (1983). Ouabain-like activity in human cerebrospinal fluid. Proc Natl Acad Sci USA 80:6101.

34. Hamlyn JM, Ringel R, Schaeffer J, Levinson PD, Hamilton BP, Kowarski AA, Blaustein MP (1982). A circulating inhibitor of ($Na^+ + K^+$)ATPase associated with essential hypertension. Nature 300:650.

35. Haupert GT Jr, Carilli CT, Cantley LC (1984). Hypothalamic sodium-transport inhibitor is a high-affinity reversible inhibitor of Na^+K^+-ATPase. Am J

Physiol 247:F919.

36. Haupert GT Jr, Chen E, Cantiello HF (1986). Hypothalamic factor regulates sodium pump activity in cultured renal tubular epithelial cells. Ann NY Acad Sci 488:540.

37. Haupert GT Jr, Stephan TR, Crabos M (1986). Target organ sensitivity to an endogenous Na,K-ATPase inhibitor from hypothalamus. Kidney Int 31:435A.

38. Haupert GT Jr, Sancho JM (1979). Sodium transport inhibitor from bovine hypothalamus. Proc Natl Acad Sci USA 76:4658.

39. Hillyard SD, Lu E, Gonick, HC (1976). Further characterization of the natriuretic factor derived from kidney tissue of volume expanded rats. Effects on short-circuit current and sodium-potassium-adenosine triphosphatase activity. Circ Res 38:250.

40. Hook JB (1969). A positive correlation between natriuresis and inhibition of renal Na-K adenosine triphosphatase by ouabain. Proc Soc Exp Biol Med 131:731.

41. Jandhyala BS, Ansari AF (1986). Elevation of sodium levels in cerebral ventricles of anaesthetized dogs triggers the release of an inhibitor of ouabain-sensitive sodium, potassium-ATPase into the circulation. Clin Sci 70:103.

42. Jones LR, Besch HR Jr (1984). Isolation of canine cardiac sarcolemmal vesicles. In Schwartz A (ed): "Methods in Pharmacology," Vol 5, Plenum Publishing Corporation, p 1.

43. Jorgensen PL (1974). Purification and characterization of $(Na^+ + K^+)$ATPase. 3. Purification from the outer medulla of mammalian kidney after selective removal of membrane components by sodium dodecylsulphate. Biochim Biphhys Acta 356:36.

44. Kaloyanides GJ, Azer M (1971). Evidence for a humoral mechanism in volume expansion natriuresis. J Clin

Invest 50:1603.

45. Kaplan MA, Bourgoignie JJ, Rosecan J and Bricker NS (1974). The effects of the natriuretic factor from uremic urine on sodium transport, water and elctrolyte content, and pyruvate oxidation by the isolated toad bladder. J Clin Invest 53:1568.

46. Kelly RA, O'Hara DS, Mitch WE, Smith TW (1986). Identification of Na,K-ATPase inhibitors in human plasma as nonesterified fatty acids and lysophospholipids. J Biol Chem 261:11704.

47. Klingmuller D, Weiler E, Kramer HJ (1982). Digoxin-like natriuretic activity in the urine of salt loaded healthy subjects. Klin Wochenschr 60:1249.

48. Kojima I, (1984). Circulating digitalis-like substance is increased in DOCA-salt hypertension. Biochem Biophys Res Commun 122:129.

49. Kojima I, Yoshihara S, Ogata E (1982). Involvement of endogenous digitalis-like substance in genesis of deoxycorticosterone-salt hypertension. Life Sci 30:1775.

50. Kramer HJ, Heppe M, Weiler E, Backer A, Liddiard C, Klingmuller D (1985). Further characterization of the endogenous natriuretic and digoxin-like immunoreacting activities in human urine:effects of changes in sodium intake. Renal Physiol 8:80.

51. Kuske R, Moreth K, Renner D, Wizemann V, Schoner W (1987). Sodium pump inhibitor in the serum of patients with essential hypertension and its partial purification from hemofiltrate. Klin Wochenschr 65:53.

52. Lichardus B, Nizet A (1972). Water and sodium excretion after blood volume expansion under conditions of constant aterial venous and plasma oncotic pressures and constant hematocrit. Clin Sci 42:701.

53. Lichtstein D, Minc D, Bourrit A, Deutsch J, Karlish SJD, Belmaker H, Rimon R, Palo J (1985). Evidence for

the presence of "ouabain-like" compound in human cerebrospinal fluid. Brain Res 325:13.

54. Lichtstein D, Samuelov S (1980). Endogenous "ouabain-like" activity in rat brain. Biochem Biophys Res Commun 96:1518.

55. London GM, Safar ME, Safar AE, Simon AC (1985). Blood pressure in the "low pressure" system and cardiac performance in essential hypertension. J Hypertens 3:337.

56. Millett JA, Holland SM, Alaghband-Zadeh, de Wardener HE (1986). Na,K-ATPase -inhibiting and glucose-6-phosphate dehydrogenase-stimulating activity of plasma and hypothalamus of the Okamoto spontaneously hypertensive rat. J Endocr 108:69.

57. Millett JA, Holland SM, Alaghband-Zadeh J, De Wardener HE (1987). Extraction and characterization of a cytochemically assayable Na^+,K^+-ATPase inhibitor/glucose-6-phosphate dehydrogenase stimulator in the hypothalamus and plasma of man and the rat. J Endocr 112:299.

58. Morgan K, Lewis MD, Spurlock G, Collins PA, Foord SM, Southgate K, Scanlon MF, Mir MA (1985). Characterization and partial purification of the sodium-potassium-ATPase inhibitor released from cultured rat hypothalamic cells. J Biol Chem 260:13595.

59. Mulvany MJ (1985). Changes in sodium pump activity and vascular contraction. J Hypert 3:429.

60. Mulvany MJ, Aalkjaer C, Petersen TT (1984). Intracellular sodium, membrane potential, and contractility of rat mesenteric small arteries. Circ Res 54:740.

61. Overbeck HW, Pamnani MB, Akera T, Brody TM, Haddy FJ (1976). Depressed function of a ouabain-sensitive sodium-potassium pump in blood vessels from renal hypertensive dogs. Circ Res 38(Suppl II):II-48.

62. Pamnani MB, Buggy J, Huot SJ, and Haddy, FJ (1981).

Studies on the role of a humoral sodium-transport inhibitor and the anteroventral third ventricle (AV3V) in experimental low renin hypertension. Clin Sci 61:57s.

63. Pamnani MB, Harder DR, Huot SJ, Bryant HJ, Kutyna FA, Haddy FJ (1982). Vascular smooth muscle membrane potential and a ouabain-like humoral factor in one kidney, one clip hypertension in rats. Clin Sci 63:31s.

64. Shimoni Y, Gotsman M, Deutsch J, Kachalsky S, Lichtstein D (1984). Endogenous ouabain-like compound increases heart muscle contractility. Nature 307:369.

65. Smith TW, Butler V Jr, Haber E, Fozzard H, Marcus FI, Bremner WF, Schulman IC, Phillips A (1982). Treatment of life-threatening digitalis intoxication with digoxin-specific Fab antibody fragments: experience in 26 cases. N Engl J Med 307:1357.

66. Smith TW, Butler VP Jr, Haber E (1969). Determination of therapeutic and toxic serum digoxin concentrations by radioimmunoassay. N Engl J Med 281:1212.

67. Somlyo AP, Broderick R, Somlyo AV (1986). Calcium and sodium in vascular smooth muscle. Ann NY Acad Sci 488:228.

68. Songu-Mize E, Bealer SL, Caldwell RW (1982). Effect of AV3V lesions on development of DOCA-salt hypertension and vascular Na^+-pump activity. Hypertension 4:575.

69. Sweadner KJ (1979). Two molecular forms of (Na^+ + K^+)-stimulated ATPase in brain. Separation, and difference in affinity for strophanthidin. J Biol Chem 254:6060.

70. Tal DM, Katchalsky S, Lichtstein D, Karlish SJD (1986). Endogenous "ouabain-like" activity in bovine plasma. Biochem Biophys Res Comm 135:1.

71. Tamura M, Kuwano H, Kinoshita T, Inagami T (1985). Identification of linoleic and oleic acids as

endogenous Na+,K+-ATPase inhibitors from acute volume-expanded hog plasma. J Biol Chem 260:9672.

72. Vanhoutte PM, Lorenz RR (1984). Na,K-ATPase inhibitors and the adrenergic neuroeffector interaction in the blood vessel wall. J Cardiovasc Pharmacol 6 (Suppl I): S88.

73. Vasdev S, Longerich L, Johnson E, Brent D, Gault MH (1985). Dehydroepiandrosterone sulfate as a digitalis-like factor in plasma of healthy human adults. Res Commun Chem Pathol Pharmacol 49:387.

74. Vatner SF, Higgins CV, Franklin D, Braunwald E (1971). Effects of a digitalis glycoside on coronary and systemic dynamics in conscious dogs. Circ Res 28:470.

OVERVIEW:

ROLE OF THE NA^+K^+-ATPase IN THE CARDIOTONIC ACTION OF CARDIAC GLYCOSIDES

Arnold Schwartz, Gunter Grupp, Earl Wallick, I.L. Grupp, and W.J. Ball, Jr.
Departments of Pharmacology and Cell Biophysics, (A.S., E.W., I.L.G., W.J.B.) Physiology and Medicine, (G.G.), University of Cincinnati College of Medicine, 231 Bethesda Avenue, Cincinnati, OH 45267-0575

INTRODUCTION

In the last three years, a considerable amount of information concerning the primary amino acid sequence and structure of the Na^+,K^+-ATPase (NKA) and its isoforms has been revealed (Shull et al., 1985 and 1986; Young and Lingrel, 1987). There appears to be no question that an important pharmacologically relevant binding site or site(s) for the cardiac glycosides exist on the catalytic (α) subunit of the Na^+,K^+-ATPase. When one considers the functions of different isoforms of the α subunit, one has to consider the specific tissue, organ, or cell the isoform is in, and the role of the Na^+,K^+-ATPase at these sites. To characterize quantitatively the sensitivity of the enzyme to an inhibitor, here a cardiac glycoside, one has to determine the specific affinity constants for inhibition of Na^+,K^+-ATPase activity. Furthermore, this affinity constant has to be correlated with the particular role or biological functions of the Na^+,K^+-ATPase in that cell type. Nowhere are the biological effects of the cardiac glycosides ("Digitalis") better seen than in the heart, an organ designed specifically for the propulsion of blood to the organs and tissues of the body. Digitalis' primary action is that of a positive inotropic agent. The clinical use of digitalis has been known since the time of William Withering (1785) and it still forms the major basis for the treatment of congestive heart failure. Digoxin, a cardiac glycoside, is still among the 15 most prescribed drugs in the world and used primarily to produce a positive inotropic effect (that is an increased force of contraction) in the failing human heart. In reviewing Withering's famous treatise, "An Account of the Foxglove," (1785) the following statement is of great interest: "That it (i.e., digitalis) has a power over the motion of the heart, to a degree yet unobserved in any other medicine, and that this power may be converted to salutary ends." The search for the mechanism of action of the cardiac glycosides has been the basis of thousands of clinical and basic science studies. By now, it appears certain that the only specific binding site for the cardiac glycosides is the Na^+,K^+-ATPase. The mechanism that has been elaborated is summarized in Figure 1.

MECHANISM

Fig. 1 $DIG + NKA \rightarrow DIG \cdot NKA \rightarrow \downarrow NKA \rightarrow$

$\uparrow Na_i \rightarrow \uparrow \otimes_{Na}^{Ca} \rightarrow \uparrow Ca_i \rightarrow + INOTROPY$

The history leading up to the recognition of the role of the Na^+,K^+-ATPase in digitalis action began with a paper by Calhoun and Harrison (1931), who observed that dogs treated with "clinical" levels of cardiac glycosides produced a very slight but significant loss of potassium from heart muscle during the development of positive inotropy. This study was followed by the demonstration of a direct action of digitalis on cat papillary muscles reported by Gold and Cattell (1940). Schatzmann (1953), Skou (1957), and Repke (1963) then established more clearly the direct glycoside-Na^+,K^+-ATPase interactions. The criteria required for an enzyme (Na^+,K^+-ATPase) to be accepted as the receptor of an agonist (digitalis) have been well established:

Characteristics of Na^+,K^+-ATPase Consistent with its Being the Receptor for Digitalis:

1. Binding is saturable and reversible.
2. K_D for binding to the Na^+,K^+-ATPase = $k_{-1}/k_1 \geq K_i$ for inhibition of Na^+,K^+-ATPase activity

3. Specific binding removed by agonist occupation (i.e., unlabelled cardiac glycoside) of receptor.
4. Stoichiometric relationship between binding and inhibition of Na^+,K^+-ATPase (1:1).
5. Direct relationship between pharmacologic effect (inotropy) and binding to and inhibition of Na^+,K^+-ATPase.
6. Contractility is increased and is related directly to inhibition of Na^+,K^+-ATPase, increase of intracellular Na^+ and Ca^{2+}.
7. Removal of digitalis from receptor, (i.e., Na^+,K^+-ATPase) restores contraction and Na^+,K^+-ATPase activity to control levels.

RESULTS

Our approach to the mechanism of activity of digitalis is an interdisciplinary one applying pharmacological, biochemical, molecular biological and immunological techniques.

In our judgment, all criteria listed above are fulfilled: Saturable binding of digitalis to Na^+,K^+-ATPase has been well established; cardiac glycoside preparations bind specifically to membrane preparations that must contain the Na^+,K^+-ATPase and the binding is modulated by both sodium and potassium (Fig. 2). The binding constants are in the same range as the inhibitor constants for enzyme activity and enzyme inhibition is reversible.

Fig. 2

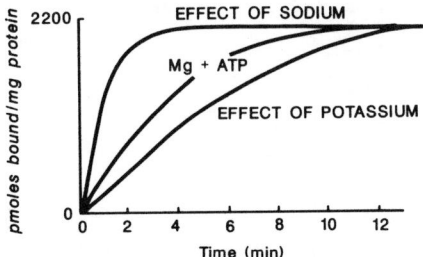

EFFECT OF ELECTROLYTES ON SPECIFIC OUABAIN BINDING TO RECEPTORS

Cardiac glycosides in many species, such as dog (Fig. 3), increase the force of contraction concomitant with an inhibition of Na^+,K^+-ATPase activity and this effect is reversible and occurs at low concentrations of digitalis.

Fig. 3

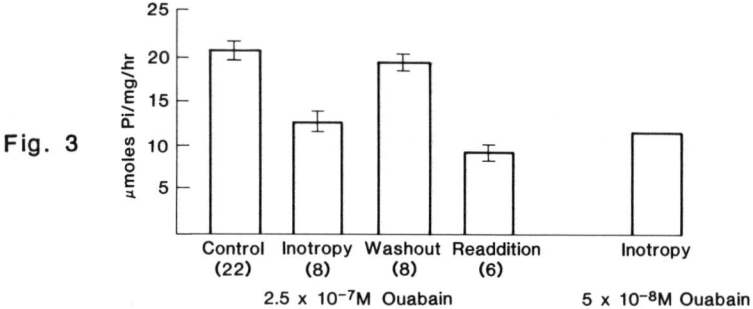

There is a relatively good relationship between the species sensitivity to cardiac glycosides with regard to positive inotropy and inhibition of the Na^+,K^+-ATPase and furthermore, there is a surprisingly good relationship between the degree of inhibition of Na^+,K^+-ATPase and the increase of contractile force in various preparations (Fig. 4).

Fig. 4

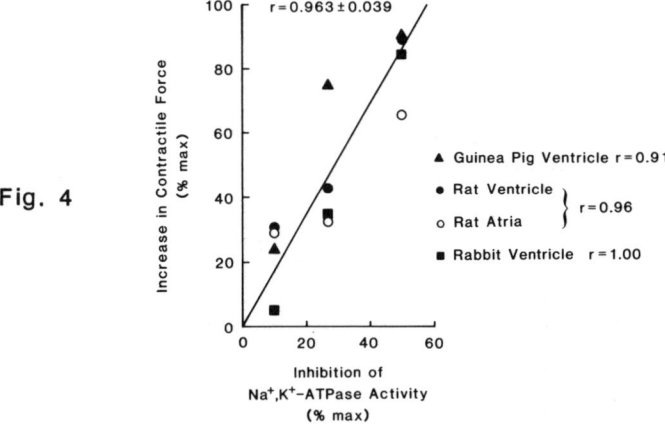

There have, however, been questions raised that Na^+,K^+-ATPase might not be the receptor of digitalis (Overview in Erdmann, 1984), but most of these issues have been satisfactorily resolved, including the issue of "pump stimulation" (G. Grupp et al., 1980, 1982, and 1985; Wehling et al., 1982). Most recently, Boyett, Hart and Levi (1986) in a re-study of strophanthidin effects in isolated stimulated sheep Purkinje fibers found a dissociation between the increase in force of contraction and an increase in intracellular sodium, particularly when the inotropic effect was washed out. The authors postulated an intracellular action of strophanthidin but without any specific evidence. The problems of ion measurement concomitant with force measurement in a "beating" preparation are considerable and render interpretation of these data very difficult (Wasserstrom, Discussion this meeting). An intracellular locus of action of digitalis was suggested by Isenberg (1984) on the basis of an interesting experiment. When injected into myocytes, ouabain and digoxin covalently bound to human serum albumin and produced a distinct and significant inotropic action. Again, interpretation is complicated by inability to ascertain exactly where the pipette for "intracellular" injection is located (i.e., is it truly intracellular or might it be located in T-system area or other extracellular sites?). These results, then, offer no significant argument against the role of the Na^+,K^+-ATPase in the pharmacological action of digitalis, particularly since no other site specifically interacts or is affected directly by digitalis.

Much is known about the relative relationships between the sensitivity of a variety of species to digitalis. It is notable that the rat ventricle, but not the atrium, responds to digitalis with an apparent biphasic inotropic action, almost as if there are "high affinity/low capacity" and "low affinity/high capacity" sites. This is illustrated in Fig. 5, adapted from I.L. Grupp et al. (1981) and Adams et al. (1982), in which a monophasic contractility response curve for the atria is contrasted with a biphasic curve of the ventricle. Our results are consistent with, but certainly do not prove, the interesting possibility of a differential distribution of isoforms of Na^+,K^+-ATPase, a concept originally suggested by Hansen (1976) and extensively developed by Sweadner (1979). Erdmann et al. (1980) brought this interesting

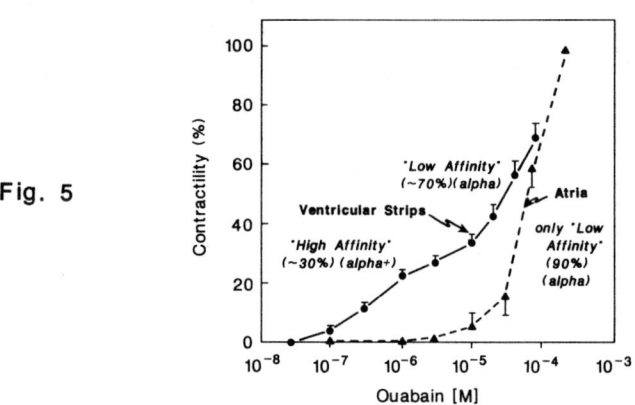

Fig. 5

discrepancy in digitalis action on the rat ventricle to our attention. He stated that a low concentration of cardiac glycoside increased contractile force of the rat ventricle but did not appear to inhibit isolated Na^+,K^+-ATPase. Moreover, the apparent affinity for the binding of ouabain to preparations of the rat ventricle was consistent with the "affinity" constant for positive inotropy but not with the I_{50} for inhibition of the Na^+,K^+-ATPase. Our experiments (I. Grupp et al., 1981 and 1984) revealed, in fact, that there were two apparent inotropic "mechanisms," a "low dose" and a "high dose" response in rat ventricle but not in atria, as illustrated in Figure 5. Careful inspection (Adams et al., 1982) of the ouabain-Na^+,K^+-ATPase inhibitory curves, however, also

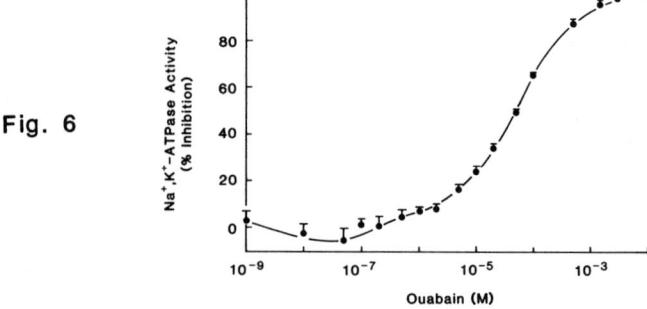

Fig. 6

revealed an approximate 10% inhibition of the isolated Na^+,K^+-ATPase by the "low dose" ouabain in purified sarcolemmal preparations from the rat ventricle (Fig. 6).

The well-established steepness of the sodium-calcium exchange mechanism (Eisner et al., 1984) is consistent with the concept that only a very small increase of internal sodium needs to occur in order for a large amount of calcium to be brought into the cell and delivered to the contractile proteins. Thus, only a partial inhibition of Na^+,K^+-ATPase, which would cause a small increase in Na^+_i, brings about a large increase in intracellular calcium and hence in contractile force. Accordingly, we (Lee et al., 1983; I. Grupp et al., 1985) carried out a study on rat ventricular preparations, notoriously "insensitive" to cardiac glycosides, in which intracellular sodium activity was measured concomitantly with contractile force. Figure 7, adapted from I.L. Grupp et al. (1985), shows that even at "low concentrations" of

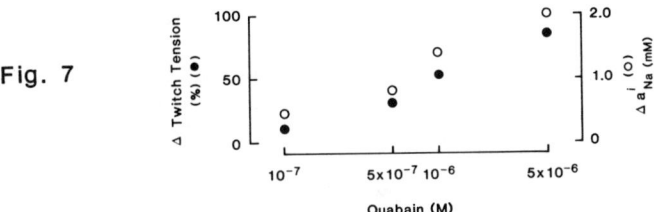

Fig. 7

OUABAIN ON TENSION AND SODIUM ACTIVITY IN RAT VENTRICLE

ouabain (10^{-7}-10^{-6}M), involving the presumed high affinity site, ouabain caused an increase in intracellular sodium activity and an increase in contraction. Thus, even in rat heart, the Na pump is inhibited by digitalis, in vivo. There is abundant evidence in other, especially glycoside sensitive tissues, that digitalis produces an increase in intracellular sodium concomitant with an increase in contractile force (e.g., Fig. 8, adapted from Eisner et al., 1984). These studies utilized sodium sensitive electrodes and isometric tension recordings (Lee et al., 1980; Eisner et al., 1983; Wasserstrom et al., 1983). All data show a clear-cut relationship between an increase in intracellular sodium produced either by digitalis or by

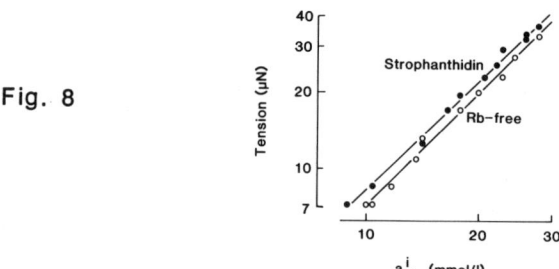

Fig. 8

removal of potassium and an increase in contractile force. Thus, in our laboratory, the concept has developed that whenever a biphasic effect occurs pharmacologically, a reasonable postulation is that multiple isoforms exist. Young and Lingrel (1987) observed the presence of α1 and α2 in rat ventricle but little or no α2 in rat atria. The genes are labeled NKAA1, NKAA2, and NKAA3 (see Lingrel et al. elsewhere in this volume), corresponding to expressed protein α1, α2, and α3. The proteins were formerly labeled α, α+, and αIII, respectively. The α2 isoform is present at lower levels in atria than in ventricle. It is predictable, therefore, that in rat atria there is primarily a "high-dose" contractile force effect, exactly what we found. Similarly, Ng and Akera (1987) recently found a 50-50 distribution of α1 and α2 isoforms in ferret heart but only a 25:75 distribution of the "low-dose/high-dose" contractile force response. Further work along these lines might reveal a distinct genetic pattern for the different functional aspects of a particular organ or tissue.

In terms of the interesting "relative insensitivity" of the rat heart and the possible explanations based upon isoform distribution, it is worthwhile to analyze and compare the α sequences at the suggested binding region for ouabain. Table 1 contains a comparison of the predicted extracellular regions H1-H2 and H3-H4 for the sheep kidney (S) α1 (Shull et al., 1985) rat kidney (R) α1, rat kidney α2 and rat kidney α3 (Shull et al., 1986). As noted by these authors, the charged amino acids, arginine and

Table 1. Comparative Sequence of Potential Ouabain Binding Regions

H1 - H2

```
           111
S NKA α₁   Gln-Ala-Ala-Thr-Glu-Glu-Glu-Pro-Gln-Asn-Asp-Asn
R NKA α₁   Arg-Ser   .   .   .   .   .   Pro   .   .  Asp
R NKA α₂   Leu   .   .  Met   .  Asp   .   .  Ser   .   .
R NKA α₃   .   .  Gly   .   .  Asp-Asp  .  Ser-Gly   .   .
```

H3 - H4

```
S NKA α₁   Glu-Tyr-Thr-Trp-Leu-Glu
R NAK α₁   .   .   .   .   .   .
R NKA α₂   Gly   .  Ser   .   .   .
R NKA α₃   Gly   .   .   .   .   .
```

aspartate, in the rat α are missing in the sheep α and there are two prolines instead of one proline in the ouabain insensitive rat α. This suggests that multiple sites of attachment as well as "conformational specificity" play roles in the binding, a concept suggested some time ago by Thomas et al. (1974). Furthermore, in comparing the sequences in regions H3-H4, the amino acid sequences at this site are identical in lamb kidney α1 (sensitive to digitalis) and the rat α1 (not sensitive to digitalis). Therefore, the primary sequence at the originally suggested ouabain binding region, namely H3-H4, cannot represent the total binding site. Consequently, we must consider other extracellular regions, such as H1-H2, H5-H6, and H7-H8 (Shull et al., 1986) as also being involved.

Immunological data from our laboratory (Ball and Lane, 1986), using a series of monoclonal antibodies raised specifically to the kidney Na^+, K^+-ATPase reveal important information. It should be remembered that it is thought that the kidney contains exclusively the α isoform. However, at this Symposium, the possibility that α+ may also be present, to a small extent, was raised (see workshop summary, Dr. Kathleen Sweadner). These mono-

Table 2. Monoclonal Antibody Binding to Rat and Lamb Kidney (Na$^+$ + K$^+$)-ATPase

Cell line: designation of clones	Lamb Kidney (Na$^+$ + K$^+$)-ATPase Binding Affinity (nM)			Rat Kidney (Na$^+$ + K$^+$)-ATPase (% Binding)	
	holoenzyme	catalytic subunit	glycoprotein	holoenzyme	catalytic* subunit
M7-PB-E9 (E5)	5.2	3	0	0	8%
M8-P1-A3	7.5	3	0	0	47%
M10-P6-B7	4.6	1.5	0	0	45%
M10-P5-C11	7.0	1,500	0	0	0%
M12-P4-E8	7.0	0.7	0	0	40%
M12-P7-F11	1.5	1.4	0	0	40%

*SDS-gel resolved catalytic subunit.
The monoclonal antibodies were raised to lamb kidney enzyme.
Antibody binding to plate-absorbed antigen was detected using the ELISA assay.

clonal antibodies, as summarized in Table 2, recognize the α subunit of the holoenzyme, which is native enzyme isolated from the lamb kidney, but show absolutely no cross-reactivity with the "native" rat kidney Na$^+$,K$^+$-ATPase. However, when the subunits are denatured ("catalytic subunit"), it is clear that most of the antibodies now recognize both the rat and the lamb α isoforms. Therefore, the tertiary structures of both external sequences and possibly intracellular sequences of the enzymes need to be taken into consideration, as well as

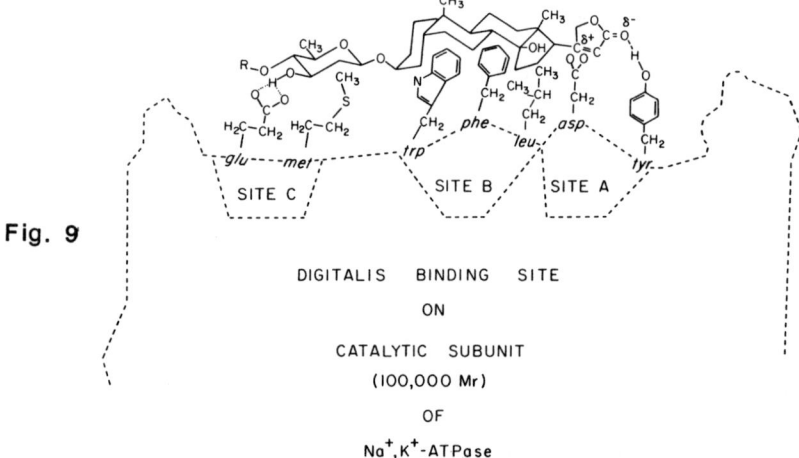

Fig. 9. DIGITALIS BINDING SITE ON CATALYTIC SUBUNIT (100,000 Mr) OF Na$^+$,K$^+$-ATPase

primary sequences when thinking about the specific binding sites for the cardiac glycosides relative to isoform differences. Figure 9 (adapted from Thomas et al., 1974, 1979) indicates a kind of model or three-dimensional diagram that was made when the structure of the Na^+,K^+-ATPase was not yet known (Thomas et al., 1974, 1979). This diagram is shown because it illustrates that it is possible that there are three sites for binding for the cardiac glycosides, namely through the sugar, through the basic steroid nucleus, and through the charged lactone group.

As additional experiments are carried out, the concept that a single region comprises the binding site for ouabain may need to be revised. For example, Kirley et al. (1986) utilizing the group specific reagents, 5,5'-dithiobis-(2-nitrobenzoic acid) and tetranitromethane have demonstrated that a sulfhydryl group is essential for ouabain binding. These two reagents react with sulfhydryl groups and cause an inhibition of ATPase activity and ouabain binding. The inactivation is specifically protected by pretreatment with ouabagenin. This suggests the possible presence of a sulfhydryl group-containing amino acid within the binding site domain(s) of the cardiac glycoside. Again, this was not suggested by the original H3-H4 domain concept (Shull et al., 1985). In contrast, Scheiner-Bobis et al. (1987) suggest that the "ouabain binding site...has not an externally accessible SH group."

Evidence concerning the existence of isoforms of Na^+,K^+-ATPase in the heart has been presented by Mansier and Lelievre (1982), Mansier et al. (1983) and Maixent (1987). They have shown high and low affinity digitalis receptor sites in rat heart and the pharmacologic action appears to become more manifest when calcium is either absent or its concentration is very low <0.25 mM (see also Grupp et al., 1986). Lelievre and his colleagues have recently presented evidence that the two isoforms, $\alpha 1$ and $\alpha 2$, exist in canine ventricle and the interesting suggestion was made that the $\alpha 2$ may be geared mostly towards the inotropic mechanism while the α might be directed towards the toxic action of the cardiac glycosides (Maixent et al., 1987). This concept, however, appears to be inconsistent with the present knowledge of isoforms.

Specific Pharmacological Correlatives of Isoforms of Na^+,K^+-ATPase in Brain

Shull et al. (1986) have shown that three classes of messenger RNA, designated NKAA1, A2, and A3, exist in rat brain. Consistent with this, Wallick et al. (submitted) have kinetically identified three forms of Na^+,K^+-ATPase which differ in affinity for ouabain. The previously unknown isoform, α3 has a binding constant of 10 nM whereas the previously identified α1 and α2 have binding constants of 40 μM and 100 nM, respectively. Furthermore, relative amounts of the isoforms detected kinetically correlate well with the amount of message.

Many interesting aspects of development and changes in isoforms are illustrated in several posters at this meeting. It is possible that the distribution of isoforms are different in the developing tissue. Further, it is possible and exciting that there may be specific changes that occur in various disease states.

A summary of changes that we have noted with respect to the pharmacological action of digitalis in different models of cardiac disease are presented in Table 3. In every case, including the atherosclerotic human, cardiomyopathic Syrian hamster, spontaneous hypertensive adult rat, and the hypertensive dog heart, there appears to be a slight shift of the concentration response curve to the left, and the maximally achievable positive inotropic effect is somewhat decreased.

Table 3. Sensitivity to Ouabain in Normal and Diseased Hearts

	ED_{50} (μM)		TD_{min} (μM)		CF_{max} (% of control)	
	Normal	Diseased	Normal	Diseased	Normal	Diseased
Human	0.135	0.035	0.3	0.1	251	154
Hamster	30	3	100	50	300	25
Rat SHR*	110	32	200	90	135	110
Dog Hypt**	0.11	0.08	2	0.5	182	115

*Spontaneous Hypertensive Rat
**Hypertension

TD_{min} = minimum toxic dose
CF_{max} = maximum contractile force

Lee et al. (1983) also found in the adult hypertensive rat a significant decrease in the number of both the low affinity and the high affinity sites with respect to inhibition of Na^+,K^+-ATPase activity, ouabain binding, and sodium dependent E/P formation. Thus the number of sites appeared to be depressed. In young prehypertensive SHR rats, there was no depression of ouabain binding sites, but a notable depression in sodium dependent E/P formation and in Na^+,K^+-ATPase activity occurred (Lee, S.W. and Wallick, E.T., unpublished observation). This evidence may point in the direction of the beginning of changes of the isoform distribution.

One of the criteria that needs to be fulfilled with respect to the importance of the Na^+,K^+-ATPase and digitalis action is the presence of natural substances within the mammalian system that may function as a regulator of the Na^+,K^+-ATPase. At the present time, there is no clear-cut evidence that there is any peptide from the brain or anywhere else that may produce an action that is as specific as digitalis. The criteria that must be fulfilled by an endogenous substance are summarized below:

1. A Concentration-dependent Positive Inotropy on the Heart.
2. A Concentration-dependent Ouabain-like Inhibition (Na/K-dependent) of Na^+,K^+-ATPase.
3. A Concentration-dependent Inhibition of ^{86}Rb-uptake.
4. No Effect on Any Membrane ATPase Other Than Na^+,K^+-ATPase.
5. No Effect on Any Non-membrane System.
6. No Effect on Any Immunoassay Other Than Digitalis Related.
7. Concentration-dependent Inhibition of ^{3}H-Ouabain Binding to Na^+,K^+-ATPase.
8. Digitalis-like Concentration-dependent Displacement of ^{3}H-Ouabain From Na^+,K^+-ATPase.

Perhaps the most important criterion is Number 8, namely, one should see a digitalis-like concentration dependent displacement of tritiated ouabain from a Na^+,K^+-ATPase (Whitmer et al., 1982). Sodium dodecylsulfate (SDS) inhibits Na^+,K^+-ATPase activity and prevents [^{3}H]ouabain binding consistent with it being an "endogenous" factor. One of the problems of this type of

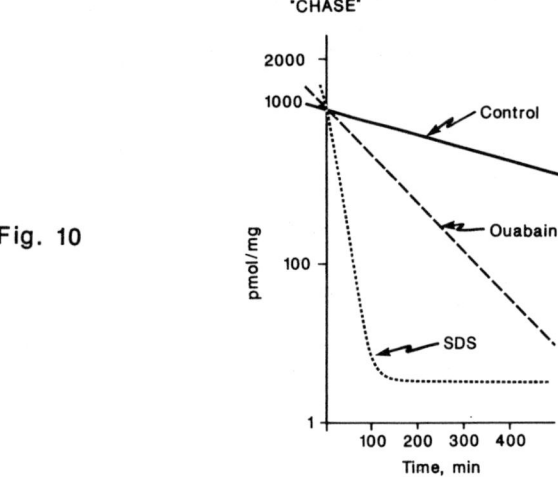

Fig. 10

data interpretation is illustrated in Fig. 10. Note that SDS causes a dissociation that is much faster than ouabain, suggesting that the effect is non-specific. Analysis of this sort has aided in eliminating fractions (which have fast off rates) from consideration for candidates for endogenous regulation (V.P. Butler, unpublished observations).

SUMMARY

In terms of evolution, the Na^+,K^+-ATPase is one of the oldest proteins. It is intriguing that this membrane-based enzyme has become biologically consistent with a multifunctional protein. One of the most interesting aspects of the Na^+,K^+-ATPase is that it contains a specific receptor domain for the oldest plant origin cardiotonic drug and that it is involved in the regulation of myocardial contractility.

REFERENCES

Adams RJ, Schwartz A, Grupp G, Grupp IL, Lee SW, Wallick ET (1982). High-affinity ouabain binding site and low dose positive inotropic effect in rat myocardium. Nature 296:167-169.

Ball WJ, Jr. and Lane, LK (1986). Immunological comparison of cardiac glycoside-sensitive (lamb) and insensitive (rat) kidney, Na^+,K^+-ATPase Biochim Biophys Acta 873:79-87.

Boyett MR, Hart G, Levi AJ (1986). Dissociation between force and intracellular Na^+ activity with strophanthidin in isolated sheep Purkinje fibres. J Physiol 381:311-331.

Calhoun JA and Harrison TR (1931). Studies in congestive heart failure: The effect of digitalis on the potassium content of the cardiac muscle of the dog. J Chem Invest 10:139-144.

Eisner DA, Lederer WJ, Vaughan-Jones RD (1983). The control of tonic tension by membrane potential and intracellular sodium activity in the sheep cardiac Purkinje fibre. J Physiol 335:723-743.

Eisner DA, Lederer WJ, Vaughan-Jones RD (1984). The quantitative relationship between twitch tension and intracellular sodium activity in sheep cardiac Purkinje fibres. J Physiol 355:251-266.

Eisner DA, Allen DG, Smith GL, Wray S (1985). The effects on an extract of the foxglove (Digitalis Purpurea) on tension and intracellular calcium concentration in ferret papillary muscle. J Physiol 365:55P.

Erdmann E, Philipp G, Scholz H (1980). Cardiac glycoside receptor, (Na^+-K^+)-ATPase activity and force of contraction in rat heart. Biochem Pharmacol 29:3219-3229.

Erdmann E (ed) (1984). Cardiac glycoside receptors and positive inotropy. Evidence for more than one receptor? Bas Res Cardiol 79(Suppl.):1-162

Gold H, Cattell M (1940). Mechanism of digitalis action in abolishing heart failure. Arch Intern Med 65:263-278.

Grupp G, Grupp IL, Schwartz A (1980). Lack of effects of neuraminidase on responses of isolated guinea pig heart preparations to ouabain. J Mol Cell Cardiol 12:1471-1474.

Grupp IL, Grupp G, Schwartz A (1981). Digitalis receptor desensitization in rat ventricle: Ouabain produces two inotropic effects. Life Sci 29:2789-2794.

Grupp G, Grupp, IL, Ghysel-Burton J, Godfraind T, Schwartz A (1982). Effects of very low concentrations of ouabain on contractile force of isolated guinea pig, rabbit, cat atria and right ventricular papillary muscles. J Pharmacol Exp Therap 220:145-151.

Grupp G, DePover A, Grupp IL, Schwartz A (1984). Analysis of the inotropic action of ouabain in rat ventricles: Two apparent ouabain inotropic response. Proc Soc Exper Biol Med 175:39-43.

Grupp G, Grupp IL, Schwartz A (1985). Reversibility of inotropic effects of ouabain on canine atrial and ventricular tissues. J Mol Cell Cardiol 15:1077-1084.

Grupp IL, Im W-B, Lee CO, Lee S-W, Pecker MS, Schwartz A (1985). Relation of sodium pump inhibition to positive inotropy at low concentration of ouabain in rat heart muscle. J Physiol 360:149-160.

Grupp G, Grupp IL, Hickerson TW, Lee S-W, Schwartz A (1986). Biphasic contractile response to ouabain. In Erdman E, Greeff JK, Skou JC (eds): "Cardiac Glycosides 1785-1985," New York: Springer, pp 99-108.

Hansen O (1976). Non-uniform populations of g-strophanthin binding sites of $(Na^+ + K^+)$-activated ATPase. BBA 433:383-392.

Isenberg G (1984). Contractility of isolated bovine ventricular myocyutes is enhanced by intracellular injection of cardioactive glycoside: Evidence for intracellular mode of action. In Erdman (ed): "Cardiac Glycoside Receptors and Inotropy." Bas Res Cardiol 79(Suppl):56-71.

Kirley TL, Lane LK, Wallick, ET (1986). Identification of an essential sulfhydryl group in the ouabain binding site of (Na,K)-ATPase. J Biol Chem 261:4525-4528.

Lee CO, Kang DH, Sokol JH, Lee KS (1980). Relation between intracellular Na ion activity and tension of sheep cardiac Purkinje fibers exposed to dihydro-ouabain. Biophy J 29:315-330.

Lee CO, Im W-B, Pecker MS, Grupp I, Lee S-W, Schwartz A (1983). Effect of ouabain on intracellular sodium ion activity and twitch tension of rat ventricular muscle. Fed Proc 42:581.

Lee SW, Schwartz A, Adams RJ, Yamori Y, Whitmer K, Lane LK, Wallick ET (1983). Decrease in Na^+,K^+-ATPase activity and [^{3}H]ouabain binding sites in sarcolemma prepared from hearts of spontaneously hypertensive rats. Hypertension 5:682-688.

Mansier P, Lelievre LG (1982). Ca-free perfusion of rat heart reveals a (Na-K)ATPase form highly sensitive to ouabain. Nature 300:535-537.

Mansier P, Cassidy PS, Charlemagne D, Preteseille M, Lelievre LG (1983). Three Na,K-ATPase forms in rat heart as revealed by K+/ouabain antagonism. FEBS Letters 153:357-360.

Maixent JM, Charlemagne D, dela Chapelle B, Lelievre LG (1987). Two Na^+K^+-ATPase isoenzymes in canine cardiac myocytes. J Biol Chem 262:6812-6818.

Ng YC, Akera T (1987). Two classes of ouabain binding sites in ferret heart and isoforms of Na^+,K^+-ATPase. Am J Physiol 252:H1016-H1022.

Repke KRH (1963). Metabolism of cardiac glycosides. In Wilbrandt W (ed): "New Aspects of Cardiac Glycosides," London: Pergamon pp 47-73.

Schatzmann HJ (1953) Herzglycoside als Hemmstoffe fuer den aktiven K^+ and Na^+ transport durch die Erythrocytenmembran. Helv Physiol Acta 11:346-354.

Scheiner-Bobis G, Zimmermann M, Kirch V, Schoner W (1987). Ouabain binding site of (Na^+,K^+)-ATPase in right-side-out vesicles has not an externally accessible SH group. Eur J Biochem 165:653-656.

Skou JC (1957). The influence of some cations on an ATP-ase from peripheral nerves. Biochim Biophys Acta 23:394-401.

Shull GE, Schwartz A, Lingrel JB (1985). Amino-acid sequence of the catalytic subunit of the (Na^+,K^+)-ATPase deduced from a complementary DNA. Nature 316:691-695.

Shull GE, Greeb J, Lingrel JB (1986). Molecular cloning of three forms of the Na^+,K^+ATPase α subunit from rat brain. Biochem 25:8125-8132.

Sweadner KJ (1979). Two molecular forms of (Na^+,K^+)-stimulated ATPase in brain: Separation and differences in affinity for strophanthidin. J Biol Chem 254:6060-6067.

Thomas R, Boutagy J, Gelbart A (1974). Cardenolide analogs V. cardiotonic activity of semisynthetic analogs of digitoxigenin. J Pharmacol Exp Ther 191:219-231.

Thomas R, Allen J, Pitts BJR, Schwartz A (1979). Cardenolide analogs. An explanation for the unusual properties of AY22241. Eur J Pharmacol 53:227-237.
Wallick ET, Kirley TL, Schwartz A (1986). Cardiac glycosides 1785-1985 biochemistry-pharmacology, clinical relevance. In Erdmann E, Greef K, Skou JC (eds): Stein Kopff, Verlag Damstadt/Springer Verlag, New York: pp 27-33.
Wasserstrom JA, Schwartz DJ, Fozzard HA (1983). Relation between intracellular sodium and twitch tension in sheep cardiac Purkinje strands exposed to cardiac glycosides. Circ Res 52:697-705.
Wehling M, Schwartz A, Whitmer K, Grupp G, Grupp IL, Wallick ET (1981). Interaction of chlormadinone with the ouabain binding site of Na^+,K^+-ATPase. Mol Pharmcol 20:551-557.
Withering W (1785). An Account of the Foxglove, and Some of its Medical Uses. Birmingham/London: M. Swinney, J. Robinson.
Whitmer KR, Wallick ET, Epps DE, Lane LK, Collins JH, Schwartz A (1982). Effects of extracts of rat brain on the digitalis receptor. Life Sci 30:2261-2273.
Young RM, Lingrel JB (1987). Tissue distribution of mRNAs encloding the α isoforms and β subunit of rat Na^+,K^+-ATPase. Biochem Biophys Res Commun 145:52-88.

ALDOSTERONE AND SODIUM INDUCE KIDNEY Na-K-ATPase
IN VITRO BY TWO DIFFERENT MECHANISMS

Catherine Barlet-Bas and Alain Doucet

Laboratoire de Physiologie Cellulaire, Collège de France, 11 Place Marcelin Berthelot, 75231 Paris Cedex 05, France

INTRODUCTION

We reported previously that the administration of a single injection of aldosterone (10 µg/kg body w.) to adrenalectomized rabbits increased within three hours the activity and the number of catalytic sites of kidney Na-K-ATPase in the collecting tubule exclusively (El Mernissi and Doucet, 1984b). It is not known, however, whether this induction of Na-K-ATPase is a primary effect of aldosterone or whether it is secondary to increased intracellular sodium concentration (Petty et al., 1981) brought about by the aldosterone-induced enhancement of luminal sodium conductance (Crabbe, 1963).

Thus, the aim of this study was to determine whether or not increasing intracellular sodium concentration may induce Na-K-ATPase in the collecting tubule, and, if it does, to compare the mechanisms underlying Na-K-ATPase stimulation in response to either aldosterone or intracellular sodium concentration. For this purpose, we developped an in vitro system in which single segments of rat cortical collecting tubules (CCT) could be maintained under "survival conditions" for over 3 hours at 37°C.

EXPERIMENTAL PROTOCOL

Single segments of CCT were microdissected under stereomicroscopic observation from collagenase-treated kidneys (Doucet et al., 1979) of either normal rats or rats adrenalectomized since 5-7 days (ADX). Segments were individually transferred into 1 µl of Eagle's MEM derived medium containing (in mM) : NaCl, 120; KCl, 5; $CaCl_2$, 1; $MgSO_4$, 1; NaH_2PO_4, 4; $NaHCO_3$,

4; glucose, 5; lactate, 10; pyruvate, 1; amino-acids, about 4; vitamins, about 3.10^{-2}; HEPES, 20; and, Dextran 40000, 3 % (w/v) and BSA 0.1 % (w/v) at pH 7.4. When necessary, the tested drugs and/or hormones were added to this medium at their required final concentration. Samples were incubated at 37°C for 2.5-3.0 hours before processing for Na-K-ATPase measurement. The maximal activity of Na-K-ATPase was determined by a microenzymatic radiochemical assay after permeabilization of cell membranes (Doucet et al., 1979). The number of catalytic sites of Na-K-ATPase was determined by the specific binding capacity of ^{3}H-ouabain (El Mernissi and Doucet, 1984a).

High intracellular sodium concentrations were obtained by adding nystatin (0.1 U/µl), an ionophore that increases the permeability to sodium, to the incubation medium. The stimulatory action of aldosterone required, to be observed, the presence of 10^{-8} M triiodothyronine in the incubation medium, as this latter hormone has been shown to facilitate in vivo as well as in vitro stimulation of Na-K-ATPase by aldosterone (Barlet and Doucet, 1987a and b).

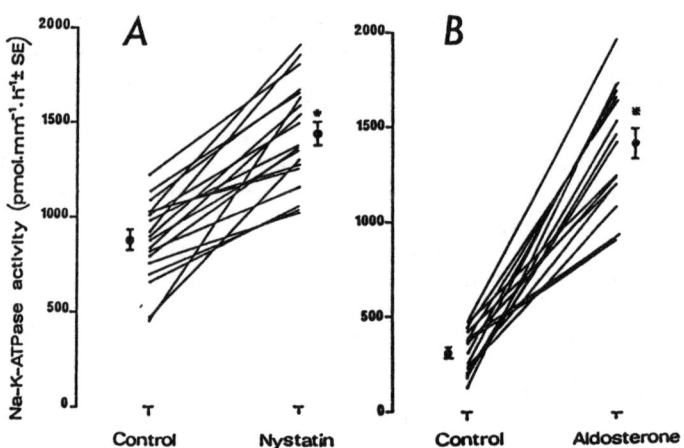

Figure 1. Effect of nystatin and of aldosterone on Na-K-ATPase activity in rat cortical collecting tubules. A- CCT from normal rats were incubated at 37°C for 3 h in the presence of nystatin (0.1 U/µl) or in its absence (control) before Na-K-ATPase measurement. B- CCT from ADX rats were incubated at 37°C for 3 h in the presence of 10^{-8} M aldosterone or in its absence (control) before Na-K-ATPase measurement. Lines join values obtained in the same animal. Points are means ± SE. Note that control values are lower in CCT from ADX rats than in CCT from normal rats. *, p<0.001, according to Student's t test.

Unless otherwise indicated, the effect of high intracellular sodium was tested on CCT from normal rats whereas that of aldosterone was evaluated on CCT from adrenalectomized rats.

RESULTS AND DISCUSSION

Results in figure 1A indicate that nystatin markedly increased Na-K-ATPase activity of CCT from normal rats within 3 hours (control : 880 $\pm$ 48 pmol.mm^{-1}.h^{-1} $\pm$ SE; nystatin : 1443 $\pm$ 63 pmol.mm^{-1}.h^{-1} $\pm$ SE; n = 19, p<0.001). Similarly (figure 1B), incubation of CCT from ADX rats for 3 hours in the presence of 10^{-8} M aldosterone stimulated Na-K-ATPase activity (ADX : 306 $\pm$ 29 pmol.mm^{-1}.h^{-1} $\pm$ SE; ADX+aldosterone : 1411 $\pm$ 78 pmol.mm^{-1}.h^{-1} $\pm$ SE; n = 16, p<0.001). The kinetics of action of nystatin and aldosterone were similar as both stimulated Na-K-ATPase activity after a 30 min period of latency and induced their maximal effect within 2-2.5 hours. Furthermore, both nystatin and aldosterone increased the specific binding of ^{3}H-ouabain in parallel with Na-K-ATPase activity, and, consequently, did not alter the turnover rate of the pump.

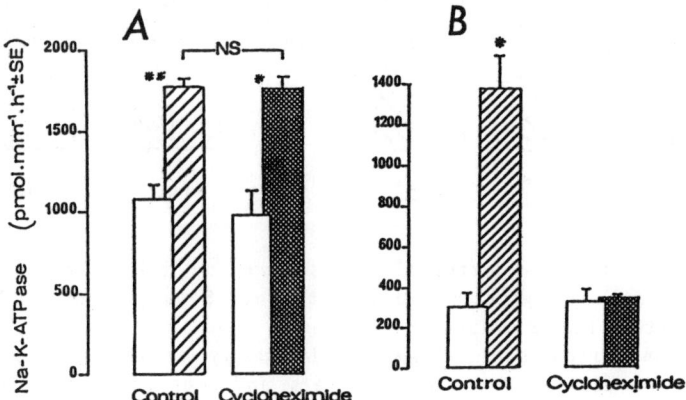

Figure 2. Effect of cycloheximide on the stimulation of Na-K-ATPase by nystatin and aldosterone. A- CCT from normal rats were incubated in the presence of nystatin (hatched and stippled bars) or in its absence (open bars), under normal conditions or in the presence of 20 µM cycloheximide. B- CCT from ADX rats were incubated in the presence of 10^{-8} M aldosterone (hatched and stippled bars) or in its absence (open bars), under normal conditions or in the presence of 20 µM cycloheximide. *, p<0.05; **, p<0.001; NS, not statistically different by variance analysis.

Although both nystatin and aldosterone increased the number of catalytic sites of Na-K-ATPase, their mechanisms of action were different since nystatin effect was independent of protein synthesis, as it was not altered by cycloheximide (figure 2A), whereas aldosterone effect was dependent of protein synthesis and completely abolished in the presence of cycloheximide (figure 2B). Similarly, actinomycin D curtailed the action of aldosterone but did not alter that of nystatin. It is concluded from these experiments that aldosterone induces the de novo synthesis of new Na-K-ATPase units whereas nystatin, by increasing intracellular sodium concentration, induces the recruitment of a preexisting pool of latent pump units.

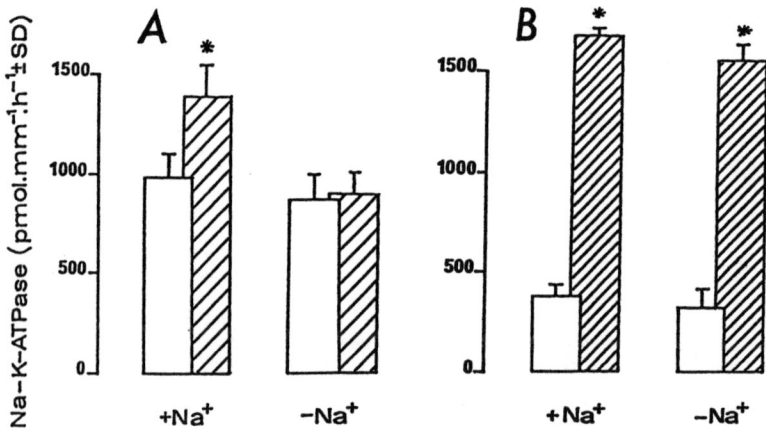

Figure 3. Effect of extracellular sodium on the stimulation of Na-K-ATPase by nystatin and aldosterone. Segments of CCT were incubated for 2 hours in a normal medium (+ Na^+) or in a medium in which sodium was replaced by choline (- Na^+). A- CCT were obtained from normal rats and were incubated in the presence of nystatin (hatched bars) or in its absence (open bars). B- CCT from ADX rats were incubated in the presence of 10^{-8} M aldosterone (hatched bars) or in its absence (open bars). *, $p<0.05$ according to Student's t test.

Aldosterone and nystatin stimulatory effects on CCT also differ by their dependence on sodium. As expected, stimulation of Na-K-ATPase activity induced by nystatin was abolished when sodium was replaced by choline in the incubation medium (figure 3A). Conversely, complete replacement of sodium by choline did not alter the stimulation of Na-K-ATPase in response to aldosterone (figure 3B). These results confirm that the observed stimulatory action of nystatin on Na-K-ATPase in the CCT is mediated by an increment of intracellular sodium concentration. They also demonstrate that the stimulation by aldosterone of Na-K-ATPase in the CCT is independent of sodium, i.e., it is not secondary to increased luminal entrance of sodium, as it had been previously suggested (Petty et al., 1981). This conclusion is in agreement with a previous report on tubular suspensions enriched in medullary collecting tubules (Rayson and Gupta, 1985).

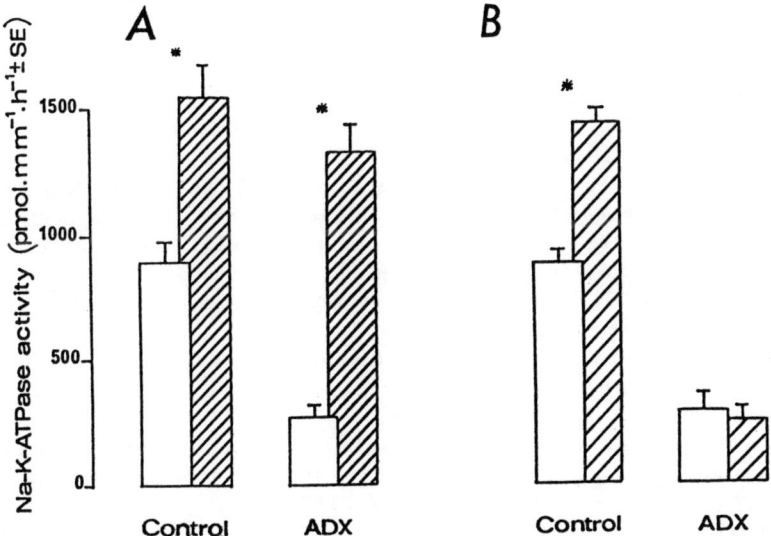

Figure 4. Effect of steroid status on the stimulation of Na-K-ATPase by nystatin and aldosterone. CCT from either normal rats (control) or rats adrenalectomized 5-7 days before study (ADX) were incubated for 2.5 h at 37°C. A- CCT were incubated in the presence of 10^{-8} M aldosterone (hatched bars) or in its absence (open bars). B- Tubules were incubated in the presence of nystatin (hatched bars) or in its absence (open bars). *, p<0.001 according to Student's t test.

Finally, induction of Na-K-ATPase by aldosterone was observed in CCT of normal as well as adrenalectomized rats (figure 4A) whereas nystatin stimulated Na-K-ATPase in CCT of normal rats exclusively (figure 4B). This last result indicates that the size of the latent pool of Na-K-ATPase, which is recruited in response to increased sodium concentration, is controled by corticosteroids.

ACKNOWLEDGEMENTS

This work was supported in part by a grant from the Centre National de la Recherche Scientifique to the Unité Associée 219.

REFERENCES

Barlet C, Doucet A (1987a). Triiodothyronine enhances renal response to aldosterone in the rabbit collecting tubule. J Clin Invest 79 : 629-631.
Barlet C, Doucet A (1987b). Interaction between triiodothyronine and aldosterone in control of kidney Na-K-ATPase. In : Biochemical Aspects of Kidney Function, Edited by Kovačević Z and Guder WG, de Gruyter, Berlin, p. 63-69.
Crabbé J (1963). Site of action of aldosterone on the bladder of the toad. Nature 200 : 787-788.
Doucet A, Katz AI, Morel F (1979). Determination of Na-K-ATPase activity in single segments of the mammalian nephron. Am J Physiol 237 : F105-F113.
El Mernissi G, Doucet A (1984a). Quantitation of ^{3}H-ouabain binding and turnover of Na-K-ATPase along the rabbit nephron. Am J Physiol 247 : F158-F167.
El Mernissi G, Doucet A (1984b). Specific activity of Na-K-ATPase after adrenalectomy and hormone replacement along the rabbit nephron. Pflügers Arch 402 : 258-263.
Petty KJ, Kokko JP, Marver D (1981). Secondary effect of aldosterone on Na-K-ATPase activity in the rabbit cortical collecting tubule. J Clin Invest 68 : 1514-1521.
Rayson BM, Gupta RK (1985). Steroids, intracellular sodium levels, and Na^+/K^+-ATPase regulation. J Biol Chem 260 : 12740-12743

REGULATION OF CARDIAC GLYCOSIDE RECEPTORS WITH DIFFERENT AFFINITIES FOR CARDIAC GLYCOSIDES IN CULTURED RAT HEART CELLS

H.J. Berger, K. Werdan, E. Erdmann

Medizinische Klinik I, Klinikum Grosshadern, University of Munich, D-8000 München 70, Germany

INTRODUCTION

The existence of receptors with different affinities for cardiac glycosides (CG) in various organs and species has been established (Erdmann et al., 1980; Shull et al., 1986; Erdmann et al., 1987). Cardiac glycoside receptors with high and low affinity for ouabain are present in cultured rat heart cells (Werdan et al., 1984). Todays most widely accepted concept is that binding of cardiac glycosides to the cell surface specifically inhibits $(Na^+ + K^+)$-ATPase and reduces Na,K-transport across the cell membrane. In consequence, the cellular Na^+-content is increased which mediates a positive inotropic effect. In the rat heart, binding of cardiac glycosides to both receptor subtypes results in inhibition of active Na,K-transport and in positive inotropic action (Grupp et al., 1986). However, toxic-arrhythmogenic effects are only seen after occupation of the low-affinity receptor. In contrast to the pharmacological consequences of the receptor occupation, the physiological role of the Na,K-ATPase-isoenzymes with high and low affinity for cardiac glycosides is unclear.

In order to answer the question whether the two distinct glycoside receptors are regulated independently from each other and to evaluate their physiological significance, we used cultured, spontaneously beating rat heart muscle cell monolayers. The influence of the modulation of growth conditions on (^{3}H)-ouabain binding, cation fluxes and contractility was studied.

METHODS

Cell Culture, Measurement of (^3H)-Ouabain Binding, $(Rb^+ + K^+)$ Influx-Rates, Contractility

Preparation and cultivation of rat heart muscle cells from 1-3 day-old SD-rats have been described in detail previously (Werdan et. al, 1984). After 36h of growth in medium (CMRL-1415) containing 10% fetal calf and horse serum, the monolayer cultures of spontaneously contracting rat heart cells were incubated in serum-free and antibiotic-free culture medium (Werdan et al.,1985). Experiments were performed after 3 days of culture in the fresh medium.

The number of CG receptors was measured as previously described (Werdan et al., 1984). Briefly, monolayer of heart muscle cells were incubated in salt solution containing 0.75 mM KCl, 0.9 mM $CaCl_2$, 5 mM HEPES and various concentrations of (^3H)-ouabain for 4h at 37 °C. For characterization of high and low affinity CG receptors, binding data have been plotted according to Scatchard. In order to estimate the number of high and low affinity receptors, binding of (^3H)-ouabain to the cells at 6×10^{-8}M and 6×10^{-6}M has been taken. At 6×10^{-8}M, 85% (15%) of specifically bound radioactivity represents binding to the high- (low-)affinity site, while at 6×10^{-6}M, 15% (85%) of the cell-bound radioactivity represents binding to the high- (low-)affinity site.

For measurement of active K^+-influx $^{86}Rb^+$ was used. To eliminate the pump reserve capacity (Werdan et al., 1983), cells have been preincubated for 4h in the presence of very low K^+-concentration (0.05 mM). For subsequent determination of $^{86}Rb^+$-uptake, cells were exposed to medium containing 0.75 mM K^+. By this procedure, transport activity of active $(Rb^+ + K^+)$-influx was proprotional to the number of active sodium pump molecules/cell.

For measurement of contraction velocity, a perfusion-stimulation-adapter has been used as previously described (Werdan and Reithmann 1987). The contraction velocity was monitored and under equilibrium conditions expressed as a percent of the maximal achievable change produced by 2 mM Ca^{2+}. The contraction velocity in cultured rat heart cells is equivalent to a positive inotropic effect .
Protein measurements are according to Lowry. All data are means from triplicate determinations. Methods have been described in detail previously (Werdan et al., 1984).

RESULTS

To answer the question whether the high and low affinity cardiac glycoside binding sites are involved in the pharmacological effect of cardiac glycosides and whether both cardiac glycoside receptors as components of the Na,K-ATPase-isoenzymes are subject to regulation, rat heart muscle cells were grown for 3 days in medium containing 3×10^{-8} M T_3 or 0.05 mM potassium. The cells did not visibly alter their morphology. They kept their ability to contract spontaneously after these 3 days. Thereafter, binding of (^{3}H)-ouabain to the high and low affinity binding sites, ($Rb^+ + K^+$)-influx-rate and contractility were measured.

Effect of Potassium Depletion on the number of High- and Low-Affinity Receptors in Cultured Rat Heart Muscle Cells

Binding studies with (^{3}H)-ouabain have demonstrated the existence of high and low affinity cardiac glycoside

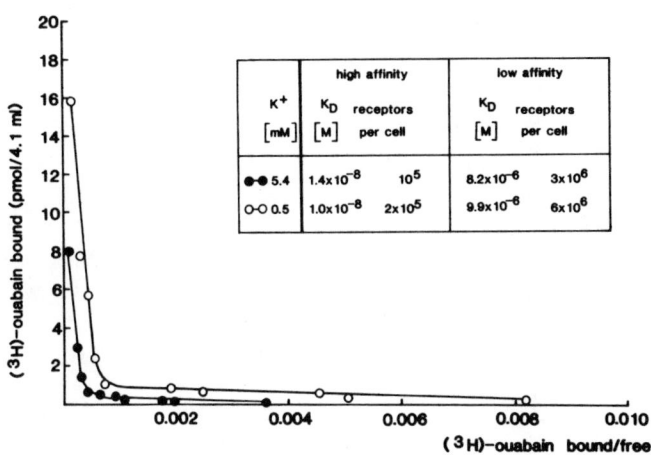

Figure 1. Effect of chronic K^+-depletion on the number of CG-receptors. Scatchard plot analysis of concentration dependent specific (^{3}H)-ouabain of control cells (5.4 mM) and cells cultured for 3 days in the presence of 0.5 mM K^+.

binding sites in cultured rat heart muscle cells. Potassium depletion increases the number of ouabain binding sites in a concentration dependent-manner. The maximal increase after a culture period of 3 days was observed at 0.5 mM K⁺ (Erdmann et al, 1987). At this K⁺-concentration Scatchard plot analysis indicates a doubling in the number of high and low affinity binding sites while both dissociation constants remain unchanged (Fig. 1).

Effect of T_3 on the Number of Cardiac Glycoside Receptors

Growth of the cells in T_3 for 3 days elevates the number of both cardiac glycoside binding sites with high and low affinity for ouabain by 50±14% and 70±14%, respectively (Fig. 2). Scatchard plot analysis has shown no change in the dissociation constants (Experiments not shown).

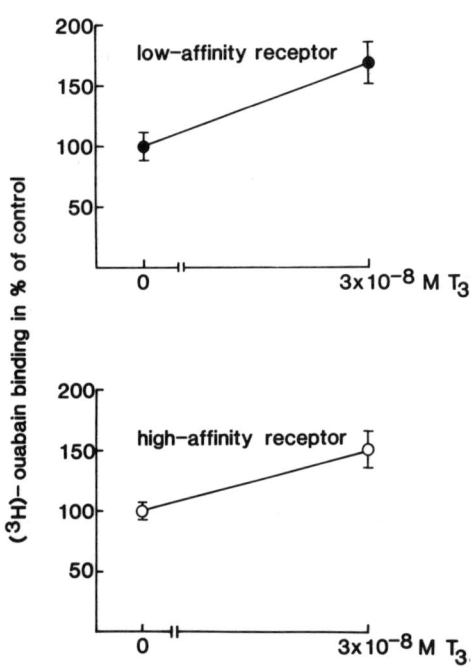

Figure 2. Effect of T_3 exposure (3×10^{-8} M, 3 days) of rat heart muscle cells in culture on high and low affinity CG receptors

Influence of Active Na⁺/K⁺-Transport on the Number of High and Low Affinity Cardiac Glycoside Receptors

The increase in the number of both cardiac glycoside receptor subtypes concerns functionally active sodium pump molecules. The increase in the number of Na,K-ATPase-isoenzymes caused by chronic K^+-depletion (Fig. 1) is accompanied by an increase in ($^{86}Rb^+ + K^+$)-influx into cells mediated by both Na,K-ATPase-isoenzymes with high as well as low affinity CG binding sites (Fig. 3).

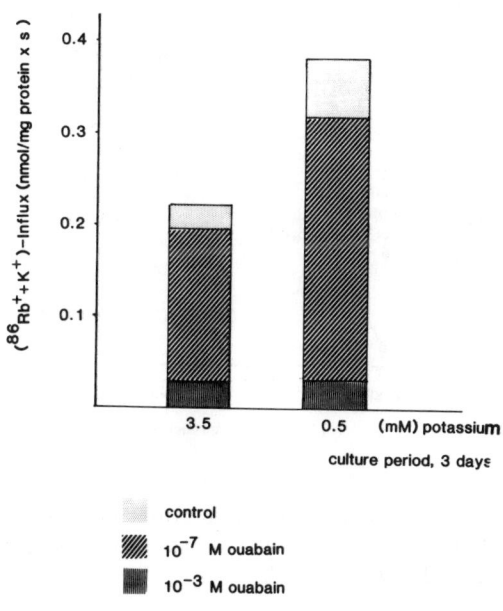

Figure 3. Effect of chronic K^+-depletion on ($^{86}Rb^+ + K^+$)-influx-rate after elimination of pump reserve capacity (see methods). A column represents total influx capacity split into the ($^{86}Rb^+ + K^+$)-influx-rate in presence of 10^{-7} M ouabain which reflects the influx after occupation of the high affinity receptor (80% (1%) high (low) affinity CG-receptor occupation) and in presence of 10^{-3} M ouabain after complete occupation of the high and low affinity receptors, which reflects ouabain-insensitive influx.

Positive Inotropic Action of Cardiac Glycosides in Rat Heart Muscle Cells in Culture

The concentration-effect curve for ouabain is biphasic in cultured rat heart muscle cells . Predominant occupation of high-affinity receptor results in 20-40% of the maximal increase in contraction velocity. High ouabain concentration with more than about 50% low-affinity occupation ($>10^{-5}$ M) finally leads to arrhythmias (Fig. 4).

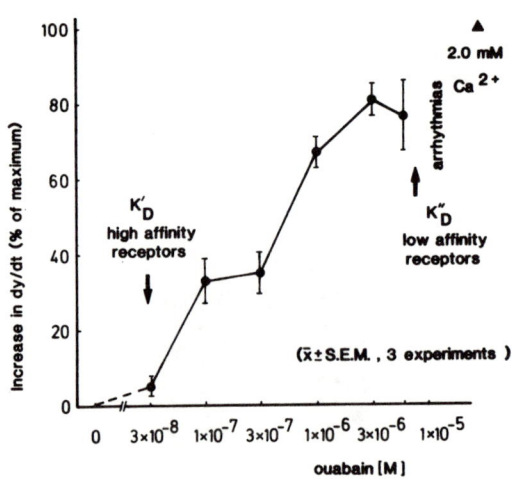

Figure 4. Effect of ouabain on contraction velocity in neonatal rat heart muscle cells in culture (modified from Erdmann et al., 1987).

An increase in the number of cardiac glycoside receptors attenuates both positive inotropic and toxic effects. This is well understood in chicken heart muscle cells with only one type of cardiac glycoside receptors (Werdan et al., 1985). In rat heart muscle cells higher numbers of high and low affinity cardiac glycoside receptors also attenuate the positive inotropic effect of ouabain at 10^{-7} M ouabain with predominant occupation of the high-affinity receptors as well as 6×10^{-6} M ouabain with occupation of high- and low-affinity cardiac glycoside receptors (Fig. 5).

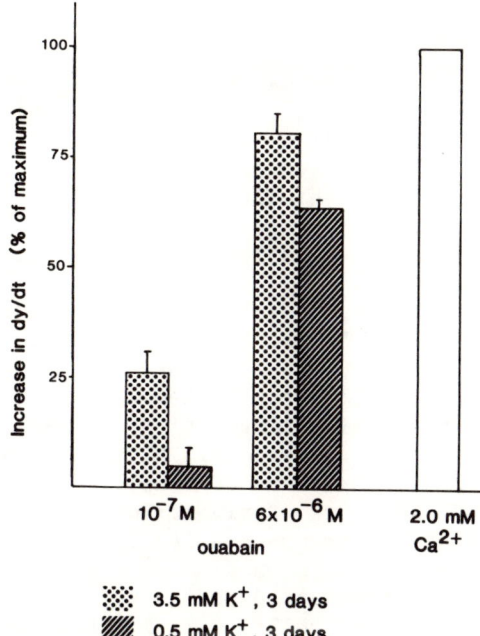

Figure 5. Effect of increased number of Na,K-ATPase isoenzyme-molecules by cultivation of rat heart muscle cells in chronic K⁺-depletion (see Fig. 1) on increase in contraction velocity by ouabain

CONCLUSIONS

Both types of Na,K-ATPase-isoenzymes with high and low affinity for ouabain participate in the regulation of Na,K-transport across the cell membrane and both mediate a positive inotropic effect by occupation of the Na,K-ATPase-isoenzymes with cardiac glycosides.

T_3 or low extracellular potassium concentration increase the number of both functional Na,K-ATPase-isoenzymes to a similar extent. This increase of active sodium pump molecules attenuates the sensitivity of the cells to cardiac glycosides.

REFERENCES

Erdmann E, Brown L, Werdan K, Berger HJ (1987). Multiple forms of the cardiac glycoside receptor with different affinities for cardiac glycosides. In Beamish RE, Panagia V, Dhalla NS (eds): "Pharmacological Aspects of Heart Disease", Boston: Martinus Nijoff Publishing, p261

Erdmann E, Werdan K, Brown L (1985). Multiplicity of cardiac glycoside receptors in the heart. Trends Pharmacol Sci 6:293-295.

Erdmann E, Philipp G, Scholz H (1980). Cardiac glycoside receptor, (Na^++K^+)-ATPase activity and force of contraction in rat heart. Biochem Pharmacol 29:3219-3229.

Grupp G, Grupp IL, Hickerson T, Lee SW, Schwartz A (1986). Biphasic contractile response to ouabain: Species specific? Calcium dependent? Altered sensitivity? In Erdmann E, Greeff K, Skou JC (eds): "Cardiac Glycosides 1785 - 1985", New York: Springer-Verlag, p 99-108.

Kim D, Marsh JD, Barry WH, Smith TW (1984). Effects of growth in low potassium medium or ouabain on membrane Na,K-ATPase, cation transport, and contractility in cultured chick heart cells. Circ Res 55:39-48.

Kim D, Smith TW (1984). Effects of thyroid hormone on sodium pump sites, sodium content, and contractile responses to cardiac glycosides in cultured chick ventricular cells. J Clin Invest 74:1481-1488.

Shull GE, Greeb J, Lingrel JB (1986). Molecular cloning of three distinct forms of the Na^+,K^+-ATPase α-Subunit from rat brain. Biochemistry 25:8125-8132.

Sweadner KJ (1979). Two molecular forms of (Na^++K^+)-stimulated ATPase in brain. J Biol Chem 254:6060-6067.

Werdan K, Reithmann C (1987). " Neonatale und embryonale Herzmuskelzellen als Myokardmodell." Stuttgart: Thieme-V

Werdan K, Reithmann C, Erdmann E (1985). Cardiac glycoside tolerance in cultured chicken heart muscle cells - a dose-dependent phenomenon. Klin Wochenschrift 63:1253

Werdan K, Wagenknecht B, Zwissler B, Brown L, Krawietz W, Erdmann E (1984). Cardiac glycoside receptors in cultured heart cells - II. Characterization of a high affinity and a low affinity binding site in heart muscle cells from neonatal rats. Biochem Pharmacol 33:1873-1886

Werdan K, Zwißler B, Wagenknecht B, Krawietz W, Erdmann E (1983). Quantitative correlation of cardiac glycoside binding to its receptor and inhibition of the sodium pump in chicken heart cells in culture. Biochem Pharmocol 32:757-760.

PERTUSSIS TOXIN MODULATES DOPAMINE INHIBITION OF Na,K-ATPase ACTIVITY IN RAT PROXIMAL CONVOLUTED TUBULE SEGMENTS

Alejandro Bertorello and Anita Aperia

Department of Pediatrics, St. Göran's Children's Hospital, 112 81 Stockholm, Sweden

INTRODUCTION

In proximal convoluted tubule (PCT) segments, locally generated dopamine (DA) inhibits Na,K-ATPase activity (Aperia et al., 1987). Hitherto we have not been able to demonstrate that the effect is mediated by cell surface receptors. Since DA is not a precursor of an inhibitor and since DA does not inhibit the purified enzyme (Bertorello and Aperia, 1987), we have assumed that DA modulation of Na,K-ATPase activity should be mediated via an intracellular signal system.

Hormones bound to all surface receptors transduce intracellular signals via GTP-dependent regulatory (G) proteins (Rodbell, 1980). G proteins have been shown to modulate CaATPase (Lotersztajn et al., 1987), potassium channels (Sasaki and Sato, 1987; Logothetis et al., 1987), and calcium channels (Hescheler, et al. 1987) but no effect related to active sodium and potassium transport has yet been reported.

In this communication we present evidence that a pertussis toxin (PT) sensitive G protein is involved in the regulation of Na,K-ATPase activity and that DA inhibits Na,K-ATPase activity via this G protein.

MATERIAL AND METHODS

The studies were performed in young male Sprague-Dawley rats weighing 150-180 g. The rats were fed synthetic rat chow ad lib (Ewos, Södertälje) and had free access to tap water. Kidney perfusion, tubule microdissection, and Na,K-ATPase assay in single tubules has been described earlier (Doucet et al., 1979; Aperia et al., 1987). Briefly, the kidney was perfused with a 0.05% collagenase solution. The PCT segments were made permeable by hypotonic shock. After preincubation with DA (Kali-Chemie, Hannover, FGR) GppNHp (Sigma, St. Louis, USA) or vehicle only (control) for 30 min at 23°C, individual tubule segments (length 0.5-1.0 mm) were incubated for 15 min at 37°C in a medium containing (in mM): NaCl 50; KCl 5; EGTA 1; Tris-HCl 100; Na_2ATP (grade II, Sigma) 10; and ^{32}P-ATP (NEN, Boston, MA, USA) 2-5 Ci/mmol in tracer amounts (5nCi/µl). For determination of ouabain-insensitive (Mg-dependent) ATPase activity NaCl and KCl were omitted, Tris-HCl 150 mM and ouabain 1 mM were added. The pH of both solutions was 7.4. The phosphate liberated by hydrolysis of ^{32}P-ATP was separated by filtration through a millipore filter after absorption of the unhydrolyzed nucleotide on activated charcoal. In each experiment (each rat) we determined total ATPase in 6-10 segments and ouabain-insensitive ATPase in 6-10 segments. The difference between the mean values for total and ouabain-insensitive ATPase was noted as Na,K-ATPase and used for further calculations.

RESULTS AND DISCUSSION

DA inhibits Na,K-ATPase activity in PCT segments in a dose-dependent way (Aperia et al., 1987). DA 10^{-5} M gives maximal inhibition. In the next experiment we evaluate DA effect in the presence and absence of GppNHp. The tubule segments were permeabilized and thus the nucleotide had full access to the interior of the cell. The enzyme assay was carried out under V_{max} condition for Na, K, and ATP.

Na,K-ATPase activity in vehicle incubated tubules was 1739 ± 266 (n = 6) pmol Pi/mm tubule/hour. In PCT segments incubated with 10^{-5}M DA Na,K-ATPase activity was 528 ± 85 pmol Pi/mm tubule/hour (n = 7). When PCT segments were incubated with DA and GppNHp 100 µM the inhibitory effect of DA was abolished. Na,K-ATPase activity was 1799 ± 77 pmol Pi/mm tubule/hour (n = 5), p < 0.001.

G proteins can be characterized with regard to toxin sensitivity (Gilman 1987). Cholera toxin interacts with a G stimulatory protein making the alpha-GTP subunit resistant to hydrolysis. Pertussis toxin interacts with G inhibitory proteins (Gi) and G proteins that have been described to induce changes in phosphatidylinositol turnover (Go). We have tested whether a pertussis toxin sensitive G protein (Gi/Go) is involved in dopamine inhibition of Na,K-ATPase activity.

We injected a sublethal dose of PT (Martinez-Olmedo et al. 1983) 10 µg/100 g BW iv into rats 3 days before the experiments. DA inhibition of PCT Na,K-ATPase was significantly attenuated. Na,K-ATPase activity in DA incubated tubules was 1259 ± 145 pmol Pi/mm tubule/hour (n = 6), which was not significantly different from the value in vehicle-incubated tubules, 1568 ± 72 pmol Pi/mm tubule/hour (n=6).

Pertussis toxin treatment had no effect on basal PCT Na,K-ATPase activity (PT control 1568 ± 72 pmol Pi/mm Pi/hour vs. 1538 ± 88 pmol Pi/mm tubule/hour in control tubules from untreated rats. Gs proteins stimulate adenylate cyclase and increases cAMP levels. The involvement of a Gs protein was ruled out by the lack of inhibitory effect of dibutyryl cAMP on Na,K-ATPase activity (db cAMP incubated tubules 2660 ± 233 pmol Pi/mm tubule/hour (n = 5) vs. vehicle-incubated tubules 1600 ± 287 pmol Pi/mm tubule/hour (n = 5)). Forskolin (10^{-3} M) which is known to activate adenylate cyclase also significantly increased Na,K-ATPase activity.

We conclude that in PCT segments a pertussis toxin sensitive G protein mediates dopamine inhibition of Na,K-ATPase activity.

REFERENCES

Aperia A, Bertorello A, Seri I (1987). Dopamine causes inhibition of Na^+-K^+-ATPase activity in rat proximal convoluted tubule segments. Am J Physiol 252:F39-F45.

Bertorello A, Aperia A (1987). Effect of L-dopa, dopamine, dihydroxyphenyl acetic acid and homovanillic acid on Na,K-ATPase activity in rat proximal tubule segments. Acta Physiol Scand 130:571-574.

Doucet A, Katz AI, Morel F (1979). Determination of Na^+-K^+-ATPase activity in single segments of the mammalian nephron. Am J Physiol 237:F105-F113.

Gilman AG (1987). G proteins: Transducers of receptor-generated signals. Ann Rev Biochem 56:615-649.

Hescheler J, Rosenthal W, Trautwein W, Schultz G (1987). The GTP-binding protein, G_o, regulates neuronal calcium channels. Nature 325:445-447.

Logothetis DE, Kurachi Y, Galper J, Neer EJ, Clapham DE (1987). The $\beta\gamma$ subunits of GTP-binding proteins activate the muscarinic K^+ channel in heart. Nature 325:321-326.

Lotersztajn S, Pavoine C, Mallat A, Stengel D, Insel PA, Pecker F (1987). Cholera toxin blocks glucagon-mediated inhibition of the liver plasma membrane (Ca^{2+}-Mg^{2+})-ATPase. J Biol Chem 262:3114-3117.

Martinez-Olmedo MA, Garcia-Sainz JA (1983). Effect of pertussis toxin on the hormonal regulation of cyclic AMP levels in hamster fat cells. Biochim Biophys Acta 760:215-220.

Rodbell M (1980). The role of hormone receptors and GTP-regulatory proteins in membrane transduction. Nature 284:17-22.

Sasaki K, Sato M (1987). A single GTPbinding protein regulates K^+-channels coupled with dopamine, histamine and acetylcholine receptors. Nature 325:259-262.

LOSS OF Na,K-ATPASE DURING SHEEP RETICULOCYTE MATURATION

Rhoda Blostein and Eva Grafova

Montreal General Hospital Research Institute,
1650 Cedar Ave., Montreal, Que., H3G 1A4, Canada.

INTRODUCTION

Sheep are dimorphic with respect to Na,K-ATPase such that the intracellular K^+ content of the mature cells of the one type (high-K^+) is high, that of the other (low-K^+) low. Their kinetic characteristics are also distinct. In contrast, reticulocytes of sheep of both genotypes have high intracellular K^+ and are kinetically similar, resembling in certain respects, mature cells of high-K^+ animals (Lee et al., 1966; Tucker and Ellory, 1971; Dunham and Blostein, 1976). During maturation, their sodium pump activity decreases and the kinetic behaviour of their pumps changes resulting in mature cells which are phenotypically either high-K^+ or low-K^+ cells (Blostein et al., 1974 and 1983; Dunham and Blostein, 1976).

We showed previously that the maturation-associated decline in ouabain binding sites of sheep reticulocytes can be retarded markedly by ATP depletion as well as by specific ligand (ouabain) binding (Weigensberg and Blostein, 1983; Blostein and Grafova, 1987). It is evident also that metabolic energy is required for the maturation-associated loss of various other distinct membrane functions including the Na/glycine cotransporter (Weigensberg and Blostein, 1983) and the nucleoside transporter (Blostein and Grafova, 1987). In the latter instance, recycling of specific nucleoside binding sites occurs and is likely to be involved in the ultimate functional loss attending maturation. In this paper we describe studies concerned with the regression of the

Na,K-ATPase catalytic subunit during sheep red cell maturation and aging.

RESULTS

Polyclonal antibodies directed against lamb kidney Na,K-ATPase were used to detect the Na,K-ATPase of plasma membranes derived from immature red cells isolated from anaemic low-K^+ and high-K^+ sheep. As shown in Fig. 1, this serum detects the kidney and brain α subunit, the brain $\alpha+$ isoform as well as the β subunit of both kidney and brain. Only the catalytic subunit of the red cell enzyme was consistently detected, its apparent mobility and relative intensity corresponding to that of a similar amount, in terms of activity, of the sheep kidney α isoform. A reactive β subunit could not be detected. The occasional appearance of a weakly-visible sharp band in the region of the β subunit is unrelated since it does not react with affinity-purified anti-α or anti-β subunit antibodies (not shown).

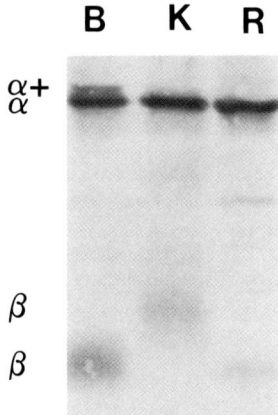

Figure 1. Western blot analysis of Na,K-ATPase. Kidney microsomes (0.62 µg, lane K), brain microsomes (3.6 µg, lane B) and reticulocyte membranes (25 µg from a high-K^+ sheep, lane R) were analyzed using antiserum diluted 1:4000.

TABLE 1. Losses in Ouabain Binding from Cells, in Immunologically Reactive α Subunit from the Plasma Membranes and Appearance of α Subunit in Shed Material.

Exp.	Culture Period (days)	Ouabain binding (sites/cell)	Immunological reactivity	
			Membranes	Shed Pellet
			(% of Control)	
A	0(control)	1643	100	–
	1	792	~100	5
	2	557	25	20
B	0(control)	1259	100	–
	1	781	40	≤10
	2	562	20	≤10

Quantitation of immunological reactivity was done by scanning the densities of xerographic images of the bands on Western blots, essentially as described by Nagata et al.[1]. For each experiment several dilutions of both membranes and shed pellets were analyzed on the same gel. Densities were directly proportional to the amounts of material added provided the amounts analyzed were no greater than ~100 ng pure kidney enzyme.

Pump Recycling: Evidence of Latent Pumps

In a series of *in vitro* experiments in which assays of [^{3}H]ouabain binding sites were carried out after relatively short intervals of culture, we observed a transient increase in ouabain binding sites in metabolically-depleted cells, either initially or after a transient decline, or when starvation was initiated after a few hours of incubation in 'fed' medium. In fed cells, lysosomotropic agents such as chloroquine, also increased the sites apparent at the surface. These observations are consistent with the notion that Na,K-ATPase recycling

[1]Nagata T, Poulsen LL and Ziegler DM, manuscript submitted

occurs in reticulocytes and may be inherent in the process(es) underlying maturation-associated regression of the sodium pump.

DISCUSSION

In this study we show that the major immunologically reactive Na,K-ATPase peptide in red cells is the full-size catalytic subunit of identical mobility as the kidney enzyme α subunit. Although the slightly higher molecular weight α+ form characteristic of axolemma Na,K-ATPase was not detected, the presence of small amounts of this and/or other isoform(s) cannot be unequivocally excluded. We were also unable to detect β subunit in sheep red cells even though the antiserum reacts with the β subunit of both kidney and brain.

Evidence of latent ouabain binding sites and of release of immunologically reactive Na,K-ATPase α subunit into the medium during in vitro culture of reticulocytes is consistent with the notion that pump recycling and exocytosis is one mechanism whereby sodium pumps are lost during maturation. Indeed, reticulocyte membrane exocytosis appears to underlie the maturation-associated loss of transferrin receptors during maturation of reticulocytes (Pan and Johnstone, 1983; Pan et al., 1985; Harding et al., 1984). In the case of the Na,K-ATPase, the extent to which regression is due to this mechanism or to one involving targeting to and degradation in intracellular organelles, or even degradation by direct cleavage at the cytoplasmic membrane surface, remains to be determined. Thus, exocytosis of Na,K-ATPase may be part of the process of surface remodelling that occurs during maturation of reticulocytes, particularly large "stress" reticulocytes (Come et al., 1974; Harding et al., 1984). As well, ATP-dependent proteolysis which is active in reticulocytes (Muller et al., 1980; Boches and Goldberg, 1982; Hersko and Ciechanover, 1982) may also effect loss of pumps. Consistent with the latter is the report that degradation of dog kidney Na,K-ATPase by dog reticulocyte lysate is enhanced by ATP addition (Inaba and Maeda, 1986).

Relative losses in transport are clearly different for distinct transport systems (Blostein and Grafova, 1987) and it is interesting to consider the evidence that ligand-mediated modulations in a specific membrane protein's conformation affects its decline. Thus, ouabain present during reticulocyte culture retards the loss in pump (ouabain binding) sites; K^+-depletion also retards pump loss, at least in cells of high-K^+ sheep (Blostein and Grafova, 1987). It is relevant that K^+-depletion or addition of sublethal amounts of ouabain retards Na,K-ATPase turnover in cultured HeLa cells (Karin and Cook, 1983). It would be interesting to determine whether distinct conformational alterations account for the higher regression of Na,K-ATPase in reticulocytes of the low-K^+ compared to high-K^+ animals. Experiments to address this question are currently underway.

REFERENCES

Blostein R, Drapeau P, Benderoff S, Weigensberg AM (1983). Changes in Na^+-ATPase and Na^+-K^+-pump during maturation of sheep reticulocytes. Can J. Biochem Cell Physiol 61:23-28.

Blostein R, Grafova E (1987). Characterization of membrane transport losses during reticulocyte maturation. Biochem Cell Biol in press.

Blostein R, Whittington ES, Kuebler ES (1974). ATPase of mammalian erythrocyte membranes: kinetic changes associated with postnatal development and following active erythropoiesis. Ann NY Acad Sci 242:305-316.

Boches FS, Goldberg AL (1982). Role of the adenosine triphosphate dependent proteolytic pathway in reticulocyte maturation. Science 215:978-980.

Castro J, Farley RA (1979). Proteolytic fragmentation of the catalytic subunit of the sodium and potassium adenosine triphosphatase. Alignment of tryptic and chymotryptic fragments and location of sites labeled with ATP and iodoacetate. J. Biol Chem 254:2221-2228.

Come SE, Shohet SB, Robinson SH (1974). Surface remodelling vs whole-cell hemolysis of reticulocytes produced with erythroid stimulation or iron deficiency anaemia. Blood 44:817-830.

Dunham PB, Blostein R (1976). Active potassium transport in reticulocytes of high-K^+ and low-K^+ sheep. Biochim Biophys Acta 455:749-758.

Harding AU, Heuser J, Stahl P (1984). Endocytosis and intracellular processing of transferrin and colloidal gold transferrin in rat reticulocytes. Eur J. Cell Biol

35:256-263.
Hersko A, Ciechanover A (1982). Mechanisms of intracellular protein breakdown. Ann Rev Biochem 51:335-364.
Inaba M, Maede Y (1986). Na,K-ATPase in dog red cells. J. Biol Chem 261:16099-16105.
Karin NJ, Cook JS (1983). Regulation of Na,K-ATPase by its biosynthesis and turnover. Curr Top Memb Transp 19:713-751.
Lee P, Woo A, Tosteson DC (1966). Cytodifferentiation and membrane transport properties in LK sheep red cells. J. Gen Physiol 50:379-390.
Muller M, Dubiel W, Rathman J, Rapoport S (1980). Determination and characteristics of energy-dependent proteolysis in rabbit reticulocytes. Eur J. Biochem 109:405-410.
Pan BT, Johnstone RM (1983). Fate of transferrin receptor during maturation of sheep reticulocyte in vitro selective externalization of the receptor cell. Cell 33:966-978.
Pan BT, Teng K, Wu C, Adam M, Johnstone RM (1985). Electron microscopic evidence for externalization of the transferrin receptor in vesicular form in sheep reticulocytes. J. Cell Biol 101:942-948.
Tucker EM, Ellory JC (1971). The M-L blood group system and active potassium transport in sheep reticulocytes. Anim Blood Grps Biochem Genet 2:77-87.
Weigensberg AM, Blostein R (1983). Energy-depletion retards the loss of membrane transport during reticulocytes maturation. Proc Natl Acad Sci USA 80:4978-4982.

ACKNOWLEDGEMENTS

This work was supported by a grant from the Medical Research Council of Canada. We thank Dr. William J. Ball for the generous gift of antiserum and Dr. Lois Lane, for purified lamb kidney Na,K-ATPase.

THE EFFECT OF EXTERNAL Na AND Ca ON THE RATE OF OUABAIN BINDING TO RESEALED HUMAN RED BLOOD CELL GHOSTS

H. Harm Bodemann* and Joseph F. Hoffman

Department of Cellular and Molecular Physiology, Yale University School of Medicine, New Haven, Connecticut, USA. *Present address: Städtisches Krankenhaus, Sindelfingen, FRG

INTRODUCTION

External Na (Na_o) as well as external Ca (Ca_o) stimulate MgATP-promoted ouabain binding to the intact red cell membrane (Bodemann and Hoffman, 1976b, 1983; Gardner and Frantz, 1974; Hobbs and Dunham, 1978). Stimulation of ouabain binding by Na_o can occur by reducing the inhibitory affinity of external K (K_o) to the Na/K-pump (Bodemann and Hoffman, 1976a) but additional binding sites are required for Na_o and Ca_o since stimulation of ouabain binding by Na_o is also possible in the presence of a saturating concentration of K_o. In addition, Ca_o does not act by changing the pump's affinity for K_o (Gardner and Frantz, 1974; Hobbs and Dunham, 1978).

This paper summarizes some preliminary results which extend previous observations concerning interactions of internal Mg (Mg_i) and Na_o and Ca_o on ATP-promoted ouabain binding to reconstituted human red blood cell ghosts (Bodemann and Hoffman, 1976a, 1983).

METHODS

The preparation of resealed ghosts, methods and analyses used are the same as those described previously (Bodemann and Hoffman, 1976a,b).

RESULTS

Figure 1 shows that the effects of Na_o and Ca_o on the ouabain binding rate to the red cell membrane are sensitive to the concentration of Mg_i. Note that a saturating concentration of K_o was present under all conditions. At low Mg_i the ouabain binding rate was low and was increased up to three-fold by Na_o. Stimulation by Ca_o was strikingly high, in the range of twenty-fold. On the other hand, at high Mg_i the rate of ouabain binding was increased with a concomitant loss in the fractional stimulation of binding by Na_o and Ca_o.

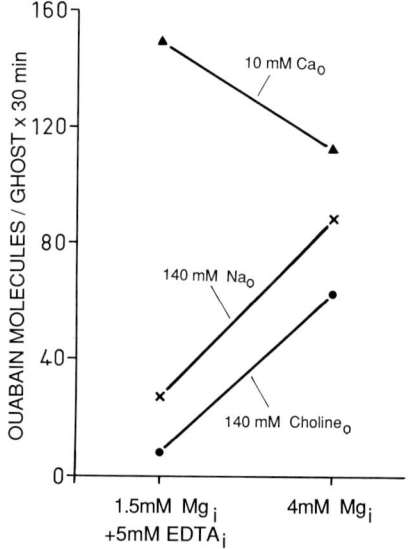

Figure 1. Stimulation of ouabain binding to resealed human red cell ghosts by Ca_o and Na_o at low and high Mg_i. The ghosts contained in addition to $MgCl_2$ and EDTA (mM): 2 ATP, 40 NaCl, 105 cholineCl, 4 KCl, 7 Tris (pH 7.4). The external solution contained in addition to $CaCl_2$ or NaCl (mM): 20 KCl, 10 Tris (pH 7.4). 125 mM cholineCl was also present when the medium contained 10 mM $CaCl_2$. Ouabain binding was measured at 37°C for 30 min in the presence of 2.5×10^{-7} M [^{3}H]-ouabain. Hematocrit 9%. Mg_i in the presence of $EDTA_i$ is estimated to be in the micromolar range.

Figure 2 extends these findings and shows that when Mg_i is 0.5 mM or greater (in the absence of $EDTA_i$) the rates of ouabain binding in the presence of 140 mM Na_o are high and relatively constant. However, in the absence of Na_o, the ouabain binding rate was increased by Mg_i, approaching the rate obtained with Na_o.

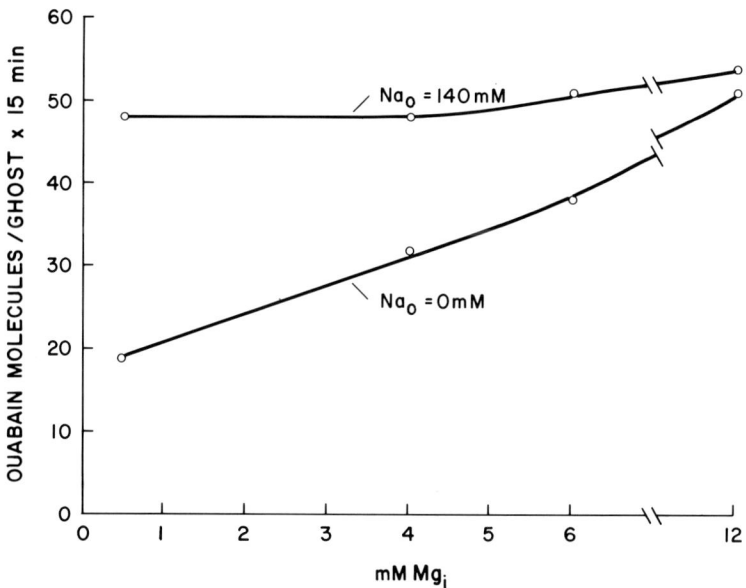

Figure 2. Effects of Na_o on ouabain binding rates at different concentrations of Mg_i. The resealed ghosts contained in addition to $MgCl_2$ (mM): 2 ATP, 44 NaCl, 105 CholineCl, 4 KCl, 7 Tris (pH 7.4). The external solution contained (mM): 20 KCl, 140 (cholineCl + NaCl), 10 Tris (pH 7.4). Ouabain binding was measured at 37°C for 15 min in the presence of 2×10^{-7} M [^{3}H]-ouabain.

Figure 3 shows the stimulation of ouabain binding by Ca_o (in the absence of Na_o) when the concentration of Mg_i was low. In this situation, the concentration of Ca_o which gave a half-maximal stimulation was less than 3 mM.

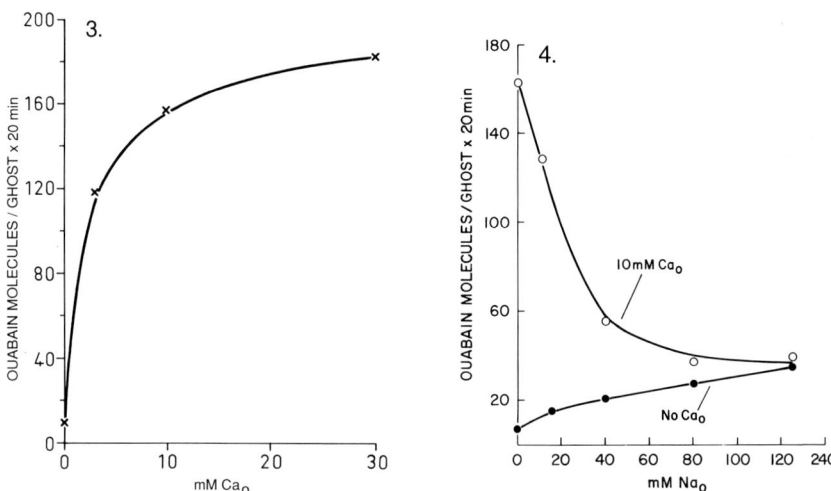

Figure 3. Stimulation of ouabain binding by Ca_o at low Mg_i. The resealed ghosts contained (mM): 1.5 $MgCl_2$, 5 EDTA (buffered with Tris), 2 ATP, 34 NaCl, 4 KCl, 7 Tris (pH 7.4), 115 cholineCl. The external solution contained in addition to $CaCl_2$ (mM): choline Cl (to maintain isotonicity), 1 KCl, 10 Tris (pH 7.4). Ouabain binding was measured at 37°C for 20 min in the presence of 2.5 x 10^{-7} M [^{3}H]-ouabain. Mg_i is estimated to be in the micromolar range.

Figure 4. Interaction between Ca_o and Na_o at low Mgi. Reconstituted ghosts contained (mM): 38 (KCl + NaCl), 5 EDTA, 1.5 $MgCl_2$, 2 ATP, 7 Tris.HCl (pH 7.4). The external solution contained (mM): 10 KCl, 10 Tris (pH 7.4) and either 0 $CaCl_2$ + 150 (NaCl + cholineCl) or 10 $CaCl_2$ + 125 (NaCl + cholineCl). Ouabain binding was measured at 37° C in the presence of 2.5 x 10^{-7} M [^{3}H]-ouabain for 20 min.

Since it is evident that either Na_o or Ca (at low Mg_i) can act to increase the rate of ouabain binding, it was important to test whether there was any interaction between Na_o and Ca_o in bringing about this effect. The results, presented in Figure 4, show that not only is there interaction but that Na_o markedly reduces the high ouabain binding rate obtained in the presence of Ca_o down to the rate obtained in the absnece of Ca_o.

DISCUSSION

It is clear that Mg_i modulates the actions that Na_o and Ca_o have on the rate of ouabain binding to red cell ghosts. This finding helps to resolve some apparently divergent effects of Na_o since it is now clear that Na_o can act either at high Mg_i (see Bodemann and Hoffman, 1976a) to alter ouabain binding rates by effecting affinity changes in K_o; or, alternatively, as indicated by the results presented in this paper, at low Mg_i, to affect the ouabain binding rate independent of K_o. Thus our results support the idea that Na_o acts on at least 2 external sites; one site is controlled by K_o (and is probably at K site), the other site is controlled transmembranally by Mg_i and is presumably not a K_o transport site.

The K_o-independent Na_o site is not known to be involved in any cation translocation step. It is possible that Ca_o acts on this site. Alternatively, Ca_o may also bind to a separate divalent cation site as had been proposed by Gardner and Frantz (1974), because, when Mg_i is low, Ca_o (in the absence of Na_o) induces ouabain binding rates higher than that obtainable with Na_o alone. If this is the case, the Ca_o site and the K_o-independent Na_o site must still interact since Na_o is able to abolish the action of Ca_o. (Parenthetically, this is the only condition known where Na_o inhibits ouabain binding.) It may be that the K_o-independent Na_o site is the same low affinity site that has been characterized by Kennedy et al. (1986) and Hobbs and Dunham (1978).

CONCLUSION

Evidence is presented that Mg_i, Na_o and Ca_o interact at sites that define ouabain binding rates. This interaction allows stimulation of ouabain binding by Na_o

and Ca_o at low Mg_i. The stimulation is gradually lost as Mg_i is increased which also results in an increase in the rate of ouabain binding. At low Mg_i, the marked stimulation of ouabain binding by Ca_o can be abolished by adding Na_o, either by a competitive-type binding of Na_o and Ca_o to the same site or by an interaction of separate Na_o and Ca_o binding sites. The results support other studies indicating the presence of a K-independent Na_o binding site (e.g. Kennedy et al., 1986; Sachs, 1987) that modulates pump conformational steps that affect its function.

Acknowledgments

This work was supported by NIH grants #HL 09906 and #AM 17433.

REFERENCES

Bodemann H H, Hoffman J F (1976a). Side-dependent effects of internal versus external Na and K on ouabain binding to reconstituted human red blood cell ghosts. J Gen Physiol 67:497-525.

Bodemann HH, Hoffman JF (1976b). Effects of Mg and Ca on the side dependencies of Na and K on ouabain binding to red blood cell ghosts and control of Na transport by internal Mg. J Gen Physiol 67:547-561.

Bodemann HH, Callahan TJ, Reichmann H, Hoffman JF (1983). Side-dependent ion effects on the rate of ouabain binding to reconstituted human red cell ghosts. Current Topics in Membranes and Transport 19:229-233. Academic Press, New York.

Gardner, JD, Frantz C (1974). Effects of cations on ouabain binding by intact human erythrocytes. J Membrane Biol 16:43-64.

Hobbs AS, Dunham PB (1978). Interaction of external alkali metal ions with the Na-K pump of human erythrocytes. J Gen Physiol 72:381-402.

Kennedy BG, Lunn G, Hoffman JF (1986). Effects of altering the ATP/ADP ratio on pump-mediated Na/K and Na/Na exchanges in resealed human red blood cell ghosts. J Gen Physiol 84:47-72.

Sachs JR (1987). Inhibition of the Na,K pump by vanadate in high-Na solutions: Modification of the reaction mechanism by external Na acting at a high-affinity site. J Gen Physiol 90:291-320.

TIME COURSE OF CHANGES IN Na,K-PUMP CONCENTRATION AND PASSIVE Na,K-FLUXES IN SKELETAL MUSCLE AFTER ADMINISTRATION OF THYROID HORMONE

Maria E. Everts, Torben Clausen and Keld Kjeldsen

Institute of Physiology
University of Aarhus
8000 Århus C, DENMARK

INTRODUCTION

The stimulatory effect of thyroid hormones on the synthesis of Na,K-ATPase has been demonstrated in a variety of tissues, but its mechanism of action is still unknown. Recent studies with rat liver slices showed that following administration of thyroid hormone (T_3) K^+-efflux was increased, before any increase in Na,K-ATPase activity could be detected (Haber and Loeb, 1984, 1986). Furthermore, it was shown in rat kidney tubules that the increase in K^+-permeability induced by valinomycin could mimick the effect of thyroid hormone in stimulating the synthesis of Na,K-pumps (Capasso et al., 1985). These studies suggested that an increase in the passive permeability for K^+ might play a role in stimulating the synthesis of Na,K-ATPase.

The present study was undertaken to explore whether similar phenomena occurred in skeletal muscle, which is one of the major target organs for thyroid hormones and contains the largest single pool of Na,K-ATPase in the body. Experiments were performed with isolated rat soleus and extensor digitorum longus (EDL) muscle which both respond to T_3 with a large (70-103%) increase in [^{3}H]ouabain-binding site concentration after 8 days of treatment (Kjeldsen et al., 1986a). The time course of the effect of T_3 treatment on the passive permeability for K^+ and Na^+ was assessed by measuring the fractional loss of ^{42}K or ^{86}Rb and the uptake of ^{22}Na in the presence of ouabain (Clausen and Kohn, 1977; Kjeldsen et al., 1986a). The concentration of Na,K-pumps was quantified in intact muscle samples by [^{3}H]ouabain binding (Nørgaard et al., 1983). Measurements of the maximum rate of active Na,K-transport have confirmed that [^{3}H]ouabain-binding site determinations quantify functional Na,K-pumps and that both parameters show parallel changes with thyroid status (Clausen et al., 1987).

RESULTS AND DISCUSSION

The time course of the effect of T_3 treatment on the concentration of [^{3}H]ouabain-binding sites and the fractional loss of ^{86}Rb as tested in the absence of ouabain or following 60 min of exposure to ouabain (1 mM) is shown in Fig. 1. Within the first 12-24 hours after the start of T_3 treatment, the increase in [^{3}H]ouabain-binding site concentration in rat soleus muscle amounted to only four per cent. In contrast, the fractional loss of ^{86}Rb measured in the contralateral muscle showed highly significant increases of 18 and 29% ($P < 0.001$) at respectively 12 and 24 hours after the first injection of T_3. When ouabain (1 mM) was added to the efflux media to prevent the return of isotope to the muscles, even larger effects were observed (31 and 42% respectively, $P < 0.001$). After 3 days of T_3 treatment, ^{86}Rb-efflux had increased by around 45% and it remained on this level for up to 8 days of daily injections of T_3. [^{3}H]ouabain-binding site concentration, however, continued to increase from a 55% rise after 3 days up to a 103% rise after 8 days.

Also when ^{42}K was used as a tracer, an early rise in K$^+$-efflux was observed, which amounted to 25% ($P < 0.001$) after 12 hours. Since the relative effects of T_3 on ^{42}K-efflux were similar to those found with ^{86}Rb over the whole time course tested, all further experiments were performed with ^{86}Rb.

In the EDL muscle the fractional loss of ^{86}Rb was around 14% larger than in the soleus muscle from the same rats. Twelve hours after a single injection of T_3, ^{86}Rb-efflux was increased by 20% ($P < 0.05$) which is the same relative rise as observed in soleus muscle. Within this short period of T_3 treatment, the [^{3}H]ouabain-binding site concentration in EDL muscle did not change whereas in an earlier study a 70% increase was observed following 8 days of T_3 treatment (Kjeldsen et al., 1986a).

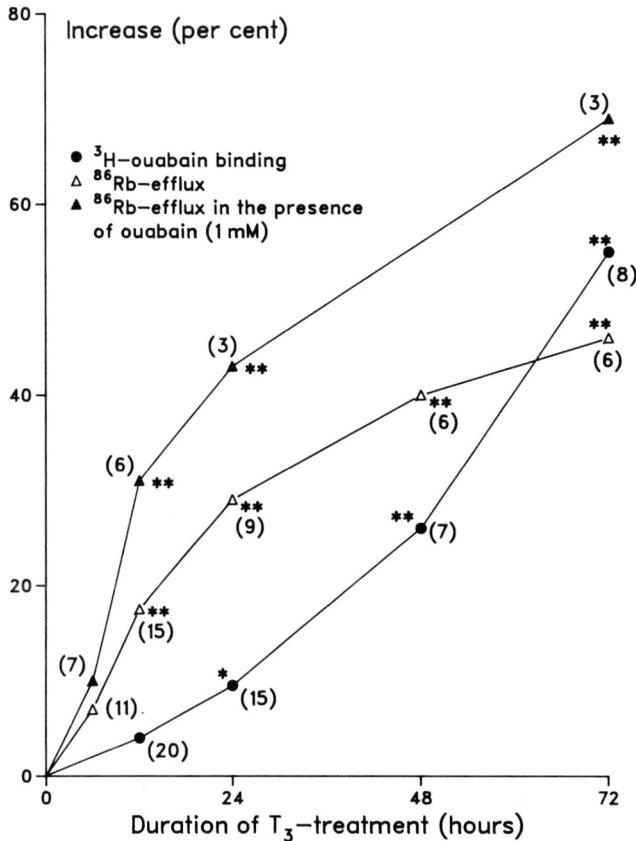

Fig. 1. Effect of T_3 treatment on ^{86}Rb-efflux in the absence and presence of 1 mM ouabain and [^{3}H]ouabain-binding site concentration in rat soleus muscle. 4 week old rats were given subcutaneous injections with T_3 (20 µg/100 g body wt.) or the solvent (1% bovine serum albumin) 6 or 12 hrs before sacrifice or at 24 hr intervals from 24 to 72 hrs before sacrifice. The [^{3}H]ouabain-binding site concentration was determined according to Nørgaard et al. (1983). ^{86}Rb-efflux was determined in the contralateral intact soleus muscle as the fraction of ^{86}Rb released per min after loading with the isotope. Values were then plotted relative to the control values. The number of animals is given in parentheses.
* $P < 0.02$ and ** $P < 0.001$ for T_3-injected compared to albumin-injected controls.

Table 1

Time course of the effect of T_3 treatment on ^{22}Na-uptake in rat soleus muscle in the absence and presence of amiloride

Experimental Conditions	^{22}Na-uptake (nmol/g wet wt./min)	
	No additions	+ Amiloride
Controls	776 ± 20 (29)	691 ± 12 (21)
12 hrs T_3	868 ± 29 (13)*	716 ± 11 (13)
24 hrs T_3	877 ± 15 (16)**	727 ± 21 (8)
2 days T_3	967 ± 32 (8)***	770 ± 19 (8)**
3 days T_3	1029 ± 42 (13)***	867 ± 41 (5)***
8 days T_3	1078 ± 44 (9)***	890 ± 16 (9)***

Values are given as means ± SE with the number of animals in parentheses. Rats were injected with T_3 (20 µg/100 g body wt.) once daily for periods lasting from 12 hours to 8 days. Euthyroid control rats received daily injections with the solvent (1% bovine serum albumin in 154 mM NaCl). Muscles were preincubated for 60 min in the presence of 1 mM ouabain without or with amiloride (0.3 mM). Thereafter they were incubated for 10 min in buffer containing ^{22}Na (1 µCi/ml) and ouabain (1 mM) without or with amiloride (0.3 mM). This was followed by four consecutive washings of 15 min at 0°C in the presence of ouabain to remove ^{22}Na from the extracellular space. ^{22}Na-activity was determined by liquid scintillation counting.
*$P < 0.02$, **$P < 0.005$ and ***$P < 0.001$ for T_3-treated compared with albumin-injected controls.

As shown in Table 1, T_3 treatment also increased the basal rate of ^{22}Na-uptake in rat soleus muscle. Similar to the time course of the effect of T_3 treatment on ^{86}Rb-efflux a significant increase in ^{22}Na-uptake was observed 12 hours after the first injection of T_3 ($P < 0.02$). ^{22}Na-uptake increased progressively with the duration of the hyperthyroid state, reached a maximum level of around 35% after 3 days and remained on this level at least up to after 8 days. To test whether the rise in ^{22}Na-uptake after T_3 treatment reflected a stimulation of the Na$^+$/H$^+$-exchange, ^{22}Na-uptake was also measured in the presence of amiloride (0.3 mM). Amiloride induced a significant decrease in ^{22}Na-uptake in all muscles tested. In contrast to the experiments without amiloride, two days of T_3 treatment were required to obtain a significant increase in ^{22}Na-uptake.

The effect of T_3 treatment for 8 days on ^{22}Na-uptake was also tested in EDL muscle. The resting ^{22}Na-uptake in euthyroid EDL muscle (722 ± 29 nmol/g wet wt./min (n=4)) was similar to that in soleus muscle (728 ± 32 nmol/g wet wt./min (n=4)). The stimulation of ^{22}Na-uptake by T_3 in EDL muscle (91%, $P < 0.001$) was even larger than that found in soleus muscle (57%, $P < 0.001$). To test whether Na$^+$-channels were involved in the response to T_3 treatment, parallel experiments were performed with tetrodotoxin (TTX) (10^{-7} M) present during preincubation and incubation. The presence of TTX did not significantly change the effect of T_3 treatment on ^{22}Na-uptake in either of the muscles suggesting that Na$^+$-channels are not involved in the increase in Na$^+$-permeability.

CONCLUSIONS

The increase in passive permeability for Na$^+$ and K$^+$ after thyroid hormone treatment was seen in both soleus and EDL muscle indicating that it probably occurs throughout the whole skeletal muscle pool. The increase in Na$^+$-influx and K$^+$-efflux showed a similar time course, indicating a common mechanism, perhaps in its later phase related to the increase in T-tubule area (Dulhunty et al., 1986). The stimulating effect of T_3 on ^{22}Na-influx was suppressed by amiloride, suggesting that a Ca^{2+}-activated Na$^+$/H$^+$-exchange is important in initiating the rise in Na$^+$-permeability. It is concluded that the early rise in the passive leaks of Na$^+$ and K$^+$ induced by thyroid hormone is a major driving force for Na,K-pump synthesis in skeletal muscle. This might constitute a general mechanism in the sense that stimulation of Na,K-pump synthesis represents an adaptation to increased Na,K-permeability. Conversely, it might be speculated that the down-regulation of Na,K-pump concentration seen in hypothyroidism and other conditions where plasma T_3 and T_4 are reduced e.g. semistarvation (Kjeldsen et al., 1986b) is preceded by a reduction of the passive permeability of Na$^+$ and K$^+$. This is currently being investigated.

REFERENCES

Capasso G, Lin JT, De Santo NG, Kinne R (1985). Short term effect of low doses of tri-iodothyronine on proximal tubular membrane Na-K-ATPase and potassium permeability in thyroidectomized rats. Pflügers Arch 403: 90-96.

Clausen T, Everts ME, Kjeldsen K (1987). Quantification of the maximum capacity for active sodium-potassium transport in rat skeletal muscle. J Physiol Lond 388: 163-181.

Clausen T, Kohn PG (1977). The effect of insulin on the transport of sodium and potassium in rat soleus muscle. J Physiol Lond 265: 19-42.

Dulhunty AF, Gage PW, Lamb GD (1986). Differential effects of thyroid hormone on T-tubules and terminal cisternae in rat muscles: an electrophysiological and morphometric analysis. J Muscle Res Cell Motil 7: 225-236.

Haber RS, Loeb JN (1984). Early enhancement of passive potassium efflux from rat liver by thyroid hormone: relation to induction of Na,K-ATPase. Endocrinol 115: 291-297.

Haber RS, Loeb JN (1986). Stimulation of potassium efflux in rat liver by a low dose of thyroid hormone: evidence for enhanced cation permeability in the absence of Na,K-ATPase induction. Endocrinol 118: 207-211.

Kjeldsen K, Everts ME, Clausen T (1986a). The effects of thyroid hormones on ^{3}H-ouabain binding site concentration, Na,K-contents and ^{86}Rb-efflux in rat skeletal muscle. Pflügers Arch 406: 529-535.

Kjeldsen K, Everts ME, Clausen T (1986b). Effects of semi-starvation and potassium deficiency on the concentration of [^{3}H]ouabain-binding sites and sodium and potassium contents in rat skeletal muscle. Br J Nutr 56: 519-532.

Nørgaard A, Kjeldsen K, Hansen O, Clausen T (1983). A simple and rapid method for the determination of the number of ^{3}H-ouabain binding sites in biopsies of skeletal muscle. Biochem Biophys Res Commun 111: 319-325.

ACKNOWLEDGMENTS

We thank Tove Lindahl Andersen, Marianne Stürup Johansen, Bente Mortensen and Susanne Olesen for skilled technical assistance and Lis Skjøt for expert secretarial assistance. M.E. Everts was supported by grants from The International Federation of University Women and The Danish Medical Research Council.

Na,K-ATPase ISOZYMES IN RAT TISSUES: DIFFERENTIAL SENSITIVITIES TO SODIUM, VANADATE AND DIHYDROOUABAIN

Gudrun Feige, Thomas Leutert and Alain De Pover

Research Department, Pharmaceutical Division, Ciba-Geigy Ltd, CH-4002 Basel, Switzerland

INTRODUCTION

The catalytic subunit of Na,K-ATPase has two main isoforms, usually denoted as α(+) and α, which were first demonstrated in rat brain (Sweadner, 1979). Their amino acid sequences have been deduced from complementary DNA (Shull et al, 1986). The most abundant isozyme is α(+) in brain and α in kidney. These two isoforms are present in cardiac myocytes and have been related to two distinct positive inotropic responses in rats (Adams et al, 1982; Finet et al, 1983; Charlemagne et al, 1987; Young and Lingrel, 1987), in mongrel dogs (Maixent et al, 1987) and in ferrets (Ng and Akera, 1987). The inotropic response associated with α(+) is of particular interest because it is sustained and is not accompanied with toxicity in vitro.

To date, there is not much evidence to distinguish the functional differences between these two isoforms. In rat adipocytes and skeletal muscle cells, insulin stimulates selectively α(+) transport activity (Lytton et al, 1985). In addition, insulin also stimulates Na^+ influx occurring through the Na/H exchanger (Rosic et al, 1985); and in turn, this could stimulate Na,K-ATPase since intracellular Na^+ concentration is the rate limiting step for Na^+ pump activity. These observations suggest that changes in Na^+ affinity may be an important regulatory mechanism in intact cells. Indeed, Lytton has reported that, in intact adipocytes, α(+) has a lower affinity for intracellular Na^+ than α, and that insulin appears to shift the α(+) $K_{0.5}$ towards a lower value (Lytton, 1985).

Purified α(+) is not stimulated by insulin and has an affinity for Na^+ similar to that of purified α (Lytton, 1985; Sweadner, 1985). In purified preparations, some regulatory factor may be missing. Therefore, in the present preliminary study, we have compared Na^+ activation curves of α(+) and α in partially purified preparations obtained from NaI and deoxycholate treated homogenates. The data suggest that, using this type of preparation, α(+) has a higher affinity for Na^+ than α.

METHODS

Na,K-ATPase was prepared essentially as described previously (Godfraind et al, 1977). Whole brains, kidneys and hearts were taken from 3-month old Wistar rats and homogenized in 5 vol. of chilled bidistilled water. The homogenate was treated with NaI solution according to Nakao (Nakao et al, 1965). The NaI mixture was centrifuged at $100000xg_{av}$ for 90 min; the pellet was suspended in 0.25 M sucrose, 10 mM Tris/HCl (pH=7.4) and centrifuged at $100000xg_{av}$ for 30 min. The resulting pellet was suspended in 10 vol. of 0.25 M sucrose, 1 mg/ml Na deoxycholate and 10 mM Tris/HCl (pH=7.4) and kept frozen overnight at -70^0. After thawing, the suspension was centrifuged successively at $20000xg_{av}$ for 30 min and at $100000xg_{av}$ for 2 h. The last pellet was suspended in 0.25 M sucrose, 10 mM Tris/HCl (pH=7.4) at 5-10 mg protein/ml. Protein content was estimated by the method of Lowry using microtiter plates according to Fryer et al (1986).

Na,K-ATPase activity was determined from released inorganic phosphate (P_i). 0.5-5 μg of microsomal protein was incubated in microtiter plates (final vol. =100 μl). The incubation medium contained, unless otherwise stated, 5 mM Tris-ATP, 5 mM $MgCl_2$, 150 mM NaCl, 20 mM KCl, 1 mM EGTA, 10 mM maleic acid, pH=7.4 adjusted with Tris. After 60 min incubation at 37^0, the reaction was stopped by addition of 100 μl of chilled solution containing 2.5% ammonium heptamolybdate / 1.7N H_2SO_4 / Fiske & SubbaRow reagent / 50% trichloroacetic acid (3/3/3/1). This mixture was prepared immediately before use. The Fiske & SubbaRow reagent contained 15% $NaHSO_3$, 0.5% Na_2SO_3 and 0.25% 1-amino-2-naphtol-4-sulfonic acid, was kept at 4^0 and was filtrated before use. Optical density was measured at 700 nm by a

Dynatech microplate reader MR600 and was linear with P_i amounts up to 0.1 μmol (O.D.=1.2). Na,K-ATPase activity was linear with time and protein provided that ATP hydrolysis did not exceed 20 %. The ATPase activities were in the range of 80-120 μmol P_i per mg protein per h for brain and kidney and 20-30 for heart. Over 90 % was ouabain sensitive.

All experimental curves were fitted to the logistic function (eq.1)

$$Y = \frac{a - d}{1 + (X/c)^b} + d$$

where a represents the response for zero concentration, b the slope factor (analogous to Hill coefficient), c the concentration for half-maximal response, and d the response for 'infinite' concentration. The method of De Lean et al (1978) adapted on a VAX8600 computer was used. Wherever relevant, curves were also fitted to the function for two independent binding sites (eq.2)

$$Y = \frac{E_{max}^1 * X}{IC_{50}^1 + X} + \frac{E_{max}^2 * X}{IC_{50}^2 + X}$$

RESULTS

The relative abundance and the catalytic turnover of α(+) and α in a Na,K-ATPase preparation from a given tissue may vary, sometimes considerably, with factors such as its age (Matsuda et al, 1984), its physiological state (Lytton et al, 1985; Charlemagne et al, 1986) and the method used for enzyme isolation. To evaluate accurately the two isoforms in the present study, we used a simple method exploiting the considerable difference in their affinities for ouabain in rat (Sweadner, 1979). Fig 1 shows inhibition curves of Na,K-ATPase by dihydroouabain. Dihydroouabain was choosen instead of ouabain because it equilibrates very rapidly with the enzyme. A low-affinity monophasic curve, which corresponds to the inhibition of α, was observed in rat kidney. On the other hand, the classical biphasic curve, which corresponds to the inhibition of both α(+) and α, was found in rat brain. The biphasic curve could be resolved by eq.2 (see METHODS), which suggests that the contribution of α(+) to Na,K-ATPase activity in this preparation of rat brain was 76%. In rat heart, an essentially monophasic curve was obtained similar to that for kidney, suggesting that α contributed to at least 90% of the enzyme activity. [^{3}H]Ouabain binding experiments confirmed these observations (data not shown).

The activation of Na,K-ATPase by Na$^+$ and K$^+$ was studied at constant ionic strength by maintaining the total Na$^+$+K$^+$ = 150 mM (Skou, 1975). Fig.2 shows that the rat brain Na,K-ATPase was activated by lower concentrations of Na$^+$ than the kidney and heart enzymes. The slope factors of Na$^+$ for the former were slightly lower than for the latter, suggesting possible differences in cooperative kinetics. Similar results were obtained when the Na$^+$ activation experiments were performed using K$^+$ = 20 mM without attempting to maintain constant ionic strength: the Na$^+$ $K_{0.5}$ were 5.4 mM for brain, 10.0 mM for heart and 12.1 mM for kidney and the respective slope factors were 1.65, 1.84 and

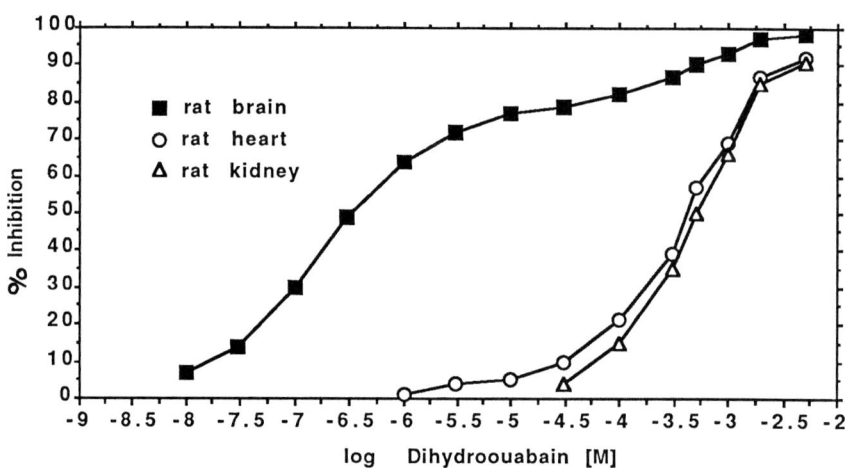

Figure 1. Na,K-ATPase inhibition by dihydroouabain. The enzyme activity was tested in the conditions given under METHODS except that the KCl concentration was 1 mM.
Parameters were estimated by using eq.1 for kidney and heart and eq.2 for brain. The low-affinity IC_{50} were 620 µM (kidney), 430 µM (heart) and 310 µM (brain). The high affinity IC_{50} was 0.16 µM for the brain. The slope factors estimated by using eq.1 were 1.06 (kidney) and 0.93 (heart).

1.86. Differences were also observed in the K⁺ activation kinetics, the brain appearing less sensitive to K⁺ than the kidney and the heart.

It has been reported that vanadate is a more potent inhibitor of Na,K-ATPase in kidney and in heart than in brain (Nechay and Saunders, 1978). This observation was confirmed in the present study (see Fig.3). Interestingly, the slope factor of vanadate inhibition was close to 1.0 in kidney and in heart, but was less than 1.0 in brain. This is consistent with the hypothesis that our rat brain preparations contain two isozymes with different affinities for vanadate. Similar data were also obtained in other

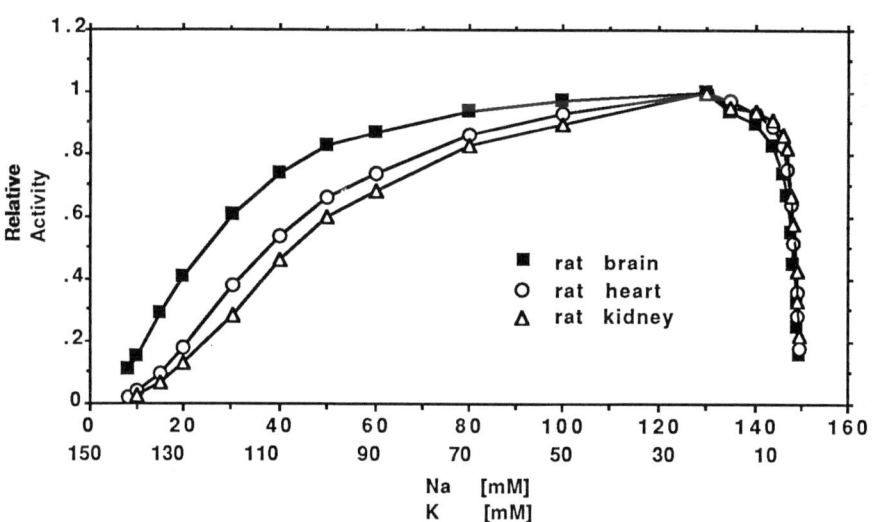

Figure 2. Na,K-ATPase activation by Na⁺ and K⁺. The enzyme activity was tested in the conditions given under METHODS except that NaCl and KCl were varied while the sum NaCl + KCl = 150 mM was kept constant. Parameters were estimated by using eq.1. The Na⁺ $K_{0.5}$ were 25 mM (brain), 40 mM (heart) and 46 mM (kidney) and the respective slope factors were 1.91, 2.22 and 2.30. The K⁺ $K_{0.5}$ were 1.83 mM (brain), 1.42 mM (heart) and 1.18 mM (kidney) and the respective slope factors were 1.25, 1.49 and 1.52.

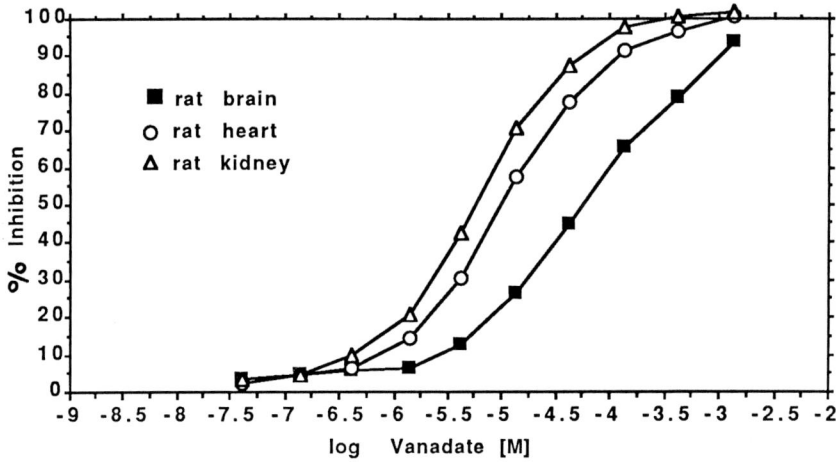

Figure 3. Na,K–ATPase inhibition by vanadate. The enzyme activity was tested in the conditions given under METHODS except that the KCl concentration was 5 mM. Parameters were estimated by using eq.1. The IC_{50} were 45 μM (brain) 7.6 μM (heart) and 4.3 μM (kidney) and the respective slope factors were 0.70, 0.88 and 0.93.

---

conditions more favourable for vanadate, e.g., at $K^+ = 20$ mM the vanadate IC_{50} were 13 μM for brain, 3.1 μM for heart and 2.2 μM for kidney and their respective slope factors were 0.74, 0.99 and 1.05.

DISCUSSION

The present data show that, using partially purified preparations ('mildly' treated membranes), Na,K–ATPase has a higher apparent affinity for Na^+ in the brain than in kidney and in heart. These differences can be accounted for by the different isozyme content of these preparations (mainly α(+) in the brain versus α in kidney and in heart) as suggested by the dihydroouabain inhibition curves (see also Sweadner, 1979). Similar data were obtained with Na,K–ATPase preparations from pig, sheep, guinea pig and rabbit tissues using the same method for enzyme isolation (data not shown).

The differences in Na^+ affinity shown in the present study are in contrast with a previous report where **pure** preparations were used (Sweadner, 1985). It is possible that Na$^+$ affinity depends on the membrane environment which may be affected by the purification procedure. Indeed, we have obtained here data (not shown) supporting this view by comparing **pure** rabbit outer medulla Na,K-ATPase (kindly provided by B. Anner, Geneva) with our partially purified preparation and with a preparation of alamethicin-treated microsomes: the Na$^+$ K$_{0.5}$ was significantly lower for the pure preparation (25 mM) than for the latter two (36 mM). In contrast, the effects of vanadate and dihydroouabain were not significantly different, suggesting that only the Na$^+$ sites were modified by the purification.

The higher apparent affinity for Na$^+$ of $\alpha(+)$ versus α in partially purified preparations may be of important significance for the regulation of Na$^+$ transport.

REFERENCES

Adams RJ, Schwartz A, Grupp G, Grupp IL, Lee SW, Wallick ET, Powell T, Twist VW, Gathiram P (1982). High affinity ouabain binding site and low dose positive inotropic effect in rat myocardium. Nature (London) 296:167-169.

Charlemagne D, Maixent JM, Preteseille M, Lelievre L (1986). Ouabain binding sites and Na,K-ATPase activity in rat cardiac hypertrophy. Expression of the neonatal forms. J Biol Chem 261:185-189.

Charlemagne D, Mayoux E, Poyard M, Oliviero P, Geering K (1987). Identification of two isoforms of the catalytic subunit of Na,K-ATPase in myocytes from adult rat heart. J Biol Chem 262:8941-8943.

De Lean A, Munson PJ, Rodbard (1978). Simultaneous analysis of families of sigmoidal curves: application to bioassay, radioligand assay, and physiological dose-response curves. Am J Physiol 235:E97-E102.

Finet M, Godfraind T (1983). The inotropic effect of ouabain and its antagonism by dihydroouabain in rat isolated atria and ventricles in relation to specific binding sites. Br J Pharmacol 80:751-759.

Fryer HJL, Davis GE, Manthorpe M, Varon S (1986). Lowry protein assay using an automatic microtiter plate spectrophotometer. Anal Biochem 153:262-266.

Godfraind T, De Pover A, Verbeke N (1977). Influence of pH

and sodium on the inhibition of guinea-pig heart Na,K-ATPase by calcium. Bioch Biophys Acta 481:202-211.
Lytton J (1985). Insulin affects the sodium affinity of the rat adipocyte Na,K-ATPase. J Biol Chem 260:10075-10080.
Lytton J, Lin JC, Guidotti G (1985). Identification of two molecular forms of Na,K-ATPase in rat adipocytes. Relation to insulin stimulation of the enzyme. J Biol Chem 260:1177-1184.
Maixent JM, Charlemagne D, de la Chapelle B, Lelievre LG (1987). Two Na,K-ATPase isoenzymes in canine cardiac myocytes. Molecular basis of inotropic and toxic effects of digitalis. J Biol Chem 262:6842-6848.
Nakao T, Tashima Y, Nagano K, Nakao M (1965). Highly specific Na,K-ATPase from various tissues of rabbit. Biochem Biophys Res Comm 19:755-758.
Matsuda T, Iwata H, Cooper JR (1984). Specific inactivation of α(+) molecular form of Na,K-ATPase by pyrithiamin. J Biol Chem 259:3858-3863.
Nechay BR, Saunders JP (1978). Inhibition by vanadium of Na,K-ATPase derived from animal and human tissues. J Environm Pathol Toxicol 2:247-262.
Ng YC, Akera T (1987). Two classes of ouabain binding in ferret heart and two forms of Na,K-ATPase. Am J Physiol 252:H1016-H1022.
Rosic NK, Staendart ML, Pollet RJ (1985). The mechanism of insulin stimulation of Na,K-ATPase transport activity in muscle. J Biol Chem 260:6206-6212.
Shull GE, Greeb J, Lingrel JB (1986). Molecular cloning of three distinct forms of the Na,K-ATPase α-subunit from rat brain. Biochemistry 25:8125-8132.
Skou JC (1975). The (Na+K)-activated enzyme system and its relationship to transport of sodium and potassium. Quart Rev Biophys 7,3:401-434.
Sweadner KJ (1979). Two molecular forms of Na,K-ATPase in brain. Separation, and difference in affinity for strophanthidin. J Biol Chem 254:6060-6067.
Sweadner KJ (1985). Enzymatic properties of separated isozymes of the Na,K-ATPase. Substrate affinities, kinetic cooperativity, and ion transport stoichiometry. J Biol Chem 260:11508-11513.
Young RM, Lingrel JB (1987). Tissue distribution of mRNAs encoding the α isoforms and β subunit of rat Na,K-ATPase. Biochem Biophys Res Comm 145:52-58.

EFFECTS OF GOSSYPOL ON THE ACTIVITY OF RABBIT KIDNEY Na,K-ATPase AND THE FUNCTIONS OF HUMAN ERYTHROCYTE MEMBRANE

Fu Yun-Feng, Zhang Shi-Lian, Lu Zhen-Min and Wang Wei.
Department of Biochemistry, Institute of Experimental Medicine, Hebei Academy of Medical Sciences, Shijiazhuang, Hebei Province, China.

INTRODUCTION

Gossypol, a polyphenolic binaphtalene-dialdehyde extracted from cotton seed oil, is a potent male antifertility agent, but it may induce a side-effect of hypokalemia, following an uncertain period of administration on rare occasions in volunteers. It was presumed that active and/or passive K^+ transport through the cell membrane must be impaired by gossypol. In this study, we have examined the inhibitory effect of gossypol on the activity of Na,K-ATPase partially purified from the outer medulla of rabbit kidney, and on the functions of erythrocyte membrane.

MATERIALS AND METHODS

Preparation of the Enzyme

Na,K-ATPase was prepared from rabbit kidney outer medulla using a discontinuous density gradient centrifugation of sodium dodecylsulfate (SDS) treated microsomal fraction as described by Jørgensen (Jørgensen, 1974). Removal of contaminating ATP, subsequent washing and storage of the resulting preparation follows the method of Schoot (Schoot et al., 1977), except for the composition of the storage medium, which contains 50 mM imidazole-HCl (pH 7.0), 0.25 M sucrose and no CDTA. The specific activity of the enzyme used in these experiments is 600-1000 $\mu mol \cdot mg^{-1}$ protein$\cdot h^{-1}$.

Preparation of Blood Samples

Blood was drawn from hemotologically healthy humans and heparinized. Red cells were washed by repetitive centrifugation and resuspension in a standard medium without K^+, i.e. 10 mM Tris-HCl (pH 7.4), 150 mM NaCl, 2 mM $MgCl_2$, 0.1 mM EDTA, 5 mM glucose, at room temperature. The buffy coat was removed carefully to minimize the loss of the lightest red cells.

Preparation of Erythrocyte Membranes

A part of the red cells isolated as described above was transferred into a certain volume of hypotonic medium to prepare the red cell ghost according to the method of Swarts (Swarts et al., 1981).

ATPase Assay

ATPase activities were assayed in terms of the liberation of inorganic phosphate. Mg,Na,K-ATPase actvity was determined in a total volume of 0.4 ml containing (mM): 5 $MgCl_2$, 5 Na_2ATP, 100 NaCl, 10 KCl and 50 imidazole-HCl (pH 7.4) at $37^\circ C$; for the Mg-ATPase activity KCl was omitted and 1 mM ouabain was added. The reaction was started by the addition of 2 μg (purified enzyme) or 20 μg (red cell ghost) protein to the prewarmed medium. After 15 min incubation the reaction was stopped by addition of 1.5 ml of ice-cold 10% (w/v) trichloroacetic acid, then 1.5 ml of a freshly prepared solution of 9.6% (w/v) $FeSO_4 \cdot 6H_2O$, 1.15% (w/v) ammonium heptamolybdate in 0.66 M H_2SO_4 was added, and after 30 min at room temperature the absorbance at 700 nm was measured. The Na,K-ATPase activity was calculated as ouabain inhibitable activity by substracting Mg-ATPase activity from Mg,Na,K-ATPase activity.

Protein determination was performed by the modified Lowry's method (Jørgensen, 1974) against bovine serum albumin as standard.

Intracellular and extracellular K^+ concentrations were determined by atomic absorption spectrophotometry.

Gossypol was dissolved in dimethylsulfoxide and adjusted to pH 7.4 with imidazole.

All kinetic data are plotted according to Lineweaver-Burk or to Dixon.

RESULTS AND DISCUSSION

Inhibition by Gossypol of Na,K-ATPase Activity

Figure 1 shows the inhibitory effect of gossypol on the Na,K-ATPase activity. The concentration required for 50% inhibition was 6.5 µM.

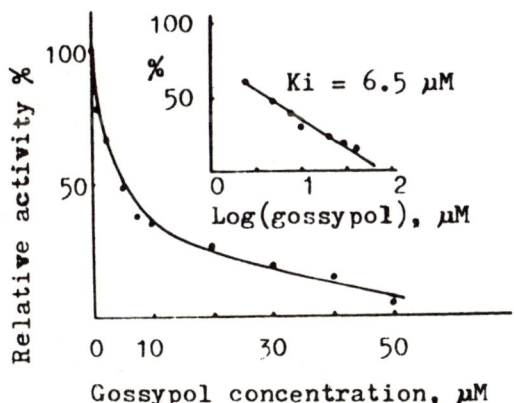

Fig. 1. Effect of gossypol on Na,K-ATPase. The assay method of the enzyme is described in the text. Gossypol was dissolved in 10 µl of dimethylsulfoxide. The control tube had the same volume of dimethylsulfoxide. In the inserted figure the concentration of gossypol is plotted logarithmically.

Kinetic Analysis of Gossypol Inhibition

The kinetic analysis of the inhibition of gossypol on Na,K-ATPase was performed. The effects of gossypol on the ATP, Na^+, K^+ and Mg^{2+} sites of the enzyme were examined. The double-reciprocal plots of the velocities vs ATP, Na^+, K^+ and Mg^{2+} concentrations were all linear in the presence or absence of gossypol, indicating non-competitive inhibition (Fig. 2).

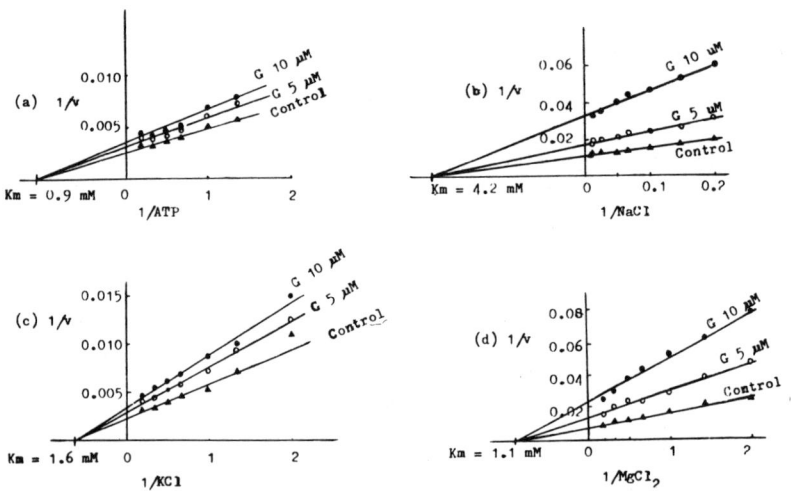

Fig. 2, a,b,c,d. Lineweaver-Burk plots of Na,K-ATPase activities as a function of ATP, Na^+, K^+ and Mg^{2+} in the presence or absence of gossypol. Gossypol was dissolved in 10 µl of dimethylsulfoxide. The concentrations of gossypol were (▲) 0, (o) 5 and (●) 10 µM.

These findings suggested that gossypol has no effects on the binding affinity, but reduces the binding capacity. The Ki values for gossypol inhibition of activation by ATP, Na^+, K^+ and Mg^{2+} were 15, 14, 4.4 and 13 µM respectively (Fig. 3a,b,c,d).

ATPase Activities of Red Cells

In table 1 the Na,K-ATPase activity of red cell ghosts was decreased by gossypol in a concentration-dependent fashion with an apparent Ki of 6.0 µM, and Mg-ATPase activity was not changed at all. It means that gossypol can inhibit Na,K-ATPase, but without affecting Mg-ATPase.

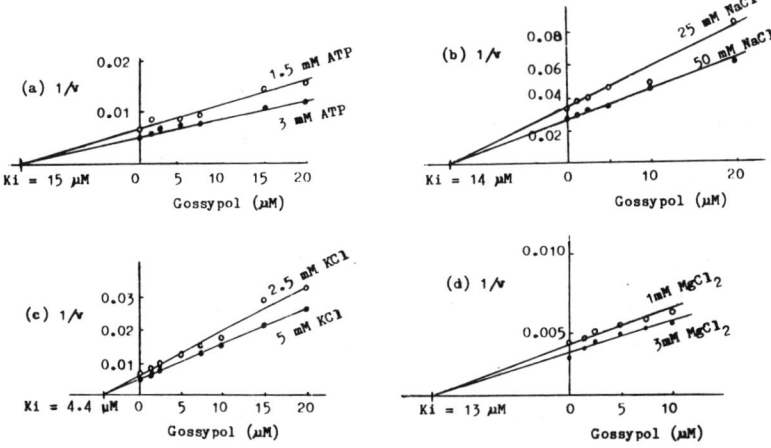

Fig. 3, a,b,c,d. Dixon plots of Na,K-ATPase activities in the presence of different concentrations of ATP, Na^+, K^+ and Mg^{2+}, which are inhibited at various concentrations of gossypol dissolved in the same volume (10 μl) of dimethylsulfoxide.

TABLE 1. ATPase Activities of Erythrocyte Membranes

Activity, μmol Pi·mg^{-1} protein·h^{-1}, M±SD

ATPase (N)	Gossypol, μM		
	0	5	10
Mg,Na,K-ATPase (11)	22.2 ± 2.5	20.9 ± 2.8	20.3 ± 2.7
Na,K-ATPase (11)	2.6 ± 0.7	1.3 ± 0.5*	0.7 ± 0.5
Mg-ATPase (11)	19.6 ± 1.8	19.6 ± 2.3	19.6 ± 2.2

* $P<0.05$

Stability of Intact Red Cells

In order to clarify whether gossypol may affect the stability of the intact red cell membrane, a certain amount of red cells was incubated with different concentrations of gossypol at room temperature for one hour. The results show that the intact red cells are lysed by gossypol in a dose (50-800 µM) dependent manner (Fig. 4).

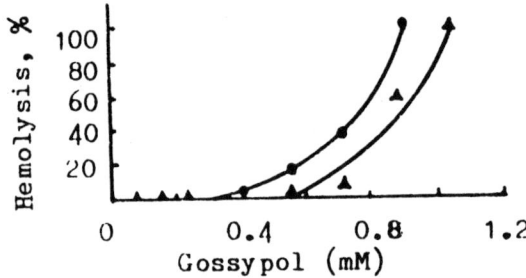

Fig. 4. Hemolysis by gossypol (●) and its antagonism by bovine serum albumin (1%, ▲).

An interesting pehenomenon is that the hemolysis by gossypol was antagonized by bovine serum albumin (BSA), as it can be seen that 1% BSA can antagonize the hemolysis by 80%, whereas 2% BSA can get to a 100% antagonism (Fig. 5). This might be the reason why hemolysis does not occur following administration of gossypol in vivo to humans or animals. We also found that gossypol increased the permeability of the red cell membrane after 30 min incubation at room temperature with a certain amount of red cells. It appears gossypol (10-40 µM) increased the K^+ efflux in a concentration-dependent manner, an effect that was blocked by 1% BSA completely (Fig. 6).

These combined data (i.e. inhibition of Na,K-ATPase and the concomittant leakage of K^+ out of the cell) may explain the mechanism of gossypol-induced hypokalemia. The difference in sensitivity of the cell membrane to gossypol in vivo and in vitro may explain the remarkable inconsistency of gossypol inhibition of the membrane functions between different individuals or species.

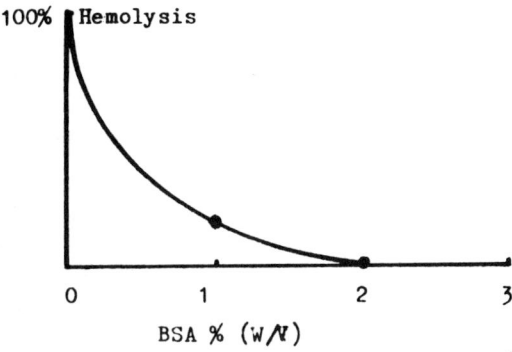

Fig. 5. Bovine serum albumin prevents gossypol (50 µM) from lysis of red cells.

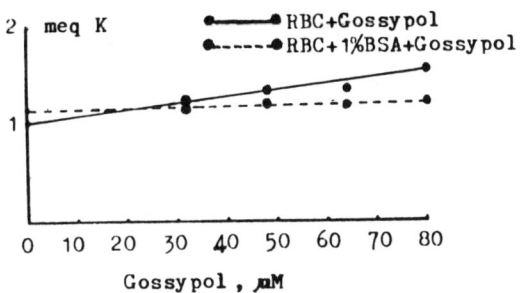

Fig. 6. Leakage of intracellular K^+ induced by gossypol and its antagonism by bovine serum albumin (1%, ●---●).

CONCLUSION

From the above results we may draw the conclusion that gossypol as a toxic material can injure and penetrate the cell membrane and cause both active and passive transport to become abnormal: Na,K-ATPase activity is inhibited and K^+ efflux is increased. This might be the main mechanism of developing hypokalemia by gossypol.

REFERENCES

Jørgensen PL (1974). Purification and characterization of (Na^++K^+)-ATPase. III. Purification from the outer medulla of mammalian kidney after selective removal of membrane components by sodium dodecylsulphate. Biochim Biophys Acta 356: 36-52.

Schoot BM, Schoots AFM, De Pont JJHHM, Schuurmans Stekhoven FMAH, Bonting SL (1977). Studies on (Na^++K^+)-activated ATPase. XLI. Effect of N-ethylmaleimide on overall and partial reactions. Biochim Biophys Acta 483: 181-192.

Swarts HGP, Bonding SL, De Pont JJHHMN, Schuurmans Stekhoven FMAH, Thien TA, Van't Laar A (1981). Cation fluxes and Na^+-K^+-activated ATPase activity in erythrocytes of patients with essential hypertension. Hypertension 3: 641-649.

INTERACTION BETWEEN PALYTOXIN AND PURIFIED
NA, K-ATPASE

E. Grell, E. Lewitzki and D. Uemura

Max-Planck-Institute for Biophysics,
Frankfurt, F.R.G., (E.G., E.L.) and
Faculty of Liberal Arts, Shizuoka University, Ohya, Shizuoka, Japan (D.U.)

INTRODUCTION

Palytoxin represents a group of extremely toxic substances which can be isolated from several Palythoa species. These toxins exhibit fairly complex structures as shown in Fig. 1 for the simplest analogue (Uemura et al., 1985). Attempts to study the molecular interactions reveal that it interacts with membranes, eg. by increasing alkali ion permeabilities (Habermann and Chhatwal, 1982) in a way which seems to be linked to membrane proteins such as Na, K-ATPase (Böttinger and Habermann, 1984; Ishida et al., 1983). Even similarities between palytoxin and ouabain concerning its inhibitory action on Na, K-ATPase have been reported (Ishida et al., 1983). In addition, as a fairly unexpected property of the toxin a promoting effect due to borate is indicated (Böttinger and Habermann, 1984). Although it seems very likely that among other effects palytoxin binds to membrane-bound Na, K-ATPase, no interactions have so far been studied on the purified enzyme.

MATERIALS AND METHODS

Membrane-bound Na, K-ATPase is isolated from red outer medulla of pig kidney acc. to Jørgensen, 1974 . The specific activity of the preparation at 37°C ranges between 30 and 40 µmole P_i

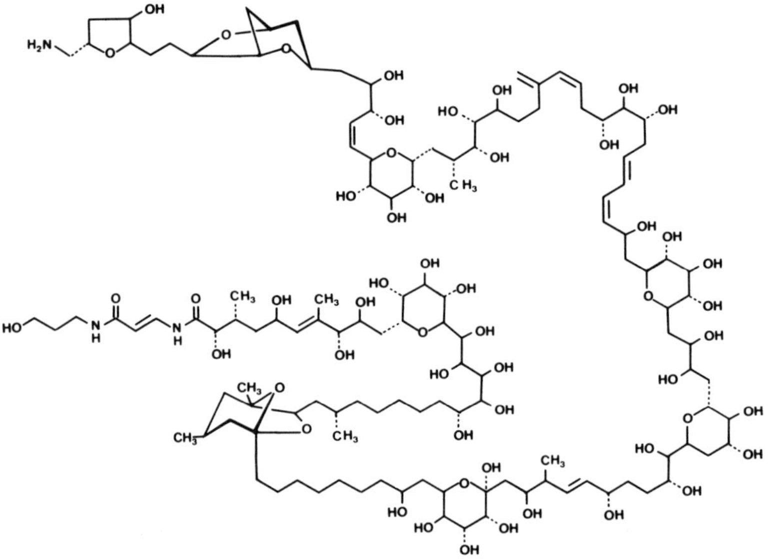

Figure 1. Chemical structure of palytoxin acc. to Uemura et al., 1985.

$mg^{-1}min^{-1}$. Fluorescein-Na, K-ATPase (FITC-Na, K-ATPase) is prepared similar to Hegyvary and Jørgensen, 1981. Palytoxin is isolated and purified similar to Hirata et al., 1979. Spectrofluorometric studies have been performed on a Spex Fluorolog 2 instrument (Model 212) equipped with a fast mixing device.

RESULTS

Appreciable inhibition of the activity of purified Na, K-ATPase and its FITC-derivative is only observed in the presence of borate which is not consistent with an earlier study (Ishida et al., 1983). The dependence of the 2,4 dinitrophenylphosphatase (DNPP) activity on palytoxin concentration at constant borate concentration is shown in Fig. 2A and allows the determination of the inhibition constant K_I^P (defined as a stability constant). The same result is obtained for the

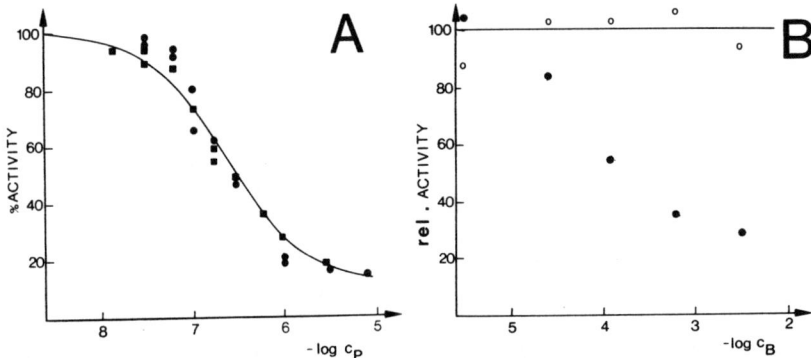

Figure 2. Inhibition of 2,4-dinitrophenylphosphatase activity of Na, K-ATPase (●) and FITC-Na, K-ATPase (■) in 40 mM imidazole-HCl, 5 mM $MgCl_2$, 2 mM KCl pH 7,5 at 37°C in dependence of palytoxin concentration (c_P) at constant 3 mM borate concentration (A); and in dependence of borate concentration (c_B) in the presence of 5×10^{-7} µM palytoxin as well as in its absence (o) (B). The theoretical curve in (A) corresponds to a log K_I^P value of 6.62 for modified (■) and unmodified enzyme (●).

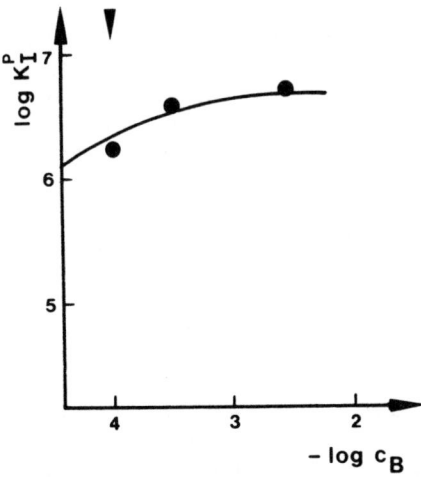

Figure 3. Inhibition of Na, K-ATPase in the presence of borate: Palytoxin inhibition constant (K_I^P) at different borate concentrations.

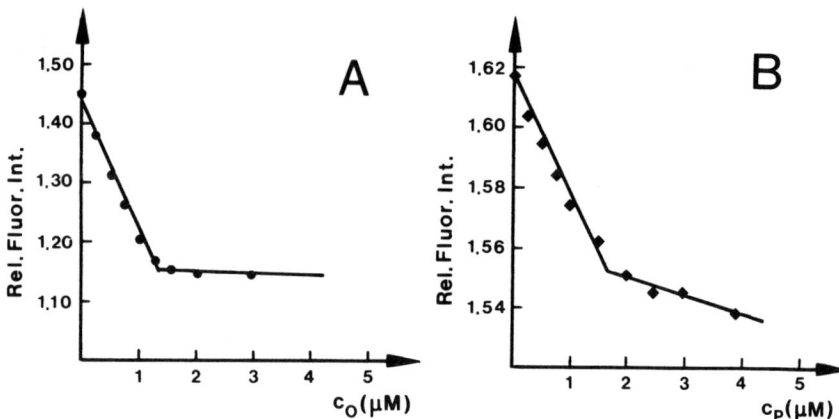

Figure 4. Saturation titration of 1.7. µM FITC-Na, K-ATPase in 25 mM imidazole-HCl pH 7.5 at 37°C containing 5 mM $MgCl_2$, 1 mM KCl and 1 mM $TrisP_i$ with ouabain (A); and in the same buffer at 37°C containing 5 mM $MgCl_2$, 1 mM KCl and 3 mM borate (B): Plot of relative fluorescence intensity at 512 nm (exc. at 489 nm) at different inhibitor concentrations. Saturation is observed for 1.4 µM ouabain (A) and 1.6 µM palytoxin.

modified and unmodified enzyme suggesting that the bound FITC-group does not essentially affect the palytoxin site. In Fig. 2B it is shown how the enzymatic activity at constant palytoxin concentration depends on the borate concentration. In the absence of palytoxin borate exhibits no observable effect on DNPP-activity (Fig. 2B). Fig. 3 conveys that the K_I^P value increases with increasing borate concentration. The stoichiometry of the enzyme-palytoxin complex in the presence of borate is obtained from a spectrofluorometric saturation titration of the FITC enzyme with palytoxin (Fig. 4).

Whereas in case of ouabain a coefficient of 0.82 is found which is larger than reported earlier (Grell et al., 1985), a value of 0.94 results for palytoxin (estimated mol. weight of enzyme: 150'000). This is in agreement with a 1:1 complex

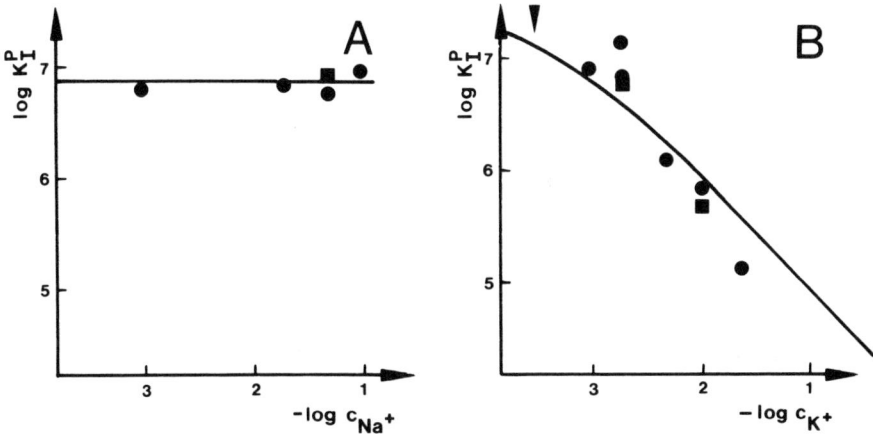

Figure 5. Effect of alkali ions on the interaction between palytoxin and Na, K-ATPase (●) as well as its FITC-derivative (■). Dependence at 37°C of palytoxin inhibition constant K_I^P in the medium 40 mM imidazole-HCl, 5 mM $MgCl_2$, 3 mM borate pH 7.5 containing 2 mM KCl as a function of Na^+ concentration (A); and in the same medium at 37°C as a function of K^+ concentration (B). The arrow in (B) indicates a log K for K^+ binding (1:1 complex formation) of 3.5.

formation reaction. Acc. to the results shown in Fig. 1-4 the following plausible reaction scheme is assumed,

$$E + PB \underset{}{\overset{K^P}{\rightleftharpoons}} EPB$$

pre-equilibrium $\quad K^B \updownarrow \quad$ P: palytoxin
$\quad\quad\quad\quad\quad\quad P + B \quad\quad$ B: borate

where prior to the interaction with the enzyme (E) a palytoxin-borate (PB) complex is formed. For formal reasons an alternative possibility consisting of an enzyme borate pre-complex formation prior to palytoxin binding cannot be excluded. The theoretical curve in Fig. 3 corresponds to a log K^B value of 4.0 and a log K^P value of 6.85.

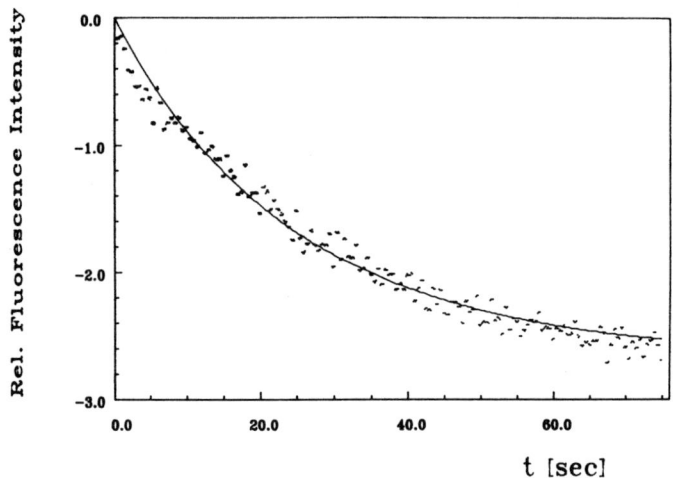

Figure 6. Kinetics of palytoxin binding to FITC-Na, K-ATPase in 25 mM imidazole-HCl containing 0.1 mM borate pH 7.5 at 37°C: Dependence of rel. fluorescence intensity at 512 nm (exc. at 489 nm) as a function of time in a rapid mixing experiment. The theoretical curve corresponds to a k_{on} value of 1.8×10^4 $M^{-1} sec^{-1}$ and a k_{off} value of $8.3 \times 10^{-3} sec^{-1}$ (log $K_I^P = 6.3$).

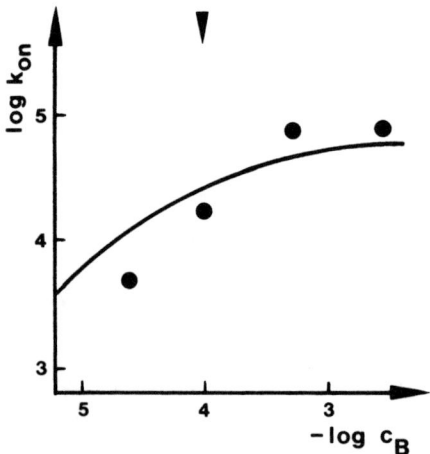

Figure 7. Plot of formation rate constant (k_{on}) of the complex in 25 mM imidazole-HCl pH 7.5 containing 1.8 µM palytoxin at 37°C at different borate concentrations. The arrow indicates a log K^B value for borate binding of 4.0.

Since binding of cardiac glycosides is diminished in the presence of K^+, the effect of Na^+ and K^+ on palytoxin binding at constant borate concentration is investigated. Fig. 5 indicates that Na^+ does not affect toxin binding but that the K_I^P value decreases with increasing K^+ concentration which would be consistent with the following competition:

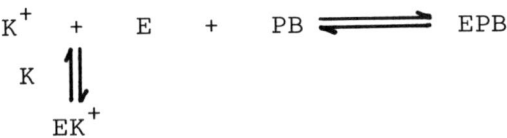

Again, similar results are obtained for the unmodified and modified enzyme (**Fig. 5**). The similarities seen between ouabain and palytoxin inhibition are further confirmed by the observation that bound palytoxin prevents the binding of the fluorescent ouabain derivative DEDO (Grell et al., 1985).

In order to get information concerning the mechanism of palytoxin binding kinetic experiments are carried out on the FITC-enzyme by employing fluorescence detection. A single kinetic reaction phase is observed (Fig. 6) which is attributed to the slow rate determining PB binding step. The formation of the intermediate state PB is assumed to be fast. The evaluation is done as given by Grell et al., 1985. Similar to the dependence of K_I^P in Fig. 3 the formation rate constant k_{on} also depends on borate concentration (Fig. 7), not however the dissociation rate constant k_{off} which is $10^{-2} sec^{-1}$. In the presence of high borate concentration a k_{on} value of 10^5 $M^{-1} sec^{-1}$ results. Since this value is similar to the one found for cardiac glycosides (Grell et al., 1985), it is assumed that a similar, slow conformational rearrangement of the enzyme may act as the rate-limiting process of toxin binding.

ACKNOWLEDGEMENT

We wish to thank Prof. E. Habermann and

Dr. H. Ruf for interesting discussions, Prof. Y. Hirata for kind support and Mrs. A. Ifftner for expert assistance. This work was supported by the German Research Council (SFB 169).

REFERENCES

Böttinger H, Habermann E (1984). Palytoxin binds to and inhibits kidney and erythrocyte Na^+, K^+-ATPase. Naunyn-Schmiedeberg's Arch Pharmacol 325: 85-87

Grell E, Lewitzki E, Ifftner A (1985). Dynamics and mechanism of cardiac glycoside binding to Na, K-ATPase. In: The Sodium Pump. The Company of Biologists Limited, Glynn, IM, Ellory E, eds., Cambridge: 289-294

Habermann E, Chhatwal GS (1982). Ouabain inhibits the increase due to palytoxin of cation permeability of erythrocytes. Naunyn-Schmiedeberg's Arch Pharmacol 319: 101-107

Hegyvary C, Jørgensen PL (1981). Conformational changes of renal sodium plus potassium ion transport adenosine triphosphatase labeled with fluorescein. J Biol Chem 256: 6296-6303

Hirata Y, Uemura D, Ueda K, Takano S (1979). Several compounds from Palythoa tuberculosa (Coelenterata). Pure Appl Chem 51: 1875-1883

Ishida Y, Tagaki K, Takahashi M, Satake N, Shibata S (1983). Palytoxin isolated from marine coelenterates. The inhibitory action on (Na, K)-ATPase. J Biol Chem 258: 7900-7902

Jørgensen PL (1974). Purification and characterization of (Na^+ + K^+)-ATPase. III. Purification from the outer medulla of mammalian kidney after selective removal of membrane components by sodium dodecylsulphate. Biochim Biophys Acta 356: 36-52

Uemura D, Hirata Y, Iwashita T, Naoki H (1985). Studies on palytoxin. Tetrahedron 41, 6: 1007-1017

CONTROL OF THE SODIUM PUMP BY LIPONUCLEOTIDES AND UN-SATURATED FATTY ACIDS: SIDE-DEPENDENT EFFECTS IN RED CELLS

W.-H. Huang, Z. Xie, S.S. Kakar and A. Askari

Department of Pharmacology, Medical College of Ohio, Toledo, Ohio 43699 (U.S.A.)

INTRODUCTION

A few years ago, in the course of our work on the $C_{12}E_8$-solubilized (Na^++K^+)-ATPase, we suspected that some differences between the properties of this detergent-treated preparation and those of the membrane-bound enzyme were not due to enzyme solubilization, but rather because of the readily reversible interaction of $C_{12}E_8$ with the enzyme in its membrane-bound form. Our subsequent experiments confirmed this, and indicated that $C_{12}E_8$ and a variety of other laboratory detergents, at concentrations (or detergent: enzyme ratios) that were too low to induce solubilization, activated the membrane-bound enzyme at suboptimal but not optimal ATP concentrations (Huang et al, 1985). In the same series of studies, we also showed that these activating effects of detergents were distinct from their well-known vesicle-opening effects that lead to apparent enzyme activation at all substrate levels in sealed vesicular preparations of the enzyme. Activation at low but not high ATP concentrations suggested the existence of hydrophobic regulatory sites on the enzyme the occupation of which by detergents mimicked the regulatory effect of ATP at its low-affinity site within the Albers-Post cycle (Huang et al, 1985). In turn, this led to the question of whether some endogenous detergents could exert physiological control over the sodium pump at these regulatory sites. Pursuing this question (Huang et al, 1986; Kakar et al, 1987) we showed that (a) unsaturated free fatty acids and CoA esters of fatty acids had activating effects on the enzyme at low substrate concentrations; (b) while the effects of free acids were biphasic (activation followed by inhibition), the CoA esters had purely activating effects on the enzyme within a reasonably physiological range of concentrations; and (c) these activating effects of CoA esters were observed not only on the ATPase activity of the purified enzyme, but also on ATP-dependent Na^+ uptake by sealed inside-out vesicles of cardiac sarcolemma. Although these findings strengthened the possibility of the physiological control of the pump by intracellular liponucleotides, it was necessary to explore the effects of the CoA esters and the free fatty acids on the pump from both sides of the plasma membrane. Since this was not possible in cardiac sarcolemmal preparations, we have used established preparations of right-side-out reconstituted ghosts (Bodemann and Hoffman, 1976) and inside-out vesicles (Mercer and Dunham, 1981) of human red cells for such studies.

RESULTS

Experiments of Fig. 1 show the effects of varying concentrations of docosahexaenoic acid (DA) and palmitoyl CoA (PCoA), added to the medium, on Na^{22} release from right-side-out ghosts that were resealed in the presence of 2 mM ATP. At concentrations up to 50 µg/ml, PCoA affected neither the ouabain-sensitive nor the ouabain-insensitive sodium efflux significantly. Increasing concentrations of the unsaturated fatty acid, however, increased the ouabain-insensitive efflux, and inhibited the ouabain-sensitive pump flux. The two components were affected in parallel within the same range of fatty acid concentrations.

When the right-side-out ghosts were resealed in the presence of suboptimal ATP (0.5 mM), PCoA again had no effect on Na^{22} efflux; and DA affected the two components of the efflux as in Fig. 1. Experiments of Fig. 2 showed that the sensitivity of the pump flux to inhibition by the unsaturated acid was about the same regardless of whether internal ATP was optimal or suboptimal.

Data of Fig. 3 show the effects of varying concentrations of PCoA, added to the medium, on ATP-dependent and ATP-independent uptake of Na^{22} by K^+-loaded inside-out vesicles. ATP concentration was suboptimal (50 µM). There was some stimulation of ATP-independent uptake at higher

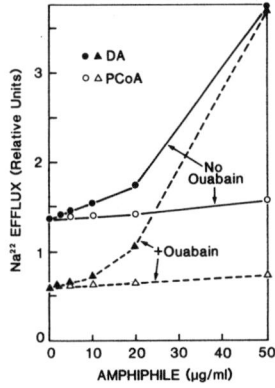 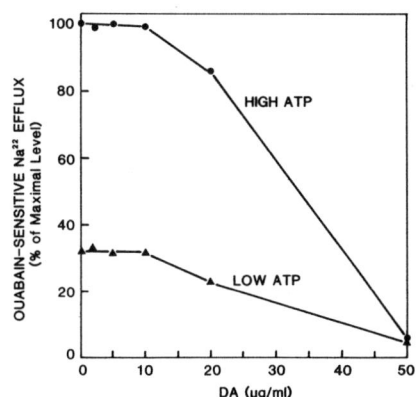

Fig. 1. (left) <u>Effects of palmitoyl CoA (PCoA) and docosahexaenoic acid (DA) on ouabain-sensitive and ouabain-insensitive release of Na^{22} from right-side-out resealed red cell ghosts.</u> Ghosts were resealed containing 0.1 mM EGTA, 3 mM $MgCl_2$, 140 mM choline chloride, 6 mM Na^{22} Cl, 2 mM ATP, and 17 mM Tris-HCl (pH 7.4). To measure efflux, they were suspended at 37°C in media containing 150 mM NaCl, 4 mM KCl, the same buffer, 0.1 mM ouabain as indicated, and the indicated amphiphiles.

Fig. 2. (right) <u>Effects of varying concentrations of DA on ouabain-sensitive Na^{22} release from right-side-out resealed ghosts loaded with high (2 mM) or low (0.5 mM) ATP.</u> Other conditions as indicated in legend to Fig. 1.

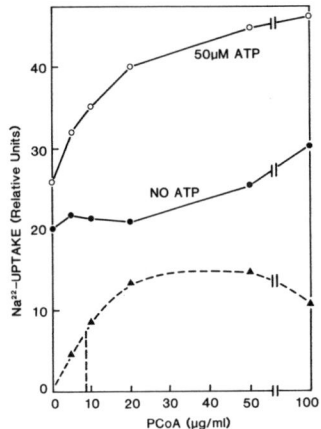 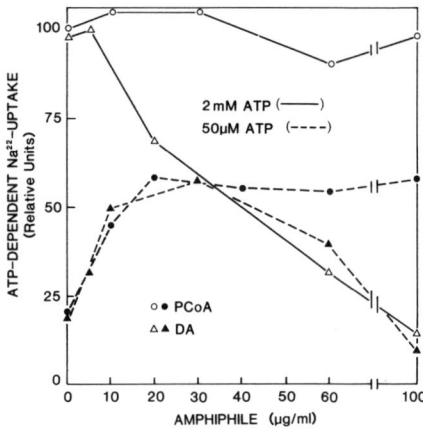

Fig. 3. (left) <u>Effects of varying concentrations of PCoA on Na^{22} uptake by K$^+$-loaded inside-out vesicles of red cells</u>. Vesicles were loaded with 0.1 mM EGTA, 25 mM KCl, 1 mM MgCl$_2$, and 2.5 mM Tris-glycylglycine (pH 6.8); and were added at 37°C to media containing 4 mM Na^{22}Cl, 16 mM KCl, 5 mM choline chloride, 1 mM MgCl$_2$, 2.5 mM Tris-glycylglycine (pH 6.8), and the indicated concentrations of PCoA. ATP concentration was 50 μM. The dotted line is the calculated PCoA-stimulated and ATP-dependent uptake.

Fig. 4. (right) <u>Effects of varying concentrations of PCoA and DA on ATP-dependent Na^{22} uptake by K$^+$-loaded inside-out vesicles in the presence of 50 μM ATP or 2 mM ATP</u>. Other conditions as indicated in legend to Fig. 3.

PCoA concentrations, but pronounced stimulation of ATP-dependent uptake was noted at all PCoA concentrations. $K_{0.5}$ value of PCoA was about 5-10 µg/ml. When similar experiments were done in the presence of 2 mM ATP, there was no effect of PCoA on the ATP-dependent Na^{22} uptake. Comparison of the effects of PCoA at optimal and suboptimal ATP are shown in one set of experiments presented in Fig. 4.

Docosahexaenoic acid, in the range of 5-100 µg/ml, had pronounced inhibitory effects on ATP-independent Na^{22} uptake (data not shown), and on the ATP-dependent uptake in the presence of 2 mM ATP (Fig. 4). When ATP was suboptimal, however, DA stimulated Na^{22} uptake at lower concentrations, but inhibited at higher concentrations (Fig. 4).

The ATP-dependent components of Na^{22} uptake by inside-out vesicles, as shown in Figs. 3 and 4, were sensitive to vanadate and digitoxigenin, but not to ouabain (data not shown).

The effects of other unsaturated fatty acids, including oleic, linoleic, and linolenic acid, were similar to those of DA (data not shown).

DISCUSSION

The present findings show that the interaction of PCoA with the intracellular domains of the red cell sodium pump activates the pump if ATP is suboptimal but not when ATP is near optimal, and that this effect is obtained at PCoA concentrations that have little or no effects on the passive fluxes. These findings agree with the results of our previous studies on the cardiac sodium pump (Kakar et al, 1987). In addition, the present data also show that from the extracellular side, PCoA has little or no effect on either the pump or the passive flux at the same concentrations that it activates from the intracellular side. This demonstration of the sidedness of PCoA effect on the pump adds further support to the proposal that CoA esters of long-chain fatty acids may indeed be intracellular regulators that protect the pump in the face of substrate depletion (Huang et al, 1986; Kakar et al, 1987).

While the effects of liponucleotides on the pump seem to be straightforward, those of the unsaturated fatty acids are rather complex. Inhibitory effects of these compounds on the isolated (Na^++K^+)-ATPase have been known for a long time (Ahmed and Thomas, 1971); and in recent years the possibility has been considered by many laboratories that unsaturated fatty acids may be the elusive endogenous digitalis-like substances (e.g., Kelly et al, 1986). The data presented here, in conjunction with our previous results (Huang et al, 1986), show that fatty acids have two distinct effects on the pump: One is similar to that of CoA esters, causing activation at low ATP levels. This is noted only when the pump is approached from the intracellular side. The other is a pronounced inhibitory effect obtained at all ATP concentrations and when the fatty acid is added to either side of the membrane. Interpretation of these findings in terms of the sidedness of fatty acid effects is difficult, however, since fatty acids freely pass through the membrane. It is also important to note that while unsaturated fatty acids do have a PCoA-like activating effect on the pump, this effect seems to be exerted at a site distinct from that of PCoA site on the enzyme (Huang et al, 1986). Another factor that complicates the study of the fatty acid effects on the pump is the observation that fatty acids, added from either side, make the membrane more leaky to Na^+. Further studies are needed to clarify the mechanisms of the different effects of unsaturated fatty acids on the pump, and to ascertain the potential physiological significance of each effect.

REFERENCES

Ahmed K, Thomas BS (1971). The effects of long-chain fatty acids on sodium plus potassium ion-stimulated adenosine triphosphatase of rat brain. J Biol Chem 246:103-109.
Bodemann HH, Hoffman JF (1976). Side-dependent effects of internal versus external Na and K on ouabain binding to reconstituted human red blood cell ghosts. J Gen Physiol 67:497-525.
Huang W-H, Kakar SS, Askari A (1985). Mechanisms of detergent effects on membrane-bound (Na^++K^+)-ATPase. J Biol Chem 260:7356-7361.
Huang W-H, Kakar SS, Askari A (1986). Activation of (Na^++K^+)-ATPase by long-chain fatty acids and fatty acyl coenzymes A. Biochem Internat 12:521-528.

Kakar SS, Huang W-H, Askari A (1987). Control of cardiac sodium pump by long-chain acyl coenzymes A. J Biol Chem 262:42-45.

Kelly RA, O'Hara DS, Mitch WE, Smith TW (1986). Identification of NaK-ATPase inhibitors in human plasma as nonesterified fatty acids and lysophospholipids. J Biol Chem 261:11704-11711.

Mercer RW, Dunham PB (1981). Membrane-bound ATP fuels the Na/K pump. J Gen Physiol 78:547-568.

ACKNOWLEDGEMENTS

This work was supported by NIH Grant P01 HL-36573 awarded by National Heart, Lung, and Blood Institute, USPHS/DHHS.

BINDING OF OUABAIN TO THE Na-K-ATPase IN OOCYTES OF XENOPUS LAEVIS IS VOLTAGE-INDEPENDENT

Andreas V. Lafaire, Bernd Schweigert and Wolfgang Schwarz

Max-Planck-Institut für Biophysik, D-6000 Frankfurt/Main (FRG)

INTRODUCTION

In the oocytes of the clawed toad Xenopus laevis a ouabain-sensitive, outwardly directed membrane current can be detected. This current has been assumed to be generated by the electrogenic Na-K-ATPase, and exhibits strong dependence on membrane potential with a characteristic maximum at about +20 to +30 mV (Lafaire & Schwarz 1985, 1986, see also Fig. 1). It has been suggested that the modulation of the

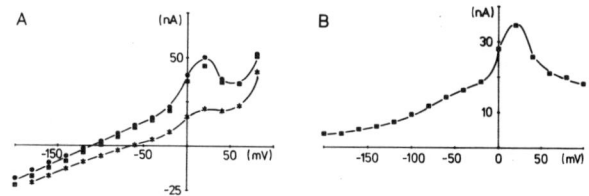

Figure 1. (A) Typical voltage dependencies of total membrane current in normal bath solution without (●,■) and with (✷) 10 μM DHO; the circles represent data obtained before application of DHO, the squares represent data obtained after washout of the DHO (note the nearly complete reversibility of the action of DHO). (B) Current component that can be inhibited by 10 μM DHO. This current was determined as the difference of total membrane current in bath solution without and with DHO.

pump current is due to modulation of the pump activity and not due to modulation of the number of pumping molecules (Schwarz & Lafaire 1986). These results were obtained from measurements of ouabain-sensitive membrane current and of ouabain binding at saturating concentrations of the inhibitors ouabain or dihydroouabain (DHO). This contribution presents further evidence for modulation of the pump activity by membrane potential. In addition, we deal with the question whether at non-saturating concentrations the inhibition of the pump current and the ouabain binding depend on membrane potential.

For the experiments described in this paper the ouabain-sensitive membrane current was determined as the difference current obtained without or with an inhibitor in the bath solution. In order to reduce influences of passive K movements, millimolar concentrations of Ba can be added to the bath solution (see e.g. Glitsch & Krahn, 1986). With this procedure we in addition examined the effect of Ba on the inhibition of the Na-K-ATPase by ouabain or DHO.

METHODS

The experiments were performed on full-grown prophase-arrested oocytes that had been preincubated in K-free medium for several hours. This treatment reversibly blocks the Na-K pump and leads to an increase of intracellular Na. The subsequently supplied bath solution usually contained (in mM) 110 NaCl, 2 $CaCl_2$, 3 KCl, and the pH was buffered to 7.2 by 5 mM morpholinopropane sulfonic acid or 5 mM N-2-hydroxyl-ethylpiperazin N'-ethansulfonic acid.

The oocytes were voltage-clamped by conventional two-microelectrodes technique. Potential-recording and current-delivering electrodes were filled with 1 and 3 mM KCl, respectively, with resistances of 10 to 30 MOhm. For stimulation and evaluation of the current-voltage dependencies rectangular voltage pulses of varing amplitude and 500 ms duration were applied every 3-5 s from a holding potential that was usually set to the respective resting potential of the oocyte (about -70 mV). The steady-state currents were determined during the last 100 ms of the test pulse.

For determination of ouabain binding, one to three oocytes that had been exposed under voltage-clamp conditions to ^{3}H-ouabain in K-free solution were washed three times in ouabain-free and K-free bath solution, and were collected in a vial for scintillation counting.

RESULTS

To demonstrate that the voltage dependence of the DHO-sensitive current as shown in Figure 1 is due to voltage-dependent activity of the sodium pump and not due to a voltage-dependent modulation of the number of pump molecules, the amount of bound ^{3}H-ouabain was determined for different voltage-clamp conditions. Figure 2 demonstrates that there are no significant differences in the amount of bound ouabain. Also, if the amount of ouabain in the bath solution was varied between 0.1 and 2 µM no significant difference in the amount of bound ouabain could be detected for cells that were kept at a constant holding potential of -70 mV, or that were additionally subjected to clamp pulses to +20 mV.

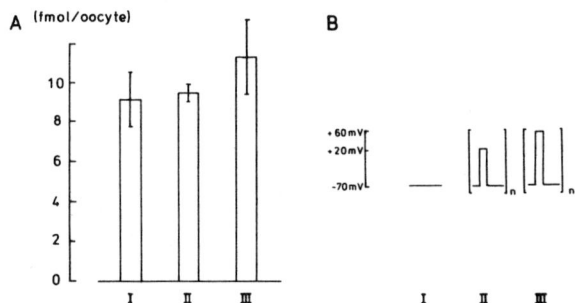

Figure 2: (A) Ouabain bound to single oocytes that were kept for 5 min under different voltage-clamp conditions (see (B)) with 0.5 µM ouabain in K^+-free bath solution. The binding of ^{3}H-labled ouabain was measured with the holding potential set to -70 mV (column I), and during application of voltage pulses (500 ms duration, 0.5 Hz) to +20 mV (the maximum in the current-voltage curve, column II) and to +60 mV (a potential were the ouabain-sensitive current is again smaller than at 20 mV, column III). The columns indicate means ± SEM (n= 10(I), 15(II) and 6(III)).

These results demonstrate that the membrane potential does not affect the binding of ouabain. Also if the reversible inhibitor DHO is used as an inhibitor of the sodium pump, no effect of membrane potential could be detected. Figure 3 demonstrates for different DHO concentrations the same voltage dependence of the DHO-sensitive membrane current.

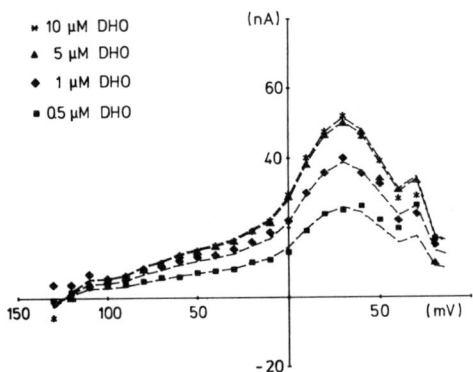

Figure 3. Voltage dependence of DHO-sensitive current for different DHO concentrations as indicated in the figure. The current was determined as the difference of total membrane current in bath solution without and with the respective DHO concentration (n= 2). The dotted lines represent the same curve obtained by scaling an average of the shown data to match the peak values for the respective DHO concentration.

Ba is known to inhibit K channels in a variety of cell membranes. Also in the oocytes of Xenopus a component of membrane current can be inhibited by Ba (Fig.4.). In order to reduce influences of the passive K-ion movements on the determination of ouabain-sensitive current, experiments were performed in the presence of 5 mM $BaCl_2$ in the bath solution. Figure 5 shows the result of the same type of experiment as shown in Figure 3 but with Ba present. Two features are apparent: in the presence of Ba (a) the characteristic voltage dependence with its maximum at +20 to +30 mV is maintained, and (b) the inhibitory potency of DHO is reduced by an order magnitude. The latter observation is demonstrated in Figure 6; in the presence of Ba the K_I value is increased from 0.5 µM to 4.4 µM.

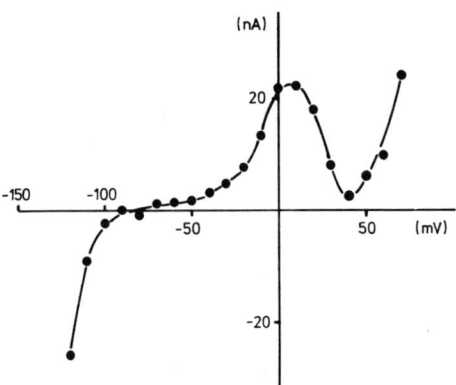

Figure 4. Voltage dependence of the current component that is inhibited by 5 mM $BaCl_2$ added to the bath solution (n= 5).

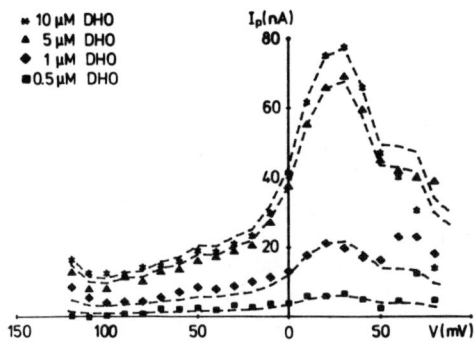

Figure 5. Voltage dependence of DHO-sensitive current for different DHO concentrations with 5 mM $BaCl_2$ in the bath solution. The current was determined as the difference of total membrane current in bath solution without and with the respective DHO concentration (n= 5). The dotted lines represent the same curve obtained by scaling an average curve to match the peak values for the respective DHO concentration.

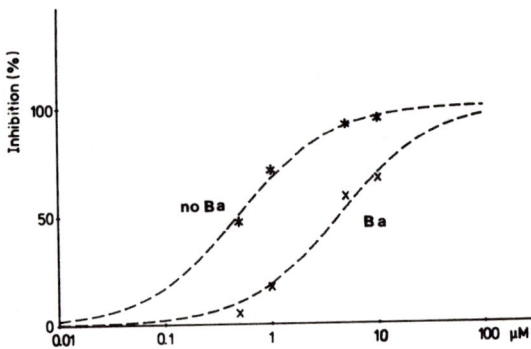

Figure 6. Dependence of pump inhibition on the concentration of DHO in the bath solution. The degree of inhibition of DHO-sensitive current at +20 mV in bath solution containing 3 mM KCl (*: without $BaCl_2$ (n= 2), X: in the presence of 5 mM $BaCl_2$ (n= 5)); lines represent least-squares fits of Michaelis-Menten kinetics to the data leading to K_I values of 0.50 ± 0.07 uM and 4.4 ± 2.1 uM for the experiments without and with Ba, respectively.

DISCUSSION

Binding of ouabain to the Na-K-ATPase leads to a specific inhibition of the enzyme activity by an one-to-one stoichiometry. For the oocytes of **Xenopus laevis** Richter et al. 1985 demonstrated a linear relationship between bound ouabain and pump inhibition. The amount of bound ouabain is given by (1) the number of binding sites and (2) the enzyme conformation. Both parameters can be modulated, and they particularly depend on the presence of ligands that modulate the enzyme activity such as external K^+ and internal Na^+.

Also the membrane potential has been demonstrated to modulate the enzyme activity (see e.g. DeWeer 1984, Gadsby et al. 1985, Glitsch & Krahn 1986, Lafaire & Schwarz 1986). In addition, the membrane potential has been suggested to affect ouabain binding by modulation of the number of binding sites in nerve fibres (Maximov & Kols 1985). In contrast to the observations by the latter authors, our measurements (Fig. 2) on the oocytes demonstrate that in these cells the number of binding sites is not affected by different voltage-clamp conditions, at least not by clamp conditions that are maintained for about 5 min. This of-course does not exclude that maintainance of longer lasting clamp conditions may alter the number of binding sites.

Since the membrane potential modulates the enzyme activity, the question may be asked whether the membrane potential can modulate ouabain binding by modulation of the dissociation constant. This contribution has demonstrated no difference in ouabain binding for oocytes that were kept at a constant holding potential or that were repetitivly subjected to depolarizing voltage clamp pulses. This could be demonstrated for different ouabain concentrations in the bath solution. The characteristic voltage dependence is also maintained at ouabain or DHO concentrations that give only partial inhibition of the pump current (Figs. 3 and 5). This again indicates voltage-independent K_I values for the inhibition of the pump current for both inhibitors.

Ba can effectively be used to reduce passive movements of K ions (Fig. 4), but it obviously also effects the Na-K pump; the sensitivity for DHO is drastically reduced (Fig. 6). Since Ba is an inhibitor of K^+-selective channels, one might speculate that Ba interfers with the K binding site at the Na-K-ATPase and by this reduces the affinity for ouabain as K binding does.

The result demonstrate that the voltage dependence is the same for different DHO concentrations and independent of the presence of the K channel blocker Ba. This is strong evidence for the assumption that the current-voltage relations of the difference currents represent the current generated by the Na-K-ATPase and that they are not disturbed by ouabain-dependent passive K currents.

REFERENCES

DeWeer P. (1984). Electrogenic pumps: Theoretical and practical considerations. In Blaustein MP, Liebermann M. (eds): "Electrogenic Transport: Fundamental principles and Physiological Implications," New York: Raven, pp. 1-15.
Gadsby DC, Kimura J, Noma A (1985). Voltage-dependence of Na/K pump current in isolated heart cells. Nature 315: 63-65

Glitsch HG, Krahn T (1986). The cardiac electrogenic Na pump. Progr Zool 33: 401-418

Lafaire AV, Schwarz W (1985). Voltage-dependent, ouabain-sensitive current in the membrane of oocytes of Xenopus laevis. In Glynn I, Ellory C (eds): "The Sodium Pump," Cambridge: The company of biologists, pp523-525.

Lafaire AV, Schwarz W (1986). Voltage dependence of the rheogenic Na^+/K^+ ATPase in the membrane of oocytes of Xenopus laevis. J Membrane Biol 91:43-51

Maximov GV, Kols OR (1985). Mechanism of activation of Na,K-ATPase in nerve fibres during rhythmic excitation. Gen Physiol Biophys 4: 279-285

Richter HP, Jung D, Passow H (1984). Regulatory changes of membrane transport and ouabain binding during progesterone-induced maturation of Xenopus ooccytes. J Membrane Biol 79: 203-210

Schwarz W, Lafaire AV (1986). Membrane potential modulates the rate of the Na^+/K^+ pump and not the number of pumping molecules. Progr Zool 33: 429-434

ACKNOWLEDGEMENT

We thank Dr. H. Passow for his comments on the manuscript, and we acknowledge the help with preparing the fugures by Heike Keim and Brigitte Lehmann. This work was supported by the Deutsche Forschungsgemeinschaft SFB 169.

AN ENDOGENOUS INHIBITOR OF Na,K-ATPase ISOLATED FROM HUMAN
PLASMA INHIBITS THE ACID PUMP OF THE STOMACH

Sven Mårdh

Department of Medical and Physiological Chemistry,
Uppsala University, Biomedical Centre, Box 575,
S-751 23 Uppsala, Sweden.

INTRODUCTION

The H,K-ATPase of the parietal cell constitutes the acid pump of the stomach. Acid secretion of the cells is controlled by chemical messengers in neurocrine, endocrine and paracrine regulatory pathways (Grossman, 1981). Acetylcholine is released near the parietal cells from postganglionic nerve cells. Gastrin is secreted from the G cells located in the mucosa of the antrum and proximal duodenum. It is released to the blood which transports it to the target cell. Finally, histamine is released from small intraglandular histamine-containing cells (Bergqvist et al., 1980) or from mast-like cells in the lamina propria of the acid-secreting mucosa (Soll, 1979). All three substances take part in the stimulatory process of acid secretion.

Acid secretion may be inhibited by somatostatin (Chew, 1983), secretin (Johnson and Grossman, 1969), epidermal growth factor (EGF, urogastron; Gonzáles et al., 1981), prostaglandins (Soll, 1980) and by catecholamines (Holton, 1973). Our present knowledge, however, about the mechanisms of inhibition of acid secretion is sparse. This paper describes an inhibitor of both Na,K-ATPase and H,K-ATPase. It is capable of producing a total inhibition of acid formation in isolated parietal cells and may represent a new type of inhibitor of acid secretion.

METHODS

The inhibitor was purified from human plasma. One hundred ml of plasma was extracted with 600 ml chloroform:methanol (1:1) plus 25 ml acetic acid and then filtered. Two phases were obtained by the addition of 100 ml of H_2O to the filtrate. The upper polar phase was evaporated to dryness. Methanol was added to the dry residue. Centrifugation removed undissolved material. Evaporation of the methanol, addition of new methanol and centrifugation was repeated twice and was then followed by a final evaporation. The extract was dissolved in 2.5 ml H_2O. It was chromatographed on a Sephadex G-10 column equilibrated and eluted with 5 mM ammonium acetate. V_o-fractions containing inhibitory activity of pNPPase were pooled, lyophilized and then dissolved in water. Rechromatography was performed on a Sephadex G-25 column. Inhibitory fractions with apparent molecular mass of 1 kDa were lyophilized and resuspended in 2 ml of H_2O (referred to as "inhibitor").

Na,K-ATPase and H,K-ATPase were prepared from outer medulla of pig kidneys (Mårdh, 1979) and from pig gastric mucosa (Ljungström et al., 1984), respectively.

Parietal cells were isolated from pig gastric mucosa (Mårdh et al., 1984). Isolated cells were stimulated by 100 µM histamine and acid formation was assayed by means of the accumulation of ^{14}C-labeled aminopyrine in their acid compartments.

RESULTS

Dose-dependent inhibition of H,K-ATPase and of Na,K-ATPase by the endogenous inhibitor was studied (Fig. 1). The ATPase as well as the pNPPase activities were inhibited. Half-maximal inhibition of the H,K-ATPase and its pNPPase was observed at 5 µl and 14 µl of the inhibitor, respectively. Corresponding values for Na,K-ATPase and its pNPPase were about 45 and 15 µl.

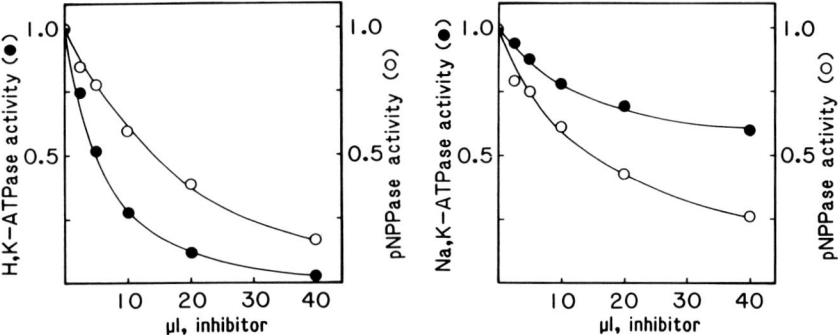

Figure 1. Effects of the endogenous inhibitor on H,K-ATPase and Na,K-ATPase and their pNPPase activities.

The endogenous factor dose-dependently inhibited acid formation in parietal cells (Fig. 2). About 11 µl of the inhibitor resulted in half-maximal inhibition. Fourty µl completely blocked the formation of acid.

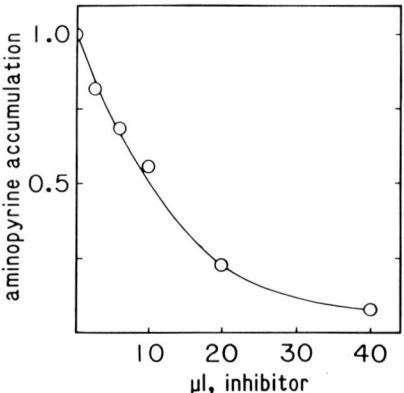

Figure 2. Inhibitition of acid formation in isolated parietal cells produced by the endogenous inhibitor.

DISCUSSION

Recently cDNA sequence analysis showed 62 % homology between H,K-ATPase and the alfa-subunit of Na,K-ATPase (Shull and Lingrel, 1987). In the present study an inhibitor of these ATPases was extracted from human plasma. The inhibitor was resistant to proteolytic attack by pronase and to heating at 100 °C for 10 min. It was effective on Na,K-ATPase and H,K-ATPase similarly to the inhibitor purified by means of size exclusion and chromatography from normal and from shock plasma of both cat and rat (Lundgren et al., 1987). SR Ca-ATPase prepared from skeletal muscle was, however, slightly stimulated by the human plasma inhibitor.

Various studies on endogenous inhibitors of Na,KATPase have been published during the last decade. The inhibitors appear to have a digitalis-like action and have been isolated from mammalian plasma (Gruber et al., 1980), and from tissues as brain, heart and hypothalamus. (For review see Schoner et al., 1986). Identification of the inhibitory component(s) appears to be a common problem. Bidard et al. (1984) concluded that possible candidates were arachidonic acid and related long chain fatty acids. Unsaturated fatty acids have been shown also to inhibit the gastric H,K-ATPase (Bin Im and Blakeman, 1982). In the present study the initial purification step included extraction with chloroform:methanol which removes most lipids and fatty acids. However, phase distribution of the partially purified inhibitor showed that 65 % of the inhibitory activity was recovered from the polar phase and the remaining 35 % from the chloroform phase. These results do not disqualify the inhibitory activity being expressed, at least partly, by 1-acyl-sn-phosphocholines as recently suggested by Tamura et al. (1987). In their assay these lyso-lechitins were inhibitory to Na,K-ATPase at 10-100 μM levels. This contrasts to the low concentrations of lysolecithins required for e.g. induction of Na^+-dependent pinocytosis in Amoeba proteus (pM levels; Arvidson and Josefsson, 1982).

The H,K-ATPase appeared to be more sensitive to inhibition than was the Na,K-ATPase and SR Ca-ATPase was slightly stimulated rather than inhibited by the presently described plasma inhibitor. Thus the inhibitor seems to have some selectivity among the transport ATPases. However, its possible physiological role and structure remain to be established.

REFERENCES

Arvidson G, Josefsson J-O (1982). Lysolecithin as a modifier of induced pinocytosis in Amoeba proteus. Eur J Cell Biol 28:34-38.

Bergqvist E, Waller M, Hammar L, Öbrink K-J (1980). Histamine as the secretory mediator in isolated gastric glands. In Schulz I, Sachs G, Forte JG, Ullrich KJ (eds): "Hydrogen ion transport in epithelia," Amsterdam: Elsevier/North-Holland Biomedical Press, pp 429-437.

Bidard J-N, Rossi B, Renaud J-F, Lazdunski M (1984). A search for an 'ouabain-like' substance from the electric organ of Electrophorus Electricus which led to arachidonic acid and related fatty acids. Biochim Biophys Acta 769:245-252.

Bin Im W, Blakeman DP (1982). Inhibition of gastric H,K-ATPase by unsaturated long-chain fatty acids. Biochim Biophys Acta 692:355-360.

Chew CS (1983). Inhibitory action of somatostatin on isolated gastric glands and parietal cells. Am J Physiol 245:G221-G229.

Gonzáles A, Garrido, J, Vial JD (1981). Epidermal growth factor inhibits cytoskeleton-related changes in the surface of parietal cells. J Cell Biology 88:108-114.

Grossman IM (1981). Regulation of gastric acid secretion. In Johnson LR (ed): "Physiology of the gastrointestinal tract," New York: Raven Press, pp. 659-671.

Gruber KA, Whitaker JM, Buckalew VM (1980). Endogenous digitalis-like substance in plasma of volume-expanded dogs. Nature 287:743-745.

Holton P (1973). Catecholamines and gastric secretion. In Holton P (ed): "Pharmacology of gastrointestinal motility and secretion," Oxford: Pergamon Press, pp. 287-315.

Johnson LR, Grossman M (1969). Characteristics of inhibition of gastric secretion by secretin. Am J Physiol 217:1401-1404.

Ljungström M, Norberg L, Olaisson H, Wernstedt C, Vega FV, Arvidson G, Mårdh S (1984). Characterization of proton-transporting membranes from resting pig gastric mucosa. Biochim Biophys Acta 769:209-219.

Lundgren O, Mårdh S, Haglind E (1987). Evidence for the existence of an endogenous inhibitor of Na,K-ATPase in plasma from cats and rats. Acta Physiol Scand 129:465-470.

Mårdh S (1979). Phosphorylation by the catalytic subunit of protein kinase of a preparation of kidney Na,K-ATPase. In Skou JC, Norby JG (eds): "Na,K-ATPase, structure and kinetics," London, New York: Academic Press, pp. 359-370.

Mårdh S, Norberg L, Ljungström M, Humble L, Borg T, Carlsson C (1984). Preparation of cells from pig gastric mucosa. Isolation of parietal cells by isopycnic centrifugation on linear gradients of Percoll. Acta Physiol Scand 122:607-613.

Schoner W, Moreth K, Kuske R, Renner D (1986). The "endogenous cardiac glycoside". In Erdmann E, Greff K, Skou JC (eds): "Cardiac glycosides 1785-1985," Darmstadt: Steinkopff Verlag, pp. 135-142.

Shull GE, Lingrel JB (1987). Molecular cloning of the rat stomach H,K-ATPase. J Biol Chem 261:16788-16791.

Soll AH (1979). Isolation of histamine-containing cells from canine fundic mucosa. Gastroenterology 77:1283-1290.

Soll AH (1980). Specific inhibition by prostaglandins E_2 and I_2 of histamine-stimulated ^{14}C-aminopyrine accumulation and cyclic adenosine monophosphate generation by isolated canine parietal cells. J Clin Invest 65:1222-1229.

Tamura M, Harris TM, Higashimori K, Sweetman BJ, Blair IA, Inagami T (1987). Lysophosphatidylcholines containing polyunsaturated fatty acids were found as Na^+,K^+-ATPase inhibitors in acutely volume-expanded hog. Biochemistry 26:2797-2806.

TWO MOLECULAR FORMS OF Na^+,K^+-ATPase IN THE FERRET HEART AND DEVELOPMENTAL CHANGES IN DIGITALIS SENSITIVITY

Yuk-Chow Ng and Tai Akera

Department of Pharmacology and Toxicology,
Michigan State University, East Lansing, MI., USA

INTRODUCTION

Two molecular forms of Na^+,K^+-ATPase, having alpha(+)- and alpha-subunits, respectively, were first reported by Sweadner (1979) in the brain tissue of several species; however, the relationship between two classes of ouabain binding sites and two molecular forms of Na^+,K^+-ATPase in the cardiac tissue is not well established. Preliminary results from our laboratory indicate that ferret heart has more than one class of ouabain binding sites. Therefore, possible existence of Na^+,K^+-ATPase isoforms and a shift in the alpha(+)/alpha ratio which may occur during development in the ferret heart were examined. A change in the relative abundance of these two forms of Na^+,K^+-ATPase would probably change the sensitivity of ferret heart to the positive inotropic and toxic actions of digitalis.

METHODS

ATPase activity was estimated from the amount of inorganic phosphate released from ATP during a 10-min incubation at $37°C$. Na^+,K^+-ATPase activity is the difference in values observed in the absence and presence of 1 mM ouabain. Ouabain binding was assayed by incubating enzyme preparations with the indicated concentration of ^3H-ouabain at $37°C$ for 60 min in the presence of 1 mM $MgCl_2$, 1 mM Tris-phosphate and 10 mM Tris-HCl buffer (pH 7.5). Specific ^3H-ouabain binding is the difference in values observed in the absence and presence of 2 mM unlabeled ouabain. Phosphoenzyme formation was examined by incubating enzyme preparations on ice for 30 sec with 100 mM NaCl, 1 mM $MgCl_2$, 10 µM [γ-^{32}P]ATP and 25 mM Tris-HCl buffer (pH 7.4).

RESULTS

Two classes of ouabain binding sites- The first indication that the ferret heart has more than one class of Na^+,K^+-ATPase having different affinities for ouabain was derived from the concentration-effect curve for ouabain-induced inhibition of Na^+,K^+-ATPase (Ng and Akera, 1987a). Inhibition of enzyme activity could be detected at ouabain concentrations as low as 1 nM, whereas complete inhibition did not occur with concentration as high as 100 µM. The fact that inhibition curve spanned over 5 log units suggests that there may be more than one class of ouabain binding sites. Specific binding of ^{3}H-ouabain to ferret heart enzyme preparations showed a slope which is significantly shallower than that predicted from Michaelis-Menten equation. Analysis of binding data using the LIGAND program revealed two classes of ouabain binding sites (Ng and Akera, 1987a). Apparent dissociation constants (K_D) for high- and low-affinity binding sites as estimated by the LIGAND program were 9.6 and 291 nM, respectively. Binding site concentrations were 22.6 pmol/mg protein for high-affinity binding sites and 22.3 pmol/mg protein for low-affinity binding sites. These results indicate that there are roughly equal abundance of the high- and low-affinity ouabain binding sites in the ferret heart.

Two molecular forms of Na,K-ATPase and two classes of ouabain binding sites- SDS-gel electrophoresis of phosphoenzyme labeled with [γ-^{32}P]ATP revealed two distinct bands with molecular weights close to 100,000 dalton (Fig. 1). That the bands were indeed the alpha subunits of Na^+,K^+-ATPase could be demonstrated by the K^+ sensitivity of the phosphoenzyme.

The relationship between two classes of ouabain binding sites observed in ferret heart and two molecular forms of Na^+,K^+-ATPase was examined from the sensitivity of two forms of the enzyme to ouabain-induced inhibition of phosphoenzyme formation. Enzyme preparations were treated with the indicated concentrations of ouabain, subsequently phosphorylated from [γ-^{32}P]ATP, and then phosphoenzymes were separated by SDS-polyacrylamide gel electrophoresis. Densitometer tracing of three separate autoradiograms of these gels from three ferret heart preparations is shown in Fig. 2. Concentrations of ouabain required to cause a 50% inhibition of phosphorylation of the alpha(+) and alpha forms were 40 and 300 nM, respectively.

These results indicate that phosphoenzyme formation of the alpha(+) form is more sensitive to inhibition by ouabain compared to that of the alpha form.

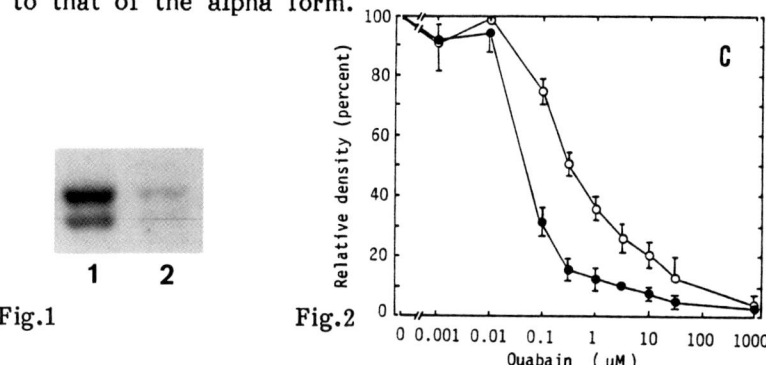

Fig.1   Fig.2

Fig. 1. Autoradiography of ^{32}P-labeled phosphoenzyme of ferret heart Na$^+$,K$^+$-ATPase preparation on SDS-polyacrylamide gel. A partially purified Na$^+$,K$^+$-ATPase preparation was phosphorylated from [γ-^{32}P]ATP. KCl (10 mM) was present during phosphorylation of the sample shown in lane 2.

Fig. 2. Relative sensitivity of alpha(+) and alpha forms of Na$^+$,K$^+$-ATPase to ouabain. Ferret heart Na$^+$,K$^+$-ATPase was preincubated with the indicated concentration of ouabain for 45 min at 37°C and then phosphorylated from [γ-^{32}P]-ATP. The ordinate represents the density of bands corresponding to alpha(+) (●) and alpha (○) forms of phosphoenzyme on ^{32}P autoradiogram as determined by scanning densitometry. (Reprinted with permission from Ng and Akera, 1987a).

Relative abundance of alpha(+) and alpha forms of Na,K-ATPase in neonatal ferret heart- To determine whether there is a developmental change in the two forms of Na$^+$,K$^+$-ATPase, the relative abundance of the alpha(+) and alpha subunits was quantified by ^{3}H-ouabain-binding studies using partially purified sarcolemmal preparations. The data were analyzed by the LIGAND program based on a two-site model (Table 1). Results indicate that the ratio of binding site concentrations for the alpha(+) and alpha forms is 1:6.7 for the 5-day-old, 1:4.1 for 7- to 10-day-old and 1:0.99 for adult ferret heart. The K_D values were not significantly different between neonatal and adult ferret hearts. These data, therefore, suggest that there is a developmental change in relative abundance of the two molecular forms of the Na$^+$,K$^+$-ATPase, with the alpha form being the predominant one in the early age.

TABLE 1. Affinity and binding site concentrations for two molecular forms of Na^+,K^+-ATPase in membrane preparations of ferret heart. Values represent mean ± approximate S.E. as calculated by the LIGAND program. (Reprinted with permission from Ng and Akera, 1987b)

Age	K_D		B_{max}	
	alpha(+)	alpha	alpha(+)	alpha
	nM		pmol/mg protein	
5 Days	10.9 ± 2.8	201 ± 31	8.1 ± 1.9	54.3 ± 2.1
7-10 days	11.9 ± 1.7	240 ± 29	25.1 ± 3.0	103.7 ± 3.1
Adult	9.6 ± 0.7	291 ± 48	22.6 ± 1.1	22.3 ± 1.0

Sensitivity of the Na,K-ATPase from newborn heart to ouabain- With the predominance of the lower affinity isoform, i.e. the alpha form, Na^+,K^+-ATPase from newborn ferret heart is expected to be less sensitive to digitalis. Indeed, low concentrations of ouabain produced substantially smaller inhibition of Na^+,K^+-ATPase isolated from neonatal heart compared

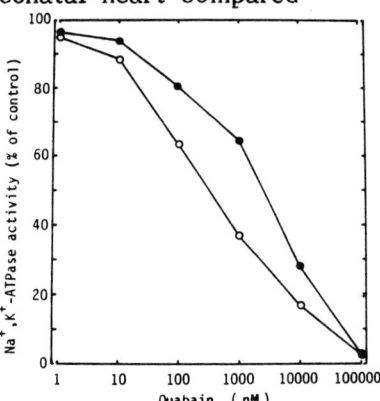

Fig. 3. Ouabain sensitivity of neonatal and adult ferret heart Na^+,K^+-ATPase preparations. Partially purified sarcolemmal preparations obtained from 7- to 10-day-old (●) or adult (○) ferrets were assayed for Na^+,K^+-ATPase activity in the presence of the indicated concentration of ouabain. (Reprinted with permission from Ng and Akera, 1987b)

to that observed with preparations obtained from adult ferret heart (Fig. 3). These data are consistent with the above finding that the ouabain-insensitive form, i.e., the alpha form, is more abundant in the neonatal preparations.

Positive inotropic effects of ouabain- In the papillary muscle preparations obtained from adult ferret heart, low concentrations of ouabain (1-30 nM) caused a modest positive inotropic effect (Fig. 4). A half-maximal increase in developed tension was observed at about 10-20 nM ouabain. At higher concentrations (0.1-10 µM), ouabain caused a further increase in developed tension with a half-maximal effect observed at about 2 µM. The biphasic positive inotropic effect in the adult ferret heart probably corresponds to inhibition of high and low affinity ouabain-binding sites at low and high doses, respectively. In the ventricular tissue of neonatal ferret heart, however, low concentrations of ouabain failed to produce a positive inotropic effect when preparations were obtained from the 1-day old ferret heart. Only at higher concentrations did ouabain cause a positive

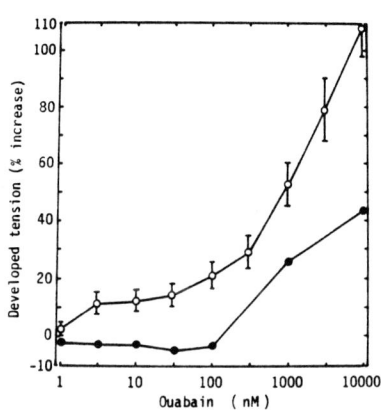

Fig. 4. Positive inotropic effects of ouabain on neonatal and adult ferret heart. Ventricular muscle were obtained from 1-day-old ferret (●) and papillary muscle preparations were obtained from the right ventricle of adult ferret heart (○). Cumulative dose response curves for ouabain were obtained by adding the glycoside at 25-min intervals. (Reprinted with permission from Ng and Akera, 1987b).

inotropic effect. The effect of ouabain was monophasic. These results indicate that neonatal ferret heart has reduced sensitivity to ouabain in concentrations ranging between 1 and 100 nM.

DISCUSSION AND SUMMARY

Our results clearly demonstrate the presence of two classes of ouabain binding sites in the ferret heart Na^+,K^+-ATPase preparations. It can also be concluded that the high- and low-affinity binding sites observed in the ferret heart are associated with alpha(+) and alpha forms of Na^+,K^+-ATPase, respectively. Furthermore, binding of ouabain to alpha(+) and alpha forms of Na^+,K^+-ATPase causes corresponding enzyme inhibition and the low- and high- dose positive inotropic effects, respectively, in the ferret heart. This is in contrast to conclusions made by Lelievre and coworkers (Maixent et al., 1987). These investigators claimed that the inotropic effect of digitalis in the dog cardiac tissue results from inhibition of the high affinity binding sites, whereas inhibition of the low affinity binding sites causes toxicity. Whether species difference can account for the difference is presently unknown; however, a direct examination of the effects of inotropic and toxic doses of digitalis on the dog heart have yet to be performed.

The relative abundance of alpha(+) and alpha forms of Na^+,K^+-ATPase changes from the time of birth to young adults in the ferret heart. The reduced inhibition of the neonatal heart enzyme preparations at low concentrations of ouabain compared to inhibition observed in adult heart enzyme preparations reflects the relatively small number of the high affinity, alpha(+) form which is present in the neonatal heart. Consistent with the small amount of the high affinity binding sites, the low dose of ouabain failed to produce a clear positive inotropic effect in the neonatal heart muscle, although the low-dose inotropic effect was apparent in the adult ferret heart.

It is well known clinically that human neonates are relatively tolerant to the therapeutic and toxic effects of the cardiac glycosides (Gorodischer et al., 1976; Kearin et al., 1980). It remains to be determined whether the present finding represents an important factor that determines digitalis sensitivity in human infants.

REFERENCE

Gorodischer R, Jusko WJ, Yaffe SJ (1976) Tissue and erythrocyte distribution of digoxin in infants. Clin. Pharmacol. Ther. 19:256-263.

Kearin M, Kelly JG, O'Malley (1980) Digoxin "receptor" in neonates: an explanation of less sensitivity to digoxin than in adults. Clin. Pharmacol. Ther. 28:346-349.

Maixent JM, Charlemagne D, Chapelle B, Lelievre LG (1987) Two Na^+,K^+-ATPase isoforms in canine cardiac myocytes. Molecular basis of inotropic and toxic effects of digitalis. J. Biol. Chem. 262:6842-6848.

Ng YC, Akera T (1987a) Two classes of ouabain binding sites in ferret heart and two forms of Na^+,K^+-ATPase. Am J. Physiol. 252:H1016-H1022.

Ng YC, Akera T (1987b) Relative abundance of two molecular forms of Na^+,K^+-ATPase in the ferret heart: Developmental changes and associated alterations of digitalis sensitivity. Mol. Pharmacol. in press

Sweadner K (1979) Two molecular forms of (Na+K)-Stimulated ATPase in brain. J. Biol. Chem. 254:6060-6067.

TWO-SIDED FUNCTIONAL Na,K-ATPase-LIPOSOMES FOR CHARACTERIZING THE PERMEABILITY AND SIDE OF ACTION OF PUMP INHIBITORS

H.G. Rey. P. Meda*, and B.M. Anner

Departments of Pharmacology and Histology*,
Geneva University Medical Center (CMU),
CH-1211 Geneva 4, Switzerland

INTRODUCTION

The molecular mechanism of action, the specificity and the side of action of many substances interacting directly or indirectly with the Na,K-ATPase of the cell membrane are often uncertain. On one hand, the target site and the selectivity of a potential pump ligand are difficult to define precisely in intact cells. On the other hand, when broken membrane fragments containing purified Na,K-ATPase molecules are used as test-system, the side at which the ligands interact with Na,K-ATPase is hardly determinable.

Therefore, we developed the ATP-filled, bi-functional Na,K-ATPase-liposomes containing co-reconstituted, randomly oriented and transport-active Na,K-ATPase molecules (Rey et al., 1987; Anner et al., 1988) into a miniaturized test-system for examining the membrane permeability and target-site of inhibitors interacting with Na,K-ATPase. It is well known that the transmembrane Na,K-ATPase molecule is asymmetric: the ATP, Mg and Na binding sites are located on the intracellular protein protusion whereas the cardioactive steroid receptor as well as the K or Rb binding sites on the extracellular protein surface. In consequence, the time required by a drug added externally to interact with the right-side-out (r-s-o) and the inside-out (i-s-o) oriented Na,K-ATPase provides information about its membrane permeability and site of action.

RESULTS

Fig. 1 shows the Rb-accumulation by the r-s-o oriented pump population in the presence of internal ATP and external RbCl. In the absence of ATP, the liposomes capture 1% of the external RbCl (Anner et al., 1984) and, thus, an entrapment above 1% reflects an inside-out Rb-gradient. The active transport catalyzed by the r-s-o pumps is blocked by the addition of external digoxin (Fig. 1). Further, digoxin prevents the extrusion of the internal RbCl by the i-s-o pumps within a 1 min incubation (Fig. 1), indicating that the drug permeates the membrane and reaches the internal receptor population within this short time period. The low passive permeability of the cholate-dialysed liposomes (Anner, 1981) prevents the leak of the accumulated Rb ions within the incubation times studied.

Fig 2. compares the time-dependence of the pump-inhibition by the hydrophilic cardioactive steroid ouabain and by the hydrophobic digoxin and the effect of the pump orientation on this inhibition. When digoxin is added externally, the active Rb-uptake by the r-s-o pumps as well as the active extrusion of the internal Rb ions by the i-s-o pumps are blocked within 1 min incubation (Fig. 2A). Conversely, ouabain does not penetrate the liposome membrane within a 7 min incubation time. Thus, it interacts only with the external receptor population on the r-s-o pumps (expressed by the block of Rb-uptake) and not with the internal receptor population, leaving the i-s-o pumps active (Fig. 2B). In control preparations, the Rb ions accumulated by the r-s-o pumps are rapidly extruded upon activation of the i-s-o oriented pumps by external ATP, despite the presence of 10% DMSO which produces only a small reduction of the transport rate (Figs. 2A, 2B).

Thus, we have now a model in which the receptor site of pure and actively transporting Na,K-ATPase molecules is accessible on the surface of liposomes. This model enables us to test potential pump-blockers. Two proposed receptor ligands, palytoxin and linoleic acid (reviewed by Schoner et al., 1986) inhibit the active Rb-uptake of the reconstituted pump only partially at high concentrations (Table 1), i.e. have not the characteristics of potent, specific Na,K-ATPase inhibitors.

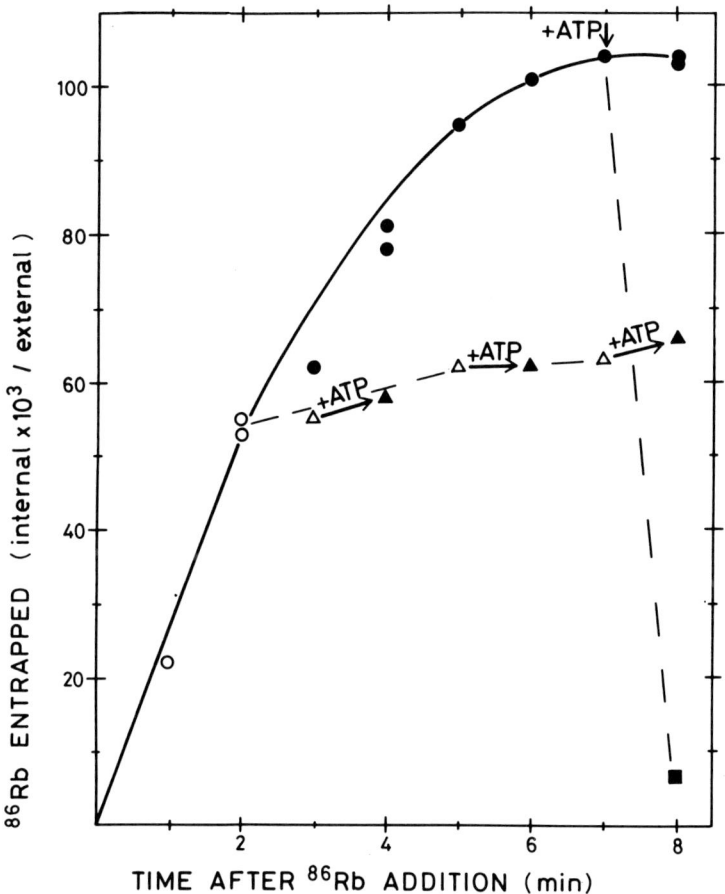

Figure 1. Rapid block of both the r-s-o and the i-s-o pump populations by 1 mM externally applied digoxin. Two μl 5 mM digoxin in 50% dimethyl sulfoxide (DMSO) (△) or 2 μl 50% DMSO alone (●) are added to 8 μl ATP-containing Na,K-ATPase-liposomes performing active Rb-uptake in the presence of 10 μM external RbCl (○). The addition of 10 mM external ATP activates the i-s-o pumps in the presence of DMSO alone (■) but has no effect if the solvent contains digoxin (▲).

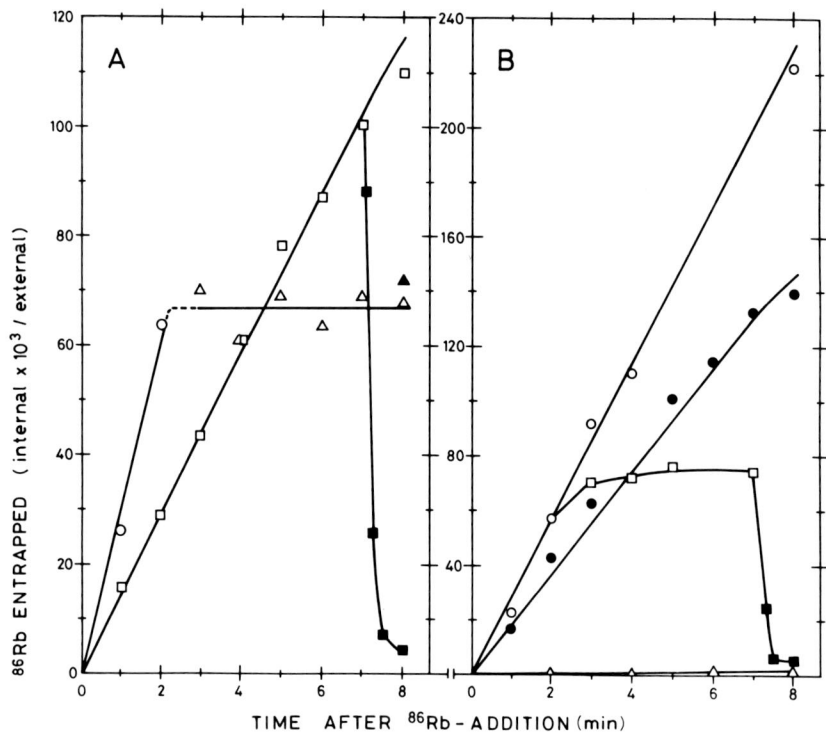

Figure 2A. Typical blockage of co-reconstituted r-s-o and i-s-o pumps by externally added digoxin. Active Rb-uptake (○) by ATP-containing Na,K-ATPase-liposomes is initiated by the addition of 4 μl 20 uM 86RbCl in transport-solution to 4 μl ATP-containing liposomes that are incubated and processed at 25°C. Digoxin (10 mM) is dissolved in DMSO. For each experimental point, 1 volume of this solution is mixed with 1 volume transport-solution immediately before use; 8 μl of the incubation mixture containing the liposomes and 10 μM 86RbCl are incubated for 2 min at 25°C. The effect of the 10% DMSO (□) used to dissolve the cardiac glycosides is controlled by mixing 1 μl of DMSO with 1 μl transport-solution and by adding it to 8 μl incubation mixture. Seven min after the solvent-addition, 2 μl of a 50 mM ATP stock solution are added to demonstrate that the 86Rb-extrusion by the i-s-o pumps is blocked by the entrapped digoxin (▲) but not by 10% DMSO (■).

(Continued)

Figure 2B. Typical selective blockage of the r-s-o oriented pump population by externally added ouabain. The experiment is performed as described in A except that, for each experimental point, 2 µl of a 5 mM ouabain solution are added, instead of digoxin, to 8 µl of the liposomes. Ouabain blocks the ^{86}Rb-uptake driven by the r-s-o pumps (□) but not the ^{86}Rb-extrusion mediated by the i-s-o pumps (■). Ouabain added together with RbCl blocks virtually the entire active transport driven by the r-s-o pumps (△); 10% DMSO decreases the active transport rate (●).

TABLE 1. Partial inhibition of r-s-o oriented pumps by high doses of palytoxin or linoleic acid.

	Concentration (µM)	Decrease of 86Rb-transport (%)*
palytoxin	10	64.8
	100	41.3
linoleic acid	500	52.7

*Mean of 2 measurements (15 min incubation at 25°C). The active Rb-uptake was measured as described in Figs. 1 and 2.

Fig. 3 illustrates the specific binding of tritiated digoxin to Na,K-ATPase molecules reconstituted in ATP-containing liposomes. The non-specific binding was determined by adding a 100-fold excess of unlabeled digoxin and was always less than 1% of the specific binding. This precise liposome binding-assay is based on the observation that the gel-filtration step at 0 C separates completely the free and the bound digoxin. The entrapped drug rapidly leaks out through the liposome membrane while the bound drug is not displaced. The same binding assay was performed with tritiated ouabain which, in contrast to the hydrophobic digoxin, does not enter the liposomes within the 10 min incubation time. Thus, ouabain binds only to the receptors exposed on the exterior liposome surface.

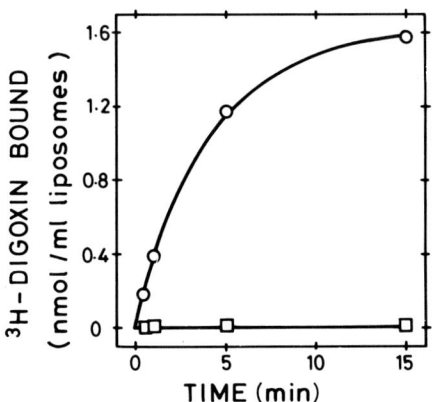

Figure 3. Specific binding of tritiated digoxin to the Na,K-ATPase reconstituted in liposomes. Na,K-ATPase-liposomes are prepared in a medium containing 50 mM ATP, 100 mM Na ions, 5 mM MgCl2, 1 mM EDTA, 30 mM histidine, pH 7.1 as described (Rey et al., 1987). They are incubated for 10 min at 25°C in the presence of 10 μM tritiated digoxin (NEN) with (O) or without (□) 1 mM cold digoxin. The unbound drug is removed by a 20 min elution in a gel column (Sephadex G-50 medium, 1 x 20 cm) at 0°C and the bound drug is measured by scintillation counting in a fraction of the eluate.

When cholate-dialysed Na,K-ATPase-liposomes are freeze-fractured, a single pump molecule appears as an intramembrane particle (Anner et al., 1984). In an attempt to visualize directly single drug molecules bound to intramembrane particles, liposomes were incubated with trititated drugs, recovered after the gel filtration step and sedimented. The pellets were processed for combined freeze-fracture and autoradiography (Carpentier et al., 1985).

Fig. 4 shows an autoradiograph of freeze-fractured
Na,K-ATPase-liposomes that were incubated in the presence
of tritiated ouabain. The location of the glycoside is
indicated by precipitates of silver nitrate filaments from
the photographic emulsion that was layered over the
freeze-fracture replica.

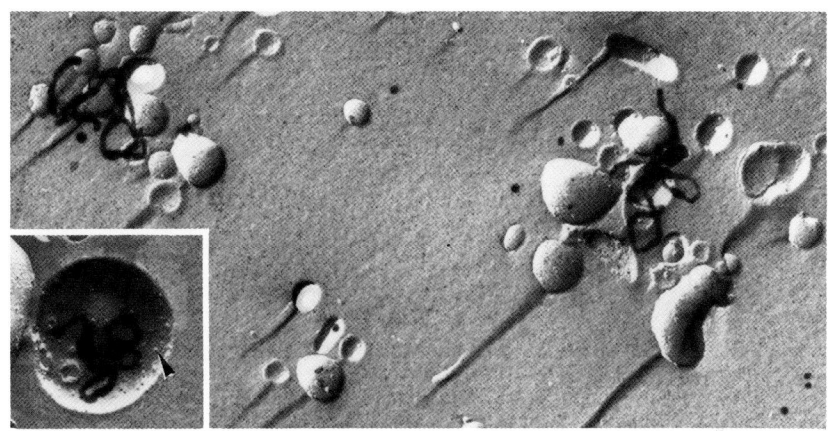

Figure 4. Autoradiograph of freeze-fractured liposomes
containing Na,K-ATPase associated with tritiated ouabain.
The r-s-o oriented Na,K-ATPase of liposomes was labeled by
a 10 min incubation at 25°C in the presence of tritiated
ouabain as described in Fig. 3. The eluted liposomes were
sedimented at 100 000 x g at 0°C and immediatly processed
first for freeze-fracture and, then, for autoradiography.
Na,K-ATPase molecules are visualized as 8-10 nm particles
(arrowhead in the inset) whereas ^{3}H-ouabain molecules are
located by deposits of curly silver filaments.

CONCLUSIONS AND PERSPECTIVES

Inside-out as well as right-side-out oriented
Na,K-ATPase molecules co-reconstituted in liposomes are
activated successively by timed addition of ATP to the
internal and external liposome compartment. Such
bi-functional liposomes represent a new tool for comparing
the membrane-permeability and molecular mechanism of known

or putative receptor ligands and other probes. In this miniaturized preparation, three essential drug characteristics, i.e., intra- and extracellular action and membrane permeability can be assessed at a time.

ACKNOWLEDGMENTS

The work was supported by the Swiss National Science Foundation, grants no. 3.536-0.83 and 3.404.86.

REFERENCES

Anner BM (1981) A K-selective channel formed by Na,K-ATPase in liposomes. Biochem. Int. 2: 365-371.
Anner BM, Moosmayer M, Rey HG (1988) Symmetric active transport in cholate-dialysed liposomes containing randomly oriented sodium pumps. This volume.
Anner BM, Robertson JD, Ting-Beall HP (1984) Characterization of (Na + K)-ATPase-liposomes I. Effect of enzyme concentration and modification on liposomes size, intramembrane particle formation and Na,K-transport. Biochim. Biophys. Acta 773: 253-261.
Carpentier JL, Brown D, Iacopetta B, Orci L (1985) Detection of surface-bound ligands by freeze-fracture autoradiography. J. Cell Biol. 101: 887-890.
Rey HG, Moosmayer M, Anner BM (1987) Characterization of (Na + K)-ATPase-liposomes. III. Controlled activation and inhibition of symmetric pumps by timed asymmetric ATP, RbCl, and cardiac glycoside addition. Biochim. Biophys. Acta 900: 27-37.
Schoner W, Moreth K, Kuske R, Renner D (1986). The "endogenous glycoside". In Erdmann E, Greef K, Skou JC (eds): "Cardiac Glycosides 1785-1985: Biochemistry- Pharmacology- Clinical Relevance," Darmstadt: Steinkopff Verlag, New York: Springer-Verlag, pp 135-142.

EFFECT OF PRIMAQUINE ON THE TOPOLOGY OF Na,K-ATPase AND THE RECEPTOR FOR ASIALOGLYCOPROTEINS.

G.J. Strous, P. van Kerkhof, A. van Bokhoven°, A.L. Schwartz*, and J.J.H.H.M. de Pont°

Laboratory for Cell Biology, University of Utrecht, °Department Biochemistry, University of Nijmegen, The Netherlands, and *Edward Mallinckrodt Departments of Pediatrics and Pharmacology, Washington University School of Medicine, Division of Pediatric Hematology-Oncology, Children's Hospital, St. Louis, USA

INTRODUCTION

Liver parenchymal cells possess a characteristic complement of membrane proteins both at the sinusoidal and at the bile canalicular membrane. Each of these proteins is constantly being inserted and removed through normal cellular processes which include membrane biogenesis, recycling and protein turnover. Membrane proteins are synthesized and transported to plasma membrane domains and at the same time are internalized and degraded. In addition, specific membrane proteins constantly recycle between their specific membrane domain and various intracellular compartments. This process, essential for the uptake of specific macromolecules via their receptors, is called receptor mediated endocytosis (Schwartz, 1984a) and requires both selective uptake mechanisms as well as sorting between proteins which recycle and those proteins which function exclusively at the cell surface.

To begin to understand the factors controlling these sorting mechanisms it is necessary to identify the exact site(s) of sorting between these classes of membrane proteins. Human hepatocytes possess two well-characterized membrane domains: basolateral (sinusoidal) and bile canalicular (apical), each with its own set of membrane proteins. In this study we have compared two sinusoidal membrane proteins: the parenchymal cell-specific receptor for asialoglycoproteins (ASGP-R) and Na,K-ATPase for their inter-organellar shuttling behavior. Na,K-ATPase catalyzes the ATP-driven pumping of Na^+ and K^+ ions and consist of a catalytic (α-chain) and a glycoprotein subunit (β-chain) in a 1:1 complex (Karin and Cook, 1983). Both proteins are abundantly present in the established hepatoma cell line HepG2, which grows as a polarized epithelium in culture and has retained essentially all specific liver functions providing an excellent model system (Geuze et al, 1983). In this contribution we provide evidence for a unique sorting mechanism, which is probably localized at the cell surface. We show that while the ASGP-R continuously recycles between the cell surface and intracellular organelles, Na,K-ATPase is maintained at the cell surface. Even under conditions in which essentially all ASGP-Rs are retained intracellularly Na,K-ATPase is retained at the plasma membrane.

METHODS

Cells. The human hepatoma cell line HepG2 (clone a16) was cultured in monolayer in minimal essential medium containing 10% decomplemented fetal bovine serum (MEM/FCS) (Schwartz et al, 1981). Confluent cultures were used in all experiments. Before the addition of primaquine (Sigma Chemical Company) or other reagents cells were incubated in serum free MEM for 30 min.
Labeling of HepG2 cells and analysis. For surface iodination near confluent HepG2 cells, grown on 60 mm plates, were chilled on ice, washed with phosphate buffered saline (PBS) and incubated on ice with 0.3 mCi/ml ^{125}I in the presence of lactoperoxidase and glucose oxidase. The reaction was quenched by addition of 0.1 M tyrosine and sodium metabisulfite. The cells were then either incubated in the presence or absence of 0.3 mM primaquine for 0 or 3h at 37°C, neuraminidase treated, and solubilized in 1% Triton X-100, 1 mM EDTA, and 0.1 mM phenylmethylsulfonyl fluoride or were solubilized without neuraminidase treatment. For metabolic labeling cells were pulse labeled with (^{35}S)methionine (800-1200 Ci/mmol, The Radiochemical Centre, Amersham) (20-30 μC/ml), chased with unlabeled methionine, and solubilized. For neuraminidase treatment the dishes were placed on ice and Vibrio cholera neuraminidase was added (30 min, 40 mU). Cells were then washed in PBS-EDTA to stop further neuraminidase action. Aliquots of the Triton X-100 soluble material were immunoprecipitated with rabbit anti-human ASGP-R (Schwartz et al, 1984b) or goat anti rabbit kidney Na,K-ATPase β-chain (Peters et al, 1984). Immunoprecipitations were carried out as previously described except that in case of the goat antiserum Staphyloccus aureus was added together with the antiserum and the immunoprecipitations were

rocked overnight at 4°C (Strous and Lodish, 1980). Isoelectric focussing (pH gradient 3-9) was done in one dimension in a minigel apparatus. Gel electrophoresis was performed in 10% polyacrylamide gels in the presence of sodium dodecylsulphate (SDS-PAGE).

RESULTS

Before we can definitively study the turnover and itinerary of Na,K-ATPase it is important to know how Na,K-ATPase is synthesized in the HepG2 cells. For this study we have focussed our attention to the β-chain of Na,K-ATPase, as its N-linked oligosaccharides can easily be used for identification purposes. Cells were labeled in the presence and absence of tunicamycin. If (^{35}S)methionine was present for 10 min, a precursor band of M_r=44,000 is present on SDS-PAGE (Fig. 1A). Labeling for 3h resulted in both precursor and mature (M_r=60,000) protein. When N-glycosylation was partly prevented by an intermediate concentration of tunicamycin it is clear that the β-chain of Na,K-ATPase is partly glycosylated. The three extra bands with apparent molecular weights of 42,000, 39,000 and 36,000 indicate that this protein normally contains 3 N-linked oligosaccharide chains.

Another feature of importance for the study of intracellular transport is the turnover of Na,K-ATPase. As can be seen in Fig. 1B the precursor polypeptide slowly disappears in the first 2-3 h after removal of radioactive methionine; but it is clear that even after 30h of chase there is still a small amount of precursor present in the cells. These polypeptides never reach the cell surface (not shown). Quantification of the radioactive mature band demonstrates that the Na,K-ATPase β-chain has a half-life of about 20h in HepG2 cells.

We wished to determine whether Na,K-ATPase is exclusively present at the cell surface or whether it also participates in endocytotic transport. Therefore, we have compared internalization of the ASGP-R with the behavior of Na,K-ATPase. Cells labeled with lactoperoxidase on ice were used for these studies. Under our conditions only proteins that are exposed to the medium are labeled. To see if these labeled proteins can enter the cell, the cells were first incubated at 37°C and thereafter returned to 0°C for treatment with neuraminidase. Under these conditions only proteins, available to the medium, are able to be de-sialylated. To visualize the fraction of Na,K-ATPase that enters the cell at 37°C the immunoprecipitated β-chain was analyzed by isoelectric focussing. As apparent

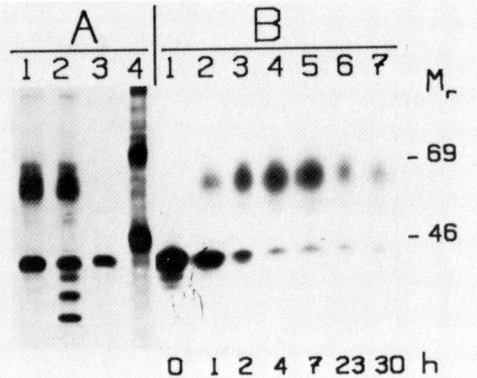

Figure 1A. *N-glycosylation of the Na,K-ATPase β-chain.* HepG2 cells were labeled for 10 min (lane 3) or 3h (lanes 1-2) with (^{35}S)methionine. The β-chain was isolated by immunoprecipitation. In lane 2 the cells were labeled and chased in 1 µg/ml tunicamycin. Lane 4, M_r standards.

Figure 1B. *Turnover of the β-chain of Na,K-ATPase.* Cells were labeled for 10 min with (^{35}S)methionine and chased as indicated.

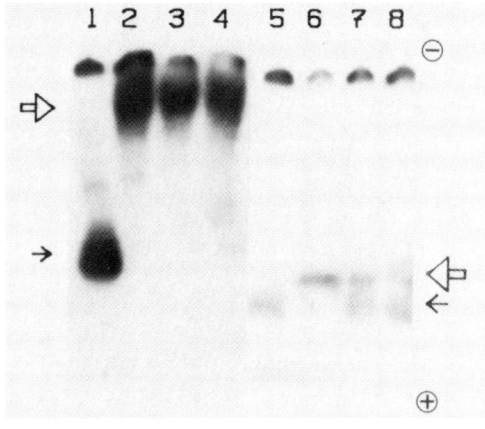

Figure 2. *Effect of primaquine on the distribution of Na,K-ATPase and ASGP-receptors.* Cells were surface-iodinated and incubated for 3h in the absence (lanes 3 & 7) or presence (lanes 4 & 8) of primaquine. Control cells (lanes 1-2, 5-6) were kept on ice after iodination. Lanes 2-4, and 6-8, cells treated with neuraminidase. The proteins were isolated by immunoprecipitation and analyzed by isoelectric focussing. Lanes 1 & 5, control cells not incubated with neuraminidase. Small arrow points to the molecules not affected by neuraminidase; big arrow points to molecules present at the outside of the cells.

from Fig. 2 the entire population of iodinated β-chain is still present at the outside of the cell after a 3h incubation period at 37°C. This time period is sufficient to reach complete equilibrium between intracellular compartments and the plasma membrane, as seen with the recycling of most cell surface receptors.

We next used the weak base primaquine to establish conditions in which at least one class of membrane proteins enters the cells without returning to the cell surface. Weak bases interfere with receptor mediated uptake of numerous ligands, including asialoglycoproteins (Tolleshaug and Berg, 1979), α_2-macroglobulin (Kaplan and Keogh, 1981), and low density lipoproteins (Basu et al, 1981). In addition, we have previously demonstrated that both receptor recycling and protein secretion is inhibited to the same extent in human hepatoma cells (Strous et al, 1985). In the present study we explore the effect of primaquine, on the behavior of Na,K-ATPase in particular as compared to the ASGP receptor. If the iodinated cells were incubated in the presence of 0.3 mM primaquine for 3h, more than 95% of Na,K-ATPase remained at the cell surface; only a trace amount was no longer accessible for neuraminidase (Fig. 2, lane 4).

We compared the effect of primaquine on the behavior of Na,K-ATPase to that of the asialoglycoprotein receptor. More than 50% of the ASGP-R is normally present inside the HepG2 cells, in equilibrium with receptors at the cell surface (Schwartz et al, 1982). Analysis of immunoprecipitated ASGP-R by isoelectric focussing after neuraminidase treatment of the cells shows that in the presence of primaquine about 50% of the ASGP-R is present inside the cell and 50% is at the plasma membrane (Fig. 2, lane 7). In the presence of primaquine almost all ASGP-R molecules are no longer accessible for neuraminidase. This result is also obtained if the cells are labeled in the presence of (^{35}S)-methionine and then chased for longer time.

We conclude that at any moment little or no Na,K-ATPase is present intracellularly. As primaquine does not interfere with the uptake of membrane proteins (Schwartz et al, 1984b) via coated pits these results indicate that Na,K-ATPase cannot enter these specialized areas of the cells, while receptor proteins have a strong preference for these membrane domains.

Supported by a grant of the Foundation for Medical Research MEDIGON (13.53.61) and NATO grant 0316/87.

REFERENCES

Basu SK, Goldstein JL, Anderson RGW, Brown MS (1981). Monensin interrupts the recycling of low density lipoprotein receptors in human fibroblasts. Cell 24:493-502.

Geuze HJ, Slot JW, Strous GJ, Schwartz AL (1983). The pathway of the asialoglycoprotein-ligand during receptormediated endocytosis: A morphological study with colloidal gold/ligand in the human hepatoma cell line HepG2. Eur J Cell Biol 32:38-44.

Kaplan J, Keogh EA (1981). Analysis of the effects of amines on inhibition of receptor-mediated fluid-phase pinocytosis in rabbit alveolar macrophages. Cell 24:925-932.

Karin NJ, Cook JS (1983). Regulation of Na,K-ATPase by its biosynthesis and turnover. Current Top. Memb Transp. 19:713-751.

Peters WHM, Ederveen AGH, Salden MHL, de Pont JJHHM (1984). Lack of immunological cross activity between the transport enzymes of Na,K-ATPase K^+/H^+-ATPase. Bioenerg Biomemb 6:223-232.

Schwartz AL, Fridovich SE, Knowles BB, Lodish HF (1981). Characterization of the asialoglycoprotein receptor in a continuous hepatoma line. J Biol Chem 256:8878-8881.

Schwartz AL, Fridovich SE, Lodish HF (1982). Kinetics of internalization recycling of the asialoglycoprotein receptor in a hepatoma cell line. J Biol Chem 257:4230-4237.

Schwartz AL (1984a). The hepatic asialoglycoprotein receptor. CRC Critical Rev Biochem 16:207-233.

Schwartz AL, Bolognesi A, Fridovich SE (1984b). Recycling of the asialoglycoprotein receptor and the effect of lysosomotropic amines in hepatoma cells. J Cell Biol. 98:732-738.

Strous GJ, Lodish HF (1980). Intracellular transport of secretory membrane proteins in hepatoma cells infected by Vesicular Stomatitis virus. Cell 22:709-717.

Strous GJ, DuMaine A, Zijderhand-Bleekemolen JE, Slot JW, Schwartz AL (1985). Effect of lysosomotropic amines on the secretory pathway on the recycling of the asialoglycoprotein receptor in human hepatoma cells. J Cell Biol 101:531-539.

Tolleshaug H, Berg T (1979). Chloroquine reduces the number of asialoglycoprotein receptors in the hepatocyte plasma membrane. Biochem Pharmacol 28:2919-2922.

NON-ESTERIFIED FATTY ACIDS AND THE CIRCULATING INHIBITOR OF Na,K-ATPase.

H.G.P. Swarts, J.A.H. Timmermans, F.M.A.H. Schuurmans Stekhoven, J.J.H.H.M. de Pont, S.J. Graafsma and T.A. Thien.

Departments of Biochemistry and Internal Medicine, University of Nijmegen, Nijmegen, The Netherlands.

INTRODUCTION

The presence of an inhibitor of Na,K-ATPase in human plasma has been reported in relation to the pathophysiology of essential hypertension (Blaustein, 1977). We have investigated the presence and the nature of such an inhibitor in plasma from pig and both normo- and hypertensive subjects.

MATERIALS AND METHODS

Isolation of the Circulating Inhibitor (CI). Pig blood in 2.85 mM EDTA, 3.4 mM EGTA and 2.65 mM gluthathione was centrifuged. The plasma was adjusted to pH 5.5, boiled for ten minutes, cooled and centrifuged. The clear supernatant was desalted on a C-18 affinity column (Waters). The CI was eluated with 50-100 % methanol and concentrated under nitrogen. Thin-layer chromatography was performed on activated HPTLC-plates, first with choloroform-methanol-acetic acid-water (40:10:10:1) and secondly, in the same demension, by the same mixture in a ratio of 40:15.3:6.3:1. Other analyses were performed as described before: Na,K-ATPase (Schoot et al., 1977), H,K-ATPase (Schrijen et al., 1980), phosphate (Broekhuyse, 1968) and non esterified fatty acids (NEFA) (Rogiers, 1977).

Blood pressure was measured with a sphygmomanometer. Mean arterial pressure (MAP) was calculated as the sum of diastolic blood pressure and 1/3 of the pulse pressure. Quetelet factor was calculated as weigth/length/length. The diagnosis of hypertension was made after exclusion of known causes according to the usual criteria.

RESULTS

A CI from pig plasma.

After deproteinating pig plasma, desalting and concentrating via a C-18 affinity column, a substance can be isolated which inhibits Na,K-ATPase in a dose dependent way.

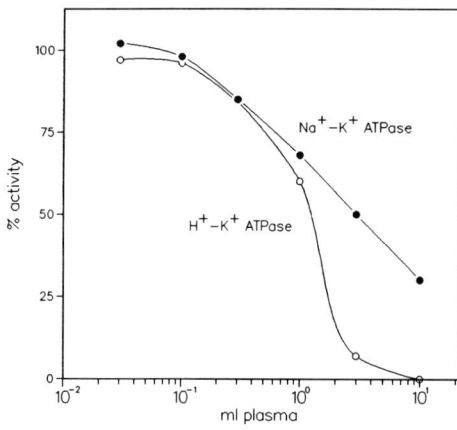

Figure 1. The effect of the CI isolated from pig plasma via a C-18 affinity column on Na,K-ATPase and H,K-ATPase activity.

Inhibition of Na,K-ATPase activity is very often correlated with the effect of ouabain and the CI is therefore often called an "ouabain-like" substance. Fig. 1 shows that this substance also inhibits the gastric mucosa enzyme H,K-ATPase. As H,K-ATPase is not affected by ouabain, the CI is not an "ouabain-like" substance.

The CI is an organic compound and not a contaminating heavy metal for after ashing (5 hours at 600 C) both the inhibition of the Na,K-ATPase and the H,K-ATPase activity are abolished. As also bovine serum albumin completly abolishes the CI effect on Na,K-ATPase (data not shown; see also Bidard et al., 1984), the data strongly indicate that this CI has also properties like NEFA.

We therefore further purified the CI by means of thin layer chromatography. Fig. 2 shows that after TLC two fractions can be isolated which inhibit Na,K-ATPase. (i). Fraction 2 contains after hydrolysis 0.59 nmol phosphate and 0.63 nmol non esterified fatty acids per ml orginal plasma, suggesting that it is a lysophospholipid. (ii). Fraction 10 contains 1.63 nmol NEFA per ml plasma.

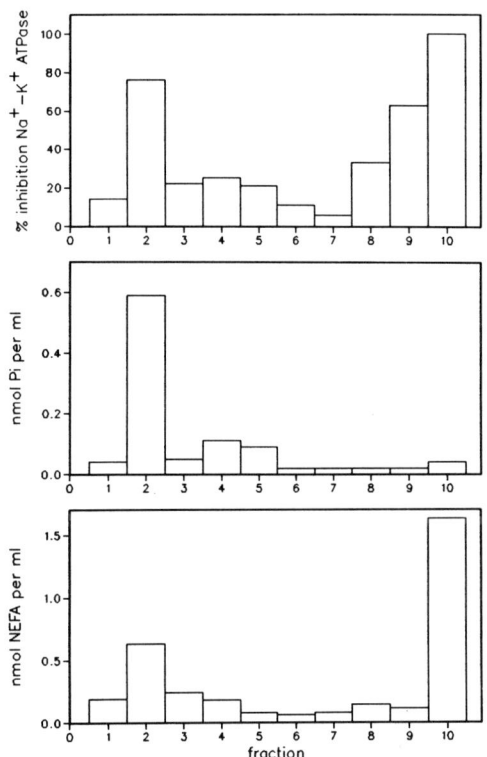

Figure 2. Purification of the CI by means of TLC. (Inhibition of Na,K-ATPase activity is tested with 14 ml orginal plasma in a total assay volume of 0.2 ml)

The effect of non esterified fatty acids on Na,K-ATPase.

From the previous data we know that NEFA can be isolated from plasma as an inhibitor of Na,K-ATPase. We have therefore studied the effect of NEFA on the Na,K-ATPase activity. NEFA inhibit the Na,K-ATPase activity in a dose dependent fashion. The degree of inhibition by the NEFA depends on the circumstances of the Na,K-ATPase assay. The data in Table 1 show that when the reaction time is increased and meanwhile the Na,K-ATPase concentration is decreased, the I-50 values decrease.

Table 1: Effect of NEFA on the Na,K-ATPase activity at 37 C. Mean values with SE are given for n=3.

Incubation time (min)		15	120
Na,K-ATPase conc. (µg per ml)		2.75	0.35
		I_{50} (µM)	
Myristic acid	(14:0)	86 ± 3	42 ± 15
Palmitic acid	(16:0)	>100	>100
Palmitoleic acid	(16:1)	26 ± 1	17 ± 1
Stearic acid	(18:0)	>100	>100
Oleic acid	(18:1)	16 ± 2	10 ± 2
Linoleic acid	(18:2)	20 ± 3	14 ± 2
Arachidonic acid	(20:4)	19 ± 3	13 ± 2

Fig. 3 shows that this shift in I-50 values of the NEFA for Na,K-ATPase is due to the change in the protein concentration during the assay. The ATP hydrolysis in the presence of oleic acid is not linear with the Na,K-ATPase concentration. The amount of inhibition decreases when Na,K-ATPase concentration is increased.

When the Na,K-ATPase concentration is kept constant and the incubation time is changed, no effect of the NEFA on the activity is observed (data not shown).

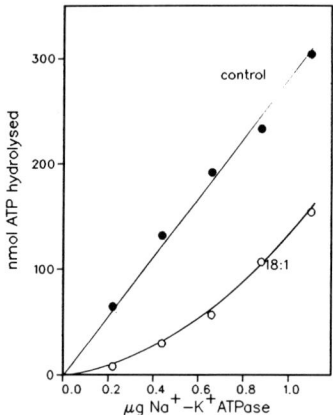

Figure 3. The effect of oleic acid (15 µM) on the amount of ATP hydrolysed by varying amounts of Na,K-ATPase. Incubation time 15 minutes.

Non-esterified fatty acids in human plasma.

From the above results we know that NEFA can inhibit Na,K-ATPase activity and that a fraction from plasma can be isolated which inhibits Na,K-ATPase and contains NEFA. We therefore studied the presence and identity of NEFA in plasma of normotensive and hypertensive individuals.

Table 2: Correlation between the mean arterial pressure (mm Hg) and the level (mM) and composition (% of total) of non-esterified fatty acids in plasma of normo- and hypertensive persons. Mean values with SE.

	NORMOTENSIVE	HYPERTENSIVE
n	18	31
age	40.4 ± 3.0	44.9 ± 1.7
Quetelet	23.0 ± 0.4	28.0 ± 1.0
MAP	85.1 ± 2.1	117.4 ± 2.4
NEFA (mM)	0.30 ± 0.02	0.49 ± 0.03
	%	
14:0	1.3 ± 0.2	1.2 ± 0.1
16:0	22.1 ± 0.7	22.7 ± 0.4
16:1	2.8 ± 0.3	3.3 ± 0.3
18:0	14.4 ± 0.7	11.9 ± 0.5
18:1	43.5 ± 0.8	45.4 ± 0.6
18:2	15.9 ± 1.2	15.4 ± 0.5

The data of Table 2 show that the plasma level of non-esterified fatty acids is significantly increased in hypertensive persons. Moreover the plasma NEFA correlates ($r=0.67$) with the mean arterial pressure (Fig. 4). The composition of the NEFA in two groups (NT and EHT) is not changed (Table 2).

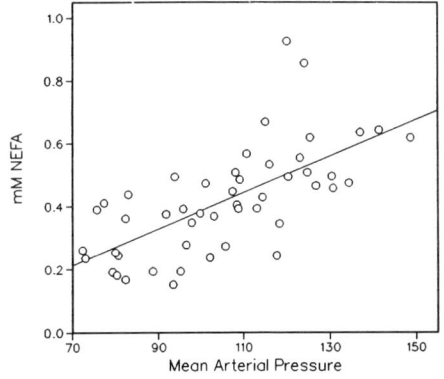

Figure 4. Relationship between the amount of NEFA in plasma and the MAP in hypertensive and normotensive persons. (NEFA = 0,00578*MAP - 0.192, r = 0.67)

CONCLUSIONS

As NEFA: i. inhibit Na,K-ATPase, ii. can be isolated from plasma, iii. are increased in plasma of patients with EHT, they can be a "circulating inhibitor" of Na,K-ATPase and may play a role in essential hypertension.

REFERENCES

Bidard JN, Rossi B, Renaud JF, Lazdunski M (1984). A search for an "ouabain-like" substance from the electric organ of electrophorus electricus which led to arachidonic acid and related fatty acids. Biochim Biophys Acta 769: 245-252.
Blaustein MP (1977). Sodium ions, calcium ions, blood pressure regulation, and hypertension: a reassessment and a hypothesis. Am J Physiol 232: C165-C173.
Broekhuyse RM (1968). Phospholipids in tissues of the eye I. Isolation characterization and quantitative analysis by two-dimensional thin-layer chromatography of diacyl and vinyl-ether phospholipids. Biochim Biophys Acta 152: 307-315
Rogiers V (1977). The application of an improved gas-liquid chromatographic method for the determination of the long chain non-esterified fatty acid pattern of blood plasma in children. Clin Chimica Acta 78: 227-233.
Schoot BM, Schoots AFM, de Pont JJHHM, Schuurmans Stekhoven FMAH, Bonting SL (1977). Studies on Na,K-ATPase. XLI Effects of N-ethylmaleimide on overall and partial reactions. Biochim Biophys Acta 483: 181-192.
Schrijen JJ, Luyben WAHM, de Pont JJHHM , Bonting SL (1980).Studies on H,K-ATPase. Essential argine residue in its substrate binding center. Biochim Biophys Acta 597: 331-344.

THE DECISIVE ROLE OF Na,K-ATPase IN STIMULATION PROCESS
OF T-LYMPHOCYTES

Márta Szamel[1,3], Mária Kövecses[1], P. Csermely[1],
L. Szollár[2], Margarete Goppelt[3], K. Resch[3], and
J. Somogyi[1]
[1]Inst. of Biochem.I., [2]Inst. of Pathophysiol.,
Semmelweis Univ. Med. Sch., Budapest (Hungary)
and [3]Div. of Mol. Pharmacol., Dept. of Pharmacol.
and Toxicol., Med. School, Hannover (F.R.G.)

INTRODUCTION

Na,K-ATPase is an ubiquitous integral constituent of cellular plasma membranes. The enzyme represents the monovalent cation pump and is the unique and specific receptor for cardiac glycosides (Schwartz and Collins, 1982). Numerous papers are dealing with the decisive role of Na,K-ATPase in stimulated T-lymphocytes or with the effect of ouabain to inhibit various aspects of lymphocyte differentiation and proliferation.

When lymphocytes are activated by antigens or mitogens a sequence of metabolic events is initiated leading to growth, differentiation and finally proliferation of the cells. Activation is initiated by the binding of antigen or mitogen to the receptors of the T-lymphocyte plasma membrane. Since the work of Quastel and Kaplan (1968,1970) it is known that activation of lymphocytes can be prevented by ouabain. Specific inhibition of several consequences of antigen or mitogen activation by ouabain may be interpreted as presumptive evidence for a requirement of the Na/K pump in these events (Quastel and Kaplan, 1968, 1970; Szamel et al., 1980; Szamel and Resch, 1981; Stoeck et al., 1983; Szamel et al., 1984). The effect of mitogens acting on intact cells to stimulate Na,K-ATPase has been abudantly documented among others by Averdunk and Lauf (1975); Averdunk (1976); Segel et al. (1979); Szamel et al. (1981,1984). However, the regulatory role of this process has not ultimately been established and other evidence was needed to determine the dependency of the activation process on activated functioning of Na,K-ATPase.

To get more direct information about the initiation of lymphocyte activation and on the possible regulatory role of Na,K-ATPase a different approach was used, e. g. to study the functional topology of the lymphocyte plasma membrane. Recently we have shown that in lymphocytes, cells with no apparent polarity, the plasma membrane is not entirely homogenous but exhibits a mosaic structure (Resch et al., 1981,1983; Szamel et al. 1987a). On the basis of affinity chromatography on Concanavalin A-Sepharose two subfractions of the plasma membrane could be separated. In the adherent fraction besides the high affinity Concanavalin A (Con A) receptor numerous enzymes were several fold enriched, including Na,K-ATPase and lysolecithin acyltransferase. The latter enzyme is responsible for the elevated fatty acid turnover of phospholipids in stimulated lymphocytes (Szamel et al. 1981; Resch et al. 1983).

In this paper further evidences are presented that lymphocytes contains specific membrane domains, furthermore that Na,K-ATPase is specifically implicated in the activation of lymphocytes and thus involved in the plasma membrane dependent regulation of cell growth and division.

EXPERIMENTAL PROCEDURES

Plasma membranes of calf thymocytes were isolated as described earlier (Resch et al. 1981). Affinity chromatography on ouabain-Sepharose, polyacrylamid gel electrophoresis and fatty acid determination were carried out according to Szamel et al. (1987a).

Mouse thymocytes were isolated and cultured in Eagle's modified essential medium as described by Szamel et al. (1987b). Permeabilisation of thymocytes was performed according to Bánfalvi et al. (1984). Isolation of serum lipoproteins and apoproteins, measurement of incorporation of ^{3}H-thymidine into DNA and that of ^{3}H-uridine into RNA were described elsewhere (Szamel et al. 1987b). Measurements of intracellular free calcium concentration with quin 2 were performed according to Csermely and Somogyi (1987).

RESULTS AND DISCUSSION

Separation of Plasma Membrane Subfractions from Calf Thymocytes on Ouabain-Sepharose

Ouabain prevents the mitogenic activation of lymphocytes in a concentration which did not affect ion transport. This suggests that Na,K-ATPase is involved in the growth regulation of lymphocytes independently of its ion pumping activity (Szamel et al. 1980).

The highly specific interaction between Na,K-ATPase and ouabain prompted further the idea to fractionate highly purified lymphocyte plasma membranes also by binding to immobilised ouabain. Indeed, using affinity chromatography on ouabain-Sepharose, plasma membranes of thymocytes could be fractionated. When highly purified plasma membrane vesicles were allowed to bind to ouabain-Sepharose in the presence of 140 mM NaCl, 0.5 mM Mg-ATP for 60 minutes, 75 per cent of the plasma membrane material eluted freely from the affinity gel, whereas 20 per cent were specifically bound to ouabain-Sepharose and could be eluted by 140 mM KCl in 20 mM HEPES, pH 7.0. Binding of ouabain to its receptor, e.g. Na,K-ATPase was an absolute requirement for fractionation as plasma membranes equilibrated with ouabain-Sepharose in the presence of 1 mM free ouabain or 140 mM KCl could be eluted in a single fraction.

The subfractions of the lymphocyte plasma membrane obtained by affinity chromatography exhibited different structural and functional properties. The specific activities of several marker enzymes, like γ-glutamyl-transpeptidase, alkaline phosphatase and Mg-ATPase were nearly identical in the unseparated membranes and in the subfractions. In contrast, some enzymes, i.e. acylCoA:lysolecithin acyltransferase and Na,K-ATPase itself were highly enriched in the adherent fraction as compared to the non-adherent one (Table 1).

These results were further substantiated by SDS-PAGE of the plasma membranes and the subfractions derived thereof. The overall polypeptide pattern of the plasma membrane and of both subfractions proved to be broadly similar by Coomassie blue staining. Among the major cytosceletal proteins actin was present at identical amounts in both subfractions, some plasma membrane proteins, however, were mar-

TABLE 1. Specific activities of membrane bound enzymes in plasma membrane subfractions isolated by affinity chromatography on ouabain-Sepharose from calf thymocytes

	activity (nmoles.mg protein^{-1}.min.$^{-1}$; 37°C)		
	plasma membrane	non-adherent fraction	adherent fraction
Mg-ATPase	132.0 ± 6.6	167.2 ± 8.3	139.6 ± 6.9
γ-GT	12.6 ± 0.6	12.8 ± 1.0	12.6 ± 0.8
AP	215 ± 10	238 ± 11	229 ± 11
Na,K-ATPase	40.3 ± 2.1	18.0 ± 0.9	67.5 ± 3.4
LAT	19.9 ± 1.1	14.0 ± 0.7	53.5 ± 2.5

(Results are means ± S.D. of three separate preparations, γ-GT, γ-glutamyl transpeptidase; AP, alkaline phosphatase; LAT, acyl-Coa: lysolecithin acyltransferase)

TABLE 2. Effect of lipoproteins on Con A-induced activation of Na,K-ATPase in mouse thymocytes

	activity (nmoles/5.10^6 cells.min^{-1}, 37°C)			
	not permeabilised		permeabilised	
10 ug/ml Con A	−	+	−	+
no addition	0.26±0.1	1.55±0.2	1.75±0.3	2.41±0.5
40 ug/ml VLDL	0.22±0.1	1.48±0.2	1.60±0.3	1.95±0.4
100 ug/ml VLDL	n. d.	1.15±0.2	1.28±0.2	1.24±0.2
25 ug/ml LDL	n. d.	1.26±0.2	0.73±0.3	1.66±0.3
50 ug/ml LDL	n. d.	1.15±0.2	1.50±0.2	1.44±0.2
25 ug/ml apo E	−	−	1.69±0.2	1.72±0.2

(Results are means±S. D. of four separate preparations, n.d., not detectable, VLDL, very low density lipoprotein; LDL, low density lipoprotein; apo E, apoprotein E)

kedly enriched in the adherent fraction compared to the non-adherent one. The most striking differences were observed in the amounts of 170, 150, 110, 94, 39 and 30 kD polypeptides. The phospholipid fatty acid composition of the plasma membrane subfractions proved to be different, as well. Whereas the amounts of the saturated fatty acids, palmitic and stearic acid were enhanced in the phospholipids of the adherent fraction, a concomitant decrease in the amounts of unsaturated fatty acids was observed in the adherent fraction as compared to the non-adherent one.

Taken together, these results supported our hypothesis on the existence of plasma membrane domains of lymphocytes, consisting a set of plasma membrane (glyco)proteins, among others some functionally important enzymes embedded in a phospholipid milieu, distinct from that of the bulk membrane. As the biochemical composition as well as the functional properties of the plasma membrane subfractions isolated by ouabain-Sepharose were nearly identical to those isolated by affinity chromatography on Concanavalin A-Sepharose, it strenghtens the idea of the domain structure of the T-lymphocyte plasma membrane.

Effect of Lipoproteins on the Con A-Induced Activation of Plasma Membrane Enzymes of Mouse Thymocytes

If one assumes that (mitogen) receptor associated enzymes, including Na,K-ATPase, are involved in the initiation of lymphocyte activation it involved the assumption that modulators (regulators) of lymphocyte activation would influence their activities. As shown in our earlier work, binding of ouabain to cell surface receptors of the lymphocyte plasma membrane inhibited the mitogen-induced activation of both Na,K-ATPase and of acylCoA: lysolecithin acyltransferase (Szamel et al., 1981) and thus interfered with the stimulation of T-lymphocytes.

TABLE 3. Effect of lipoproteins on Con A-induced activation of acylCoA: lysolecithin acyltransferase in mouse thymocytes

	nmol ^{14}C-oleate incorporated / 10^7 cells	
10 ug/ml Con A	−	+
No addition	0.38 ± 0.05	0.72 ± 0.09
10 ug/ml VLDL	0.38 ± 0.07	0.66 ± 0.08
25 ug/ml VLDL	0.38 ± 0.07	0.64 ± 0.08
50 ug/ml VLDL	--	0.37 ± 0.07
100 ug/Ml VLDL	0.32 ± 0.07	0.30 ± 0.07
1 ug/ml LDL	0.38 ± 0.06	0.66 ± 0.08
10 ug/ml LDL	0.37 ± 0.04	0.41 ± 0.06
50 ug/ml LDL	0.32 ± 0.04	0.35 ± 0.05
10 ug/ml apo E	0.35 ± 0.07	0.41 ± 0.05

(Results are means ± S.D. of four separate preparations, cells were preincubated for 6 hours before determination of enzyme activity)

Similarly, serum lipoproteins, well known inhibitors of lymphocyte activation interfered with the Con A-induced activation of both Na,K-ATPase and of acylCoA: lysolecithin acyltransferase in mouse thymocytes (Table 2 and 3).

TABLE 4. Effect of serum lipoproteins on the activities of plasma membrane-bound enzymes of Con A-stimulated mouse lymphocytes

	activity (nmoles.5×10^6 cells^{-1}.min, 37°C)			
	not permeabilised		permeabilised	
10 ug/ml Con A	−	+	−	+
alkaline phosphatase				
no addition	1.4±0.1	1.2±0.08	1.1±0.09	1.1±0.06
100 ug/ml VLDL	−	−	1.2±0.08	1.3±0.04
50 ug/ml LDL	−	−	1.2±0.03	1.3±0.05
γ-glutamyl-transpeptidase				
no addition	0.87±0.02	0.68±0.06	0.89±0.01	0.88±0.08
100 ug/ml VLD1	−	−	1.25±0.08	1.07±0.03
50 ug/ml LDL	−	−	1.27±0.04	1.08±0.03
Mg-ATPase				
no addition	6.31±1.61	6.07±1.04	7.31±1.97	6.21±1.04
100 ug/ml VLDL	−	−	7.11±0.91	6.51±1.40
50 ug/ml LDL	−	−	7.06±0.96	5.94±0.72
25 ug/ml apo E	−	−	7.36±1.12	6.51±0.97

(Results are means ± S.D. of four separate preparations)

While the activation of the enzymes associated with the high affinity mitogen receptors were prevented by very low density and low density lipoproteins as well as by their apoprotein component, non of the enzymes, randomly distributed in the plasma membranes e.g. Mg-ATPase, nor alkaline phosphatase or γ-glutamyl transpeptidase were influenced by the lipoproteins (Table 4).

Similarly, the Con A-induced rapid elevation of intracellular free calcium concentration in mouse thymocytes was not either affected by lipoproteins. All these findings further support the role of receptor associated domains, specially the decisive function of Na,K-ATPase in the physiological as well as in the pharmacological regulation of lymphocyte activation.

REFERENCES

Averdunk R (1976). Early changes of "leak flux" and the cation content of lymphocytes by Concanavalin A. Biochem Biophys Res Comm 70: 101-109.
Averdunk R, Lauf PK (1975). Effects of mitogen on sodium-potassium transport, ^{3}H-ouabain binding, and adenosine triphosphatase activity in lymphocytes. Exp. Cell. Res 93: 331-342.
Bánfalvi G, Soóki-Tóth Á, Sarkar N, Csuzi S, Antoni F (1984). Nascent DNA chains synthesized in reversibly permeable cells of mouse thymocytes. Eur J Biochem 139: 553-559.
Csermely P, Somogyi J (1987). The possible pitfalls of the measurement of intracellular Ca concentration in T lymphocytes with the fluorescent dye, quin 2. Immunobiology 174: 980-987.
Quastel MR, Kaplan GJ (1968). Inhibition by Ouabain of Human Lymphocyte Transformation Induced by Phytohaemagglutinin in vitro. Nature 219: 198-200.
Quastel MR, Kaplan GJ (1970). Lymphocyte Stimulation: The Effect of Ouabain on Nucleic Acid and Protein Synthesis. Exp Cell Res 62: 407-420.
Resch K, Schneider S, Szamel M (1981). Separation of Right-side-Out-Oriented Subfractions from Purified Thymocyte Plasma Membranes by Affinity Chromatography on Concanavalin A-Sepharose. Anal Biochem 117: 282-292.
Resch K, Schneider S, Szamel M (1983). Characterization of functional domains of the lymphocyte plasma membrane. Biochem Biophys Acta 733: 142-153.
Schwartz A, Collins JH (1982). Na^+/K^+-ATPase, structure and mechanism of action of digitalis. In Martonosi AN (ed) "Membranes and Transport" Vol 1 New York: Plenum Press pp 521-527.
Segel GB, Kovach G, Lichtman MA (1979). Sodium-potassium adenosine triphosphatase activity of human lymphocyte membrane vesicles: kinetic parameters, substrate specificity, and effects of phytohemagglutinin. J. Cell Physiol 100: 109-118.
Stoeck M, Northoff H, Resch K (1983). Inhibition of mitogen-induced lymphocyte proliferation by ouabain: interference with interleukin 2 production and interleukin 2 action. J. Immunol 131: 1433-1437.
Szamel M, Somogyi J, Csukás I, Solymossy F (1980). Effect of ouabain on macromolecular synthesis during the cell cycle

in mitogen-stimulated human lymphocytes. Biochem Biophys Acta 633: 347-360.

Szamel M, Resch K (1981). Inhibition of lymphocyte activation of membrane phospholipid metabolism. Biochem Biophys Acta 647: 297-301.

Szamel M, Schneider S, Resch K (1981). Functional interrelationship between $(Na^+ + K^+)$-ATPase and lysolecithin acyltransferase in plasma membranes of mitogen stimulated rabbit thymocytes. J. Biol Chem 256: 9198-9204.

Szamel M, Somogyi J, Seebass R, Schneider S, Resch K (1984). Functional interrelationship between (Na+K)ATPase and lysolecithin acyltransferase in the plasma membrane domains of lymphocytes. In Trón L, Damjanovich S, Fonyó A and Somogyi J (eds) "Membrane dynamics and transport of normal and tumor cells" Budapest: Akadémiai Kiadó, pp. 181-193.

Szamel M, Goppelt M, Bessler W, Wiesmüller KH, Resch K (1987a). Separation of plasma membrane domains of calf thymocytes by affinity chromatography on ouabain Sepharose. Biochem Biophys Acta in press.

Szamel M, Kövecses M, Csermely P, Szollár L, Somogyi J (1987b). Serum lipoproteins inhibit the early activation of plasma membrane bound enzymes in mouse thymocytes. Biochem Biophys Acta submitted for publication.

MEMBRANE PHOSPHORYLATION IS INVOLVED IN THE INHIBITORY EFFECT OF MICROMOLAR Ca^{2+} ON RAT MYOMETRIAL Na^+K^+-ATPase

A. Turi and J. Somogyi

Institute of Biochemistry I., Semmelweis Medical School, P.O.B. 260. H-1444 Budapest, 8. (Hungary)

INTRODUCTION

Ca^{2+} is a potent inhibitor of Na^+K^+-ATPase in the millimolar range in all tissues investigated, but sensitivity of Na^+K^+-ATPase activity against micromolar Ca^{2+} was reported only in erythrocytes so far (Yingst and Marcovitz 1983).

We have previously reported the inhibition of the rat myometrial Na^+K^+-ATPase by micromolar Ca^{2+}. This effect could be abolished by SDS pretreatment of microsome preparation. Dimethyl sulfoxide (DMSO) proved to protect the Ca^{2+} sensitivity of SDS treated enzyme (Turi and Török 1985).

cAMP has been shown to influence the Na^+K^+-ATPase in a number of tissues (Lingham and Sen 1982, Giesen et al. 1984). Our results presented here indicate that cAMP dependent-protein kinase is involved in the modulation of Na^+K^+-ATPase activity by Ca^{2+} in the myometrial plasma membrane.

EXPERIMENTAL PROCEDURES

SDS as well as DMSO+SDS treated microsome fraction was prepared as described previously (Jørgensen 1974, Turi and Török 1985). Plasma membrane fraction was purified from microsomes by the method of Pitts (1979). Na^+K^+-ATPase activity was determined in terms of the ATP hydrolysis inhibited by 1 mM ouabain in the presence of 50 mM Tris-HCl (pH 7.4), 5 mM ATP, 5 mM $MgCl_2$, 20 mM KCl, 100 mM NaN_3, 1 mM EGTA and Ca^{2+} in a concentration indicated. The free Ca^{2+} content at different Ca^{2+}/EGTA ratios was measured with ionselective Ca-electrode (Radelkis, Hungary). Preparations were phosphorylated in the presence of 0.1 mM (gamma-^{32}P)ATP, 5 mM $MgCl_2$ and 30 mM Tris-HCl (pH 7.4) at $0°C$ for 10 min. SDS polyacrylamide gel electrophoresis was performed according to Laemmli (1974). ^{32}P incorporation was visualised by autoradiography using Kodak X-ray films.

RESULTS AND DISCUSSION

Na^+K^+-ATPase activity determined in the presence of SDS could be inhibited by micromolar Ca^{2+} both in the microsome and plasma membrane fractions of rat myometrium. (Table 1). This effect of Ca^{2+} proved to be reversible, since it was abolished by excess of EGTA (Fig. 1).

TABLE 1. Effect of Ca^{2+} on the Na^+K^+-ATPase activity of myometrial microsome and plasma membrane fractions

Addition	$\bar{x}$ Na^+K^+-ATPase activity (umol P_i/mg protein/hour)		Inhibition by Ca^{2+}
	1 mM EGTA	3 uM Ca^{2+}	%
microsome fraction	6.7	3.7	45
plasma membrane fraction	15.1	9.2	39

$\bar{x}$ 1 mg of microsome or plasma membrane fraction was incubated in the presence of 0.5 mg SDS at 20 °C for 10 min before the enzymatic assay.

When the microsome fraction was treated by SDS, about 50 per cent of proteins was solubilized and the specific activity of Na^+K^+-ATPase enhanced twice. However, the Ca^{2+} sensitivity of enzyme disappeared almost completely. On the other hand, the presence of DMSO during the SDS treatment did not influence significantly the solubilization, but preserved the Ca-sensitivity of the Na^+K^+-ATPase (Table 2).

To investigate what kind of solubilized membrane component(s) could be responsible for the Ca^{2+} sensitivity, some suitable compounds were tested (Table 3). The calmodulin antagonist, calmidazolium remained without effect on DMSO+SDS treated preparation. However, the heat stable cAMP dependent-protein kinase inhibitor completely abolished the

Ca^{2+} - sensitivity of the Na^+K^+-ATPase.

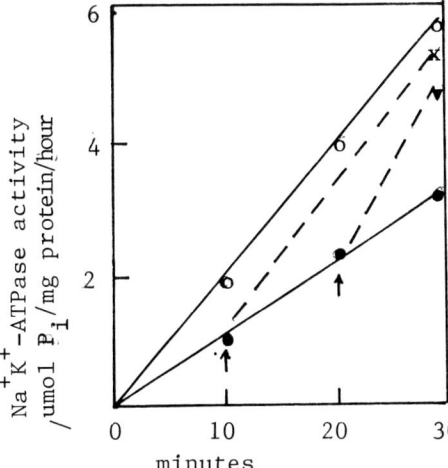

Na^+K^+-ATPase activity of microsome fraction was measured in the presence of (o) 1 mM EGTA, and (•) 3 µM free Ca^{2+}. (x, ▼) 5 mM EGTA (final concentration) was added in 10 µl volume when the arrowes indicate.

Figure 1. Reversibility of the Ca^{2+} inhibition

TABLE 2. Ca^{2+} sensitivity of SDS and DMSO+SDS treated preparations

Addition	Na^+K^+-ATPase activity (µmol P_i/mg protein/hour)		Inhibition by Ca^{2+}
	1 mM EGTA	3 µM Ca^{2+}	%
microsome fraction	6.75 0.56	3.65 0.54	46
SP fraction	15.6 1.1	14.6 1.3	7
DSP fraction	17.0 0.9	11.0 1.1	35

(SP and DSP: Pellets obtained by centrifugation of microsomes at 100 000 x g for 60 min followed treatment with 0.5 mg SDS or 40 per cent DMSO plus 0.5 mg SDS)

TABLE 3. The effect of calmidazolium, cAMP dependent-protein kinase inhibitor and calmodulin on the Ca^{2+} sensitivity of myometrial Na^+K^+-ATPase activity

Addition	Na^+K^+-ATPase activity (umol P_i/mg protein/hour)		Inhibition by Ca^{2+}
	1 mM EGTA	3 uM Ca^{2+}	%
DSP	18.9	9.6	49
DSP + 5×10^{-6} M calmidazolium	17.8	9.8	45
DSP + 50 ug cAMP dependent protein kinase inhibitor	15.1	13.7	9.3
SP	14.4	12.47	13.4
SP + 5 ug calmodulin	13.8	12.38	10.4

These results suggest that the Ca^{2+} sensitivity of the myometrial Na^+K^+-ATPase activity is dependent on the presence of functional cAMP dependent-protein kinase. Indeed, in the DMSO+SDS treated preparation active protein kinases existed as demonstrated in phosphorylated gel patterns, while the incorporation of ^{32}P in the SDS treated microsomes proved to be very poor (Fig. 3). It is meaning, that Ca^{2+} enhanced the ^{32}P incorporation in a few protein bands. As show in Fig. 4 phosphorylation of certain proteins of the DMSO+SDS treated preparation were reduced by cAMP dependent-protein kinase inhibitor. Furthermore, two distinct substrates could be phosphorylated by exogenously added catalytic subunit of cAMP dependent-protein kinase in the SDS treated preparation.

The proteins phosphorylated both by endogenous and exogenous cAMP dependent-protein kinase are supposed to be involved in the mechanism of low concentration Ca^{2+} inhibition. Recently, many Ca^{2+} binding proteins were demonstrated in the myometrial plasmamembrane (Grover 1986). The further steps of our investigations are the isolation of this substrate proteins and examination of its Ca^{2+} bindig abilities.

Phosphorylation is Involved in Ca^{2+} Sensitivity / 461

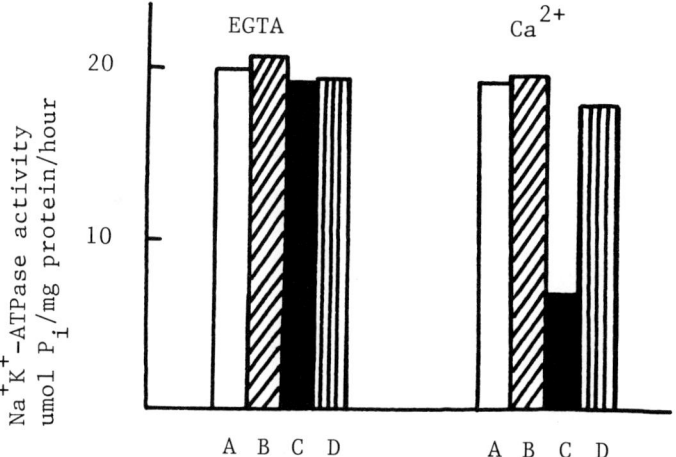

Figure 2. Effect of catalytic subunit of cAMP dependent-protein kinase and its inhibitor on the Ca^{2+} sensitivity of Na^+K^+-ATPase activity
A: control SP preparation; B: plus 50 ug cAMP dependent-protein kinase inhibitor; C: plus 5 ug catalytic subunit of cAMP dependent-protein kinase; D: plus 5 ug catalytic subunit and 50 ug inhibitor

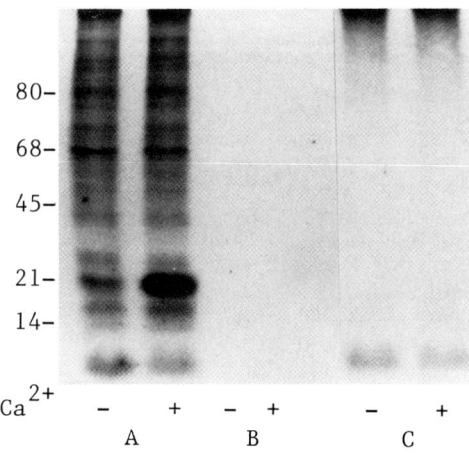

Figure 3. Phosphorylation of (A) DMSO+SDS treated preparation, (B) 100000xg supernatant after SDS treatment and (C) SDS treated preparation

The reaction mixtures contained 5.3 uM free Ca^{2+} as indicated.

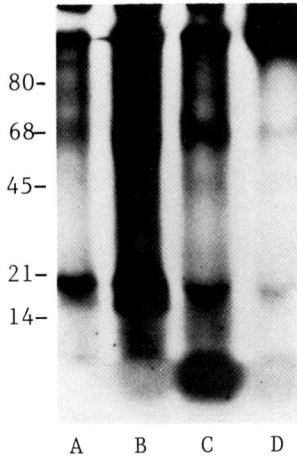

Figure 4. Effect of catalytic subunit of cAMP dependent-protein kinase and its inhibitor on the phosphorylation of DMSO+SDS and SDS treated preparation
(A) DMSO+SDS treated preparation plus cAMP dependent-protein kinase inhibitor (20 ug)
(B) DMSO+SDS treated preparation
(C) SDS treated preparation plus catalytic subunit of cAMP dependent-protein kinase (5 ug)
(D) SDS treated preparation

The samples were phosphorylated in the presence of 5.3 uM Ca^{2+}

REFERENCES

Giesen EM, Imbs JL, Grima M, Schmidt M and Schwartz J (1984). Modulation of renal ATPase activities by cyclic AMP. Biochem Biophys Res Commun 120: 619

Grover AK (1986). Calcium binding proteins in smooth muscle membrane. Cell Calcium 7:101

Jorgensen PL (1974). Purification and characterization of $(Na^{+}+K^{+})$-ATPase: III. Purification from the outer medulla of mammalian kidney after selective removal of membrane components by sodium dodecylsulphate. Biochim Biophys Acta 356:35

Laemmli UK (1970). Cleavage of structural proteins during assembly of the head of Bacteriophage T4. Nature 227:680

Lingham R and Sen AK (1982). Regulation of rat brain $(Na^{+}+K^{+})$-ATPase by cyclic AMP. Biochim Biophys Acta 688:475

Pitts BJR (1979). Stoichiometry of sodium calcium exchange in cardiac sarcolemmal vesicles. J Biol Chem 254:6232

Turi A and Török K (1985). Myometrial $(Na^{+}+K^{+})$ activated ATPase and its Ca^{2+} sensitivity. Biochim Biophys Acta 818:123

Yingst DR and Marcovitz MJ (1983). Effect of hemolysate an calcium inhibition of the $(Na^{+}+K^{+})$-ATPase activity of human red blood cells. Biochem Biophys Res Commun 111:970

EFFECTS OF ALDOSTERONE ON Na,K-ATPase TRANSCRIPTION, mRNAs, AND PROTEIN SYNTHESIS, AND ON TRANSEPITHELIAL Na+ TRANSPORT IN A6 CELLS.

F. Verrey 1), E. Schaerer 2) P. Fuentes 3)
J.P. Kraehenbuhl 2) and B.C. Rossier 1)

Institute of Pharmacology and Biochemistry,
University of Lausanne, Rue du Bugnon 27,
CH-1005 Lausanne, Switzerland

INTRODUCTION

Mineralocorticoid hormones promote sodium reabsorption across tight epithelia such as the distal nephron, the colon and the toad urinary bladder (Rossier et al., 1985).

These steroid hormones bind to intracellular receptors which are in turn activated and regulate the transcription of certain genes. The product of these genes and other indirect induced or repressed proteins constitute a hormonal domain. This modulation of certain proteins results in an increased transepithelial sodium reabsorption.

Na,K-ATPase has previously been identified as one of the mineralocorticoid (aldosterone) induced proteins. In the toad-bladder system its biosynthesis is increased two- to three fold after a lag time of three to six hours. Its induction is mediated by the occupancy of the type I (high affinity) mineralocorticoid receptor and is inhibited by the competitive antagonist spironolactone, by the transcription inhibitor actinomycin D, but not by the apical sodium channel blocker amiloride (Geering et al., 1982), suggesting that its induction is mediated by transcriptional regulation of its genes or, alternatively of a regulatory protein.

Na,K-ATPase is an integral membrane protein composed by two subunits (alpha- or catalytic- and beta-subunit). In tight epithelia, sodium ions enter the cell via apical, amiloride-sensitive sodium channels, and are extruded from the cells into the basolateral fluid by the Na,K-ATPase. This enzyme generates, therefore, the driving force for transepithelial sodium transport.

A6 cells, derived from the kidney of Xenopus laevis, are an appropriate system to study the action of aldosterone on transepithelial sodium transport. They form a monolayer and display a mineralocorticoid responsive transepithelial sodium transport when plated on a collagen coated filter (Handler et al., 1983, Paccolat et al., 1987). To study the regulation of Na,K-ATPase in this system at the molecular level, we cloned and sequenced cDNAs coding for both alpha and beta subunit of A6 cells (Verrey et al., manuscript in preparation). We have demonstrated by Northern blot analysis and immunoprecipitation of biosynthetically labeled proteins, that the induction of Na,K-ATPase is mainly mediated by the accumulation of both alpha and beta-subunit mRNAs (Verrey et al., 1987). In this study we show that these changes in specific mRNA pools are preceded by an increase in Na,K-ATPase gene transcription.

RESULTS

A6 cells were grown 14 days on collagen coated polycarbonate filters and incubated 6 and 96 hours with 300nM aldosterone or 6 hours with 300uM spironolactone, a competitive inhibitor of mineralocorticoid hormones. Sodium transport was measured in a modified Ussing chamber and cells were collected for the measurement of RNA levels, or pulse labeled 45 min. with 35S-methionine before collection for the assessment of the protein synthesis rate. The results of 2 to 4 independent experiments for the biochemical data and 7 to 12 filters for the values of short circuit current are summarized in figure 1.

On Northern blots, α-subunit appears as a single 3.9 kb band. On the same blots, cytoplasmic actins were measured as a negative control for induction. β-subunit mRNA appears as a major band at 2.6 kb. On SDS polyacrylamide gel electrophoresis, immunoprecipitated α-subunit forms a single 98kD band and β-subunit a 42 kD core-glycosylated and a 49 terminally glycosylated band.

During the first 6 hours of aldosterone stimulation, the biosynthesis of the α- and the β-subunit increases about

2-fold relative to the total protein synthesis. We now show that this change in the rate of protein synthesis is mainly mediated by an accumulation of the respective mRNAs. This induced level of Na,K-ATPase mRNAs and protein synthesis is maintained over 96 hours under these culture conditions.

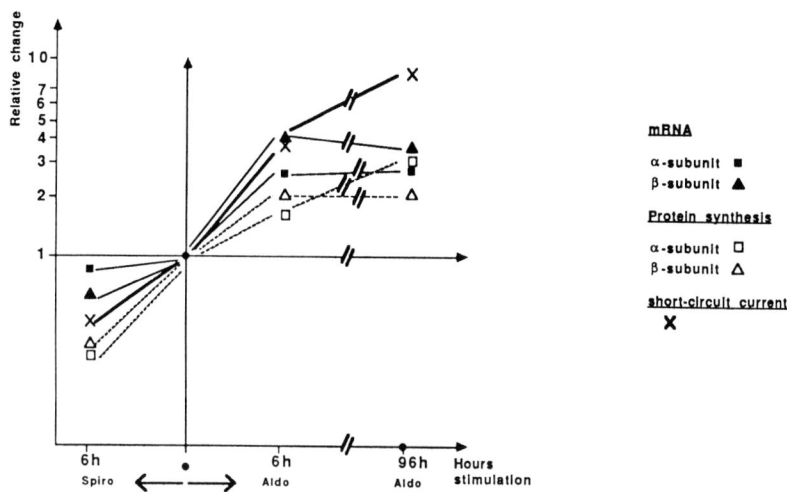

Figure 1. Effect of aldosterone and spironolactone on A6 cells grown on filters. The values obtained by scanning the autoradiographs of Northern blots and SDS-page and by measuring the short circuit current are expressed as fractional changes. Control values (zero time) are set arbitrarily to zero. Aldosterone was 300nM and spironolactone 300uM

The aldosterone-induced biochemical changes are paralleled during the first six hours by an increase in transepithelial Na^+ transport. A further increase during prolonged stimulation (96 hrs) suggests that there are additional sodium transport regulatory mechanisms, which involve other effectors, such as the apical, amiloride sensitive sodium channel.

We have demonstrated that the same cells, when plated on plastic dishes, respond initially to aldosterone by the same type of biochemical changes at the level of the sodium pump. But in contrast to the cells grown on filters, the response is not maintained over longer periods of stimulation (24 to 96 hours, data not shown).

We have investigated whether protein synthesis is necessary for the induction of Na,K-ATPase mRNAs by aldosterone. High doses of cycloheximide, which inhibit over 90% of protein synthesis, cause drastic changes in the cell monolayer, as documented by a complete fall in the transepithelial resistance. At 0.5 µg/ml cycloheximide, about 75% of the total protein synthesis is inhibited, without any change in the transmural resistance. Over 90% of the aldosterone-dependent increase in Na^+ transport is inhibited by 0.5 µg/ml cycloheximide but the accumulation of specific mRNAs in response to aldosterone is fully maintained (data not shown). From these data, we can conclude that aldosterone regulates the level of Na,K-ATPase mRNAs even when the overall rate of protein synthesis is severely decreased. We nevertheless can not yet decide whether the aldosterone-induced elevation of Na,K-ATPase mRNA levels is independent of protein synthesis or not.

To determine whether the observed changes in specific mRNAs induced by aldosterone are due to an increase in gene transcription or to a change at a posttranscriptional level (e.g. mRNA stability), we adapted a nuclear$Rtn-gn transcription assay (Schibler et al., 1983) to the A6 cells plated on filters. Approximately 10^6 isolated, RNase treated nuclei are submitted to in vitro RNA chain elongation. Equal amounts of extracted labeled RNA (0.15 or 0.3×10^6 cpm) are hybridized to filter-bound cDNAs. Hybridization and washes are performed according to the manufacturers protocols (Genescreen, du Pont) with some modifications: hybridization temperature is 50°C and RNAse A and RNAse T1 are used in the washing procedure to decrease background and non-specific hybridization. Hybridized RNA is visualized by autoradiography and quantified by densitometric scanning.

As shown in table 1, treatment for 45 minutes with 300nM aldosterone increases the relative number of RNA polymerase II transcribing the tested genes. Type 3 cytoskeletal actin (Mohun et al., 1984) transcription is decreased by 25-65% and the transcription of Na,K-ATPase α- and β-subunits is increased by 164 and 224% respectively. These changes in the rates of gene transcription could fully account for the changes observed at the level of the specific mRNA accumulation.

TABLE 1. Effect of aldosterone on Na,K-ATPase gene transcription

	type 3 actin	Na,K-ATPase	
		α-subunit	β-subunit
Control	1	1	1
Aldosterone * (45 min)	0.56 (0.37-0.74)	1.64 (1.52-1.76)	2.24 (2.12-2.36)

*Mean of 2 independent experiments expressed as fractional change: test/control (range is indicated in parenthesis). Control is arbitrarily set to 1.0. 300 nM aldosterone was added to the media 45 min. prior to cell collection.

DISCUSSION

Na,K-ATPase is basolaterally-located, and generates the driving force for transepithelial Na^+ transport. The induction of its biosynthesis by Aldosterone has been reported earlier (Geering et al.,1982, Paccolat et al.,1987). We have now demonstrated that the changes in the rate of α- and β-subunit synthesis are mainly mediated by an accumulation of specific mRNAs. We have addressed the question of whether this aldosterone induced change in the Na,K-ATPase mRNA pools could be accounted for by a change in the rate of specific gene transcription. Performing nuclear run-on transcription assays, we could show that indeed Na,K-ATPase gene transcription is increased and could fully explain the observed change at the level of the mRNAs. This early change in gene transcription could be due to a direct effect of the activated hormone-receptor complex on the Na,K-ATPase genes or, alternatively, to an earlier induced (or repressed) intermediate. Nuclear run-on transcription assays performed in the presence of stringent inhibition of protein synthesis will help answer this question. The availability of Na,K-ATPase genes (Shull and Lingrel, 1987) and of a cDNA encoding the putative mineralocorticoid receptor (Arriza et al.,1987) opens new perspectives in the study of transepithelial sodium transport regulation by mineralocorticoid hormones.

REFERENCES

Arriza JL, Weinberger C, Cerelli G, Glaser TM, Handelin BL, Housman DE, Evans RM (1987). Cloning of human mineralocorticoid receptor complementary DNA: Structural and functional kinship with the glucocorticoid receptor. Science 237:268-275.

Geering K, Girardet M, Bron C, Kraehenbühl JP, Rossier BC (1982). Hormonal regulation of (Na^+,K^+)-ATPase biosynthesis in the toad bladder. J Biol Chem 257:10338-10343.

Handler JS (1983). Use of cultured epithelia to study transport and its regulation. J Exp Biol 106:55-69.

Mohun TJ, Brennan S, Dathan N, Fairman S, Gurdon JB (1984). Cell type-specific activation of actin genes in the early amphibian embryo. Nature 311:716-721.

Paccolat MP, Geering K, Gaeggeler HP, Rossier BC (1987). Aldosterone regulation of Na^+ transport and Na^+-K^+-ATPase in A6 cells: role of growth conditions. Am J Physiol 252:C468-C476.

Rossier BC, Paccolat MP, Verrey F, Kraehenbühl JP, Geering K (1985). Mechanism of action of aldosterone: a pleiotropic response. In Dumont JE, Hamprecht B, Nunez J (eds): "Hormones and Cell Regulation, Vol. 9, INSERM European Symposium," Amsterdam: Elsevier, p 209-225.

Schibler U, Hagenbüchle O, Wellauer PK, Pittet AC (1983). Two promoters of different strengths control the transcription of the mouse alpha-amylase gene Amy-1[a] in the parotid gland and the liver. Cell 33:501-508.

Shull MM, Lingrel JB (1987). Multiple genes encode the human Na^+,K^+-ATPase catalytic subunit. Proc Natl Acad Sci USA 84:4039-4043.

Verrey F, Schaerer E, Zoerkler P, Paccolat MP, Geering K, Kraehenbühl JP, Rossier BC (1987). Regulation by aldosterone of Na^+,K^+-ATPase mRNAs, protein synthesis, and sodium transport in cultured kidney cells. J Cell Biol 104:1231-1237.

INOTROPIC AND TOXIC ACTIONS OF SEVERAL CARDIAC STEROIDS IN SHEEP CARDIAC TISSUES*

J. Andrew Wasserstrom and David E. Farkas

Department of Medicine, Northwestern University Medical School, Chicago, IL, 60611, and The Cardiac Electrophysiology Laboratory, The University of Chicago, Chicago, IL, 60637.

INTRODUCTION

The most widely held view of the basis for the positive inotropic effect of cardiac steroids relies on inhibition of the Na^+,K^+-ATPase as the singular mechanism of drug action (Repke, 1964). Na^+,K^+-pump inhibition causes an increase in Na^+_i, which secondarily increases contractile Ca^{2+} via Na-Ca exchange (Reuter and Seitz, 1968). Both positive inotropy and the toxicity of cardiac steroids are attributed to this increase in Ca^{2+}. However, the margin of safety between positive inotropy and arrhythmogenesis differs for several cardiac glycosides in vivo (Acheson et al., 1964; Mendez et al., 1973) which is inconsistent with a single mechanism of action.

The present study examines the actions of several cardiac steroids, including the semi-synthetic agent, actodigin, to produce their inotropic and toxic effects in sheep cardiac tissues. Actodigin differs from other cardiac glycosides only in its position of attachment of the lactone ring (C2 instead of C3) to the steroid nucleus (Deghenghi, 1970) which reportedly improves its therapeutic ratio (Mendez et al., 1973).

*Supported by USPHS HL-30724 and HL-20592.

METHODS

One end of a sheep papillary muscle (diameter $\leq$1mm) was pinned to the chamber floor and the other was hooked to a force transducer. The test chamber was superfused with modified Tyrodes solution (6-8 ml/min, pH=7.3, 37°C) with the following composition (in mM): NaCl, 137; KCl, 5.4; $CaCl_2$, 1.8; $NaHCO_3$, 22; NaH_2PO_4, 2.4; $MgCl_2$, 1.0; glucose, 5.5. A Purkinje fiber was pinned to the chamber floor next to the papillary muscle. Both tissues were stimulated at 1 Hz (for more details, see Wasserstrom et al., 1983; Brill and Wasserstrom, 1986).

Na^+-sensitive microelectrodes were fabricated according to methods described elsewhere (Wasserstrom et al., 1983; Brill and Wasserstrom, 1986). The difference potential $V_{Na}-V_m$ was calculated from digitized voltage measurements (Apple II computer). Intracellular Na^+ activity (a_{Na}^i) was calculated using the Nikolsky equation with the appropriate values for slope and selectivity derived during calibration. Transmembrane potentials were measured using conventional microelectrodes (3M KCl, 12-20 megohms). The Purkinje fiber was impaled with voltage and Na^+-sensitive electrodes within 1 mm of each other. a_{Na}^i was measured by briefly interrupting the stimulus for 5 to 10 sec. This allowed the Na^+-sensitive electrode (time constant of about 200 msec) to reach a stable potential for computer measurement.

Actodigin ($2-4 \times 10^{-6}$M), ouabain (5×10^{-7}M) or acetylstrophanthidin (AS, $3-4 \times 10^{-7}$ M), was added to the Tyrodes solution. Measurements of a_{Na}^i and action potentials in Purkinje fibers and twitch tension in muscle were obtained every minute during drug superfusion and washout. Sotalol ($1-5 \times 10^{-7}$M) was present to block Beta-adrenoceptors but had no effect on the actions of the cardiac steroids.

Data are presented as mean $\pm$ SEM. Significance was determined at $p<0.05$ unless otherwise specified using a one-way analysis of variance.

RESULTS

Figure 1 shows typical effects of actodigin (first row) and ouabain (second row) on action potentials and a_{Na}^i in Purkinje fiber and twitch tension in muscle at 1 Hz. Actodigin decreased maximum diastolic potential (MDP) from -86 mV in control to -82 mV at peak tension with no transient depolarizations (TDs, middle column). In contrast, ouabain

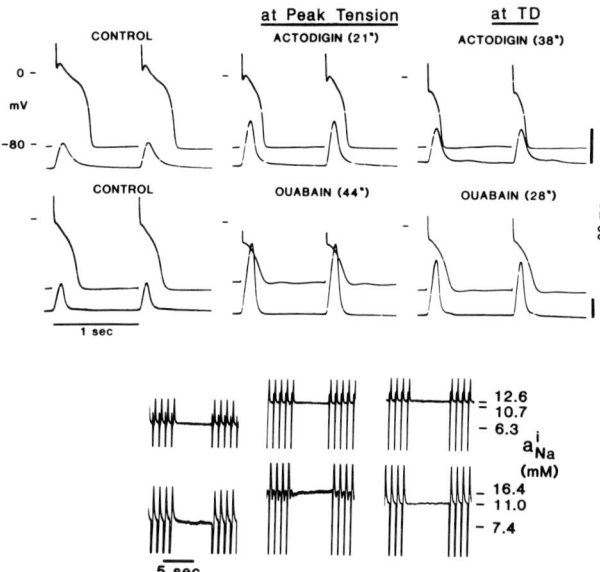

Figure 1. Effects of Actodigin and Ouabain on Action Potentials and a_{Na}^i in Purkinje Fibers and Twitch Tension in Muscles (1Hz). Top panels show recordings of transmembrane potential from Purkinje fiber and twitch tension from muscle. Bottom panels show recordings of a_{Na}^i corresponding to times and conditions indicated above.

reduced MDP from -82 mV in control to -71 mV at peak tension and caused prominent TDs. The first TDs occurred after peak inotropy with actodigin (right column) and were associated with additional depolarization, action potential shortening and a decline in twitch tension. On the other hand, TDs appeared with ouabain before peak inotropy.

The bottom panels of Figure 1 show measurements of a_{Na}^i that coincide with the recordings above. With actodigin, a_{Na}^i increased from 6.3 mM in control to 10.7 mM at peak inotropy and to 12.6 mM when TDs occurred. In contrast, ouabain increased a_{Na}^i from 7.4 mM to 11.0 mM when TDs occurred and yet peak inotropy was not achieved until a_{Na}^i had increased to 16.4 mM.

The time course of the effects of actodigin is shown in Figure 2 (left panel). Actodigin rapidly increased a_{Na}^i and twitch tension and TDs occurred at higher a_{Na}^i than and after peak inotropy. With ouabain, TDs appeared (right panel) before and at lower a_{Na}^i than the peak inotropic effect.

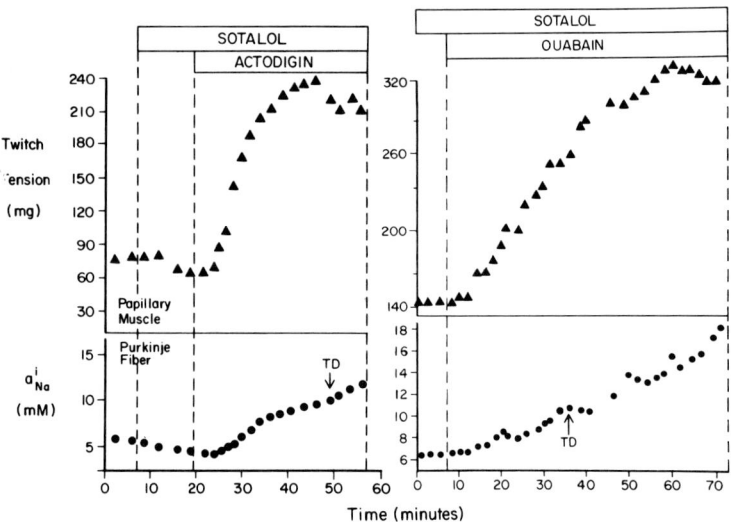

Figure 2. Time Course of Effects of Actodigin and Ouabain on Twitch Tension in Muscles and a_{Na}^i in Purkinje Fibers.

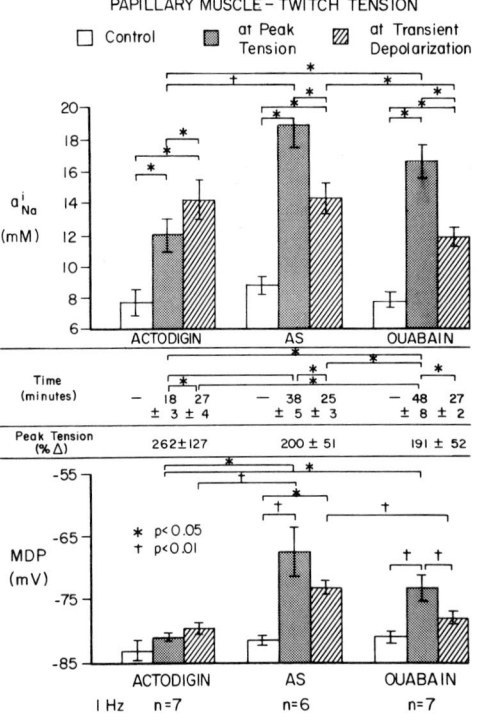

Figure 3. Summary of Effects of Actodigin, AS and Ouabain on Purkinje Fibers and Muscles (1 Hz).

A summary of the effects of all three agents is shown in Figure 3. Actodigin, AS and ouabain produce TDs after the same duration of drug exposure and at the same a_{Na}^i, yet actodigin produced its peak inotropic effect earlier, at lower a_{Na}^i and with less depolarization than did the other cardiac steroids.

These results suggest that there are differences between drugs on the relation between Na^+_i and twitch tension. Figure 4 shows a Na^+_i-tension curve from one of six experiments in which both actodigin and AS were added to the superfusate. The order of drug exposure was distributed randomly between preparations. Despite nearly identical initial a_{Na}^i, the slope and peak of this relationship differ for the two agents. Actodigin produced greater force than AS at the same a_{Na}^i and the peak inotropic effect for actodigin occurred at lower a_{Na}^i than with AS.

The effects of these three cardiac steroids on action potentials in Purkinje fibers are summarized in Table 1. Actodigin caused little change in MDP whereas AS and ouabain caused depolarization both at peak tension and TDs. At peak tension, overshoot (OS) was reduced by AS and ouabain but not by actodigin. Actodigin shortened the action potential at both peak inotropy and TD while ouabain had little effect on duration. Abbreviation of the action potential with AS occurred at a lower MDP and greater aNai than with actodigin.

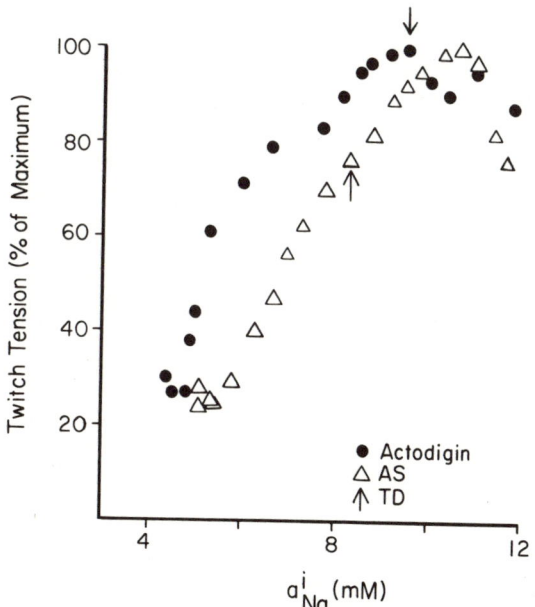

Figure 4. Na^+-Tension Relationship for Actodigin and AS in a Purkinje Fiber-Muscle Preparation (1 Hz).

TABLE 1: EFFECTS OF CARDIAC STEROIDS ON ACTION POTENTIALS IN SHEEP CARDIAC PURKINJE FIBERS

	ACTODIGIN (3×10^{-6}M)			AS (3×10^{-7}M)			OUABAIN (5×10^{-7}M)		
	CONTROL	AT PEAK TENSION	at TD	CONTROL	AT PEAK TENSION	at TD	CONTROL	AT PEAK TENSION	at TD
MDP (mV) ± n	-83.0 1.53	-80.8 0.65 6	-79.5 0.72	-81.4 0.61	-67.4** 4.19 7	-73.3* 1.10	-81.1 0.66	-73.3** 1.89 18	-78.1+ 0.94
OS (mV) ± n	29.7 2.72	26.2 2.87 6	23.3* 2.54	34.5 3.15	14.8** 5.82 6	24.0*** 6.28	31.3 1.88	17.5** 3.12 15	24.6++ 2.24
AMP (mV) ± n	112.7 3.30	107.0 2.92 6	103.0** 2.85	115.8 3.40	85.8** 7.97 6	97.3*** 7.22	112.1 2.39	93.3** 4.03 15	102.9*++ 2.67
APD$_{50}$ (msec) ± n	258.8 38.24	184.7* 22.14 6	171.2* 14.11	140.8 37.13	68.8* 14.11 5	104.8 16.99	177.9 13.77	143.3 15.64 15	148.2 11.45
APD$_{90}$ (msec) ± n	331.2 36.58	253.7** 27.01 6	235.3** 16.12	251.7 35.79	167.3* 5.59 6	196.7* 17.18	268.4 15.61	243.3 21.94 15	257.5 15.97

* $p<0.05$; ** $p<0.01$ significant difference from control
\+ $p<0.05$ significant difference from value at peak tension

MDP, maximum diastolic potential; OS, overshoot; AMP, amplitude;
APD$_{50}$/APD$_{90}$, action potential duration at 50% and 90% of repolarization

DISCUSSION

The idea that both therapeutic (inotropic) and toxic (arrhythmic) actions of cardiac glycosides result from inhibition of the Na^+,K^+-ATPase cannot account for differences in margin of safety for various cardiac steroids (Acheson et al., 1964; Mendez et al., 1973). We used an in vitro model to compare cardiac glycoside effects in the two tissue types involved clinically. Ouabain and AS increased a_{Na}^i in Purkinje fibers coincident with positive inotropy in ventricular muscle but also caused depolarization and electrophysiological manifestations of toxicity, TDs, prior to the achievement of peak inotropy. Their therapeutic efficacy is limited to the positive inotropy that occurs in working myocardium before TD-based arrhythmias occur, probably in the ventricular conducting system. TDs arise as a consequence of Ca^{2+} overload of sarcoplasmic reticulum, resulting in cytoplasmic oscillations of Ca^{2+} that activate a non-specific cation conductance (Kass et al., 1978). Inhibition of the Na^+,K^+-ATPase also promotes arrhythmias through depolarization, which directly increases TD amplitude (Ferrier 1980; Wasserstrom and Ferrier, 1981) and brings TDs closer to threshold potential.

Actodigin increases a_{Na}^i but also produces positive inotropy that is distinct from that of the other two cardiac steroids. We have not yet identified a Na^+-independent positive inotropic effect (Wasserstrom et al., 1983 and the present study), and thus there may be another inotropic mechanism superimposed upon the accumulation of a_{Na}^i. Previously, we could not identify a second mechanism, probably because the range of drug action was limited to a 4-5 mM increase in a_{Na}^i where tissue variability obscured changes in the Na^+-tension relationship (Wasserstrom et al., 1983).

There are several important consequences of this second positive inotropic mechanism. First, the Na^+-tension relationship is altered such that there is a greater positive inotropic effect at lower a_{Na}^i than with other glycosides. Second, MDP is depolarized very little which helps to suppress the appearance and amplitude of TDs until after peak inotropy. Third, the peak inotropic effect of actodigin occurs at a lower a_{Na}^i than with conventional cardiac glycosides, thus reducing the occurrence of Ca^{2+} overload induced toxicity. These actions probably underly the improved therapeutic index reported for this agent in situ (Mendez et al., 1973).

In conclusion, direct measurements of $a_{Na}{}^i$, force and action potentials in cardiac tissues suggest that actodigin has a second positive inotropic action that may not be related to changes in $a_{Na}{}^i$. The result is that TDs are suppressed until a higher $a_{Na}{}^i$ is achieved than at peak inotropy. These factors contribute to a positive inotropic potency of actodigin that occurs in the relative absence of toxicities.

REFERENCES

Acheson GH, Kahn JB Jr., Lipicky RJ (1964). A comparison of dihydro-ouabain, dihydro-digoxin, dihydro-digitoxin, 3-acetylstrophanthidin, erysimin and ouabain given by continous infusion into dogs. Naunyn-Schmiedeberg's Arch Pharmacol 248:247.

Brill DM, Wasserstrom JA (1986). Intracellular sodium and the positive inotropic effect of veratridine and cardiac glycoside in sheep Purkinje fibers. Circ Res 58:109.

Deghenghi R (1970). Synthetic cardenolides and related products. Pure Appl Chem 21:153.

Ferrier GR (1980). Effects of transmembrane potential on oscillatory afterpotentials induced by acetylstrophanthidin in canine ventricular tissues. J Pharmacol Exp Ther 215:332.

Kass RS, Lederer WJ, Tsien RW, Weingart R (1978). Role of calcium ions in transient inward currents and aftercontractions induced by strophanthidin in cardiac Purkinje fibres. J Physiol (London) 281:187.

Mendez R, Pastelin G, Kabela E (1973). The influence of the position of attachment of the lactone ring to the steroid nucleus on the action of cardiac glycosides. J Pharmacol Exp Ther 188:189.

Repke, K (1964). The biochemical action of digitalis. Klin Wochenschr 42:157.

Reuter, H, Seitz N (1968). The dependence of calcium efflux from cardiac muscle on temperature and external ion composition. J Physiol (London) 195:451.

Wasserstrom JA, Ferrier GR (1982). Voltage dependence of digitalis afterpotentials, aftercontractions and inotropy. Amer J Physiol 241:H646.

Wasserstrom JA, Schwartz DJ, Fozzard HA (1983). Relation between intracellular sodium and twitch tension in sheep cardiac Purkinje strands exposed to cardiac glycosides. Circ Res 52:697.

EFFECT OF APAMIN ON THE Na,K PUMP AND Na,K-ATPase

Hana Zemková, Jan Teisinger[+] and František Vyskočil

Institute of Physiology, Czechoslovak Academy of Sciences, and [+]Institute of Hygiene and Epidemiology, Prague, Czechoslovakia

INTRODUCTION

The similarity between the K^+ binding site on the Na,K-ATPase and that on the voltage-dependent K^+ channel is shown by the similar effects of Tl^+, Rb^+, Cs^+ and NH_4^+ ions on the two sites (Edwards, 1982) and by the inhibitory effects of quarternary ammonium cations on both the voltage-dependent K^+ channels and active Na^+ and K^+ transport (Sachs, 1967). In the present work we examined the effects of another K^+ channel blocker, apamin, on the Na,K pump. Apamin has been recently found to inhibit one type of Ca^{2+}-dependent K^+ conductance which is present in many cell types and which is responsible for prolonged afterhyperpolarization of action potentials in neurons and skeletal muscle cells (Schmid-Antomarchi et al., 1985). We found that apamin also inhibits the K^+-induced hyperpolarization of Na-loaded skeletal muscles and the activity of Na,K-ATPase, i.e. it inhibits the electrogenic pump. A possible connection between the apamin-sensitive channel and Na,K-ATPase is discussed.

METHODS

Electrophysiological Measurements

Experiments were performed on strips of Na-loaded diaphragm muscle of female white mice. Freshly dissected diaphragms were incubated for 3-5 h in an oxygenated (95 % O_2 + 5 % CO_2) K^+-free solution of the following composition (in mM): NaCl, 136.8; $CaCl_2$, 2.0; $MgCl_2$, 1.0; $NaHCO_3$, 12.8; NaH_2PO_4, 1.0; glucose, 11.0; pH = 7.4 at room temperature $20\pm2\ °C$. Resting membrane potentials of surface muscle fibres were measured by means of glass intracellular microelectrodes filled with 3 M KCl (resistance 5-10 MΩ) before and 5 min after addition of 5 mM KCl to the bath. The difference between these two values was taken as a measure of electrogenic Na,K pump activity (Kernan, 1962; Dlouhá et al., 1981).

Estimation of Na,K-ATPase Activity and ^{3}H-ouabain Binding

Enzymatic activity of the Na,K-ATPase was measured as the production of P_i in membrane fractions from the rat cerebral cortex. The reaction was started by addition of 2.5 mM ATP to the microsomes (0.03-0.05 mg membrane protein per milliliter) which were preincubated for 5 min at 37 °C in 1.25 ml of 100 mM NaCl, 20 mM KCl, 5 mM $MgCl_2$, 100 mM TRIS-HCl, pH = 7.4. The basal ouabain-insensitive activity of ATPase was measured in a K^+-free solution with 2×10^{-4} M ouabain.

The binding of ^{3}H-ouabain to the rat brain microsomal fractions was carried out in 50 mM TRIS buffer, pH = 7.4. Membrane fractions were incubated with 100 nM ^{3}H-ouabain for 1 h at 37 °C in the presence of 5 mM TRIS-P_i and 1 mM V^{5+} and 0, 1, 5 or 10 mM K^+. Non-specific binding of ouabain was determined in the presence of unlabelled ouabain (0.1 mM).

RESULTS

Electrogenic Effect of the Na,K pump

Incubation of the mouse diaphragm for 3-5 h

in K^+-free solution depolarized the average membrane potential from -75.3 ± 1.1 mV to -65.9 ± 1.0 mV (the mean $\pm$ S.E.M.). The addition of 5 mM K^+ to the bath with Na-loaded muscle hyperpolarized the membrane to -75.6 ± 1.0 mV. Apamin inhibited this K^+-induced hyperpolarization in a dose-dependent manner; 50 % inhibition was found with about 10^{-8} M apamin (Table 1.). Apamin alone was without effect on the resting membrane potential in either freshly isolated muscles in a normal 5 mM K^+ solution (-75.6 ± 0.5 mV before and -74.5 ± 0.9 mV 10 min after 10^{-7} M apamin, 3 muscles) or in Na-loaded muscles before K^+ application (Table 1.).

TABLE 1. Effect of Apamin on the K^+-induced Hyperpolarization in Na-loaded Mouse Diaphragm Muscles

	Membrane potential (mV)^x		Difference	n
	K^+-free	After 5 mM K^+		
Control	64.9 ± 1.0 (70)	75.6 ± 1.0 (72)	10.6	4
Apamin				
5×10^{-9} M	66.1 ± 1.3 (42)	74.3 ± 1.1 (40)	8.2	2
10^{-8} M	66.8 ± 1.3 (56)	71.9 ± 1.0 (50)	5.1	3
10^{-7} M	65.3 ± 0.9 (51)	69.1 ± 0.8 (63)	3.8	3
10^{-6} M	68.2 ± 1.3 (42)	69.6 ± 1.2 (38)	1.4	2

^{x}Average resting membrane potential $\pm$ S.E.M. from n muscles; the number of muscle fibres whose potential was measured is given in parentheses.

It is well known that the plasma membrane of skeletal muscle cells can be hyperpolarized by insulin (Zierler, 1957), catecholamines (Clausen and Flatmann, 1977) and extracellular NADH (Zemková et al., 1984). Examination of the effects of apamin on the hyperpolarization induced by insulin, noradrenaline or NADH showed that only insulin- and NADH-induced hyperpolarizations were completely inhibited by 10^{-7} M apamin (Table 2.). The noradrenaline-induced hyperpolarization was not inhibited even by 10^{-6} M apamin.

TABLE 2. Effect of Apamin on Insulin, NADH and Noradrenaline(NA)-induced Hyperpolarization in Na-loaded Muscles

	Membrane potential (mV)		Difference	n
	K^+-free	After insulin 100 mU/ml		
Control	65.5±1.8 (45)	71.8±1.2 (56)	6.3	3
Apamin 10^{-7} M	62.8±0.8 (54)	60.2±1.2 (60)	-2.6	3
	K^+-free	After NADH 10^{-5} M		
Control	64.0±1.3 (45)	69.6±1.1 (51)	5.6	2
Apamin 10^{-7} M	59.7±1.0 (53)	58.9±1.2 (63)	-0.8	3
	K^+-free	After NA 10^{-5} M		
Control	61.8±1.0 (57)	82.8±1.2 (60)	21.0	3
Apamin 10^{-6} M	60.6±1.1 (65)	81.5±1.9 (54)	20.9	3

Na,K-ATPase Activity and ^{3}H-ouabain Binding

The effect of apamin on the enzymatic activity of Na,K-ATPase was measured in the rat brain microsomes. The Na,K-ATPase was inhibited by nanomolar amounts of apamin; half-maximal inhibition was obtained with about 1.5×10^{-8} M apamin (Fig. 1.). The binding of ^{3}H-ouabain was not inhibited by apamin either in the presence of K^+ ions which have a stimulatory effect on specific apamin binding

(Seagar et al., 1986) or in its absence (data not shown).

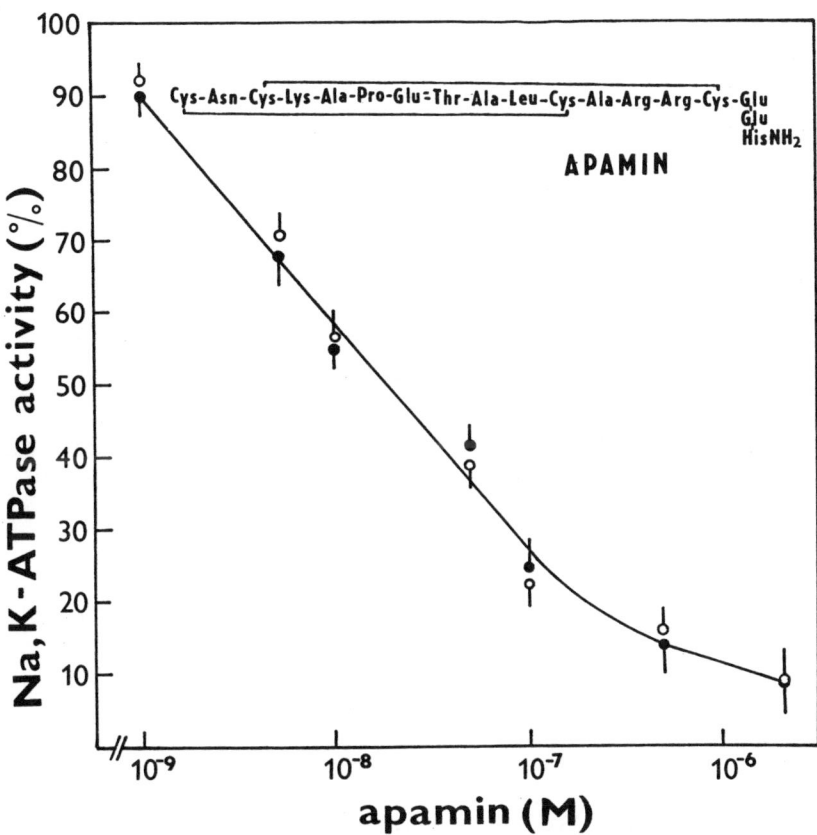

Figure 1. The effect of apamin on the Na,K-ATPase under normal assay conditions (o) and in solutions with decreased ATP (0.2 mM, ●). Identical inhibitory effect of apamin under these two conditions supports the idea that apamin does not act on the cytoplasmic side of the enzyme. Each value represents the mean ± S.E.M. of 5-6 enzymatic measurements.

DISCUSSION

The inhibition of K^+-induced hyperpolarization of Na-loaded muscle as well as of Na,K-ATPase by nanomolar concentrations of apamin suggests that it inhibits both the small conductance Ca^{2+}-activated K^+ channel (Blatz and Magleby, 1986) and the electrogenic Na,K pump with the same efficiency. The apamin block of the insulin- and NADH- but not the noradrenaline-induced hyperpolarization is in keeping with other evidence that the effects of insulin and NADH are mediated by the Na,K pump, while noradrenaline hyperpolarizes muscle cells by a different mechanism (Zemková et al., 1985).

There is no direct evidence that apamin, a polypeptide of 18 amino acids (Habermann, 1972) can cross the membrane. Therefore we assume that it inhibits the Na,K-ATPase from the outside of the membrane by binding to a site which must be different from that for ouabain, since apamin did not inhibit the binding of ^{3}H-ouabain. This suggests that apamin apparently decreases the turnover rate of the pump without changes in the number of pumping sites.

The sensitivity of the Na,K-ATPase to the K^+ channel blockers indicates that there might be a functional or structural interrelationship between these two proteins in the membrane. With the exception of apamin no specific, high-affinity ligands for K^+ channels have, as yet, been identified. The isolation of an apamin-binding protein from the rat skeletal muscle (Schmid-Antomarchi et al., 1985) and brain microsomes (Seagar et al., 1986) is thus the first attempt to characterize the structure of any K^+-channel. Differential photoaffinity labelling revealed that specific apamin binding is localized at the border between 3-4 polypeptide chains of M_r = 22-115 kDa (Seagar et al., 1986) which are supposed to belong either to putative subunits of the apamin-sensitive channel or to different apamin receptors. The molecular size of the rat brain Na,K-ATPase alpha subunit, M_r = 112 kDa (Shull et al., 1986), is close to that of the largest polypeptide of apamin-binding site. It is possible that

Na,K-ATPase forms one of these subunits of the apamin-sensitive channel. But probably only one form of three Na,K-ATPase isoenzymes recently reported in many cells, including skeletal muscles (Lytton, 1985), might be associated with the apamin binding site. The density of ouabain binding sites is greater by one order of magnitude than the density of apamin receptors (Clausen, 1986; Seagar et al., 1986). The possibility that disabled Na,K-ATPase can function as an apamin-sensitive channel can apparently be excluded for the following reason: The leakage pathway through the Na,K pump molecule incorporated into lipid bilayer is equal to a channel conductance of about 40-50 pS (Reinhardt et al., 1984) which is about 3 times higher than the conductance of the single apamin-sensitive Ca^{2+}-activated K^+ channel (10-14 pS) in cultured skeletal muscle cells (Blatz and Maglegy 1986).

Possible functional connection between the apamin-sensitive channel and Na,K-ATPase should be proven by more experiments. But if the pump and the channel are two different proteins the two different apamin receptor molecules might have some parts which are similar structurally and therefore are equally accessible for apamin binding.

REFERENCES

Blatz AL, Maglegy KL (1986). Single apamin-blocked Ca-activated K^+ channels of small conductance in cultured rat skeletal muscle. Nature 323: 718-720.
Clausen T (1986). Regulation of active Na^+-K^+ transport in skeletal muscle. Physiol Rev 66: 542-580.
Clausen T, Flatman JA (1977). The effect of catecholamines on Na-K transport and membrane potential in rat soleus muscle. J Physiol 270: 383-414.
Dlouhá H, Teisinger J, Vyskočil F (1981). The effect of vanadate on the electrogenic Na^+/K^+ pump, intracellular Na^+ concentration and electrophysiological characteristics of mouse skeletal muscle fibre. Physiol bohemoslov 30: 1-10.
Edwards C (1982). The selectivity of ion channels in nerve and muscle. Neuroscience 7: 1335-1366.
Habermann E (1972). Bee and wasp venoms. Science

177: 314-322.

Kernan RP (1962): Membrane potential changes during sodium transport in frog sartorius muscle. Nature 193: 986-987.

Lytton J (1985). The catalytic subunits of the (Na^+,K^+)-ATPase α and $\alpha(+)$ isoenzymes are the products of different genes. Biochem Biophys Res Comm 132: 764-769.

Reinhardt R, Linemann B, Anner BM (1984): Leakage-channel conductance of single (Na^+,K^+)-ATPase molecules incorporated into planar bilayers by fusion of liposomes. Biochim Biophys Acta 774: 147-150.

Sachs JR (1967). Competitive effects of some cations on active potassium transport in the human red blood cell. J clin Invest 46: 1433-1441.

Schmid-Antomarchi H, Renaud JP, Romey G, Hugues M, Schmid A, Lazdunski M (1985). The all-or-none role of innervation in expression of apamin receptor and of apamin-sensitive Ca^{2+}-activated K^+ channel in mammalian skeletal muscle. Proc Natl Acad Sci USA 82: 2188-2191.

Seagar MJ, Labbé-Julié C, Granier C, Goll A, Glossmann H, Van Rietschoten J, Couraud F (1986). Molecular structure of rat brain apamin receptor: Differential photoaffinity labeling of putative K^+ channel subunit and target size analysis. Biochemistry 25: 4051-4057.

Shull GE, Greeb J, Lingrel JB (1986). Molecular cloning of three distinct forms of the Na^+,K^+-ATPase α-subunit from rat brain. Biochemistry 25: 8125-8132.

Zemková H, Svoboda P, Teisinger J, Vyskočil F (1985). On the mechanism of catecholamine-induced hyperpolarization of skeletal muscle cells. Naunyn-Schmiedeberg's Arch Pharmacol 329: 18-23.

Zemková H, Teisinger J, Vyskočil F (1984). Hyperpolarization of mouse skeletal muscle plasma membrane induced by extracellular NADH. Biochim Biophys Acta 775: 64-70.

Zierler KL (1957). Increase in resting membrane potential of skeletal muscle produced by insulin. Science 126: 1067-1068.

Index

Acetylstrophanthidin, 199, 469–474
Acidification, intracellular, 176, 177, 178
Acid pump, plasma inhibitor and, 417–421
Actin, 160
Active transport
 exchange diffusion, 6
 flux ratios, 7–8
 frog skin, isolated, 6–7
 isotope studies, 4–5
 maximum capacity for, in heart, 233–238
 physiology, 3–4
 secondary, 12–13, 213–214, 215–216
 short-circuit criterion, 8–10
 single-file diffusion, 8
 solvent drag, 12
 stoichiometry, 10–12
Actodigin, 469–476
Acyl CoA: lysolecithin acyltransferase, 453
Adenylate cyclase system, 277–278
Adipocytes, 281, 377
Adipose tissue, 106
Adrenalectomy, 281, 284, 339–344
Adrenaline, 204
Adrenergic receptors
 alpha, catecholamine binding, 279
 beta, 470
 beta-2, 204
AISC, 113–117
Aldosterone, 281–282
 kidney enzyme regulation, 222, 339–344
 toad bladder cells (A6), 463–467
Alpha-subunits. *See* Subunits, alpha-
Amiloride, 176, 375
 heterogeneity of pump, 188, 189
 kidney enzyme regulation, 224, 226
 and pump rate regulation, 183
Amino acid decarboxylase, 179
Amino acid sequences, ouabain binding sites, 329–331
Ankyrin, 57, 245–249
Antidiuretic hormone, 277–278

Antigen activation, lymphocytes, 449
Antiports, 157–161, 173
Apamin, 477–483
Arrhythmias, 254, 469
Artemia salina, 35–41
Asialoglycoprotein receptor, 437–441
A6 cells, 282, 463–467
Asparagine residues, glycosylation sites, 22–23
Assembly, subunits, 20–26
ATP
 heart enzyme, 3-O-methylfluorescein phosphate activity, 257–260
 kidney enzyme regulation, 222, 223
 and ouabain binding, 323
N-Azidobenzoyl-ouabain, 54
Azido-iodophenethylamidosuccinyl-cymarin (AISC), 113–117

Beta-subunits. *See* Subunits, alpha- and beta-; Subunits, beta-
Biosynthesis, 174
 beta-subunit, *Artemia salina*, 35–41
 in vitro. *See* In vitro systems
 subunits, 20–26
Biphasic sensitivity
 inotropic action, 325–327
 ouabain binding, 379
Bladder
 cell lines
  aldosterone effects, 463–467
  tunicamycin effects, 164–168
 regulation
  aldosterone, 282–283
  vasopressin, 277–278
Boyle-Conway theory, 4
Brain
 alpha-subunit isoforms
  vs. heart, 43–48
  photoaffinity labeling studies, 113–117
  tissue specificity in rat, 120–125

beta-subunit gene expression, 127–133
differential sensitivity to Na, vanadate, and dihydroouabain, 377–383
intraindividual variations in gene organization, 136, 137, 140
regulation
inhibitors, 306–307
see also Hormones
tissue-specific gene expression, 105, 109, 111
Brine shrimp, 22; *see also Artemia salina*

Calcium, 300
cardiac steroids and, 469
cardiomyopathy and, 269–270
and ouabain binding, red cell ghosts, 365–370
myometrium, rat, 457–462
thymocytes, 454
Calcium channels, 353, 458, 460
Calmidazolium, 458, 460
Calmodulin antagonist, 458, 460
cAMP, 457, 458, 460, 461, 462
Cardiac glycosides, 20
azido-iodophenethylamidosuccinyl-cymarin, 113–117
cardiac alpha-subunit, immunological and electrophoretic differences, 149–154
and contractility, 324–325
isoform differences in sensitivity, 105
ouabain-binding regions, 328–334
amino acid sequences, 328–329
isoform differences, 330–332
monoclonal antibody studies, 329–331
normal vs. diseased hearts, sensitivity in, 332–333
SDS effects, 333–334
receptor function, criteria for, 322–323
receptor regulation in cultured cells, 345–351
sheep heart, inotropic and toxic actions, 469–476
sodium-calcium exchange, 327
species differences in sensitivity, 325–326
tissue differences, 108
Cardiac muscle. *See* Heart
Cardiomyopathy, 332
human cardiac enzyme, 269–273

3-O-methylfluorescein phosphate activity, 260
Catalytic subunit. *See* Subunits, alpha-
Catecholamines, 279
muscle hyperpolarization, 479
pump stimulation by, 204
cDNA
alpha-subunit isoforms
in vitro expression, 71–75
tissue specificity, 119–126
beta-subunit, 29, 106
ouabain resistance, transfer of, 98–102
Cell culture
aldosterone effects in A6, 463–467
coupling ratio in, 176
heart cell glycoside receptors, 345–351
kidney enzyme regulation, 226–227
localization in epithelial cells, 51–57
monensin effects, 157–161
mRNA, pump-specific, 59–64
neuronal, isozyme appearance in, 91–95
pertussis toxin effects, 353–356
primaquine effects, 437–442
pump rate measurement vs. isolated membranes, 184–185
regulation
aldosterone, 282
insulin, 280
long-term, 180–184
short-term, 179–180
renal tubular epithelial cell cytoskeleton, 245–249
reticulocyte maturation, 360, 361–362
sodium influx, 172
toad bladder cells, 164–168, 282, 463–467
Cell-free translation
deletion plasmids, 79–84
protein synthesis, 71–75
see also In vitro systems
Cell surface
renal tubular cells, 211
see also Plasma membrane; Sidedness
Cellular cooperation, 185–186
Cerebrospinal fluid, 105, 308
Channel blockers
apamin, 477–483
calmidazolium, 458, 460

Chloride transport, 7
Choline, 366
Chromosomal assignment, mouse, 102–103
Chymotrypsin, 131
Circulating inhibitor. *See* Inhibitors
Citrate buffer, 143–146
Cloning. *See* cDNA; Molecular cloning
Codistribution, renal tubular epithelial cells, 245–249
Coenzyme A esters, 401–406
Collagen, 248
Concanavalin A, 453–455
Conformational changes, 20, 163
Contractile force, cardiac glycoside sensitivity, 324–325
Coreglycosylation
   maturation process, 22, 24, 26–30
   tunicamycin and, 163–168
Cortical collecting tubules, kidney enzyme regulation, 339–344
Corticosterone, 222
Cotranslational coreglycosylation, 26–30
Cotranslational insertion, 21
   Artemia salina, 39–41
   brain, rat, 127, 130–132
   deletion mutants, 83–84
Cotranslational translocation, 25
Cotransport, 173, 174, 215
Counter transport, 173, 174
Coupling ratio, 175, 176
C-termini, 25
CV1 cells, 176, 188
Cyclic AMP, 457, 458, 460, 461, 462
Cycloheximide, 341
Cymarin, azido-iodophenethylamidosuccinyl-cymarin, 113–117
Cytoplasmic side, 19, 131
Cytoskeleton, 57, 245–249

Deletion mutants, 77–84
Development
   alpha-subunit isoforms
      ferret heart enzyme, 423–428
      molecular cloning, 119–126
      rat heart, 151, 152–153
      tissue specificity in rat, 120–125
      intraindividual variations in gene organization, 136, 138

   neuronal, appearance in cell culture, 91–95
   reticulocyte maturation, 357–363
   skeletal muscle enzyme regulation, 253
Dexamethasone, 222, 284
Diabetes mellitus, 253, 260
Diacylglycerol, 279
Diaphragm muscle
   apamin effects, 477–483
   tissue specificity in rat, 120–125
Dietary potassium, 217, 218, 219
Differentiation
   alpha-subunit, rat
      heart, 151, 152–153
      tissue specificity, 120–125
   intraindividual variations in gene organization, 136, 138
   neuronal, appearance in cell culture, 91–95
   pump rate regulation, 181–183
   reticulocyte maturation, 357–363
Diffusion
   exchange, 6
   single-file, 8
Digitalis, 267, 300–301
   alpha-subunit isoforms
      dog heart, 263–267
      heart vs. brain, 46–47
   ferret heart, developmental changes in sensitivity, 423–428
   see also Cardiac glycosides
Digoxin
   radioimmunoassay, 301, 305
   two-sided liposomes, 430, 431, 432
Dihydroouabain
   differential sensitivities in rat tissues, 377–383
   ouabain-binding in oocytes, voltage independence of, 409–415
Distal convoluted tubules. *See* Kidney
Distribution, in heart, 237–238
DMSO, 459, 460, 461, 462
DNP, 11
DOCA, 222
Docking protein, 24, 25–26
Docosahexaenoic acid, 402–405
Dog heart, alpha-subunit isoforms, 263–267
Dog kidney. *See* MDCK cells

L-Dopa, 179
Dopamine, 179–180, 353–356
Dorsal root ganglia, 185
Downhill transport, 8

Electrical potential
  apamin and, 477–483
  cardiac steroids, 469–476
  frog skin, 7
  renal tubular cells, 211
  short-circuit criterion, 8–10
  voltage-independence of oocyte ouabain-binding, 409–415
Electron probe analysis, 175, 176
Electrophoretic mobility, 105, 149–154
Embryos
  intraindividual variations in gene organization, 136, 138
  see also Fetal development
Endocytosis, receptor mediated, 437
Epithelial cells
  respiratory, tissue specificity in rat, 122–123
  tubular, subcellular localization, 51–57, 245–249
Exchange diffusion, 6
Extracellular sites. See Subunits, beta-
Eye, 105, 143–146

Fatty acids, 301, 401–406
  and circulating inhibitor, 443–448
  T-lymphocytes, 452
Ferret heart, 423–428
Fetal development
  alpha-subunit, rat
    heart, 151, 152–153
    tissue specificity, 120–125
  intraindividual variations in gene organization, 136
  skeletal muscle enzyme regulation, 253
Fibroblasts
  coupling ratio in, 176
  monensin effects, 157–161
  sodium influx, 172
Flux ratios, 7–8, 12
Fodrin, 57, 245
Freeze-fractured liposomes, 435
Frog skin, 6–7, 11
Functional analysis. See Gene mapping

Gastrointestinal tract, tissue specificity, 120–125
Gelatin, 248
Gene expression
  aldosterone effects in toad bladder cells (A6), 463–467
  cDNA clone, rat brain, 127–133
  in vitro, 71–75
  in *Saccharomyces cerevisiae*, 85–90
  thyroid hormones and, 286–290
  tissue-specific, 65–69, 105–111
  tunicamycin and, 163–168
Gene mapping
  chromosomal assignment in mouse, 102–103
  transfer of ouabain resistance, 98–102
    alpha 1-subunit cDNA, 98–102
    chromosome-mediated, 98
Gene organization, intraindividual variations in, 135–142
Gene transfer. See Gene mapping
Glucocorticoids, 222, 283–284
Glycosides, cardiac. See Cardiac glycosides
Glycosylation, 440
  Artemia salina, cell-free system, 37–39
  maturation of subunits, 22, 24, 26–30
  rat brain, 128
  sites of, 22–23
  tunicamycin and, 163–168
  see also Subunits, beta-
N-Glycosylation, 440
Gossypol, 385–391
G-proteins, 353
Guinea pig heart
  maximum capacity for active transport, 233–238
  3-O-methylfluorescein phosphate activity, 257–260

H,K-ATPase, 417–421
Hamster heart, 257–260
Heart
  alpha-subunit isoforms, 109, 110, 263–267
  vs. brain, 43–48
  ferret, 423–428
  immunological and electrophoretic differences, 149–154

3-O-methylfluorescein phosphate
  activity, 257-260
  relative abundance, 108
  tissue specificity in rat, 120-125
beta-subunit, 109, 110
  mRNA species, multiple, 130
  thyroid hormone and, 133
cardiac glycosides. *See* Cardiac
  glycosides
cultured cells, glycoside receptors,
  345-351
differential sensitivity to Na, vanadate,
  and dihydroouabain, 377-383
ferret, molecular forms and developmental changes, 423-428
human, 269-273
K, extracellular, 197-198
maximum capacity, right vs. left ventricle, 233-238
pump quantification, 257-260
pump rate, maximum, 200, 201, 202
regulation. *See* Hormones; Inhibitors
sheep cardiac tissues, cardiac steroids,
  469-476
tissue differences in subunit distribution,
  109, 110
HeLa cells
  deletion plasmids, 77-84
  monensin effects, 157-161
  ouabain effects, 363
  pump-specific mRNA, 59-64
Henle's loop, 209-210
Hepatocytes
  asialoglycoprotein receptors, 437
  coupling ratio, 176
  regulation, insulin, 280
  sodium influx, 172
Hepatoma cells, primaquine effects, 438, 439
HEPG2, 438, 439
Hereditary hypertension, 299
HK cells, reticulocyte maturation, 357-363
Hormones
  and acid secretion, 417
  catecholamines, 279
  glucocorticoids, 283-284
  insulin, 280-281
  kidney enzyme regulation, 222, 226-227
  mineralocorticoids, 281-283

pump rate regulation
  long-term, 180-184
  short-term, 179-180
pump stimulation by, 204
sheep cardiac tissues, cardiac steroids,
  469-476
skeletal muscle enzyme regulation, 253
thyroid, 284-290
toad bladder cells (A6), 463-467
vasopressin, 277-278
HPLC, 303-312
Humans
  genomic library, 106-111
  heart enzyme
    3-O-methylfluorescein phosphate
      activity, 257-260
    quantification, 269-273
  intraindividual variations in gene organi-
    zation, 135-142
  membrane insertion in, 77-84
  plasma inhibitor, acid pump effects,
    417-421
  red cell membranes
    gossypol effects, 385-391
    ouabain binding, 365-370
Hydrogen ion pump, 417-421
Hydrophobic domain, beta-subunit, 129
Hyperpolarization, apamin and, 477-483
Hypertension, 298-299, 332
  inhibitor in, 443-448
  skeletal muscle enzyme regulation, 253
Hyperthyroidism, 253, 375
Hypothalamus, 306-307, 309
Hypothyroidism, 253, 260, 375

Immunochemical studies
  alpha-subunit immunological and electro-
    phoretic differences, 149-154
  renal tubular epithelial cells, 245-249
  tunicamycin effects, 164-168
Inhibition
  calcium effects, 457-462
  differential sensitivity to Na, vanadate,
    and dihydroouabain, 377-383
  and sodium leak, 172
Inhibitors
  concept and status, 297-303
  experimental evidence, 297-300
  and gastric acid pump, 417-421

gossypol, 385–391
identification, 300–303
kidney enzyme, 222
non-esterified fatty acids and, 443–448
palytoxin, 393–399
purification, 303–312
two-sided liposomes, 429–436
Inotropic agents
ferret heart, developmental changes in sensitivity, 423–428
receptor regulation in cultured heart cells, 345–351
sheep cardiac tissues, 469–476
see also Cardiac glycosides
Inotropy
alpha-subunit, immunological and electrophoretic differences, 149–154
biphasic action, 325–327
Insertion signal, 132
Insulin, 280–281, 377
isoform differences in sensitivity, 106
muscle hyperpolarization, 479, 480
Intracellular ions. See Physiological role; *specific ions*
Intracellular localization, in kidney, 211
Intracellular signals, 353
Intracellular transport, 26–30, 163
In vitro systems
kidney enzyme regulation by aldosterone and sodium, 339–344
subunit synthesis and assembly, 20–26
alpha-subunit, 28–29
Artemia salina, 35–41
deletion plasmids, 79–84
membrane insertion, 79–84, 130–131
run-on transcription assays, 69
translation, 22–24
Ion fluxes, 175, 176
glycoside receptor regulation in cultured heart cells, 345–351
skeletal muscle, 239–241
Ionic coupling ratio, 175, 176
Ionophores
calcium, calmodulin antagonist, 458, 460
kidney enzyme regulation, 224
potassium, apamin, 477–483
Ischemia, cardiac, 263–267
Isoforms

brain vs. heart, rat, 43–48
ferret heart, 423–428
pump rate regulation, hormonal, 183
see also Subunits, alpha-
Isoproterenol, 204
Isotope studies, 4–5
Isozymes
alpha-subunit, immunological and electrophoretic differences, 149–154
neuronal, appearance in cell culture, 91–95
in rat tissues, 377–383

Jorgensen procedure, 151

Kidney
alpha-subunit isoforms
relative abundance, 108
tissue specificity in rat, 120–125
beta-subunit
mRNA species, multiple, 130
thyroid hormone and, 133
cell culture systems. See MDCK cells
differential sensitivity to Na, vanadate, and dihydroouabain, 377–383
gossypol effects, 385–391
intraindividual variations in gene organization, 136, 137, 140, 141
ouabain sensitivity, 265, 377–383
pertussis toxin and, 353–356
pump rate in, 178, 181–183
regulation
by aldosterone, 282–283, 339–344
differentiation and, 181–183
inhibitors. See Inhibitors
by sodium, 339–344
see also Hormones
renal tubular epithelial cell cytoskeleton, 245–249
tissue-specific differences
in gene expression, 65–69, 105
in subunit distribution, 109, 111
Kidney function, 207–227
function of Na,K-ATPase, 212–221
potassium reabsorption, 219–221
potassium secretion, 216–219
quantitative correlations, 214–215
secondary active transport, 215–216

sodium reabsorption and secondary
active transport, 213–214
localization, 208–212
distribution profile, 208–211
heterogeneity of, 211–212
intracellular, 211
regulation of, 221–227
Laminin, 248
Latent pumps, 361–362
Leaks, 172
pump activity and, 187–188
pump rate and, 173, 174, 181–183
Linolenic acid, 433
Lipids, 301
nonesterified fatty acids, 443–448
receptor-mediated uptake, 441
T-lymphocytes, 452
two-sided liposomes, 433
Liponucleotides, 401–406
Lipoproteins, T-lymphocytes, 452
Liposomes
two-sided, 429–436
see also Vesicles
Lithium, 160, 161
Liver
alpha-subunit isoforms, 108
beta-subunit
mRNA species, multiple, 130
thyroid hormone and, 132–133
intraindividual variations in gene organization, 141
regulation, by hormones, 279
tissue-specific differences, 65–69, 109
LK cells, 357–363
Localization, renal epithelial cells
cytoskeleton and, 245–249
distribution profile, 208–212
sorting in, 51–57
Low-density lipoproteins, 441, 452
Lung
alpha-subunit isoforms, 108
intraindividual variations in gene organization, 136, 137, 139–140
tissue-specific gene expression, 105, 109, 111
Lymphocytes, 449–455
Lysolecithin acyltransferase, 450, 453
Lysolecithins, 420

Lysophospholipids, 301
Madin Darby canine kidney cell line. *See* MDCK cells
Magnesium
and ouabain binding, red cell ghosts, 366, 367, 369
skeletal muscle enzyme regulation, 253
Magnesium ATP, and ouabain binding, 323
Malignancy, intraindividual variations in gene organization, 139–140, 141
Mass spectroscopy, 305–306, 307
Matrix coatings, MDCK cells, 248
Maturation, 26–30
alpha-subunit, rat heart, 151, 152–153
intraindividual variations in gene organization, 136, 138
kidney enzyme regulation, 222
and localization in epithelial cells, 51–57
pump rate regulation, differentiation and, 181–183
red cells, 357–363
tunicamycin and, 163
Maximum capacity, right vs. left ventricle, 233–238
Maximum pump activity, skeletal muscle, 200–201, 202, 241–243
MDCK cells
coupling ratio in, 176
cytoskeleton, 245–249
pump sharing, 185–186
pump-specific mRNA, 59–64
regulation, by vasopressin, 278
subcellular localization in, 51–57
Membrane insertion
beta-subunit
Artemia salina, 39–41
rat brain, 130–132
human enzyme, 77–84
subcellular localization in epithelial cells, 51–57
subunits, 20–26
tunicamycin and, 163–168
Membranes, cell
electrical potential. *See* Electrical potential
isolated vs. whole cells, 184–185
lymphocytes, 450
myometrium, calcium effects, 457–462

primaquine effects, 437–441
renal tubular cells, 211
subcellular localization in epithelial cells, 51–57
two-sided liposomes, 429–436
voltage independence of oocyte ouabain binding, 409–415
Metabolism, and active transport, 10–12
Mineralocorticoids, 281–283
kidney enzyme regulation, 222, 225
toad bladder cells (A6), 463–467
Mitogen activation, lymphocytes, 449, 453–455
Molecular cloning
alpha-subunit isoforms, 119–126
beta-subunit, rat brain, 130–131
deletion plasmids, 79–84
Saccharomyces cerevisiae, 85–90
see also cDNA
Molecular forms, ferret heart enzymes, 423–428
Molecular weight, rat brain alpha-isoforms, 115
Monensin
cell culture effects, 157–161
kidney enzyme regulation, 224
Monoclonal antibodies, ouabain binding sites, 329–331
Mouse, 29
gene mapping and transfer techniques, 98–102
intraindividual variations in gene organization, 135–136
kidney distribution profiles, 210
MRC5, monensin effects, 157–161
mRNA
aldosterone effects in toad bladder cells (A6), 463–467
alpha-subunit, monensin and, 157–161
beta-subunit, rat brain, 130
pump-specific, from HeLa and MDCK cells, 59–64
tissue-specific differences in gene expression, 65–69
translation, in vitro. *See* In vitro systems
Muscle
alpha-subunit isoforms, tissue specificity in rat, 120–125

cardiac. *See* Heart
myometrium, calcium effects, 457–462
skeletal. *See* Skeletal muscle
Myometrium, calcium effects, 457–462
Myosin, 154

NAB-ouabain, 54, 55–57
Neonatal development
alpha-subunit, rat heart, 151, 152–153
pump rate regulation, 183
skeletal muscle enzyme, 253
Nephron. *See* Kidney function
Nerve growth factor, 91–95, 185
Neural tissue
isozyme appearance in, 91–95
pump rates, maximum, 200, 201
regulation. *See* Hormones
tissue specificity in rat, 120–125
see also Brain
NIH 3T3 cells, 172
N-linked glycosylation, 38
Non-esterified fatty acids, 301, 443–448
Noradrenaline, 479, 480
Norepinephrine, 300
N-termini, 84, 119
brain, rat, 128, 131–132
membrane organization, 22–23, 24, 25
Nuclear magnetic resonance, 305
Nucleotide sequences, 128, 129
Number of pumps, 174
Nystatin, 224, 339–344

Oocytes. *See Xenopus*, oocytes
Osmoregulation, 5
Ouabain
apamin and, 478, 479, 480–481
binding site characterization, 328–334
cellular cooperation and pump sharing, 185–186
differential sensitivities in rat tissues, 377–383
gene mapping and gene transfer techniques, 98–103
glycoside receptor regulation in cultured cells, 345–351
heart
vs. brain, 43–48
ferret, 423–428
human, 269–273

vs. kidney, 265
maximum capacity for active transport, 233-238
rat, 153-154
sheep, 469-474
insulin and, 280
and pump turnover, 363
red cell ghosts, binding to, 365-370
and short-circuit current, 11
subunit interactions and, 29-30
tissue differences, 108
two-sided liposomes, 429-436
Xenopus oocytes, 183, 409-415

Palmitoyl CoA, 402-405
Palytoxin, 393-399, 433
Permeability, two-sided liposomes, 429-436
pH, intracellular, 176, 177, 178
Pheochromocytoma cells, 91-95
Phosphate ions, renal tubular cells, 215
Phosphorylation, myometrium, 457-462
Physiological role, 3-4
cellular cooperation and sharing of pumps, 185-186
heterogeneity of pump, 188-189
leaks, 187-188
Na,K-ATPase measurement and pumping activity, 184-185
pump rate
measurement, 173-176
rapid variation of, 176-179
regulation
ontogenic and hormonal, 180-184
short-term, 179-180
Placenta, 136, 137
Plasma inhibitor
and hydrogen ion pump, 417-421
nonesterified fatty acids and, 443-448
see also Inhibitors
Plasma membrane. *See* Membranes, cell
Plasma potassium
skeletal muscle enzyme regulation, 254
see also Potassium, extracellular
Plasmids
beta-subunit, rat brain, 130-131
Saccharomyces cerevisiae, 85-90
Polyadenylation, 128
Polysomes, 21, 35-36
Postnatal development

intraindividual variations in gene organization, 138
pump rate regulation, hormonal, 183
Posttranslational control, 20-21, 289
Posttranslational membrane insertion, 24
Potassium
extracellular
kidney enzyme regulation, 223, 224
normal variation in, 197-198
pump density, 157-161
and pump rate, 196-197
skeletal muscle enzyme regulation, 254
glycoside receptor regulation in cultured heart cells, 345-351
heart enzyme, 3-O-methylfluorescein phosphate activity, 257-260
intracellular
excitation-related changes in muscle, 200-203
kidney enzyme regulation, 223, 224
normal variation in, 203
pump response, 198-200
kidney function
reabsorption, 219-221
secretion, 216-219
and ouabain binding, 323
physiological role of Na,K-pump. *See* Physiological role
skeletal muscle, basal and maximal uptake, 241-243
Potassium adaptation, 217, 218, 219
Potassium channels, apamin, 477-483
Potassium depletion, 220
heart enzyme, 3-O-methylfluorescein phosphate activity, 260
kidney enzyme regulation, 226
and pump loss, 363
skeletal muscle enzyme regulation, 253
Potassium efflux
muscle, thyroid hormone effects, 371
skeletal muscle, 239-243
Potassium influx, ratio to Na influx, 175
Potassium load, 217, 218, 219, 253
Potential difference. *See* Electrical potential
Prenatal development. *See* Fetal development
Primaquine, 437-441
Progesterone, 183

Protein kinase C, 279
Protein kinases, 279, 457, 458, 460, 461, 462
Protein synthesis
    aldosterone effects in toad bladder cells (A6), 463–467
    cell-free, 71–75; *see also* In vitro systems
Proteolysis, subunits, 164
    alpha-subunit, nascent, 28–29
    beta-subunit, rat brain, 130–131, 132
Protomers, 19, 163
Proximal convoluted tubules, 176, 177, 178, 209
    pertussis toxin and, 353–356
    pump-leak uncoupling, 187–188
Pump concentration
    cell culture, monensin and, 157–161
    muscle, thyroid hormone effects, 371–375
Pump rate, 173, 175
    maximum, 200–201, 202
    rapid variation in, 176–179
    sodium influx and, 175
Pump sites, in heart, 237–238
Purification, inhibitors, 303–312
Purified enzyme
    gossypol effects, 385, 387–388
    palytoxin and, 393–399
Pyrithiamin, 153

Quantification
    cardiac enzyme
        human, 269–273
        3-O-methylfluorescein phosphate activity, 257–260
    in skeletal muscle, 252–254
    skeletal muscle enzyme, 251–252

Rabbit
    intraindividual variations in gene organization, 135–136
    kidney
        distribution profiles, 210
        gossypol effects, 385, 387–388
Radioimmunoassays, 301, 305
Rat
    alpha-subunit isoforms
        heart vs. brain, 43–48

        immunological and electrophoretic differences, 149–154
        photoaffinity-labeling studies, 113–117
        tissue specificity, 119–126
    beta-subunit gene expression, 127–133
    brain enzyme, 43–48
    differential sensitivities in, 377–383
    heart enzyme
        vs. brain, 43–48
        glycoside receptor regulation in cultured cells, 345–351
        immunological and electrophoretic differences, 149–154
        maximum capacity for active transport, 233–238
        3-O-methylfluorescein phosphate activity, 257–260
    kidney, distribution profiles, 210
    myometrium, calcium effects, 457–462
    neuronal enzyme appearance in cell culture, 91–95
    tissue-specific differences, 65–69
Receptor-mediated endocytosis, 437
Recombinant plasmids, 85–90
Recycling, pump, 361–362, 437–441
Red cells
    gossypol effects, 385–391
    lipids and liponucleotides, 401–406
    ouabain binding, 365–370
    reticulocyte maturation, 357–363
    sodium influx, 172
Regulation
    aldosterone effects in toad bladder cells (A6), 463–467
    hormonal. *See* Hormones
    inhibitors. *See* Inhibitors
    inotropic action. *See* Inotropic agents; Inotropy
    kidney function, 221–227
    monensin and, 157–161
    in skeletal muscle, 252–254
Renal carcinoma, 141
Renal function. *See* Kidney function
Respiration, transport and, 10–11
Respiratory epithelium, 122–123
Restriction fragment length polymorphism, 135–142
Restriction maps, 107–108, 135–142

Reticulocyte maturation, 357–363
Retina, 143–146
RPTC, coupling ratio in, 176
Run-on transcription assays, 69

Saccharomyces cerevisiae, 85–90
SDS, 333–334, 457, 459, 460, 461
Secondary active transport, 12–13, 213–216
Serum inhibitors. *See* Inhibitors
Sharing of pumps, 185–186
Sheep
    cardiac tissues, cardiac steroids, 469–476
    reticulocyte maturation, 357–363
Short-circuit criterion, 8–10
Short-circuit current, 12
Sidedness
    beta-subunit, rat brain, 131
    lipids and liponucleotides, 401–406
    renal tubular cells, 211
    two-sided liposomes, 429–436
Signal recognition particle, 22, 23, 24
Single-file diffusion, 8
Skeletal muscle
    alpha-subunit isoforms, 108
    apamin effects, 477–483
    homeostasis in striated muscle, 195–204
        catecholamines and, 204
        intracellular ions, excitation-related changes, 200–203
        intracellular ions, normal variation in, 203
        intracellular ions, pump response, 198–200
        normal variation in extracellular K, 197–198
        pump response to extracellular K, 196–197
    quantification, 251–252
    regulation, 252–254
    significance of changes, 254
    sustained work, 239–244
        basal and maximal K uptake, 241–243
        ion changes per contraction, 239–244
    thyroid hormone effects, 371–375
    tissue differences in subunit distribution, 109
    tissue-specific gene expression, 106
Skin
    frog, 6–7, 11

regulation. *See* Hormones
Sodium
    differential sensitivities in rat tissues, 377–383
    extracellular, 223, 224
    intracellular, 175, 176
        cardiac steroids and, 469
        excitation related changes in muscle, 200–203
        kidney enzyme regulation, 222, 223, 224, 225, 226–227, 339–344
        normal variation in, 203
        and pump density, 157–161
        and pump rate, 176–177, 195, 196
        pump response, 198–200
    kidney enzyme, 213–214, 223, 224, 339–344
    and ouabain binding, red cell ghosts, 365–370
    physiological role of Na,K-pump. *See* Physiological role
Sodium-calcium exchange, 327, 469
Sodium dodecyl sulfate, 333–334, 457, 459, 460, 461
Sodium efflux, maximum, 200, 201, 202
Sodium flux, pump rate variation, 176–179
Sodium gradient, 174
Sodium-hydrogen ion exchanger, 176–178, 188, 377
Sodium influx, 174
    ratio to K influx, 175
    skeletal muscle, 239–241
Sodium leaks, 172
Sodium-potassium ratio, 223, 224
Solvent drag, 12
Somatic rearrangements, 139–140
Sorting
    primaquine effects, 437
    subcellular localization in epithelial cells, 51–57
Sotalol, 470
Species differences
    cardiac glycoside sensitivity, 325–326
    ouabain sensitivity, 98–103
Spectrin, 245–247
Spinal cord, 120–125
Squid axon, 200, 201
Starvation, 253, 375

Steady state rates, sodium pumping, 172
Stoichiometry, metabolism and, 10–12
Stomach
    alpha-subunit isoforms, 108
    plasma inhibitor and, 417–421
    tissue differences in subunit distribution, 109, 111
Striated muscle. *See* Muscle, striated
Strophanthidin, 93, 199, 279, 325
Strophanthidin derivatives, 469–474
Structural processing, 26–30, 163–168
Subcellular localization, 51–57, 245–249
Subunits, alpha-
    biosynthesis and membrane insertion, 80–82
    ferret heart, developmental changes in, 423–428
    gene characterization, 106–107
    gene mapping, 98–103
    heart
        vs. brain, 43–48
        immunological and electrophoretic differences, 149–154
    intraindividual variations in gene organization, 135–142
    isoforms, 71–75
        cardiac glycoside action, 330–332
        dog heart, 263–267
        gene mapping and gene transfer techniques, 98–103
        molecular cloning, 119–126
        rat brain vs. heart, 43–48
        retinal, citrate pretreatment and, 143–146
        thyroid hormone and, 288–290
        tissue-specific gene expression, 105–106
    monensin and, 157–161
    mRNA purification, 61
    in rat tissues
        brain vs. heart, 43–48
        differential sensitivities, 377–383
    regulation
        thyroid hormone and, 288–290
        see also Hormones
    tissue distribution, 107–110
Subunits, alpha-3, 71–75, 93, 106, 150
    isoforms, photoaffinity-labeling studies, 113–117

    tissue distribution, 108
    tissue-specific expression, fetal rat, 125
Subunits, alpha- and beta-
    aldosterone effects in toad bladder cells (A6), 463–467
    deletion mutants, 77–84
    interactions in synthesis and assembly, 20–26
    intraindividual variations in gene organization, 135–142
    in vitro expression, 71–75
    monensin effects, 157–161
    primaquine effects, 437–441
    reticulocyte maturation, 357–363
    structural arrangements, 163–168
    synthesis and assembly, 20–26
    tissue-specific differences, 65–69
    tissue-specific gene expression, 105–111
    tunicamycin and, 163–168
Subunits, beta-
    biosynthesis and membrane insertion, 82–84
    differences in electrophoretic mobility, 106
    gene expression
        rat brain, 127–133
        in *Saccharomyces cerevisiae*, 85–90
    hybrid, 29
    in vitro synthesis, *Artemia salina*, 35–41
    tissue distribution, 109
    tunicamycin and, 163–168
Symports, 173

Tetrodotoxin, 143, 375
3T3 cells, 172
Thymocytes, 450–453
Thyroid gland
    intraindividual variations in gene organization, 136, 137, 141
    skeletal muscle enzyme regulation, 253
Thyroid hormones, 284–290
    brain enzyme beta-subunit, 132–133
    heart enzyme
        glycoside receptor regulation in cultured cells, 348
        3-O-methylfluorescein phosphate activity, 260
    kidney enzyme regulation, 222

skeletal muscle enzyme regulation, 253, 371–375
Tissue-specific expression, 105–111
   alpha-subunit isoforms, molecular cloning, 119–126
   differential sensitivities in rat, 377–383
   intraindividual variations in gene organization, 135–142
T-lymphocytes, 449–455
   Con A-activated, 453–455
   membrane subfractions, separation of, 451–453
Toad bladder
   cell lines, 164–168, 463–467
   regulation
      aldosterone, 282–283
      vasopressin, 277–278
Topology, primaquine effects, 437–441
Toxins
   cardiac steroids, 469–476
   palytoxin, 393–399, 433
   pertussis toxin effects, 353–356
Transcription
   aldosterone effects in toad bladder cells (A6), 463–467
   in vitro. *See* In vitro systems
Transcription analysis, 69
Transfection, heterogeneity of pump, 188, 189
Transformation, *Saccharomyces cerevisiae*, 85–90
Translation, 21–22
   beta-subunit, *Artemia salina*, 36–37
   in vitro. *See* In vitro systems
Transmonolayer resistance, 248
Transport
   aldosterone effects in toad bladder cells (A6), 463–467
   intracellular, 26–30
   tunicamycin and, 163
Triiodothyronine. *See* Thyroid hormones
Trypsin, subunit studies, 164
   alpha-subunit, nascent, 28–29
   beta-subunit, rat brain, 130–131

T3
   glycoside receptor regulation in cultured heart cells, 348
   see also Thyroid hormones
Tumor, 139–140, 141
Tunicamycin, 26, 163–168
Turnover
   beta-chain, 440
   pump, 363

Uncoupling, pump-leak, 187–188
Untranslated region, beta-subunit, 128
Uremia, 301

Valinomycin, 371
Vanadate, 279
   cardiac enzyme
      human, 269
      maximum capacity for active transport, 233–238
   differential sensitivities in rat tissues, 377–383
Vanadyl ribonuclease complex, 35
Vasopressin, 277–278
Ventricle, 109–110; *see also* Heart
Very low density lipoprotein, 452
Vesicles
   lipids and liponucleotides, 401–406
   renal tubular cells, 211
   two-sided liposomes, 429–436
Voltage-dependence. *See* Electrical potential

Wheat germ system, 22–24
Wilms' tumor, 141

Xenopus, 28
   cell lines, 164–167, 282, 463–467
   glycosylation patterns, 28
   oocytes
      ouabain-binding, voltage independence of, 409–415
      pump rate regulation, 183

Yeast, gene expression in, 85–90